Volatile Biomarkers

Non-Invasive Diagnosis in Physiology and Medicine

Volatile Biomarkers

Non-Invasive Diagnosis in Physiology and Medicine

Edited by

Anton Amann

Breath Research Institute
Austrian Academy of Sciences
Rathausplatz 4, A-6850 Dornbirn
Austria

Univ.-Clinic for Anesthesia
Innsbruck Medical University
Anichstr 35, A-6020 Innsbruck
Austria
ETH-Zentrum
Laboratory of Physical Chemistry
CHN E24, CH-8092 Zürich
Switzerland

David Smith

Institute for Science and
Technology in Medicine
Keele University, Stoke-on-Trent, ST4 7QB
UK

AMSTERDAM • BOSTON • HEIDELBERG • LONDON • NEW YORK • OXFORD
PARIS • SAN DIEGO • SAN FRANCISCO • SINGAPORE • SYDNEY • TOKYO

ELSEVIER

Elsevier
The Boulevard, Langford Lane, Kidlington, Oxford OX5 1GB, UK
Radarweg 29, PO Box 211, 1000 AE Amsterdam, The Netherlands
225 Wyman Street, Waltham, MA 02451, USA
525 B Street, Suite 1900, San Diego, CA 92101-4495, USA

First edition 2013

British Library Cataloguing in Publication Data
A catalogue record for this book is available from the British Library

Library of Congress Cataloging-in-Publication Data
A catalog record for this book is available from the Library of Congress

For information on all Elsevier publications
visit our website at **store.elsevier.com**

Printed and bound in China
13 14 15 16 17   10 9 8 7 6 5 4 3 2 1

ISBN: 978-0-444-62613-4

Contents

PART B REAL-TIME ANALYSIS OF EXHALED BREATH

PART C PHYSIOLOGICAL AND CLINICAL STUDIES

List of Contributors

Agapios Agapiou
National Technical University of Athens (NTUA), School of Chemical Engineering, Sector I, 9 Iroon Polytechniou Street, 15773 Athens, Greece

Clemens Ager
Breath Research Institute, Austrian Academy of Sciences, Rathausplatz 4, 6850 Dornbirn, Austria; University Clinic for Anesthesia, Innsbruck Medical University, Anichstraße 35, 6020 Innsbruck, Austria

Alexander A. Aksenov
Department of Mechanical and Aerospace Engineering, One Shields Avenue, University of California, Davis, CA 95616, USA

Anton Amann
Breath Research Institute, Austrian Academy of Sciences, Rathausplatz 4, 6850 Dornbirn, Austria; University Clinic for Anesthesia, Innsbruck Medical University, Anichstraße 35, 6020 Innsbruck, Austria

Hamzeh Bardaweel
Department of Mechanical and Aerospace Engineering, One Shields Avenue, University of California, Davis, CA 95616, USA

Peter J. Barnes
Section of Airway Disease, National Heart and Lung Institute, Imperial College London, London, UK

Maria Baur
Breath Research Institute, Austrian Academy of Sciences, Rathausplatz 4, 6850 Dornbirn, Austria; University Clinic for Anesthesia, Innsbruck Medical University, Anichstraße 35, 6020 Innsbruck, Austria

A.M. Biaggi-Labiosa
NASA Glenn Research Center, 21000 Brookpark Road, Cleveland, OH 44135, USA

Bogusław Buszewski
Department of Environmental Chemistry and Bioanalytics, Faculty of Chemistry, Nicolaus Copernicus University,7 Gagarin St., 87-100 Toruń, Poland

C.W. Chang
ASRC /NASA Glenn Research Center, 21000 Brookpark Road, Cleveland, OH 44135, USA

William H.K. Cheung
Department of Mechanical and Aerospace Engineering, One Shields Avenue, University of California, Davis, CA 95616, USA

Massimo Corradi
Department of Clinical and Experimental Medicine, University of Parma, Via Gramsci 14, I-43126 Parma, Italy

Ben de Lacy Costello
Centre for Research in Biosciences, University of the West of England, Coldharbour Lane, Frenchay, Bristol BS16 1QY, UK

Martin Daniaux
Department of Radiology, Medical University Innsbruck, Anichstrasse 35, 6020 Innsbruck, Austria

Cristina E. Davis
Department of Mechanical and Aerospace Engineering, One Shields Avenue, University of California, Davis, CA 95616, USA; Clinical and Translational Sciences Center (CTSC), UC Davis Medical Center, Sacramento,CA 95817, USA

Jürgen Dunkl
Oncoytyrol-Center for Personalized Cancer Medicine GmbH, Eduard Bodem Gasse 3, 6020 Innsbruck, Austria; Institute of Ion Physics and Applied Physics, Leopold-Franzens University Innsbruck, Technikerstrasse 25, 6020 Innsbruck, Austria

Prabir K. Dutta
Ohio State University, Department of Chemistry, 100 West 18th Avenue, Columbus, OH 43210, USA

Raed A. Dweik
Cleveland Clinic, Respiratory Institute & Pathobiology/Lerner Research Institute, Desk A90, 9500 Euclid Ave., Cleveland, OH 44195, USA

Ashley Esarik
Northern Ontario School of Medicine and Department of Biology, Lakehead University, Thunder Bay, Ontario, Canada P7B5E1

Anna Filipiak
Breath Research Institute, Austrian Academy of Sciences, Rathausplatz 4, 6850 Dornbirn, Austria

Wojciech Filipiak
Breath Research Institute, Austrian Academy of Sciences, Rathausplatz 4, 6850 Dornbirn, Austria; University Clinic for Anesthesia, Innsbruck Medical University, Anichstraße 35, 6020 Innsbruck, Austria

Hossam Haick
The Department of Chemical Engineering and Russell Berrie Nanotechnology Institute, Technion-Israel Institute of Technology, Haifa 32000, Israel

Armin Hansel
Institute of Ion Physics and Applied Physics, Leopold-Franzens University Innsbruck, Technikerstrasse 25, 6020 Innsbruck, Austria

Jens Herbig
Oncoytyrol-Center for Personalized Cancer Medicine GmbH, Eduard Bodem Gasse 3, 6020 Innsbruck, Austria; Ionimed Analytik GmbH, Eduard Bodem Gasse 3, 6020 Innsbruck, Austria

Hartmann Hinterhuber
Univ.-Clinic for Psychiatry, Innsbruck Medical University, Anichstr. 35, 6020 Innsbruck, Austria

Marieann Högman
Department of Medical Sciences, Respiratory Medicine and Allergology, Uppsala University, SE 75105 Uppsala, Sweden; Centre for Research and Development, Uppsala University/County Council of Gävleborg, SE 801 88 Gävle, Sweden

Ildikó Horváth
Department of Pulmonology, Semmelweis University, Budapest, Hungary

Michael Hubalek
Department of Gynecology and Obstetrics, Medical University Innsbruck, Anichstrasse 35, 6020 Innsbruck, Austria

Gary W. Hunter
NASA Glenn Research Center, 21000 Brookpark Road Cleveland, OH 44135, USA

Tadeusz Jezierski
Institute of Genetics and Animal Breeding of Polish, Academy of Sciences, Department of Animal Behaviour, Jastrzębiec, 05-552 Wólka Kosowska, Poland

Gennadii Kamarchuk
B. Verkin Institute for Low Temperature Physics and Engineering, National Academy of Sciences of Ukraine, 47 Lenin Avenue, Kharkov 61103, Ukraine

Julian King
Breath Research Institute, Austrian Academy of Sciences, Rathausplatz 4, 6850 Dornbirn, Austria; University of Vienna, Faculty of Mathematics, Nordbergstr. 15, 1090 Wien, Austria

Helin Koc
Breath Research Institute, Austrian Academy of Sciences, Rathausplatz 4, 6850 Dornbirn, Austria; Vorarlberg University of Applied Sciences, Hochschulstr. 1, 6850 Dornbirn, Austria

Ingrid Kohl
Oncoytyrol-Center for Personalized Cancer Medicine GmbH, Eduard Bodem Gasse 3, 6020 Innsbruck, Austria; Ionimed Analytik GmbH, Eduard Bodem Gasse 3, 6020 Innsbruck, Austria

Ievgeniia Kushch
Institute for Children and Adolescents Health Care, National Academy of Medical Sciences of Ukraine, 52-A 50 let VLKSM Avenue, Kharkov 61153, Ukraine

Jae Kwak
Monell Chemical Senses Center, 3500 Market Street, Philadelphia, PA 19104, USA

Alice M. Kwan
Department of Mechanical and Aerospace Engineering, One Shields Avenue, University of California, Davis, CA 95616, USA

D. Laskowski
Cleveland Clinic, Respiratory Institute & Pathobiology/Lerner Research Institute, Desk A90, 9500 Euclid Ave., Cleveland, OH 44195, USA

Tomasz Ligor
Department of Environmental Chemistry and Bioanalytics, Faculty of Chemistry, Nicolaus Copernicus University, 7 Gagarin St., 87-100 Toruń, Poland

Chung-Chiun Liu
Department of Chemical Engineering, Case Western Reserve University, 10900 Euclid Ave., Cleveland, OH 44106, USA

D.B. Makel
Makel Engineering, Inc., 15505 Neo Parkway, Cleveland, OH 44128, USA

Rosa Margesin
Institute of Microbiology, Leopold-Franzens University of Innsbruck, Technikerstraße 25, 6020 Innsbruck, Austria

Pekka Meriläinen
Department of Medical Sciences, Respiratory Medicine and Allergology, Uppsala University, SE 75105 Uppsala, Sweden

Wolfram Miekisch
University of Rostock, Department of Anaesthesia and Intensive Care, Schillingallee 35, D-18057 Rostock, Germany

Pawel Mochalski
Breath Research Institute, Austrian Academy of Sciences, Rathausplatz 4, 6850 Dornbirn; Univ.-Clinic for Anesthesia, Innsbruck Medical University, Anichstr, 35, 6020 Innsbruck, Austria; Institute of Nuclear Physics PAN, Radzikowskiego 152, 31342 Kraków, Poland.

Anil S. Modak
Cambridge Isotope Laboratories Inc., 50 Frontage Road, Andover, MA 01810, USA

Suvra P. Mondal
Ohio State University, Department of Chemistry, 100 West 18th Avenue, Columbus, OH 43210, USA

Paolo Montuschi
Department of Pharmacology, Faculty of Medicine, Catholic University of the Sacred Heart, Largo F. Vito, 1, 00168 Rome, Italy

Antonio Mutti
Department of Clinical and Experimental Medicine, University of Parma, Via Gramsci 14, I-43126 Parma, Italy

Markus Nagl
Department of Hygiene, Microbiology and Social Medicine, Division of Hygiene and Medical Microbiology, Innsbruck Medical University, Fritz-Pregl-Straße 3, 6020 Innsbruck, Austria

Anna-Carin Olin
Occupational and Environmental Medicine, Sahlgrenska Academy, Box 414, SE-405 30 Gothenburg, Sweden

Daniel J. Peirano
Department of Mechanical and Aerospace Engineering, One Shields Avenue, University of California, Davis, CA 95616, USA

Joachim D. Pleil
National Exposure Research Laboratory, Office of Research and Development, US Environmental Protection Agency, Research Triangle Park, NC 27711, USA

Alexander Pospelov
National Technical University "Kharkov Polytechnic Institute", 21 Frunze Str., Kharkov 61002, Ukraine

George Preti
Monell Chemical Senses Center, 3500 Market Street, Philadelphia, PA 19104, USA; Department of Dermatology, School of Medicine, University of Pennsylvania, PA 19104, USA

Norman M. Ratcliffe
Centre for Research in Biosciences, University of the West of England, Coldharbour Lane, Frenchay, Bristol BS16 1QY, UK

Terence H. Risby
Bloomberg School of Public Health, Johns Hopkins University, 615 North Wolfe Street, Baltimore, MD 21205, USA

Brian M. Ross
Northern Ontario School of Medicine and Department of Biology, Lakehead University, Thunder Bay, Ontario, Canada P7B5E1

Joanna Rudnicka
Department of Environmental Chemistry and Bioanalytics, Faculty of Chemistry, Nicolaus Copernicus University,7 Gagarin St., 87-100 Toruń, Poland

Stefan W. Ryter
Pulmonary and Critical Care Medicine, Brigham and Women's Hospital, Harvard Medical School, 75 Francis Street, Boston, MA 02115, USA

Michael Schivo
Center for Comparative Respiratory Biology and Medicine, 451 Health Sciences Drive, #6517, Davis, CA 95616, USA; Clinical and Translational Sciences Center (CTSC), UC Davis Medical Center, Sacramento, CA 95817, USA

Alex Schmid
Breath Research Institute, Austrian Academy of Sciences, Rathausplatz 4, 6850 Dornbirn, Univ.-Clinic for Anesthesia, Innsbruck Medical University, Anichstr, 35, 6020 Innsbruck, Austria

Jochen K. Schubert
University of Rostock, Department of Anaesthesia and Intensive Care, Schillingallee 35, D-18057 Rostock, Germany

David Smith
Institute for Science and Technology in Medicine, School of Medicine, Keele University, Thornburrow Drive, Hartshill, Stoke-on-Trent ST4 7QB, UK

Jon R. Sobus
National Exposure Research Laboratory, Office of Research and Development, US Environmental Protection Agency, Research Triangle Park, NC 27711, USA

Steven F. Solga
St. Luke's Hospital, 801 Ostrum Street, Bethlehem, PA 18015, USA School of Medicine, Johns Hopkins University, 733 North Broadway, Baltimore, MD 21205, USA

Patrik Španěl
J. Heyrovský Institute of Physical Chemistry, Academy of Sciences of the Czech Republic, Dolejškova 3, 182 23, Prague 8, Czech Republic

Andreas Sponring
Breath Research Institute, Austrian Academy of Sciences, Rathausplatz 4, 6850 Dornbirn, Austria; University Clinic for Anesthesia, Innsbruck Medical University, Anichstraße 35, 6020 Innsbruck, Austria

Gerald Teschl
University of Vienna, Faculty of Mathematics, Nordbergstr. 15, 1090 Wien, Austria

Susanne Teschl
University of Applied Sciences Technikum Wien, Höchstädtplatz 5, 1200 Wien, Austria

Ulrike Tisch
The Department of Chemical Engineering, Technion-Israel Institute of Technology, Haifa 32000, Israel

Jakob Troppmair
Daniel-Swarovski Research Laboratory, Department of Visceral-, Transplant- and Thoracic Surgery, Innsbruck Medical University, Innrain 66, 6020 Innsbruck, Austria

Claire Turner
The Open University, Department of Life, Health and Chemical Sciences, Walton Hall, Milton Keynes, MK7 6AA, UK

Karl Unterkofler
Breath Research Institute, Austrian Academy of Sciences, Rathausplatz 4, 6850 Dornbirn, Austria; Vorarlberg University of Applied Sciences, Hochschulstr. 1, 6850 Dornbirn, Austria

Marta Walczak
Institute of Genetics and Animal Breeding of Polish Academy of Sciences, Department of Animal Behaviour, Jastrzębiec, 05-552 Wólka Kosowska, Poland

Benjamin J. Ward
Makel Engineering, Inc., 15505 Neo Parkway, Cleveland, OH 44128, USA

Anna Wenda-Piesik
Department of Plant Growth Principles and Experimental Methodology, University of Technology and Life Sciences, 20 Kordeckiego St., 85-225 Bydgoszcz, Poland

Helmut Wiesenhofer
Breath Research Institute, Austrian Academy of Sciences, Rathausplatz 4, 6850 Dornbirn, Austria

Jennifer C. Xu
NASA Glenn Research Center, 21000 Brookpark Road, Cleveland, OH 44135, USA

Konstantin Zamuruyev
Department of Mechanical and Aerospace Engineering, One Shields Avenue, University of California, Davis, CA 95616, USA

Yuriy Zrodnikov
Department of Mechanical and Aerospace Engineering, One Shields Avenue, University of California, Davis, CA 95616, USA

Foreword

Volatile compounds, the majority being organic in nature, VOCs, are continuously generated in the human body and are partially emitted via exhaled breath and through the skin. VOCs are also released by blood, urine, and fecal samples. It stands to reason that in some way these VOCs carry information on the physiological and metabolic status of the individual and it is because of this that these VOCs, often referred to as volatile metabolites or biomarkers, have attracted the attention of physicians, physiologists, and surgeons as potential aids to clinical diagnosis and therapeutic monitoring. Hence, the detection, identification, and quantification of these volatile biomarkers have stimulated the development of sophisticated analytical methods and instrumentation by which individual biomarker molecules, or combinations of different biomarker molecules, can be analyzed to sufficient accuracy to be clinically useful. The expectation is that particular patho-physiological conditions can be signaled by individual volatile biomarker molecules, and this has been realized by the close association of exhaled NO with asthma, a much researched phenomenon that is now finding utility in clinical practice. However, this exemplar is, as yet, a rare circumstance. Much research is indicating that if complex and very variable conditions, such as cancer, are to be recognized by VOC detection in exhaled breath, they are much more likely to be signaled by a combination of VOCs, some specific to the condition, but also by unusual distributions of common breath metabolites that are present in the breath of all individuals.

The status of breath research up to the year 2005 was well documented and described in the book "Breath Analysis for Clinical Diagnosis and Therapeutic Monitoring" edited by A. Amann and D. Smith (World Scientific Publishing Company, Singapore). The present text reports the continued progress of this topic, not least in the development of new instrumental analytical techniques. It expands the topic toward volatiles released by other biological fluids, including urine and stool, in the last of which bacterial action plays an important, even dominating role in the generation of volatiles, and also the release of volatiles from *in vitro* cell and bacterial cultures that can provide valuable pointers for *in vivo* studies.

This book comprises chapters written by authorities in particular areas of volatile biomarkers research. Some of the chapters are focused, others are more broadly based. There are chapters on the development of analytical instrumentation, including mass spectrometry-based devices and solid state sensors, on physiological and clinical studies via breath, urine, and stool VOC analyses, on volatile compounds emitted by pathogenic microorganisms and on breath condensate and exhaled particulates. Standardization of sampling methodologies for VOCs are described throughout the chapters, a topic that is more and more being realized as critical to breath analysis, since it will allow meaningful comparisons of results obtained in different laboratories. There is little point in attempting to subdivide, order, and categorize too rigorously the chapter topics in this book, but some generalizations are attempted, as can be seen by the sectioning of the contents list.

In one sense, we have "put the cart before the horse" by including as the first chapters those topics that focus on interpretation and analysis of breath concentrations data, including mathematical and statistical approaches, the physiological models of the transport of volatiles from blood to breath, and other issues and challenges in human breath research. The rationale for this is the need for these important aspects to be considered when designing new techniques for breath analysis, notably of solid state sensors that must be very sensitive, yet quite small, low in costs and easy to use by both health professionals and patients alike. Such devices contrast with the larger mass spectrometry-based analytical instruments, including gas chromatography-mass spectrometry (GC-MS) proton transfer reaction mass spectrometry (PTR-MS) and selected ion flow tube mass spectrometry (SIFT-MS) instruments, which are more research tools than monitoring instruments and are generally unsuitable for the clinical environment and point-of-care application. The important advances made in these analytical methods and instrumentation for VOC analysis are impressive. It is becoming clear that combinations of these various analytical methods offer the most profitable approach to biomarker research and discovery. This is now contributing to the new field of metabolomics that is represented in several chapters in this book.

A group of chapters deal with specific physiological and clinical studies, both in the clinic/hospital environment and in the workplace, which exploit the well-established analytical methods referred to above. A major advance has been made in sampling methodology of exhaled breath and the headspace of other biological fluids. In this sense, those techniques like GC-MS that require sample collection using bags and traps, such as solid phase micro extraction (SPME) and adsorption/thermal desorption (ATD) are not ideal, and direct sampling techniques together with immediate analyses are most desirable. The direct sampling techniques include PTR-MS and SIFT-MS and some sensors, notably those specifically designed to detect and quantify the diatomic molecules NO and CO in exhaled breath. These direct sampling techniques are described in separate chapters, which also indicate how they are utilized to monitor and study wide-ranging conditions and diseases, including asthma.

Real-time analysis of exhaled breath has been developed to a high degree in that it is now possible to directly analyze single breath exhalations for several compounds and separately analyze the end-tidal portion of exhaled breath. Even so, the composition of exhaled breath passing along the airways and through the oral cavity can become modified, which can be a serious consideration when systemic biomarkers at the parts-per-billion by volume (ppbv) and lower are intended as indicators of disease. However, mouth-exhaled breath and nose-exhaled breath can now be separately analyzed for some VOCs in just a few seconds and so it can be determined which VOCs originate in the oral cavity (by bacterial and enzymatic activity) and those which largely originate at the alveolar interface and are most likely to be truly systemic. Similarly, these *real-time* analyses can equally be utilized to analyze volatile compounds in the humid gas/vapor headspace of urine, cell and bacterial cultures and stool samples. The results of such *in vitro* studies can assist in directing the researcher toward specific biomarkers generated *in vivo* by mammalian cells and bacterial cells, the last analyses, for example, greatly contributing to the identification of harmful respiratory and gut bacteria. Much effort has been given to the search for volatile cancer biomarkers and there are chapters dedicated to this topic, including a chapter that describes how dogs can "sniff out" tumors. Progress has been limited, but the prize of discovery of a non-invasive biomarker of early-stage tumors is driving this research ever forward.

While strictly speaking not directly concerned with volatile biomarkers, exhaled breath condensate and particulates/aerosols are closely allied to breath volatiles research and discussions of these topics are properly included as separate chapters in this book. Knowledge of the types of VOCs emitted by human urine is being exploited in the detection and location of persons trapped in collapsed buildings following natural disasters such as earthquakes. A separate chapter is given to this important topic.

Some final remarks are in order here to indicate the way forward for VOC research and for breath research in particular. Accurate on-line, *real-time* measurements of the concentrations of trace metabolites in exhaled breath and other biological media can now be made down to ppbv concentration in the laboratory, but a major desire is to move direct analysis from the laboratory toward the clinical setting. This offers a real challenge, but direct analyses of exhaled breath of ventilated patients in the operating theatre and intensive care units are already being explored. Yet an even greater challenge lies ahead, which is the identification and accurate quantification in real time of new biomarkers of disease and infection, many, if not most, likely to be present in exhaled breath at sub-ppbv levels. This calls for greater analytical sensitivity than is currently available by on-line analytical techniques, and so further instrumentation developments are required. When these challenges in breath analysis have been met and overcome, the analyses of volatiles above other biological media important in clinical diagnosis (viz. urine, stool, blood, open infected wounds) should be straightforward. What is certain is that only thorough experimental work and innovation will promote this difficult and challenging topic such that trace VOC analysis will become a valuable diagnostic method that will enhance and extend conventional clinical diagnostics.

We are grateful to all the authors of the chapters in this book. Without their expert research work in advancing the topic of volatile biomarkers and exploring their relevance to physiology and medicine, coupled with their skill in reporting their research, this book could not have come into being. We are especially grateful to Hossam Haick for the depiction of an exhaling person that appears on the book cover, to Venelin Chernogorov for his editorial help and to Patrik Španěl for his invaluable comments in response to the many queries posed during the editing of the chapters.

Anton Amann
David Smith

PART

A

Interpretation of Breath Analysis Data

CHAPTER

1 Mathematical and Statistical Approaches for Interpreting Biomarker Compounds in Exhaled Human Breath

Joachim D. Pleil and Jon R. Sobus

National Exposure Research Laboratory, Office of Research and Development, US Environmental Protection Agency, Research Triangle Park, NC 27711, USA

1.1 INTRODUCTION

The various instrumental techniques, human studies, and diagnostic tests that produce data from samples of exhaled breath have one thing in common: they all need to be put into a context wherein a posed question can actually be answered. Exhaled breath contains numerous compounds; just the volatile organic fraction alone has been estimated to represent in excess of 500 different chemical species. In addition, the aerosol fraction contains proteins, signaling molecules, dissolved inorganic compounds, and even bacteria and viruses adding to the complexity of the total sample.

No single technique can detect everything in breath, in fact, even the most broadly designed breath measurements result in suites of compounds restricted by the methods used. For example, reactive oxygen species may be observed using real-time sensors or real-time mass spectrometry (MS), but not by gas chromatography-MS (GC-MS), whereas GC-MS can discriminate among a variety of hydrocarbons, alcohols, and ketones that may overlap completely in a real-time MS instrument without benefit of chromatographic separation. Furthermore, the fraction of the breath (gas-phase or aerosol phase) also determines what measurements can be made; for example proteins and signaling molecules could be detected in exhaled breath condensate via enzyme-linked immunosorbent assay (ELISA), nuclear magnetic resonance (NMR), or liquid chromatography (LC) MS, but not with any gas-phase instruments such as those based on optical spectroscopy or gas chromatography.

The first issue that the data analyst faces is that all data he or she sees, regardless how complex or detailed, is not comprehensive but always stratified (restricted) by the chemistry, instrumentation, and thermodynamics of the choices made for sampling and analysis. The second issue is that a suite of compounds measured in any given sample does not represent an independent set of variables; sub-groups of biomarker compounds often have appreciable covariance reflecting similar metabolic pathways or exogenous sources. The third issue relates to variability; any given sample is unique and it can never be taken for granted that the constituents of the breath are the same between people nor within one person over time. Finally, any individual compound

Volatile Biomarkers. http://dx.doi.org/10.1016/B978-0-44-462613-4.00001-5
2013 Published by Elsevier B.V.

in breath can have a wide range of concentrations that are all considered "normal" or "unremarkable" in the apparently healthy general population. This has implications for assessing health or exposure status based on just a few data points; under such constraints, the analyst can only interpret a measurement as a statistical probability that the concentration is probative.

1.2 DATA INTERPRETATION

In this chapter, we describe a series of mathematical and statistical approaches geared specifically to the interpretation of volatile organics in exhaled breath that can be implemented to address the four issues outlined above. Although all methods interact, we have assigned them to five categories for the purposes of this discussion as follows:

1. Data visualization and summary statistics.
2. Variable independence and clustering.
3. Population statistics and variance components.
4. Stochastic models and meta-data.
5. Dynamic models and longitudinal data.

Each of these approaches can provide distinct information about a breath data set, generally increasing in detail in the order listed. We note that the broader interpretations gleaned from categories 1–3 feed the modeling processes in 4 and 5. We further note that not all of these procedures need to be performed; often it is sufficient to answer a question with a simple analysis based on a graph or a summary table. In the following, we describe these categories and show how they can progressively tell a story about acquired data. We assume that multiple compounds have been measured across a number of people, subjects may be grouped as cases and controls, host factor meta-data have been collected, and measurements may have been repeated, either with or without intervention or treatment.

1.2.1 Data visualization and summary statistics

The first step in any analysis of newly acquired data is to get a "feel" for how the experiment turned out. Generally, we make graphs and calculate averages, standard deviations, data ranges, etc. We have found that it is very helpful to get all of the data into some form of visual representation, either as bar graphs, cluster plots, or heat maps, depending on complexity of the data structure. For example, in a recent publication, we described measurements of a series of polar volatile organic compounds (PVOCs) in exhaled breath condensate (EBC) made during an intervention study of diesel exhaust exposure.[1] Human subjects (3 males and 6 females) were each exposed for 2 h to a dilute diesel exhaust atmosphere and to a purified air atmosphere on separate occasions.[2] A variety of samples, including EBC, were collected immediately before and after the exposures, and again 24 h later; EBC means data for eight of the most prevalent compounds were plotted by nine subjects grouped by

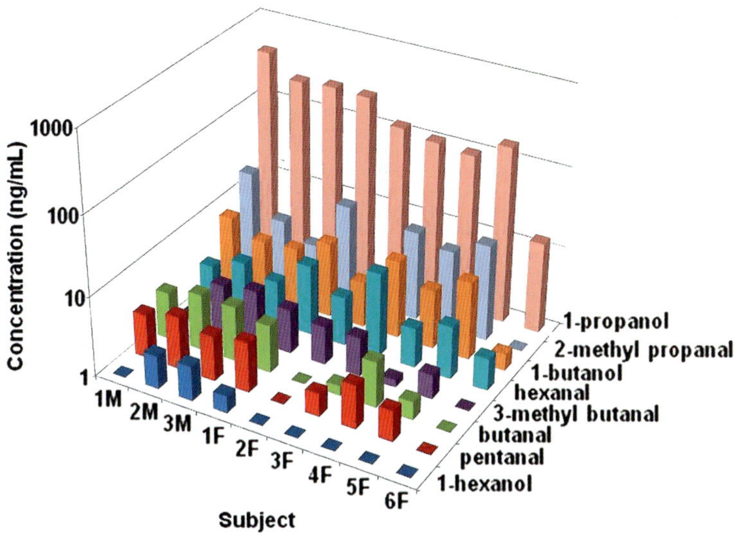

FIGURE 1.1
Bar graph visualization of balanced exhaled breath data from an intervention study of diesel exhaust exposure. Summary breath data for nine subjects and eight compounds.

gender (Figure 1.1). We noted a slight gender bias for some compounds, but otherwise the results were unremarkable and we went on to assess the data from a statistical perspective. In a subsequent publication, however, we developed methodology for data visualization using heat map style graphics.[3] Briefly, heat maps are visual representations of quantitative data on two axes; the *x*-axis reflects individual samples and the *y*-axis consists of groups of measured parameters. The field between the axes is comprised of an array contiguous boxes color coded to reflect quantitation. The term "heat map" is derived from the convention that the higher levels tend toward red, the lower levels tend toward blue.

We revisited the data set from Hubbard et al.[1] and found that heat map visualization was capable of showing more pattern detail. This technique had to be restricted to seven subjects with complete data to avoid blank spots on the heat map as we lost a few samples to follow-up. Figure 1.2 shows this alternate approach; note that we now have access to results from individual subjects' samples, for all 16 measured PVOCs, and the flexibility for grouping samples by gender and longitudinal parameters. Here, the gender effect becomes obvious; males were expressing much higher levels of many of the measured PVOCs, especially 2-methyl propanal, 1-heptanol, butanal, pentanal, 1-hexanol, and 3-methyl-3-pentanol. Only hexanal and heptanal reverse this trend. We further see that there is no apparent treatment effect or longitudinal time effect, that is, there are no obvious pattern differences between diesel exposures and air exposures, nor pre- and post-exposure data. These simple

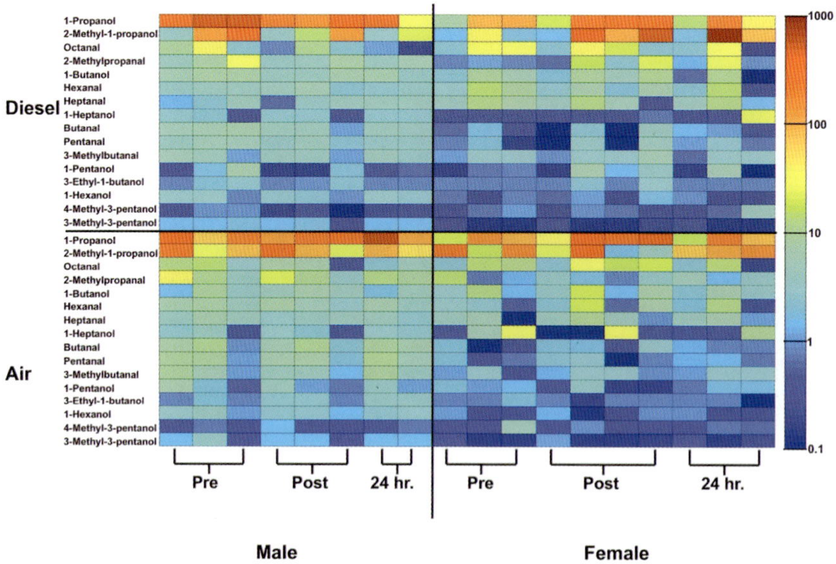

FIGURE 1.2

Heat map visualization of balanced exhaled breath data from an intervention study of diesel exhaust exposure. Individual samples grouped by longitudinal time frame (pre-exposure, post- exposure, and 24 h post-exposure) as well as by gender.

visualization approaches, coupled with summary statistics, can provide hints as to how more detailed mathematical approaches could be implemented to quantify these observations. Also, we note that the visualization methods for complex data make the subsequent statistical calculations more accessible to the readership.

1.2.2 Variable independence and clustering

One of the most vexing problems encountered in data interpretation procedures is the determination of actual independence of what we generally denote as "independent variables." Subsequent statistical evaluations rely on the notion that measurements that are used in models are not overly correlated. For example, consider the height and weight parameters of human subjects as host-factor data in a complex breath experiment. We know that taller people tend to be heavier, and so there is significant correlation expected. If one were to treat height and weight as independent and place them both into a model for predicting a health outcome (along with a series of breath biomarkers), one can get completely different interpretations as to their respective importance depending upon which was entered first. This is why one often sees body-mass index (BMI) used as a composite parameter instead. Now consider unexpected correlations among variables, for example, among compounds measured in the exhaled breath of subjects in an intervention or case-control study. If certain compounds are repeatedly behaving the same way, then their inclusion in even simple multivariate regression models will result in mathematical instability.

We cannot know *a priori* if different analytes are tracking the same outcome, however, we can perform various correlation tests among presumed independent variables to assess the degree of independence. The most fundamental method is the correlation matrix wherein the regression between pairs of variables (V_i, V_k) is calculated as the Pearson correlation coefficient "r" which has possible values ranging from -1 to 1. Positive r-values indicate that a larger V_i is associated with a larger V_k, whereas a negative r indicates that a larger V_i is associated with a smaller V_k. The closer the r-value is to 0, the more independent the two variables. For example, in a study measuring height (H), weight (W), and percent body fat (BF), $r = 0.486$ for W vs. H, $r = 0.074$ for H vs. BF, and $r = 0.613$ for W vs. BF.[4] Based on the results of this simple correlation matrix, one would probably not include both parameters H and W, nor both W and BF, but would keep H and BF as host factors in a resulting model.

The correlation matrix approach is considered a semi-quantitative measure for overall data independence interpretation because it treats variables two at a time; one cannot discern directly how multiple variables, or combinations of variables, correlate. A more sophisticated approach is available wherein Eigenvector projections are calculated for all variables in n-dimensional space. Using statistical software such as proc VARCLUS from SAS Inc. (Cary, NC, USA), it is possible to develop a "dendrite" diagram that can be used to create clusters of variables that have certain levels of co-linearity. This is similar to principle components analysis (PCA), but rather than grouping samples, this approach groups variables. Variable clustering serves two purposes, it improves model stability by identifying/removing collinearity, and reduces the total number of variables for a more parsimonious model.[5]

As an example, consider the dendrite diagram (dendrogram) in Figure 1.3 where we show a generic example of both host factor and measurement data clustering.[5] In this case, we use a hypothetical data structure with six host factors (gender, BMI, age, etc.) and seven environmental/biomarker variables (e.g. concentration measurements in blood or urine). This methodology has been applied to complex environmental dioxin congener data[6] and to studies of jet fuel exposure in the US Air Force.[7]

By collapsing the more highly correlated variables into clusters, we do not lose much explanatory power, but improve the interpretive power of subsequent models by increasing the ratio (n/m) between number of samples "n" and the number of independent variables "m." We have observed that $n/m \geqslant 10$ is a good rule of thumb for assessing the importance of individual variables.[6] We caution that this procedure requires a certain amount of judgment on the part of the investigator. The first issue is choosing the variance level vs. number of clusters. This is very dependent on the eventual n/m parameter. The second issue is how to combine information from variables collapsed into clusters. There are a number of choices that depend on the nature of the variables; in the above example, we use a simple sum, but other strategies can be employed as well. For example, given highly correlated variables such as HF3 and HF4 (especially if they are not continuous), one can just discard one or the other.

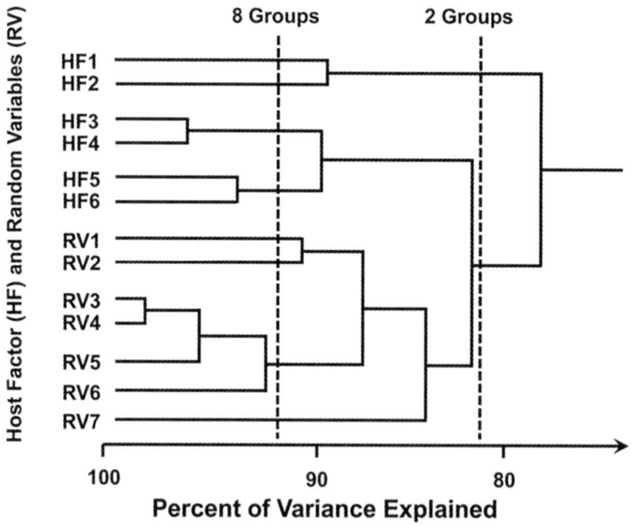

FIGURE 1.3

Example dendrogram of variable cluster analysis. Starting with a total of $m = 13$ independent variables ($p = 6$ host factors and $q = 7$ environmental variables), forming 8 clusters (HF1, HF2, HF3 + HF4, HF5 + HF6, RV1, RV2, RV3 + RV4 + RV5 + RV6, and RV7) results in explained variance of ~93%. Collapsing the clusters further to 2 Groups explains ~83% of the variance, but now is difficult to interpret.

For measured concentrations, it is sometimes better to normalize each measurement to a total to avoid having a particular variable overwhelm the sum.

1.2.3 Population statistics and variance components

Traditional mathematical analyses of breath biomarkers, both for clinical and environmental research applications, utilize measurement statistics (e.g. mean and median) to evaluate differences between groups. For example, it is common to evaluate measurements statistics to compare cases vs. controls, exposed vs. unexposed subjects, or males vs. females (as shown in our earlier example of a diesel intervention study). Furthermore, it is common to employ stratified data analyses—that is, analyses using multiple levels of organization—to reduce the impacts of confounding variables. Consider the evaluation of smoking effects on breath biomarker levels; one could perform a single evaluation using all subjects (smokers vs. non-smokers), two evaluations after stratifying by sex (male smokers vs. male non-smokers; female smokers vs. female non-smokers), four evaluations after stratifying by sex and disease status (male smoker [case] vs. male smoker [control]; male non-smoker [case] vs. male non-smoker [control]; female smoker [case] vs. female smoker [control]; female non-smoker [case] vs. female non-smoker [control]), and so on. These stratified analyses can narrow down the potential origins of an observed effect, but require larger sample numbers with each level of stratification. Thus, an investigator must balance the desired outcome of a given analysis with costs required to achieve sufficient statistical power and sensitivity.[6]

Given adequate sample numbers for stratified group analyses, it is prudent to first investigate the underlying distribution(s) of the biomarker data in question. Our earlier examples for data analysis (i.e. data visualization and variable clustering) can be considered qualitative or semi-quantitative, and generally have no *a priori* conditions for data structure. Hypothesis testing on the other hand, is entirely quantitative, and in many cases relies on distributions of measurement data being approximately Gaussian, or "normal." Simple diagnostic procedures exist in most software packages to evaluate data distributions. Often times, simple histograms, normal probability plots, or quantile-quantile plots can be visually inspected to evaluate normality assumptions. More advanced statistical packages offer statistical tests (e.g. Shapiro-Wilk and Kolmogorov-Smirnov) to confirm results from visual inspection.

Measurements of analytes in biological media are often "lognormally" distributed; that is, the distribution of the logged values is approximately normal. Lognormal data can be identified by a right-skewed distribution of the original (non-transformed) data, reflecting few values at exceedingly high levels, and many values near, but not below, a lower threshold (generally zero). Additional signs of lognormally distributed data include: (1) an observed arithmetic mean value that is larger than the median (reflecting the differential effects of extremely high values on these statistical parameters) and (2) an observed confidence interval that includes negative values (it is impossible to have negative amounts of a biomarker). When one encounters these signs, it is best to evaluate data transformation approaches before pursuing quantitative hypothesis testing.

"Parametric" testing procedures are generally used to evaluate normally distributed data and lognormally distributed data that have been log-transformed. Commonly used parametric tests include the Student's *t*-test, the paired *t*-test, and analysis of variance (ANOVA). A Student's *t*-test evaluates differences in biomarker measurements across two groups (e.g. men vs. women); a paired *t*-test evaluates differences in biomarker measurements between paired samples from individuals (e.g. pre-intervention vs. post-intervention); and ANOVA compares biomarker measurements across multiple groups (e.g. children vs. adolescents vs. adults). In the event that biomarker data are not Gaussian, or cannot be mathematically transformed to approximate normality, "non-parametric" methods exist for hypothesis testing. For example, the Mann-Whitney test, Wilcoxon test, and Kruskal-Wallis test are common substitutes for the Student's *t*-test, paired *t*-test, and ANOVA, respectively. These tests generally approximate the results of the parametric tests given a large enough sample size. However, the exact data requirements (e.g. sample number, sample independence, random selection) for any given test should always be considered prior to hypothesis testing.

The statistical tests discussed thus far generally utilize a single observation for each subject. (While the paired *t*-test uses two observations per subject, the difference between the two values [i.e. a single value] is used for hypothesis testing.) Therefore, different approaches to data analysis are required when multiple measurements exist for each subject. At the most basic level, it is of interest to evaluate the spread of the data, or the "variance," between and within subjects. To do this, the total variance across all measurements is first partitioned into that which is observed across repeated measurements of individual subjects, known as "within-person (intra-individual) variance," and that which is observed across average levels of all subjects known as "between-person (inter-individual) variance." These between- and within-person variance components can be calculated using a number of techniques.[8] The simplest of these methods is analysis of variance (ANOVA) which is suitable for balanced data sets where the same number of measurements exists for each subject. More complex methods, such as restricted maximum likelihood estimation (REML), may be required when working with unbalanced data sets; these methods are typically available only in more advanced statistical software packages.

Once the variance components are established, they become instrumental for identifying the origins of exposure or disease, and in turn, the best opportunities for mitigation or intervention. Large within-person variance and small between-person variance indicates little difference between subjects on average, but large differences over time for each subject. This result may point to a temporal event, random or otherwise, that affects each subject equally. Alternatively, small within-person variance and large between-person variance indicates little change over time, but marked differences between subjects. This result may point to a host-specific parameter (e.g. genotype, fitness-level) that dictates biomarker response. In the first example, a general intervention strategy, equally applicable to all subjects, might be suitable to reduce an exposure or limit a biological response. In the second example, a targeted strategy would likely be necessary to first identify the cause of elevated biomarker levels for certain individuals, and then for intervention.

1.2.4 Stochastic models and meta-data

After identifying group-based differences in biomarkers levels, and within- and between-subject variance components (for repeated-measures studies only), the next step is to develop statistical models using study meta-data. Statistical models serve two functions for breath research: (1) they allow investigators to simultaneously identify multiple significant predictors of breath biomarker levels and (2) they provide a platform for predicting breath biomarker levels in other studies where only meta-data exist.

An important decision for model development is identifying a dependent variable; this is not as easy as it sounds. Often, the dependent variable is obvious by design and represents the "outcome" for a subject; it can be binomial, that is, cancer/not cancer, infected/not infected, or it can be a continuous health outcome variable such as cholesterol level, total urinary protein, pulmonary function (e.g. forced expiratory volume in 1 s (FEV1), forced vital capacity (FVC)), or DNA damage (e.g. sister chromatid exchange (SCE), strand breaks), among others.

However, there are many occasions when the choice of the dependent variable is not obvious, especially in environmental or cross-sectional public health studies for which the analyst was not consulted in sampling design. For example, suppose one has measured a suite of exogenous chemicals and metabolites in exhaled breath, and has acquired meta-data concerning recent activity, job type, height, weight, gender, ethnicity, etc., but no health effects information was collected or observed. What could the dependent variable be for interpretation purposes? In such cases, the first question to be addressed is: What do we want to know? Generally, we would like to explore the linkages between suspected exposure sources and the resulting internal dose in human subjects. Suppose that in the totality of all measurements, we observe benzene, toluene, ethyl-benzene, and xylenes (BTEX), plus many other organic compounds in breath. If one of the suspected sources for all exposures is automobile exhaust, we could sum the BTEX numbers into one composite parameter and use this to represent the dependent variable for overall fuels exposure. This is a bit of a bootstrapping approach, especially if we keep the individual BTEX compounds as independent continuous variables, but often this is the only way to build a stochastic model in the absence of a designed dependent variable.

Given that we have a number of "independent variable" measurements and related meta-data and some consensus dependent (outcome) variable, then the next step is to determine which independent variables and data actually tell a story relating to the changes in the identified outcome variable. This requires some form of a modeling approach, and if the data include both host factors and continuous variables, the best approach is a generalized linear model, or a "mixed effects" model.[8,9]

A general form of such a model is:

$$Y_{hij} = \left\{\beta_1 X_{1_{hij}} + \beta_2 X_{2_{hij}} + \cdots + \beta_p X_{p_{hij}}\right\} + \alpha_h + b_{hi} + \epsilon_{hij},$$

where Y_{hij} is the value of some relevant biological parameter for the jth observation of the ith subject in the hth group; $X_{1_{hij}}, X_{2_{hij}}, \ldots, X_{p_{hij}}$ are the values of the fixed effect variables such as environmental chemical concentrations (in air, water, food, dust, etc.), and host factors such as age, health state, gender, genetic polymorphisms, ethnicity, etc.; p is the total number of fixed effects (note: the host factors may be fixed for all j within a given i); $\beta_1, \beta_2, \ldots, \beta_p$ are the corresponding modeled coefficients for the fixed effects and host factors; α_h is the random effect for the hth group; b_{hi} is the random effect for the ith subject from the hth group; and ϵ_{hij} is the residual (unexplained) error for the jth observation of the ith subject from the hth group.

Software applications for this style of approach are commercially available (e.g. proc MIXED, SAS). Upon calculation, the coefficients and their p-values can be interpreted to determine the effect of including the particular fixed effect or random effect variable in the final model for explaining the variance in the biological parameters' values. This can be done with iterative steps of forward addition or reverse elimination with the eventual objective being a parsimonious model without appreciable loss of modeling power. Once the final model is established, we can observe which exposure parameters and fixed effects are more likely to cause perturbations to the systems biology.[5]

In an earlier example, based on published results from a diesel exhaust intervention study, we demonstrated how visualization tools can be used to identify likely predictors of breath biomarker levels. Specifically, we used a 3D bar graph and a heat map to show the likely effect of gender, and the unlikely effects of the exposure intervention and sample time, on PVOC levels in EBC. Following from this qualitative work, we now demonstrate a statistical assessment of this data using mixed models. We note that these methodologies and results have been published alongside the aforementioned graphical work.[1]

PVOC levels in breath were treated individually as model dependent variables, whereas model independent variables included fixed effects for the time of sample collection (pre-intervention vs. post-intervention vs. 24 h post-intervention), the type of intervention (diesel exhaust vs. purified air), and gender (male vs. female), as well as random effects for subject and residual error. Results of the models are shown in Table 1.1, and confirm a significant gender effect ($p < 0.1$) for 4 out of 9 modeled PVOCs, with men having higher levels than women in all cases. While not shown in Table 1.1, results also confirm no significant effects ($p > 0.1$) of collection time or exposure intervention on these PVOCs. These quantitative results corroborate those of the earlier qualitative analyses. Furthermore, these results include variance components estimates (listed under "Random effects" in Table 1.1) that can be used to explore subject susceptibilities and possible intervention strategies.[10,11]

1.2.5 Dynamic models and longitudinal data

Dynamic models are used to interpret time dependence. In some study designs, the breath data structure investigates certain applied or observed external conditions. In diagnostic medicine, this could be a pre/post-drug treatment study, or a time-dependent study to monitor post-operative recovery status. In environmental studies, this could be a sample time series to determine the rates of uptake and elimination from different profiles of deliberate or incidental exposures to chemicals. Regardless of the exact design, the primary focus is temporally resolved data. Such studies are invaluable for deducing the time constants of absorption, distribution, metabolism, and elimination of exogenous compounds like pharmaceuticals and environmental pollutants, and for assessing different metabolic pathways. The underlying assumptions of such a study are that kinetic and physiological parameters measured under controlled conditions will be consistent for other exposure or treatment profiles. This allows us to model different scenarios to assess internal dose and metabolism.

For example, consider that we have gathered a series of breath samples pre-, during-, and post-exposure to a well-defined concentration of a specific compound.[12–14] There are a number of ways of interpreting such data using classical pharmacokinetic principles. Researchers including Wallace from US Environmental Protection Agency (EPA), Weisel from Rutgers University, and Raymer from Research Triangle Institute constructed closed form solutions from dynamic measurements to assess the time constants and internal (hypothetical) compartments of the human body such as blood and lymphatic fluids, highly perfused tissues including organs

Table 1.1 Mixed models results for PVOCs in EBC from subjects enrolled in a diesel exhaust intervention study.

PVOC	Fixed Effects						Random Effects					
	Intercept			Gender			$\hat{\sigma}_b^2$			$\hat{\sigma}_w^2$		
	Est.	*(SE)*	*p-Value*	*Est.*[a]	*(SE)*	*p-Value*	*Est.*	*(SE)*	*p-Value*	*Est.*	*(SE)*	*p-Value*
1-Hexanol	−0.592	(0.281)	0.07	1.27	(0.483)	0.01	0.438	(0.251)	0.04	0.138	(0.033)	<0.0001
Butanal	0.135	(0.355)	NS	1.31	(0.602)	0.04	0.583	(0.406)	0.08	0.710	(0.170)	<0.0001
Pentanal	0.151	(0.347)	NS	1.27	(0.588)	0.04	0.545	(0.382)	0.08	0.734	(0.175)	<0.0001
2-Methylpropanal	1.33	(0.273)	0.002	0.787	(0.453)	0.09	0.206	(0.240)	NS	1.04	(0.248)	<0.0001
1-Propanol	4.86	(0.304)	<0.0001			NS	0.733	(0.411)	0.04	0.450	(0.107)	<0.0001
1-Butanol	1.63	(0.191)	<0.0001			NS	0.274	(0.161)	0.05	0.239	(0.058)	<0.0001
Hexanal	1.32	(0.220)	0.0003			NS	0.325	(0.209)	0.06	0.498	(0.118)	<0.0001
1-Pentanol	0.079	(0.225)	NS			NS	0.338	(0.299)	0.1	1.14	(0.275)	<0.0001
2-Methyl 1-propanol	3.95	(0.373)	<0.0001			NS	0.542	(0.640)	NS	3.34	(0.796)	<0.0001

$\hat{\sigma}_b^2$: *Estimated between-subject variance.*

$\hat{\sigma}_w^2$: *Estimated within-subject variance.*

NS: Not significant.

[a] *Parameter estimates for male subjects; female subjects were the reference group (i.e. Est = 0).*

and muscle, poorly perfused tissues including tendons, bone marrow, and connective tissues, and finally adipose tissue.[15–17] These models were based on simple exponential uptake and decay behaviors with different time constants assigned empirically to the hypothetical compartments. This provided important information regarding the internal distribution of specific compounds, but did not provide organ-specific dose, nor did it allow generalizing to random intermittent exposures.

Subsequently, new methods were developed for interpreting observational kinetic data and estimating response to input functions for the general case. These included very complex physiologically based pharmacokinetic models (PBPK) that require a great deal of specific data, and also prior knowledge about organ dose kinetics, partition functions, and diffusion parameters that are generally only available from invasive animal studies.[18,19] Although PBPK models are powerful tools, they are beyond the scope for this discussion.

A hybrid method was developed by Pleil et al.,[12] wherein the closed form solution was replaced with an iterative solution. This approach requires a starter set of empirical data linked precisely to well-characterized exposure data. Typically, this is generated in a human exposure chamber.[2,20] The methodology is based on conversion of differential equations to difference equations that can then be easily evaluated in standard spread sheet software as follows.

In the simplest form, consider that one has concentration measurements of the central compartment (blood), $C_{\text{blood}}(t)$, and matching environmental air measurements, $C_{\text{air}}(t)$ for some exogenous compound. One further assumes some stable (first order) rate constant for inhalation uptake (k_u) and some aggregate first order elimination rate constant k_e in 1/time units. Under the assumption of linear kinetics,

$$\mathrm{d}C_{\text{blood}}(t)/\mathrm{d}t = k_u C_{\text{air}}(t) - k_e C_{\text{blood}}(t)$$

For a well-controlled air concentration profile (e.g. $C_{\text{air}}(0) = 0$ and $C_{\text{air}}(t > 0) = C_0$), this differential equation is easily solved with the closed form solution:

$$C_{\text{blood}}(t) = C_0(k_u/k_e)[1 - \exp(-k_e t)],$$

when the boundary conditions are applied. This function can then be fitted to the real data of the proposed step function exposure as could be generated in an environmental chamber, and the rate constants k_u and k_e can be estimated. Regrettably, such closed form solutions are only possible in the simplest conceptual models. In general, there are multiple compartments (blood, highly perfused tissues, poorly perfused tissues, adipose (fatty) tissues, connective tissues, etc.) that all have different capacities and time constants. Each compartment may further exhibit different elimination pathways and time constants as well. As such, the only way to model behavior is via iterative computational methods. We have developed a difference equation approach to compute biomarker concentrations from essentially any complex system given initial conditions, some kinetic data, and a reasonable conceptual model of the compartmental structure.

Using the simple example above, one can rewrite the differential equation as a difference equation in the form of:

$$\Delta C_{\text{blood}}(t)/\Delta t = k_u C_{\text{air}}(t) - k_e C_{\text{blood}}(t),$$

which can be expanded for any arbitrary time increment as:

$$[C_{\text{blood}}(t + \Delta t) - C_{\text{blood}}(t)]/[(t + \Delta t) - t] = k_u C_{\text{air}}(t) - k_e C_{\text{blood}}(t)$$

and subsequently rearranged as:

$$C_{\text{blood}}(t + \Delta t) = C_{\text{blood}}(t) + [k_u C_{\text{air}}(t) - k_e C_{\text{blood}}(t)]\Delta t,$$

which is a form wherein each future concentration can be calculated from the previous point for any complex exposure function $C_{\text{air}}(t)$ as long as the rate constants and initial conditions are known. This iterative procedure is successful for more complex models for which we can draw a conceptual diagram because we never need to deduce a closed form solution. Although beyond the scope of this discussion, Pleil[21] has presented a detailed discussion and application of these techniques for blood and breath biomarkers, multiple compartments, and human metabolites.

Figures 1.4a and 1.4b shows applications of this procedure to time-dependent blood and breath data for methyl tertiary butyl ether (MTBE) chamber exposures and the resulting phase-1 human metabolite tertiary butyl alcohol (TBA). Note that breath data have the same kinetic behavior as the blood data and that the TBA metabolite exhibits biological damping (delayed response) making it a longer time-frame marker for MTBE exposure.

Once the rate constant parameters for the known chamber exposures are calculated, the model can then be applied to any arbitrary exposure input function under the assumption that the concentration lie within the regime of linear kinetics. For environmental levels, this is almost always the case. As such, a solid longitudinal data set from known (controlled) exposures with precise empirical biomonitoring data can be used to implement a generalized model for any exposure profile.

1.3 CONCLUSIONS AND RECOMMENDATIONS

The premise of this chapter is simple: the value of breath analysis does not end with making a series of measurements. It is critically important to understand the underlying structure of the data, and then to apply the appropriate mathematical and statistical interpretations to be able to use the data to actually answer a particular question. Certainly, the level of detail and the exact techniques are dependent on the available data and meta-data (as well as the question). If one only wants to know what kind of compounds may be found in breath using a chosen sampling and analytical scheme, then simply listing a set of chemicals and their means and ranges may be sufficient. If one wants to determine how fast an anesthetic is eliminated after a surgery or an environmental compound is metabolized, then a simple time series

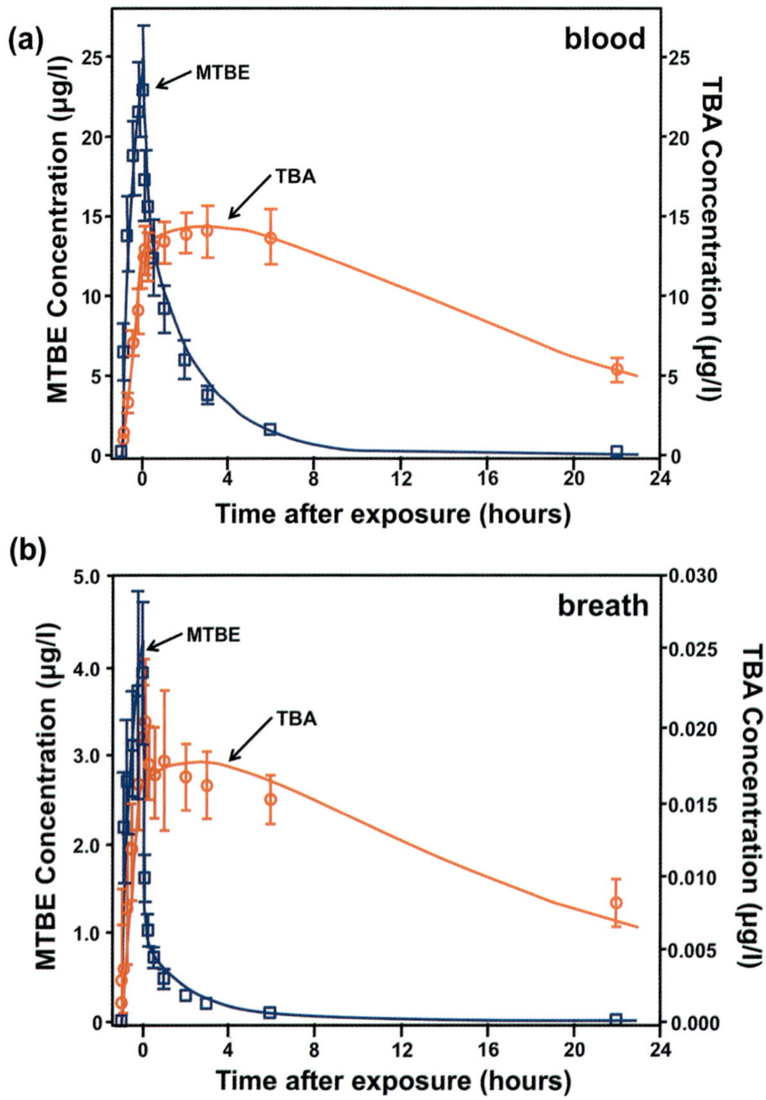

FIGURE 1.4
Observed concentrations and empirical models of MTBE and TBA biomarkers in (a) human blood and (b) human exhaled breath resulting from a 1 h exposure to 3 ppmv MTBE. Note that the phase-1 metabolite response of TBA is damped with half-life of ~12 h in contrast to the native compound MTBE with half-life of ~1 h.

analysis based on classical pharmacokinetics (rate constant) analysis will suffice. If, however, one wants to use breath analysis for discerning the reasons for a particular adverse health outcome, then a much more complex analysis that includes host-factor data, suites of compounds, repeated measurements, and a sufficiently large subject pool is required. In short, the data tell a story only if the mathematics and statistics are sound, and also if the sampling design is suited to actually answer the posed question in the first place.

DISCLAIMER

The United States Environmental Protection Agency through its Office of Research and Development funded and managed the research described here. This research has been subjected to Agency review and approved for publication.

Acknowledgments

The authors thank Terence Risby from Johns Hopkins University, Baltimore, MD, USA, Anton Amann from the Austrian Academy of Sciences, Innsbruck, Austria, Michael Phillips from Menssana Research Inc., Newark, NJ, USA, Wolfram Miekisch from Rostock University Hospital, Rostock, Germany, and Jens Herbig from Ionimed Analytik, Innsbruck Austria for many insightful discussion regarding the use of human breath biomonitoring data. We also thank Linda Sheldon, Tzipporah Kormos, Karen Oliver, and Myriam Medina-Vera of US EPA, and Matthew Stiegel from University of North Carolina, Chapel Hill for their expert advice.

References

1. Hubbard HF, Sobus JR, Pleil JD, Madden MC, Tabucchi S. Application of novel method to measure endogenous VOCs in exhaled breath condensate before and after exposure to diesel exhaust. *J Chromatogr B: Analyt Technol Biomed Life Sci* 2009;**877**:3652–8.
2. Sobus JR, Pleil JD, Madden MC, Funk WE, Hubbard HF, Rappaport SM. Identification of surrogate measures of diesel exhaust exposure in a controlled chamber study. *Environ Sci Technol* 2008;**42**:8822–8.
3. Pleil JD, Stiegel MA, Sobus JR, Liu Q, Madden MC. Observing the human exposome as reflected in breath biomarkers: heat map data interpretation for environmental and intelligence research. *J Breath Res* 2011;**5**:037104.
4. Burmaster DE, Murray DM. A trivariate distribution for the height, weight, and fat of adult men. *Risk Anal* 1998;**18**:385–9.
5. Pleil JD, Sheldon LS. Adapting concepts from systems biology to develop systems exposure event networks for exposure science research. *Biomarkers* 2011;**16**:99–105.
6. Pleil JD, Lorber MN. Relative congener scaling of Polychlorinated dibenzo-*p*-dioxins and dibenzofurans to estimate building fire contributions in air, surface wipes, and dust samples. *Environ Sci Technol* 2007;**41**:7286–93.

7. Pleil JD, Stiegel MA, Sobus JR. Breath biomarkers in environmental health science: exploring patterns in the human exposome. *J Breath Res* 2011;**5**:046005.
8. Rappaport SM, Kupper LL. *Quantitative exposure assessment*. El Cerrito, California, USA: Stephen Rappaport; 2008.
9. Singer JD. Using SAS PROC MIXED to fit multilevel models, hierarchical models, and individual growth models. *J Educ Behav Stat* 1998;**23**:323–55.
10. Sobus JR, McClean MD, Herrick RF, Waidyanatha S, Nylander-French LA, Kupper LL, et al. Comparing urinary biomarkers of airborne and dermal exposure to polycyclic aromatic compounds in asphalt-exposed workers. *Ann Occup Hyg* 2009;**53**:561–71.
11. Sobus JR, Pleil JD, McClean MD, Herrick RF, Rappaport SM. Biomarker variance component estimation for exposure surrogate selection and toxicokinetic inference. *Toxicol Lett* 2010;**199**:247–53.
12. Pleil JD, Kim D, Prah JD, Rappaport SM. Exposure reconstruction for reducing uncertainty in risk assessment: example using MTBE biomarkers and a simple pharmacokinetic model. *Biomarkers* 2007;**12**:331–48.
13. Pleil JD, Lindstrom AB. Exhaled human breath measurement method for assessing exposure to halogenated volatile organic compounds. *Clin Chem* 1997;**43**:723–30.
14. Pleil JD, Fisher JW, Lindstrom AB. Trichloroethene levels in human blood and exhaled breath from controlled inhalation exposure. *Environ Health Perspect* 1998;**106**:573–80.
15. Wallace L, Pellizzari E, Gordon S. A linear model relating breath concentrations to environmental exposures: application to a chamber study of four volunteers exposed to volatile organic chemicals. *J Expo Anal Environ Epidemiol* 1993;**3**:75–102.
16. Weisel CP, Kim H, Haltmeier P, Klotz JB. Exposure estimates to disinfection by-products of chlorinated drinking water. *Environ Health Perspect* 1999;**107**:103–10.
17. Raymer JH, Pellizzari ED, Thomas KW, Cooper SD. Elimination of volatile organic compounds in breath after exposure to occupational and environmental microenvironments. *J Expo Anal Environ Epidemiol* 1991;**1**:439–51.
18. Kim D, Andersen ME, Pleil JD, Nylander-French LA, Prah JD. Refined PBPK model of aggregate exposure to methyl tertiary-butyl ether. Toxicol Lett 2007;**169**:222–35.
19. Kim D, Nylander-French LA. Physiologically based toxicokinetic models and their application in human exposure and internal dose assessment. *Exs* 2009;**99**:37–55.
20. Prah J, Ashley D, Blount B, Case M, Leavens T, Pleil J, et al. Dermal, oral, and inhalation pharmacokinetics of methyl tertiary butyl ether (MTBE) in human volunteers. *Toxicol Sci* 2004;**77**:195–205.
21. Pleil JD. Influence of systems biology response and environmental exposure level on between-subject variability in breath and blood biomarkers. *Biomarkers* 2009;**14**:560–71.

CHAPTER

2 Issues and Challenges in Human Breath Research: Perspectives from Our Experience

Steven F. Solga* **and Terence H. Risby**†

*St. Luke's Hospital, 801 Ostrum Street, Bethlehem, PA 18015, USA School of Medicine, Johns Hopkins University, 733 North Broadway, Baltimore, MD 21205, USA

†Bloomberg School of Public Health, Johns Hopkins University, 615 North Wolfe Street, Baltimore, MD 21205, USA

2.1 INTRODUCTION

Breath analysis enjoys obvious intuitive appeal. There are many potential advantages of breath tests over conventional laboratory tests including that they are safe and painless, allow repeat sampling and analysis, attractive and easy, and can be performed anywhere from a pediatric clinic to an intensive care unit. Results can be available quickly, even immediately, allowing for on the spot therapeutic assessments. Breath analysis may also be inexpensive; after the investment of the monitor, there is little cost to an actual sample analysis. Perhaps the most exciting advantage is the potential capacity to evaluate disease and wellness states using molecules not readily measurable in the blood or by any other means. Some examples of molecules identified in breath, their biochemical foundations, and possible disease correlates are offered in Tables 2.1 and 2.2.

Given these positive advantages, it is surprising that breath analysis has not lived up to its potential. In principle, the requirements for the successful development of breath biomarkers should not be hard to achieve. Moreover, real-time monitors can be based on a variety of technologies, including mass or optical spectroscopy, fast gas chromatography, electrochemistry, etc. and these fields have experienced significant progress over the last decade on many fronts, including improved speed, size, sensitivity, and accuracy. Therefore, many investigators have anticipated that their breath research would lead to clinical success. Unfortunately, however, these aspirations have rarely been actualized. The reasons for these failures, and a path forward, can be understood in a brief hypothetical example.

Volatile Biomarkers. http://dx.doi.org/10.1016/B978-0-44-462613-4.00002-7

Table 2.1 Biochemical bases of breath molecules.

Molecule	Physiological Basis	Molecule	Physiological Basis
Water	Respiration	Ethanol	Gut bacterial metabolism
Carbon dioxide	Respiration	Acetaldehyde	Ethanol metabolism
Hydrogen	Carbohydrate metabolism	Isoprene	Cholesterol biosynthesis
Hydrocarbons	Lipid peroxidation/ metabolism	Methanol	Fruit metabolism
Carbon monoxide	Heme catabolism (HO-1)	Formaldehyde	Methanol metabolism
Nitric oxide	Biological mediator (iNOS)	Methylamine	Protein metabolism
1-Pentane	Lipid peroxidation	Ethene	Lipid peroxidation
Ethane	Lipid peroxidation	Methane	Carbohydrate metabolism
Ammonia	Protein metabolism	Methanethiol	Methionine metabolism
Acetone	Decarboxylation of acetoacetonate	Carbon disulfide	Gut bacterial metabolism
Hydrogen sulfide	Protein metabolism	Carbonyl sulfide	Gut bacterial metabolism
		Methyl sulfide	Methionine metabolism

Table 2.2 Current breath tests at various stages of development.

Clinical Test	Molecule
Capnography	Carbon dioxide
Neonatal jaundice	Carbon monoxide
Gastrointestinal diagnosis	Hydrogen, methane
Airway reactivity	Nitric oxide
Oxidative stress	Ethane, ethene, 1-pentane, aldehydes
Cholesterol biosynthesis	Isoprene
Renal function	Ammonia, alkylamines
Hepatic function	Ammonia, carbonyl sulfide, methanethiol, methyl sulfide
Infections (metabolites, host response)	Carbon monoxide, ethane, pentane, ethene, nitric oxide, hydrogen cyanide,
Heart transplant rejection	Branched chain hydrocarbons

2.2 DEFINING NORMAL IN CLINICAL PRACTICE: THE CASE OF A ROUTINE LIVER BLOOD TEST

Alanine aminotranferase (ALT) is a commonly ordered blood assay used as a marker of liver inflammation and/or injury. In a mid-sized community hospital, it would be measured on dozens of inpatients per day, with many more being performed in an outpatient setting. ALT is not specific to the liver, and may be abnormal (elevated) in disease states that do not involve that organ at all. Over the years, there have been promising breath assays proposed which may also measure liver inflammation. Might there be an important opportunity here?

Normal ALT ranges, like those of other blood assays, were defined years ago by using a presumably "healthy" reference population, often blood donors or other large databases. Usually, this involved evaluation of hundreds if not many thousands of samples. Commonly, "abnormal" is simply defined as plus two standard deviations. However, the upper limit of normal (ULN) remains debated; as of April 2011, at St. Luke's Hospital in Bethlehem, Pennsylvania, USA, the ULN is 65 IU/L. However, the large, commercial vendors, LabCorp and Quest Diagnostics, define the ULN as 55 and 40 IU/L, respectively. However, it is recognized that many of those presumed "healthy" may have had significant undiagnosed serious liver disease, especially hepatitis C and fatty liver (e.g. the blood assay for hepatitis C was not available when many of these studies were done, *and* the population at that time was not obese). Moreover, these ranges often do not take in account age, gender, race, and physiological conditions such as pregnancy which have been known to affect ALT.[1] A more recent analysis, supported by the American Association for Liver Disease (AASLD), endorses a new upper limit of 30 IU/L for men, and 19 IU/L for women.[2] Even more importantly, even if the above issues are settled with clarity, it is rarely certain that just because a patient has an abnormal ALT, they have a (liver) disease, and the opposite is also true. These issues are ubiquitous in medicine; doctors are tasked with interpreting data in a context, which requires experience and knowledge of biology.

2.3 DEVELOPING A BREATH TEST: CAN THE BLOOD ASSAY BE UNSEATED?

While breath analysis enjoys many potential advantages over other diagnostic modalities such as blood assays, important drawbacks must be acknowledged. Imagine that your lab, after much effort and investment, has developed a Breath-LFT, or b-LFT, that perfectly mirrors the blood assay in your lab using reference gases. Your team, with input from basic scientists in liver physiology, is convinced that b-LFT measures liver inflammation as certain as the blood assay. Still, your team must address several concerns to bring this to clinical utility.

First, and perhaps most importantly, you must interface your monitor reliably with a human being. This means there must be an instrument to precisely and reproducibly collect exhaled breath and perform the analysis. Your strategy must control for variations in the breath cycle and the rather unpredictable way in which human subjects breathe when tested. This may be particularly challenging if b-LFT relies on volatile, hard to measure compounds, especially if background, ambient air contamination is an issue.

Only when this step has been thoroughly mastered can you then move to the next step: namely, the evaluation for potential confounders that may affect b-LFT measurement. While some of these confounders may also influence the standard blood assay (e.g. gender, obesity, etc.), many others may impact breath only (e.g. poor dentition, pulmonary disease, tobacco use, etc.). Thus, your burden to probe b-LFT for confounding variables will be much greater than for the blood assay.

Next, you will need to scale it up. That is, in order to convincingly define normal ranges, you will need to evaluate hundreds or, better, several thousand healthy volunteers. Practically speaking, in order to do this, you will need several monitors with several testing sites, all of them working equally well and in exactly the same way (again, potentially controlling for background air contamination). Only after this is accomplished can you contemplate the evaluation of b-LFT in subjects with known liver disease, and then you will need to study a variety of liver diseases to be really convincing.

2.4 CONVINCING CLINICIANS

Imagine you have enjoyed great success testing b-LFT. You have proceeded through all of the above steps, managing to solve unforeseen problems, defining normal ranges and disease states precisely, and you are still convinced b-LFT reflects liver inflammation as well as its blood counterpart. Will the clinicians and thought leaders of the AASLD endorse it? Will they incorporate it into their own research, and use it to manage patient care? Probably not. Despite all of its known problems, clinicians will continue to use blood ALT for several good reasons: the biology is reasonably well understood, the ALT blood assay itself is highly reproducible with little if any sampling error, there are decades of experience of evaluating liver disease using blood ALT, it is used worldwide, and it is part of a chemistry panel (e.g. the comprehensive metabolic panel, CMP) that is routinely measured in both in- and outpatients. To elaborate more on this last point, the CMP inexpensively measures multiple important blood chemistries of interest, including serum sodium, potassium, creatinine, bilirubin, etc. It is almost always the case that clinicians want these other chemistries in addition to ALT; thus, they will order a CMP regardless of the existence of a perfect surrogate b-LFT. Thus, the b-LFT would be an extra test for the patient to undergo and expense, even if it is nearly free.

Finally, episodic sampling for blood ALT, for almost all purposes, is useful and the results are usually available within a couple of hours. Thus, despite flaws of blood ALT and *the initial promise of b-LFT*, perhaps it was not a good idea from the beginning. Your efforts are lost. Many lesser failures populate the breath analysis literature. At times, this has dampened enthusiasm for serious investment, leading to unhappy clichés like: *Breath Analysis: The Greatest Clinical Tool of the Future, and Always Will Be.*

2.5 BREATH MARKERS

The hypothetical blood ALT *versus* b-LFT example introduces another critical issue: almost always, the onset of a disease can be captured in changes in the concentration of a given marker, not its presence or absence. There are very few yes/no markers in medicine; in general, laboratory and clinical interpretation are required. For example, tumor markers such as alpha fetoprotein (AFP) for liver cancer are *present* in all humans; this blood test can only *suggest* liver cancer when elevated outside a defined range and *only then* within a context (i.e. benign viral hepatitis can elevate AFP 20× the ULN). Even highly accurate over-the-counter urine pregnancy tests, which rely on the measurement of beta-human chorionic gonadotropin (β-HCG), must distinguish physiologic β-HCG from pregnancy, and still can be falsely positive in women (and men) who are not pregnant but rather have a β-HCG secreting tumor. This is true of essentially all other tests; e.g. with ALT, while the "normal" range is debated, a result of 50, 100, 1000, or 10,000 (all which are routinely possible) must be placed in a clinical context to have any significance.

There has been a longstanding but misguided belief that breath analysis can identify and measure unique markers of disease. Unique biomarkers in breath *can only* originate from the ingestion, inhalation, or dermal absorption of foreign substances or from bacterial, viral, or fungal metabolism. Parenthetically, it may be worth noting that some of the only yes/no blood tests in medicine are similar (i.e. tests which measure the presence or absence of, for example, the viruses hepatitis B or C). Thus, the onset of disease results in changes in the concentrations of breath molecules and not the production of unique breath biomarkers. While there may be many applications in clinical medicine where breath analysis may outperform blood analysis (e.g. ammonia) and other applications where breath analysis holds promise in measuring a molecule not readily measurable in the blood (e.g. ethane), these molecules are *not unique* to breath, and must be interpreted like any other laboratory test (e.g. ALT), with all of the associated issues articulated above.

2.6 CONCLUSION

Breath analysis holds untold potential, and with the dramatic advances in instrumentation and a worldwide surge of research, it is an exciting time. However, in order to "*Make Breath Analysis a Clinical Reality,*" the following issues must be understood:

i. *Choose a target compound and disease state very carefully. Enlist a clinician from the start.* This is important not only from standpoint of basic biological plausibility, but also to ask "even if this project works perfectly, will anyone actually use it?" There are three generic risks to any project development with a commercial ambition: scientific risk, regulatory or legal risk, and market risk. This point requests that the last risk be considered with the first.
ii. *Carefully interface the monitor with human breath; very compulsively probe for variation and confounders.* This point mandates expertise in the complex physiology of breath collection. Some of the requirements for diagnostic breath analyses are shown in Table 2.3.
iii. *Anticipate the need to evaluate very many subjects, including healthy volunteers as well as those with the disease of interest. This will require heavy statistical*

Table 2.3 Requirements for a useful clinical diagnostic breath test.

Identities of molecules found in the breath of healthy subjects must be well documented for the following:

- The biochemical pathways for their production and the relationships between these biochemistries and disease
- The normal ranges and limiting concentrations for the breath molecules for "healthy" subjects
- The variations of normal ranges and limiting concentrations for the breath molecules with age, gender, and ethnicity
- Other sources of these molecules that may confound the analysis

Concentrations of molecules that are markers of disease states, or early indicators of disease must be well documented for the following:

- The minimum concentrations of breath biomarkers indicative of disease
- The relationship between concentrations of breath biomarkers and disease state or staging of disease
- The variations of abnormal concentrations of breath biomarkers with age, gender, and ethnicity

input. Perhaps the most common shortfall in the published human breath literature is small sample size. Given all the variables and uncertainties of human biology, there can be little confidence in conclusions that involve only a few subjects. As noted above, publications on standard blood ALT, a far from perfect marker of liver inflammation, generally have 20–100× the number of subjects that are cited in most human breath publications. This is achievable and must be done. And it should be noted, on the upside, that research in monitor development and clinical research in human breath analysis can be done on a large scale for a small fraction of the time and cost of drug development.

In summary, it is an extremely exciting time in the field of breath analysis, and there is real reason for enthusiasm. Breath analysis research must mature, in the ways presented herein, in order for it to realize its long-held potential.

References

1. Dufour DR, Lott JA, Nolte FS, Gretch DR, Koff RS, Seeff LB. Diagnosis and monitoring of hepatic injury. II. Recommendations for use of laboratory tests in screening, diagnosis, and monitoring. *Clin Chem* 2000;**46**:2050–68.
2. Prati D, Taioli E, Zanella A, Della Torre E, Butelli S, Del Vecchio E, et al. Updated definitions of healthy ranges for serum alanine aminotransferase levels. *Ann Intern Med* 2002; **137**:1–10.

PART

B

Real-Time Analysis of Exhaled Breath

CHAPTER

3 Physiological Modeling for Analysis of Exhaled Breath

Julian King[*,†], Helin Koc[*,‡], Karl Unterkofler[*,‡], Gerald Teschl[†], Susanne Teschl[§], Pawel Mochalski[*,], Hartmann Hinterhuber[††], and Anton Amann[*,‡‡]**

[*]*Breath Research Institute, Austrian Academy of Sciences, Rathausplatz 4, 6850 Dornbirn, Austria*

[†]*University of Vienna, Faculty of Mathematics, Nordbergstr. 15, 1090 Wien, Austria*

[‡]*Vorarlberg University of Applied Sciences, Hochschulstr. 1, 6850 Dornbirn, Austria*

[§]*University of Applied Sciences Technikum Wien, Höchstädtplatz 5, 1200 Wien, Austria*

[**]*Institute of Nuclear Physics PAN, Radzikowskiego 152, 31342 Krakow, Poland*

[††]*Univ.-Clinic for Psychiatry, Innsbruck Medical University, Anichstr. 35, 6020 Innsbruck, Austria*

[‡‡]*Univ.-Clinic for Anesthesia, Innsbruck Medical University, Anichstr. 35, 6020 Innsbruck, Austria*

3.1 INTRODUCTION

There is nothing more practical than a good theory (David Hilbert[§§]).

Due to its broad scope and applicability, breath gas analysis holds great promise as a versatile framework for general bio-monitoring applications. As a biochemical probe, volatile organic compounds (VOCs) in exhaled breath are unique in the sense that they can provide both *non-invasive* and *continuous* information on the metabolic and physiological state of an individual. Apart from diagnostics and therapy control, this information might potentially be used for dynamic assessments of normal physiological function (e.g. by a stress test on a cycle ergometer, in an intra-operative setting, or in a sleep lab), pharmacodynamics (drug testing), or for quantifying body burden in response to environmental exposure (e.g. in occupational health).[1–8] Furthermore, as has been impressively demonstrated in the Nobel prize-winning work of Furchgott, Ignarro, and Murad on the small inorganic molecule nitric oxide, trace gases can actively participate in the regulation of physiological events.[9,10] This renders breath VOCs as an intriguing tool for examining more fundamental endogenous processes.

While sufficiently accurate and fast instrumental techniques are vital for fully exploiting this huge potential, also several methodological issues need to be addressed before exhaled breath analysis can supplement blood tests. Specifically, a crucial yet underestimated aspect of exhaled breath analysis concerns the blood-gas kinetics of the VOCs under scrutiny as well as their systemic distribution and physiological

[§§]Also attributed to Immanuel Kant.

Volatile Biomarkers. http://dx.doi.org/10.1016/B978-0-44-462613-4.00003-9

flow within the human body. Much can be learnt in this context from the cumbersome development of the nitric oxide breath test, which nowadays ranks among the most successful breath tests based on an endogenous compound.[1] This paradigmatic example shows that a purely explorative search for VOC biomarkers is not sufficient: in order to make breath tests operational, a thorough quantitative understanding of the underlying gas exchange mechanisms and their sensitivity with respect to physiological factors is required. In particular, such knowledge can then contribute to the implementation of unified sampling protocols, similar to the ATS/ERS guidelines for determining exhaled lower respiratory tract nitric oxide and nasal nitric oxide.[11]

Due to the current lack of definite sampling recommendations for other trace gases, *end-tidal* breath collection is usually adopted as best sampling practice. Accordingly, the first part of the exhalation—corresponding to the gas volume filling the anatomical dead space and hence not participating in gas exchange according to classical pulmonary inert gas elimination theory—is discarded and only the last exhalation segment (supposed to contain "alveolar air") is collected for analysis. In more sophisticated sampling systems, end-tidal extraction is triggered by virtue of simultaneously measured CO_2-concentrations and/or respiratory flow.[12,13] While such setups are an important contribution to current standardization efforts in breath sampling, they generally give little insight into a series of major experimental questions:

- What is the quantitative relationship between breath VOC levels and their underlying endogenous (e.g. blood) VOC levels?
- Does the end-tidal VOC concentration level reflect the alveolar level?
- How does the variability of breath VOC concentrations relate to varying physiological conditions?

In this context, mathematical modeling and simulation can be employed to establish mechanistic descriptions of the gas exchange processes governing the VOC under scrutiny. Once such a model has been developed it becomes possible to understand the key relationships underlying the physiological behavior of the compound and to identify its defining parameters. Such a quantitative treatment provides a sound conceptual framework for experiment design as well as decision making and can aid substantially in preventing misinterpretations of empirical results.[14]

Hence, in the following we shall try to answer the above questions by modeling the exhalation patterns of two highly abundant, endogenous trace gases found in human breath:

- isoprene, which has a low solubility in blood (as reflected by a small blood:gas partition coefficient $\lambda_{\mathrm{b:air}} \approx 0.95$),[56] is lipophilic and presumed to be correlated with cholesterol biosynthesis[15] and
- acetone, which is relatively much more soluble in blood ($\lambda_{\mathrm{b:air}} \approx 340$) and has been linked to fat catabolism.

Both isoprene and acetone can be viewed as prototypical examples for the analysis of two major classes of VOCs, even though they cannot cover the whole spectrum of different physico-chemical properties. Furthermore, these two compounds rank among the most notable VOCs studied in the context of breath gas analysis and thus represent natural choices for a more detailed assessment of observable trace gas behavior in exhaled breath.

3.2 REAL-TIME MEASUREMENTS: EXPERIMENTAL BASICS

The range of measurement techniques employed for breath gas analytical investigations is extremely diverse and each method comes with its specific strengths and weaknesses.[2,16] Within the present setting, we will mainly focus on direct mass spectrometry, particularly proton transfer reaction mass spectrometry (PTR-MS).[17–20] In brief, this technique has proven to be a sensitive method for the quantification of volatile molecular species on the basis of chemical ionization within a drift reactor tube. More specifically, it exploits the proton transfer reaction of precursor hydronium ions, H_3O^+, originating from an adjoint hollow cathode discharge, with molecules M, viz.,

$$H_3O^+ + M \rightarrow MH^+ + H_2O$$

Specifically, this reaction process is selective to VOCs with proton affinities higher than water (166.5 kcal/mol), thereby avoiding the ionization of the bulk components of exhaled air, N_2, O_2, and CO_2. Count rates of the nascent product ions MH^+ or fragments thereof appearing at specified mass-to-charge ratios m/z can subsequently be converted into absolute concentrations of the investigated compounds. Further details on quantification can be found in Ref. 21.

A major hallmark of PTR-MS is its *real-time* capability, allowing for concentration measurements of VOCs with breath-by-breath resolution (i.e. on a time-scale of <0.5 s). The possibility of generating high sample rate data can be viewed as an essential requirement for relating short-term changes in breath VOC concentrations to quick physiological variations (e.g. in blood or ventilatory flow). An experimental setup combining real-time PTR-MS trace gas measurements with data streams reflecting hemodynamic and respiratory parameters is shown in Figure 3.1. For a detailed description the interested reader is referred to Ref. 22 (see also Ref. 23 for cross-validations by means of independent GC-MS data).

The above-mentioned instrumentation allows for the parallel evaluation of *end-tidal* VOC concentration profiles and several decisive physiological factors during rest, exercise challenges on a cycle ergometer, and in a sleep laboratory setting.[24,25] Such synchronized data streams are shown in Figure 3.2 and represent an indispensable phenomenological basis pertinent to any quantitative modeling approach.

3.3 MODELING

In classical pulmonary inert gas elimination theory the relationship between the blood concentrations of VOCs and their respective concentrations in the gaseous phase is captured by the well-known Farhi equation.[26] It predicts a direct proportionality

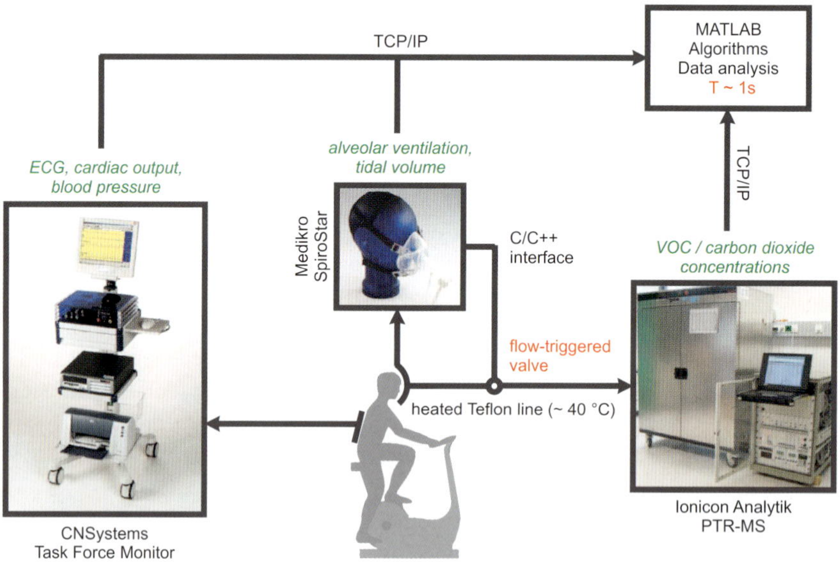

FIGURE 3.1
Experimental setup used for obtaining VOC profiles in conjunction with a number of physiological parameters (reproduced with permission from Ref. 22). Items in italic correspond to measurable variables. A selective analysis of predefined breath segments is achieved by flow-triggered sample extraction. More specifically, an automatic shutter valve installed along the gas sampling line guarantees that breath samples are drawn into the PTR-MS device exclusively during the end-tidal phase of each exhalation.

between the alveolar concentration C_{A} of a VOC (presumed to be accessible from end-tidal exhalation segments) and its concentration C_{a} and $C_{\bar{\mathrm{v}}}$ in arterial and mixed venous blood, respectively. Here, the associated proportionality factor depends on the substance-specific blood:gas partition coefficient $\lambda_{\mathrm{b:air}}$ (describing the diffusion equilibrium at the alveolar-capillary interface according to Henry's law), alveolar ventilation $\dot{V}_{\mathrm{A}}$ (governing the transport of the compound through the respiratory tree), and cardiac output $\dot{Q}_{\mathrm{c}}$ (controlling the rate at which the VOC is delivered to the lungs), viz.,

$$C_{\mathrm{measured}} = C_{\mathrm{A}} = \lambda_{\mathrm{b:air}} C_{\mathrm{a}} = \frac{C_{\bar{\mathrm{v}}}}{\lambda_{\mathrm{b:air}} + \dfrac{\dot{V}_{\mathrm{A}}}{\dot{Q}_{\mathrm{c}}}} \tag{3.1}$$

An elementary but particularly useful tool for deriving such formulae is to employ the fundamental physical concept of *mass balance* coupled with a *compartmental* model design. Within a physiological framework, the compartmental approach to mass transfer traditionally consists in dividing the body into an ensemble of localized anatomic structures (like lung, liver, kidney, etc.) or roughly homogenous tissue volumes (lumping together tissue groups of a given type such as fat, muscle, viscera, bone, etc.)

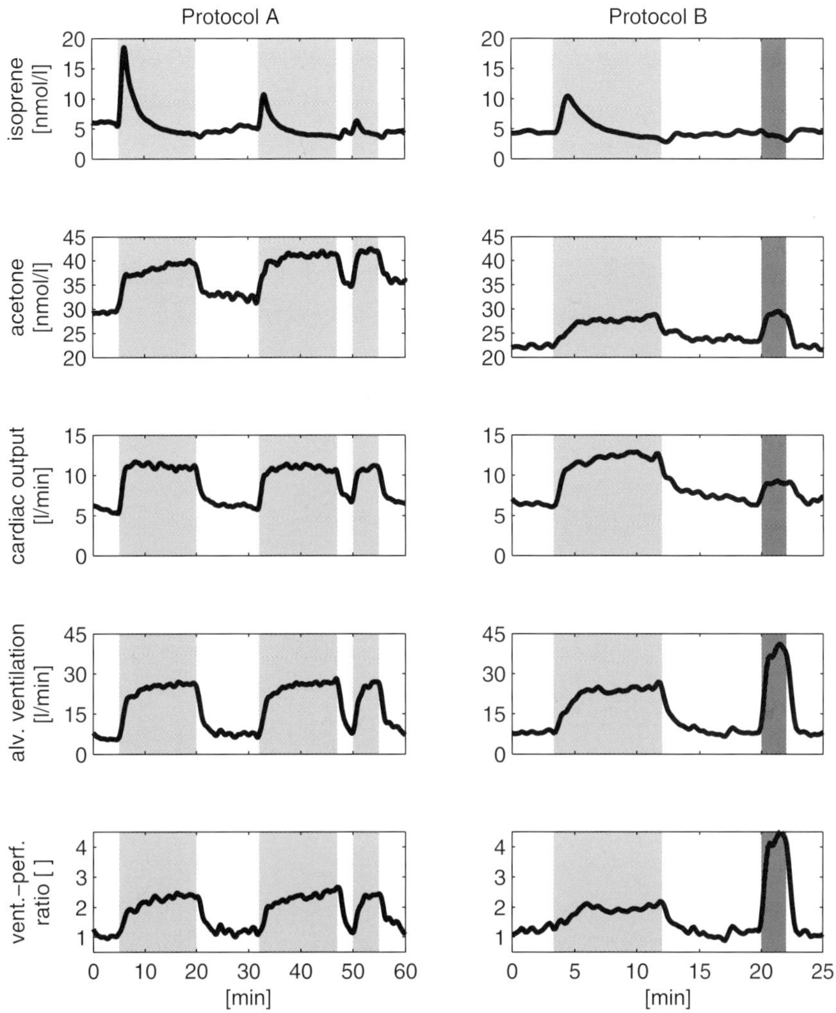

FIGURE 3.2

Typical smoothed profiles of end-tidal isoprene and acetone concentrations as well as hemodynamic and respiratory parameters in response to two distinct workload regimes. Segments shaded in light and dark gray indicate exercise and hyperventilation, respectively. *Protocol A*: rest (0–5 min), ergometer challenge at 75 W (5–20 min), rest (20–32 min), ergometer challenge at 75 W (32–47 min), rest (47–50 min), ergometer challenge at 75 W (50–55 min), rest (55–60 min). *Protocol B*: rest (0–3 min), ergometer challenge at 75 W (3–12 min), rest (12–20 min), hyperventilation with increased tidal volume (20–22 min), rest (22–25 min). Data correspond to one representative healthy male volunteer from the study cohort in Ref. 22.

in which the investigated VOC can reasonably be postulated to behave uniformly. These compartments are subsequently interconnected via the arterial and venous network, yielding a simplified representation of the human organism that reflects known or hypothesized physiology and biochemistry.[27–31] A typical example of such a model is shown in Figure 3.3.

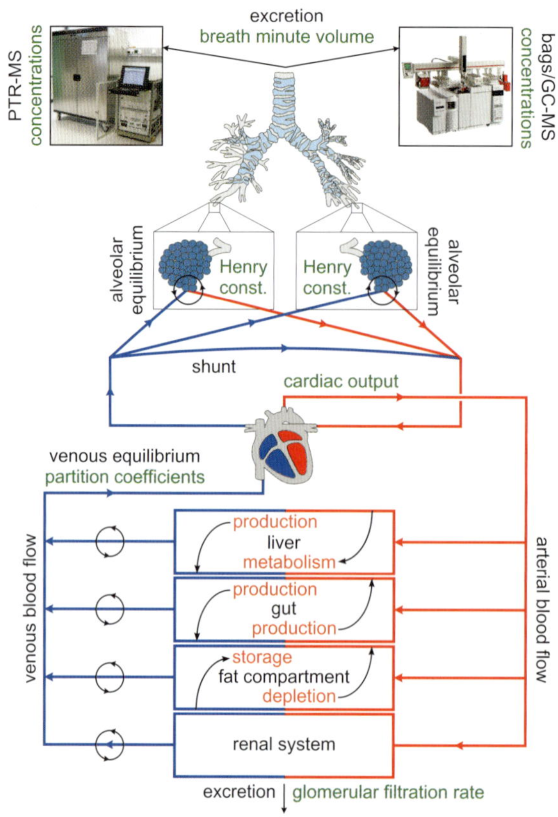

FIGURE 3.3

Physiologically based compartment model for the distribution of an endogenous volatile organic compound within the human body. The spectrum of model parameters can roughly be divided into fixed, measured, and unknown variables. The first class typically comprises quantities that have been determined a priori and are assumed to remain constant during the course of experimentation, such as approximate tissue volumes or partition coefficients. Information of this type has been compiled by several investigators (see, e.g. Refs. 27,32, and references therein). Measured parameters include quantities that are accessible by virtue of the given experimental setup (blood flow, ventilation), while unknown parameters (e.g. production, metabolism, storage, and depletion rates) have to be extracted by calibrating the model using available experimental data.

Since their introduction in the 1930s,[33] compartment models have received widespread attention in the biomedical sciences. The reason for this popularity is that the model structures bear a direct correspondence with the underlying physiology, thereby maintaining a clear-cut interpretation of the involved parameters. Moreover, by writing down the associated mass balance relations[***] the underlying model structure can directly be expressed as a set of ordinary differential equations (ODEs) of the form

$$\dot{\mathbf{x}}(t) = \mathbf{g}(\mathbf{x}(t), \mathbf{u}(t), \vartheta), \quad \mathbf{x}(t_0) = \mathbf{x}_0$$

These equations describe the evolution of a *state vector* $\mathbf{x}(t)$, belonging to an open set $X \subseteq \mathbb{R}^n$, within an observation interval $[t_0, t_{\max}]$. The right-hand side $\mathbf{g}$ usually is a continuously differentiable function with respect to its arguments and depends on a set of *measured*, possibly time-varying variables $u \in U \subseteq \mathbb{R}^q$ that are postulated to be uniformly bounded over $[t_0, t_{\max}]$ as well as on a vector $\vartheta \in \Theta \subseteq \mathbb{R}^p$ lumping together a set of *constant*, unknown model parameters. The latter will have to be estimated from available process data (possibly together with some components of the initial conditions $\mathbf{x}_0$). In the present setting, $\mathbf{x}$ typically reflects a vector of molecular concentrations of the investigated trace gas within the tissue compartments introduced; $\mathbf{u}$ stands for external inputs (blood/respiratory flows, temperature, etc.) that can be modified by experimentation; ϑ includes kinetic constants such as endogenous production or metabolization rates. Furthermore, information on the model dynamics can be collected via the (scalar) *measurement equation*

$$y(t) = h(\mathbf{x}(t), \vartheta),$$

where h is some function defining the observable model output (i.e. the measurable breath concentration). Specific examples for the framework introduced above can be found in Sections 3.3.1 and 3.3.2.

It is instructive to note that the Farhi equation (3.1) represents a simple form of such a compartment model, involving only one gas exchange unit (the lungs) and assuming steady state conditions throughout the measurement (i.e. the rate of gas build-up/depletion in the lungs is 0). It is therefore not surprising that this formulation might be inadequate for explaining non-steady state behavior, e.g. in response to relatively quick physiological transitions (for instance during exercise). Indeed, from the data in Figure 3.2 the profiles of isoprene and acetone appear to considerably depart from the trend predicted by Eq. (3.1). More specifically, at the onset of exercise the ventilation-perfusion ratio $\dot{V}_A/\dot{Q}_c$ increases drastically and hence, other factors being equal, a more or less pronounced drop in the respective breath concentrations could be expected.[†††]

As we shall illustrate in the following, in the case of isoprene this discrepancy can be explained by reference to a peripheral (extrahepatic) production of the compound,

[***] Stating that the rate at which the compartmental amount of a VOC changes per time unit equals the rate of mass transfer into the compartment less the rate of removal from the compartment.

[†††] Such a drop can in fact be observed for some endogenous breath VOCs, e.g. butane (cf. Figure 2 in Ref. 23).

which is realized by complementing the basic Farhi model with two systemic compartments affecting the time evolution of the mixed venous concentration $C_{\bar{v}}$ according to their fractional perfusion. For acetone, a mechanistic description of the profiles in Figure 3.2 can be achieved by taking into account pre- and post-alveolar gas exchange. This amounts to dividing the pulmonary compartment into two parts, the alveolar tract and the conducting airways.

3.3.1 Isoprene

3.3.1.1 Background

Due to its volatility and low affinity for water and blood, isoprene exchange occurs mainly in the alveolar region. Physical activity causes marked changes in exhaled breath isoprene concentrations of humans.[22,34,35] Extensive real-time measurements during moderate workload ergometer challenges have been conducted by Karl et al.[34] and King et al.,[22] showing an exercise-induced increase of breath isoprene concentrations by a factor of 2–3 in mixed exhaled breath,[34] and by a factor of 3–4 in end-tidal breath.[22] Typical profiles are shown in Figure 3.2. While Karl et al. have explained this behavior by reference to a delay in the rise of ventilation with respect to the rise in perfusion at the start of exercise (i.e. by assuming a temporary drop in the average pulmonary ventilation-perfusion ratio), recent evidence points toward a substance-specific wash-out from an isoprene buffer tissue within the human body.

3.3.1.2 Some insightful experiments

As can be deduced from Figure 3.2, exercise bouts at constant workload interrupted by breaks of variable duration lead to markedly different heights of the characteristic breath isoprene concentration peak at the onset of pedaling, despite an almost identical behavior of cardiac output and alveolar ventilation throughout all workload segments.[22] Additional information can be distilled from these profiles by performing one-legged ergometer experiments at moderate workloads,[36] which appear to favor the hypothesis of a peripheral tissue source affecting breath isoprene output. More specifically, a switch of the working leg after a short break following an ergometer exercise segment at 50 W results in an immediate recovery of the initial peak height, whereas continuing the exercise with the same leg leads to a wash-out effect similar to the two-legged case.

Experiments carried out by stimulating muscle groups with distinct masses provide some hints with respect to the tissues involved in this behavior.[37] Breath isoprene concentration profiles during repeated arm-crank exercises with intermediate pauses qualitatively resemble the trends observed during ergometer challenges. However, compared to two-legged ergometer exercise, arm-crank exercise induces much lower peak isoprene concentrations, despite similar time courses of the ventilation-perfusion ratio within both exercise regimes. From these experiments it might hence be concluded that a major part of isoprene variability during exercise phases can be ascribed to an increased fractional perfusion of the working locomotor muscles, eventually leading to higher isoprene concentrations in mixed venous blood at the onset of physical activity. This rationale is in accordance with the predominant physiological role of working muscles during exercise. Collectively, the skeletal muscles constitute up to 40–45% of body weight, which is more than any other single organ. At rest, about 10–15% of cardiac output is distributed to skeletal muscle, while during strenuous exercise the latter may receive more than 80% of the total blood flow. This underlines the importance of skeletal muscle as one of the predominant factors in systemic hemodynamics.[38]

3.3.1.3 A physiological model for breath isoprene dynamics

In the light of the experiments outlined above the mathematical model of isoprene distribution to be presented here adopts the assumption of a peripheral source of isoprene in the body. Following the methodological steps described at the beginning of Section 3.3, the lung is modeled by one single alveolar unit and the systemic part consists of two homogenous compartments: a richly perfused tissue compartment (including intestines, liver, brain, connective muscles, and skin) and a peripheral tissue compartment. The model structure is sketched in Figure 3.4. For the associated compartmental mass balance equations as well as model validation and estimation we refer to the original publication.[36] In order to correlate the model output to experimental outcomes, it is postulated that the measurable breath isoprene concentration corresponds to the alveolar level C_A.

While production and metabolic elimination of isoprene occurs in both systemic compartments, peripheral tissue takes the role of an isoprene buffer and is postulated to contain the working muscle groups, which receive a disproportionally high fraction of the systemic blood flow during exercise. According to the simulation in Figure 3.5, at rest the peripheral compartment is characterized by high isoprene concentrations resulting from extrahepatic production at a constant rate. However, due to the relatively small fraction of total blood flow to these tissues, the mixed venous concentrations are mainly governed by the lower isoprene content of the venous blood returning from the richly perfused tissue group. At the start of exercise, the fractional perfusion q_{per} to the periphery increases and consequently the mixed venous concentrations become dominated by peripheral venous return. The isoprene concentration peak visible in breath can thus be interpreted as a direct result of the associated increase in mixed venous blood concentrations.

3.3.1.4 Practical implications

Dynamic isoprene breath profiles corroborate the view that the breath output of certain VOCs can quickly vary with the physiological status of an individual. While a high sensitivity with respect to hemodynamics, respiration, food intake, etc. imposes strict regulations on the breath sampling protocol for such compounds, it can also provide further insights into their metabolic and physiological fate within the human body. In particular, for VOCs stemming from a specific endogenous (tissue) source, a change in the respective breath output should be observable if that source is activated, stimulated, or depressed. Using an analogous line of argument, by employing

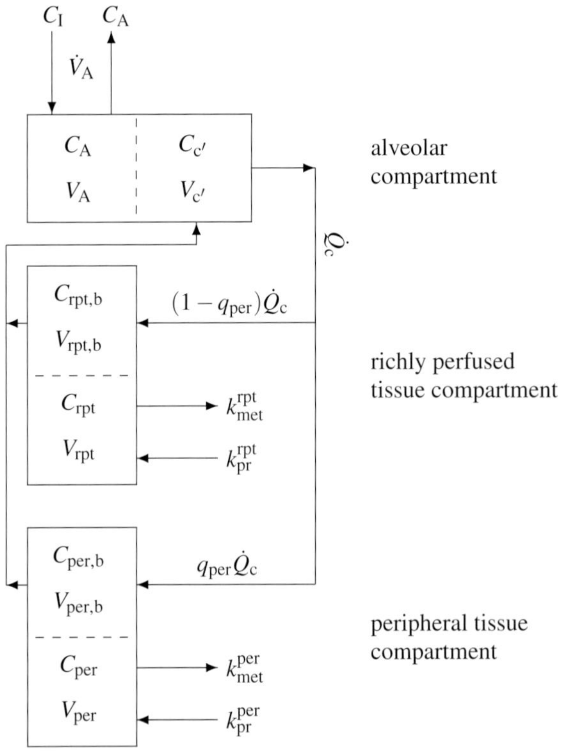

FIGURE 3.4

Schematic representation of the model structure used for describing the dynamics of isoprene concentrations (C) in various parts of the human body. The latter is divided into three distinct functional units (with volumes V): alveolar/end-capillary compartment (gas exchange), richly perfused tissue (metabolism and production), and peripheral tissue (storage, metabolism, and production). Fractional blood flow to the periphery is denoted by q_{per}, while k_{pr} and k_{met} represent constant production and (linear) metabolism rates, respectively. Dashed boundaries indicate a diffusion equilibrium. Subscripts denote as follows: I—inhaled; A—alveolar; c'—end-capillary; rpt—richly perfused tissue; per—peripheral tissue; b—blood. Reproduced with permission from Ref. 36.

exercise scenarios on an ergometer we could give support to the idea that some peripheral (muscle) tissue may play a role in isoprene synthesis. Recently, this viewpoint has further been strengthened by observing drastically reduced breath and blood isoprene levels in late stage Duchenne muscle dystrophy patients.[57] While these findings contrast previous work postulating a purely hepatic origin of isoprene,[15,39] they provide interesting new perspectives on the age dependence of isoprene output[40–42] as well as its reduction under statin therapy.[15,34] Furthermore, the above modeling study opens up the possibility for a mechanistic interpretation of future experimental findings and challenges new hypotheses.

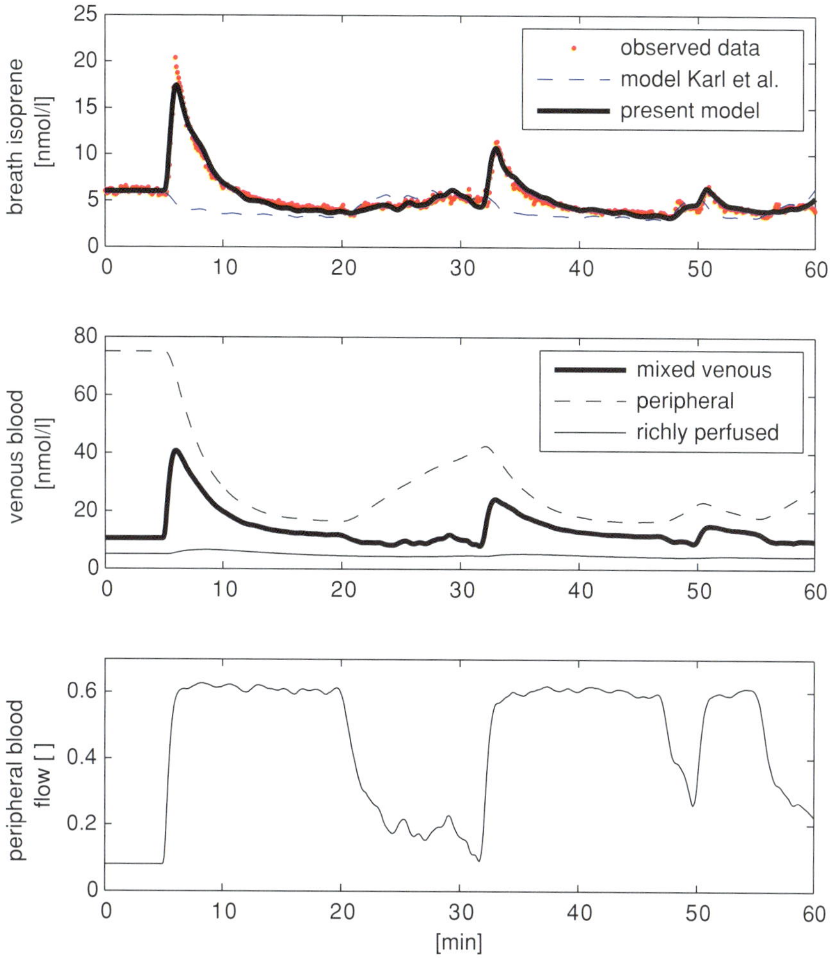

FIGURE 3.5
First panel: model response fitted to the isoprene breath data in Figure 3.2, Protocol A. Second panel: predicted concentrations in mixed venous blood and venous blood returning from the peripheral and richly perfused tissue groups. Third panel: predicted profile of fractional blood flow q_{per} to peripheral tissue. Reproduced with permission from Ref. 36.

3.3.2 Acetone

3.3.2.1 Background

In contrast to isoprene, acetone is both highly blood and water soluble. Hydrophilic compounds such as acetone have been demonstrated to interact with the water-like mucus membrane lining the conductive airways, an effect which has become known as *wash-in/wash-out* behavior. Extensive details can be found, e.g. in Refs. 43–46.

Briefly, the term *wash-in/wash-out* refers to the fact that while fresh air is being inhaled, it becomes enriched with hydrophilic VOCs stored in the airway surface walls of the peripheral bronchial tract, thus leading to a decrease of the respective pressure/tension gradient between the gas phase and capillary blood in the alveolar space. This causes an effective reduction of the driving force for gas exchange in the alveoli and minimizes the unloading of the capillary VOC level. Correspondingly, during exhalation the aforementioned diffusion process is reversed, with a certain amount of hydrophilic VOCs being stripped from the air stream and redepositing onto the previously depleted mucus layer. As a phenomenological consequence, exhaled breath concentrations of such VOCs tend to be diminished on their way from the deeper respiratory tract to the airway opening. The resulting discrepancies between the "true" alveolar and measured breath concentrations can be substantial (even if breath samples are drawn in a strictly standardized manner, e.g. by employing CO_2-controlled sampling) and will depend on a variety of factors such as airway temperature profiles, airway perfusion, and volumetric flow.

3.3.2.2 Some insightful experiments

As in the case of isoprene, end-tidal acetone breath profiles during moderate ergometer challenges contradict the qualitative behavior anticipated from classical pulmonary inert gas elimination theory. From Figure 3.2, end-tidal acetone levels in response to exercise generally resemble the profile of ventilatory flow, showing abrupt increases and decreases in the range of 10–40% at the onsets and ends of the individual workload periods, respectively.[22] Similarly, during hyperventilation breath acetone levels tend to increase, whereas a (slight) decrease might be expected from Eq. (3.1) (due to a drastic rise of the ventilation-perfusion ratio). Further evidence for the inadequacy of classical pulmonary inert gas elimination theory when it comes to describing the blood-breath relationships of hydrophilic compounds can be derived from isothermal rebreathing experiments.[47] The heuristic intention leading to isothermal rebreathing is to create an experimental situation diminishing the influence of the *wash-in/wash-out* mechanisms discussed above. This can be accomplished by *closing the respiratory loop*, i.e. by continuous re-inspiration and re-exhalation of a fixed mass of air from a rebreathing receptacle (e.g. a Tedlar bag), causing the air stream to equilibrate with the mucosa linings over the entire respiratory cycle.[48–50] Additionally, warming the rebreathing volume to body temperature will ensure a similar solubility of these VOCs in both regions of the pulmonary tract, alveoli, and airways. End-tidal breath profiles of highly water soluble VOCs such as acetone and methanol may increase by a factor of up to 1.5 during isothermal rebreathing,[47] until approaching a plateau value that corresponds to the underlying alveolar concentration. Again, this behavior sharply contrasts with the formalism underlying Eq. (3.1).

3.3.2.3 A physiological model for breath acetone dynamics

The experimental findings described above strongly suggest that airway gas exchange should be taken into account as a major determinant affecting the breath levels of hydrophilic VOCs. From a modeling viewpoint, this can be achieved by dividing the pulmonary tract into two separate compartments, an airway (bronchial) space, and the alveolar space. The former comprises both a gas phase and a mucus membrane, which is assumed to have the physical properties of water and acts as a reservoir for acetone.[51,52] Part of the acetone dissolved in this layer is transferred to the bronchial circulation, whereby a major fraction of the associated venous drainage is postulated to join the pulmonary veins via the post-capillary anastomoses.[53] The alveolar region itself is again represented by one single homogeneous alveolar unit. Similar to Section 3.3.1, the systemic part of the model consists of two functional units: a liver compartment, where acetone is endogenously produced and metabolized, as well as a passive tissue compartment representing an effective storage volume. The latter basically lumps together tissue groups with similar blood:tissue partition coefficient $\lambda_{\mathrm{b:tissue}} \approx 1.38$, such as richly perfused tissue, muscles, and skin.[52,54] A schematic representation of the model structure is presented in Figure 3.6. For the associated compartmental mass balance equations we again refer to the original publication.[55] In order to correlate the model output to experimental outcomes, it is postulated that the measurable breath acetone concentration corresponds to the bronchial level C_{bro}.

A direct implication of the *wash-in/wash-out* mechanisms discussed at the beginning of this section is that during free tidal breathing at rest the amount of hydrophilic VOCs reaching the alveolar region with each inhalation will be approximately equal to the amount extracted with each exhalation. In other words, under these circumstances the bronchio-alveolar net exchange of such compounds over one breath cycle will be close to 0 and a major part of the overall pulmonary gas exchange (e.g. up to 96% as in the case of acetone[54]) can be ascribed to the conducting airways. As a consequence, a concentration gradient develops along the respiratory tree, with higher levels of the respective VOC found in the proximal (alveolar) parts. Increasing ventilatory flow rates and volume minimizes equilibration times with the more distal mucus layers and hence counteracts the saturation and depletion of the air stream during inhalation and exhalation, respectively. Consequently, the location of gas exchange is continuously shifted toward the alveolar region and the contribution of the latter becomes more pronounced. In particular, the end-exhaled (measured) breath concentration approaches values closer to the alveolar level.

This rationale can be incorporated into the model by including a diffusion term of the form $D(C_{\mathrm{bro}} - C_{\mathrm{A}})$ into the underlying mass balance equations. This expression reflects the above-mentioned net gas exchange between the alveolar and the bronchial tract over a complete breath cycle (i.e. the conductance parameter D is close to 0 during free tidal breathing at rest and increases with ventilatory flow).

The relationships outlined in the previous paragraphs are summarized in Figure 3.7, showing a calibrated version of the model with the observable model output (i.e. the bronchial concentration) fitted to the experimental data in Figure 3.2, Protocol B. It is interesting to note that the alveolar acetone concentration stays relatively constant throughout the entire experiment, i.e. the short-term variability observable in breath acetone concentrations can almost exclusively be attributed to airway gas exchange.

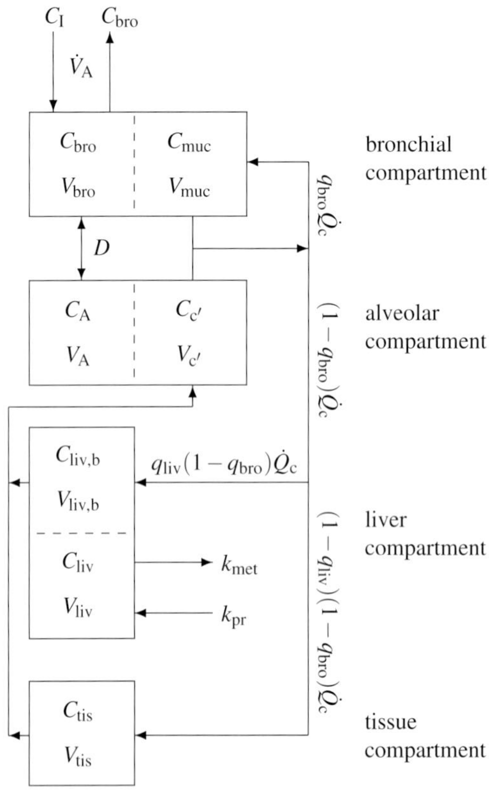

FIGURE 3.6

Schematic representation of the model structure used for describing the dynamics of acetone concentrations (C) in various parts of the human body. The latter is divided into four distinct functional units (with volumes V): bronchial/mucosal compartment (gas exchange), alveolar/end-capillary compartment (gas exchange), liver (metabolism and production), and tissue (storage). Fractional blood flows are denoted by q, while k_{pr} and k_{met} represent constant production and (linear) metabolism rates, respectively. Dashed boundaries indicate a diffusion equilibrium. Subscripts denote as follows: I—inhaled; bro—bronchial; muc—mucosal; A—alveolar; c'—end-capillary; liv—liver; tis—tissue; b—blood. Reproduced with kind permission of Springer Science+Business Media from Ref. 55.

Similarly, only minute changes in the underlying blood and tissue concentrations are expected to occur.

For the sake of completeness, it should be mentioned that another important factor affecting the gas exchange of hydrophilic VOCs is airway temperature. In particular, changes in the latter will affect the solubility of acetone and similar compounds in the mucus surface of the respiratory tree.[43,48] For example, the local solubility of acetone in the mucus layer increases from $\lambda_{muc:air}(37°C) \approx 392$ in the lower respiratory tract

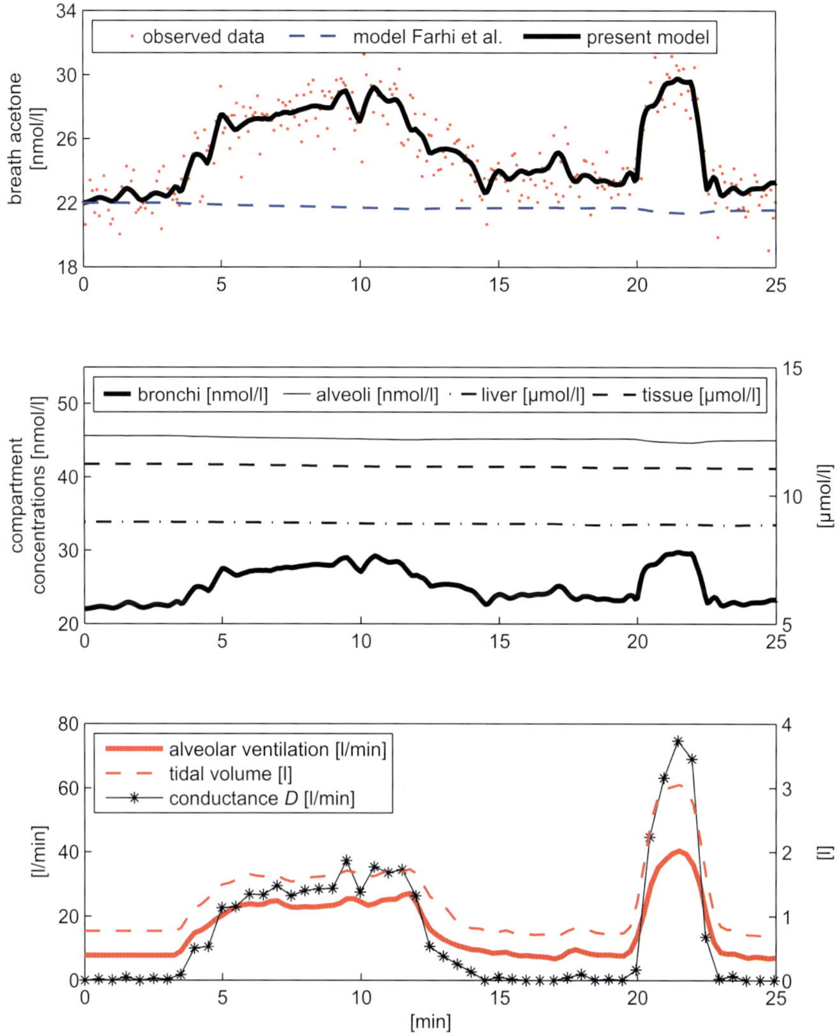

FIGURE 3.7

First panel: Model response fitted to the acetone breath data in Figure 3.2, Protocol B. Second panel: corresponding concentrations in the bronchial, alveolar, liver, and tissue compartments. Third panel: measured respiratory parameters and the estimated time course of the conductance parameter D, reflecting the net gas exchange between the alveolar and the bronchial tract as described in the text. Note that D is close to 0 during free tidal breathing at rest (i.e. the measurable breath acetone levels are almost exclusively determined by the respective levels in the conducting airways) and varies in parallel with ventilatory flow. In particular, higher values of D correspond to an increased contribution of the alveoli to total pulmonary gas exchange. Reproduced with kind permission of Springer Science+Business Media from Ref. 55.

to $\lambda_{muc:air}(32°C) \approx 498$ at the mouth, thereby accentuating the *wash-out* of air stream acetone during the exhalation phase, cf. Section 3.3.2.2. Warming the airways to body core temperature can minimize this effect by ensuring a uniform solubility distribution throughout the airways. This is consistent with an increase of the measured breath concentrations during isothermal rebreathing experiments.[47,49,50] For further details on how to incorporate explicit temperature dependences into the mechanistic model presented above, the reader is referred to Ref. 55. This paper also further reports on the validation of the model, e.g. during exposure scenarios.

3.3.2.4 Practical implications

The major conclusion that can be drawn from quantitatively modeling exhalation dynamics of hydrophilic VOCs is that for this class of compounds the measured end-exhaled breath concentrations and the underlying alveolar levels must *not* be taken as identical. For perspective, the experimental evidence outlined above indicates that for acetone during free tidal breathing at rest the associated values can differ by a factor of more than 1.5. In particular, this implies that multiplying end-tidal breath concentrations with the substance-specific *in vitro* blood:gas partition coefficient $\lambda_{b:air}$ will generally result in an underestimation of the true arterial blood concentration of hydrophilic VOCs. *In vivo*, the (arterial) blood-breath ratio strongly depends on a variety of additional factors that can vary considerably among individuals, including airway temperature and bronchial blood flow (both exerting an inversely monotonous influence on the blood-breath ratio). Modeling can give a good mechanistic account of these relationships. In particular, it can be used to predict the impact of conditions/pathologies that are known to alter airway gas exchange (e.g. in asthma or bronchiectasis) and can provide hints for designing and evaluating sampling protocols aimed at improving the reliability and information content of breath measurements involving hydrophilic VOCs.

3.4 CONCLUDING REMARKS

In summary, we believe that breath gas analysis would greatly profit from a more causal understanding of VOC kinetics in response to distinct physiological states. However, the development of quantitative formulations relating breath concentrations of trace gases to their underlying systemic levels clearly lags behind the enormous analytical progress in the field. This situation calls for the establishment of a proper modeling framework, striving for validated mechanistic descriptions of VOC dynamics in different parts of the organism. As an outlook, from an operational point of view such modeling studies can:

- help to identify potential confounding factors affecting the interpretation of breath test results and propose corrective measures;
- guide the ongoing standardization of sampling procedures and provide formal means for evaluating the information content of newly proposed experimental settings;

- enhance the fundamental understanding of the physiological role of VOCs, e.g. by relating phenomenological features of observable breath concentration profiles to specific distribution, transport, bio-transformation, and excretion processes of these substances; and
- stimulate biochemical investigations exploring the endogenous (e.g. cellular) sources of VOCs.

In particular, the last two points are directed towards further consolidating the potential role of breath gas analysis in general bio-monitoring applications.

Acknowledgments

J. King, H. Koc, and G. Teschl gratefully acknowledge support from the Austrian Science Fund (FWF) under Grant No. Y330. J. King, K. Unterkofler, and P. Mochalski gratefully acknowledge support from the Austrian Science Fund (FWF) under Grant No. P24736-B23. We greatly appreciate the generous support of the government of Vorarlberg, Austria. We appreciate funding from the Austrian Federal Ministry for Transport, Innovation and Technology (BMVIT/BMWA, project 836308, KIRAS).

References

1. Amann A, Smith D. *Breath analysis for clinical diagnosis and therapeutic monitoring*. Singapore: World Scientific; 2005.
2. Amann A, Spanel P, Smith D. Breath analysis: the approach towards clinical applications. *Mini Rev Med Chem* 2007;**7**:115–29.
3. Amann A, Poupart G, Telser S, Ledochowski M, Schmid A, Mechtcheriakov S. Applications of breath gas analysis in medicine. *Int J Mass Spectrom* 2004;**239**:227–33.
4. Buszewski B, Kesy M, Ligor T, Amann A. Human exhaled air analytics: biomarkers of diseases. *Biomed Chromatogr* 2007;**21**:553–66.
5. Rieder J, Lirk P, Ebenbichler C, Gruber G, Prazeller P, Lindinger W, et al. Analysis of volatile organic compounds: possible applications in metabolic disorders and cancer screening. *Wien Klin Wochenschr* 2001;**113**:181–5.
6. Miekisch W, Schubert JK. From highly sophisticated analytical techniques to life-saving diagnostics: technical developments in breath analysis. *TrAC—Trends Anal Chem* 2006;**25**:665–73.
7. Pleil JD. Role of exhaled breath biomarkers in environmental health science. *J Toxicol Environ Health B: Crit Rev* 2008;**11**:613–29.
8. Amorim LCA, de L Cardeal Z. Breath air analysis and its use as a biomarker in biological monitoring of occupational and environmental exposure to chemical agents. *J Chromatogr B: Analyt Technol Biomed Life Sci* 2007;**853**:1–9.
9. Koshland Jr DE. The molecule of the year. *Science* 1992;**258**:1861.
10. Barnes PJ. Nitric oxide and airway disease. *Ann Med* 1995;**27**:389–93.
11. American Thoracic Society, European Respiratory Society recommendations for standardized procedures for the online and offline measurement of exhaled lower respiratory nitric oxide and nasal nitric oxide. *Am J Respir Crit Care Med* 2005;**171**:912–30.

12. Herbig J, Titzmann T, Beauchamp J, Kohl I, Hansel A. Buffered end-tidal (BET) sampling—a novel method for real-time breath-gas analysis. *J Breath Res* 2008;**2**:037008.
13. Birken T, Schubert J, Miekisch W, Noldge-Schomburg G. A novel visually CO_2 controlled alveolar breath sampling technique. *Technol Health Care* 2006;**14**:499–506.
14. Wolkenhauer O, Mesarovic M, Wellstead P. A plea for more theory in molecular biology. *Ernst Schering Research Foundation workshop*; 2007. p. 117–37.
15. Stone BG, Besse TJ, Duane WC, Evans CD, DeMaster EG. Effect of regulating cholesterol biosynthesis on breath isoprene excretion in men. *Lipids* 1993;**28**:705–8.
16. Cao WQ, Duan YX. Current status of methods and techniques for breath analysis. *Crit Rev Anal Chem* 2007;**37**:3–13.
17. Lindinger W, Hansel A, Jordan A. On-line monitoring of volatile organic compounds at PPTV levels by means of proton-transfer-reaction mass spectrometry (PTR-MS)—medical applications, food control and environmental research. *Int J Mass Spectrom* 1998;**173**: 191–241.
18. Lindinger W, Hansel A, Jordan A. Proton-transfer-reaction mass spectrometry (PTR-MS): on-line monitoring of volatile organic compounds at pptv levels. *Chem Soc Rev* 1998;**27**:347–54.
19. Blake RS, Monks PS, Ellis AM. Proton-transfer reaction mass spectrometry. *Chem Rev* 2009;**109**:861–96.
20. Herbig J, Muller M, Schallhart S, Titzmann T, Graus M, Hansel A. On-line breath analysis with PTR-TOF. *J Breath Res* 2009;**3**:027004.
21. Schwarz K, Filipiak W, Amann A. Determining concentration patterns of volatile compounds in exhaled breath by PTR-MS. *J Breath Res* 2009;**3**:027002.
22. King J, Kupferthaler A, Unterkofler K, Koc H, Teschl S, Teschl G, et al. Isoprene and acetone concentration profiles during exercise on an ergometer. *J Breath Res* 2009;**3**:027006.
23. King J, Mochalski P, Kupferthaler A, Unterkofler K, Koc H, Filipiak W, et al. Dynamic profiles of volatile organic compounds in exhaled breath as determined by a coupled PTR-MS/GC-MS study. *Physiol Meas* 2010;**31**:1169–84.
24. King J, Kupferthaler A, Frauscher B, Hackner H, Unterkofler K, Teschl G, et al. Measurement of endogenous acetone and isoprene in exhaled breath during sleep. *Physiol Meas* 2012;**33**:413–28.
25. Amann A, Telser S, Hofer L, Schmid A, Hinterhuber H. Exhaled breath gas as a biochemical probe during sleep. In: Amann A, Smith D, editors. *Breath analysis for clinical diagnosis and therapeutic monitoring*. Singapore: World Scientific; 2005. p. 305–16.
26. Farhi LE. Elimination of inert gas by the lung. *Respir Physiol* 1967;**3**:1–11.
27. Reddy MB. *Physiologically based pharmacokinetic modeling: science and applications*. Hoboken, NJ: Wiley-Interscience; 2005.
28. Leung HW. Development and utilization of physiologically based pharmacokinetic models for toxicological applications. *J Toxicol Environ Health* 1991;**32**:247–67.
29. Andersen ME. Physiological modelling of organic compounds. *Ann Occup Hyg* 1991;**35**:309–21.
30. Gerlowski LE, Jain RK. Physiologically based pharmacokinetic modeling: principles and applications. *J Pharm Sci* 1983;**72**:1103–27.
31. Wagner PD. In: Farhi LE, Marsh Tenney S, editors. *Handbook of physiology: Section 3: the respiratory system, Volume IV: Gas Exchange*, New York: Oxford University Press; 1988. p. 257–81.
32. Fiserova-Bergerova V. *Modeling of inhalation exposure to vapors: uptake, distribution, and elimination*. Boca Raton, FL: CRC Press; 1983.

33. Teorell T. Kinetics of distribution of substances administered to the body: I. The extravascular modes of administration. *Arch Int Pharmacod Ther* 1937;**57**: 205–25.
34. Karl T, Prazeller P, Mayr D, Jordan A, Rieder J, Fall R, et al. Human breath isoprene and its relation to blood cholesterol levels: new measurements and modeling. *J Appl Physiol* 2001;**91**:762–70.
35. Turner C, Spanel P, Smith D. A longitudinal study of breath isoprene in healthy volunteers using selected ion flow tube mass spectrometry (SIFT-MS). *Physiol Meas* 2006;**27**:13–22.
36. King J, Koc H, Unterkofler K, Mochalski P, Kupferthaler A, Teschl G, et al. Physiological modeling of isoprene dynamics in exhaled breath. *J Theor Biol* 2010;**267**:626–37.
37. Koc H, King J, Teschl G, Unterkofler K, Teschl S, Mochalski P, et al. The role of mathematical modeling in VOC analysis using isoprene as a prototypic example. *J Breath Res* 2011;**5**: 037102.
38. Mohrman DE, Heller LJ. *Cardiovascular physiology*. 6th ed. New York: Lange Medical Books/McGraw-Hill; 2006.
39. Deneris ES, Stein RA, Mead JF. Acid-catalyzed formation of isoprene from a mevalonate-derived product using a rat liver cytosolic fraction. *J Biol Chem* 1985;**260**:1382–5.
40. Smith D, Spanel P, Enderby B, Lenney W, Turner C, Davies SJ. Isoprene levels in the exhaled breath of 200 healthy pupils within the age range 7–18 years studied using SIFT-MS. *J Breath Res* 2009;**4**:017101.
41. Risby TH. Volatile organic compounds as markers in normal and diseased states. In: Marczin N, Yacoub M, editors. *Disease markers in exhaled breath: basic mechanisms and clinical applications*. Washington, DC: IOS Press; 2002. p. 418.
42. Kushch I, Arendacka B, Stolc S, Mochalski P, Filipiak W, Schwarz K, et al. Breath isoprene-aspects of normal physiology related to age, gender and cholesterol profile as determined in a proton transfer reaction mass spectrometry study. *Clin Chem Lab Med* 2008;**46**:1011–8.
43. Tsu ME, Babb AL, Ralph DD, Hlastala MP. Dynamics of heat, water, and soluble gas exchange in the human airways: 1. A model study. *Ann Biomed Eng* 1988:**16**:547–71.
44. Tsu ME, Babb AL, Sugiyama EM, Hlastala MP. Dynamics of soluble gas exchange in the airways: II. Effects of breathing conditions. *Respir Physiol* 1991;**83**:261–76.
45. Anderson JC, Babb AL, Hlastala MP. Modeling soluble gas exchange in the airways and alveoli. *Ann Biomed Eng* 2003;**31**:1402–22.
46. Anderson JC, Hlastala MP. Breath tests and airway gas exchange. *Pulm Pharmacol Ther* 2007;**20**:112–7.
47. King J, Unterkofler K, Teschl G, Teschl S, Mochalski P, Koc H, et al. A modeling-based evaluation of isothermal rebreathing for breath gas analyses of highly soluble volatile organic compounds. *J Breath Res* 2012;**6**:016005.
48. Jones AW. Role of rebreathing in determination of the blood-breath ratio of expired ethanol. *J Appl Physiol* 1983;**55**:1237–41.
49. Ohlsson J, Ralph DD, Mandelkorn MA, Babb AL, Hlastala MP. Accurate measurement of blood alcohol concentration with isothermal rebreathing. *J Stud Alcohol* 1990;**51**:6–13.
50. O'Hara ME, O'Hehir S, Green S, Mayhew CA. Development of a protocol to measure volatile organic compounds in human breath: a comparison of rebreathing and on-line single exhalations using proton transfer reaction mass spectrometry. *Physiol Meas* 2008;**29**: 309–30.
51. Kumagai S, Matsunga I. A lung model describing uptake of organic solvents and roles of mucosal blood flow and metabolism in the bronchioles. *Inhal Toxicol* 2000;**12**:491–510.
52. Mork AK, Johanson G. A human physiological model describing acetone kinetics in blood and breath during various levels of physical exercise. *Toxicol Lett* 2006;**164**:6–15.

53. Lumb A. *Nunn's applied respiratory physiology*. Oxford: Butterworth-Heinemann; 2005.
54. Anderson JC, Lamm WJ, Hlastala MP. Measuring airway exchange of endogenous acetone using a single-exhalation breathing maneuver. *J Appl Physiol* 2006;**100**:880–9.
55. King J, Unterkofler K, Teschl G, Teschl S, Koc H, Hinterhuber H, et al. A mathematical model for breath gas analysis of volatile organic compounds with special emphasis on acetone. *J Math Biol* 2011;**63**:959–99.
56. Mochalski P, King J, Kupferthaler A, Unterkofler K, Hinterhuber H, Amann A. Measurement of isoprene solubility in water, human blood and plasma by multiple headspace extraction gas chromatography coupled with solid phase microextraction. *J Breath Res* 2011;**5**:046010.
57. King J, Mochalski P, Unterkofler K, Teschl G, Klieber M, Stein M, et al. Breath isoprene: Muscle dystrophy patients support the concept of a pool of isoprene in the periphery of the human body. *Biochem Biophys Res Commun* 2012;**423**:526–30.

PART

C

Physiological and Clinical Studies

CHAPTER

4

Recent SIFT-MS Studies of Volatile Compounds in Physiology, Medicine and Cell Biology

Patrik Španěl* **and David Smith**†

*J. Heyrovský Institute of Physical Chemistry, Academy of Sciences of the Czech Republic, Dolejškova 3, 182 23, Prague 8, Czech Republic

†Institute for Science and Technology in Medicine, School of Medicine, Keele University, Thornburrow Drive, Hartshill, Stoke-on-Trent ST4 7QB, UK

4.1 INTRODUCTION

Since its inception in 1996,[1–4] the major focus and rational for the development of selected ion flow tube mass spectrometry, SIFT-MS, has been to realize a technique that can be exploited for on-line, real-time analysis of volatile trace gases of biological origin, especially with breath analysis in mind,[5] but extending to the analysis of trace compounds in the headspace of biological fluids, particularly urine[5–13] and blood, and trace compounds emitted by cell[14–16] and bacterial[17–21] cultures of potential clinical significance. Much work has now been carried out in these areas, which demonstrates the value and potential of SIFT-MS, as reported in the cited references. However, SIFT-MS has now been applied to the analysis of volatile trace compounds emitted by other diverse media, including food and food products,[22,23] animal waste[24] and gases present in rumen gas,[25,26] and those generated by incineration[27] and by explosions.[28,29] These wider applications of SIFT-MS and the key results obtained have been summarized in previous review papers,[30,31] while the historical development of SIFT-MS and subsequent applications are summarized in a quite recent review.[5] The ion chemistry underpinning SIFT-MS and the details of the analytical method as an ambient chemistry analytical tool are given in a recent tutorial review.[6] Hence, the details of the physics and ion chemistry involved in SIFT-MS are largely excluded from this chapter and the focus is on the new advances made in its application to *in vivo* breath analysis for clinical diagnosis and therapeutic monitoring and for relevant *in vitro* studies of urine and blood headspace and mammalian cell and bacterial cultures carried out during the last 5 years or so, i.e. beyond those studies reported in the book *Breath Analysis for Clinical Diagnosis and Therapeutic Monitoring* published in 2005.[32]

SIFT-MS allows real-time analysis of trace gases down to parts-per-billion by volume, ppbv, concentrations, avoiding sample collection into bags or onto surface

Volatile Biomarkers. http://dx.doi.org/10.1016/B978-0-44-462613-4.00004-0

or cryogenic traps that can compromise the sample. It has also been shown by very elegant experiments to produce accurate analyses at the sub-ppbv level.[33] It is now considered as one of the several techniques capable of ambient analysis[6] that are rapidly developing and being applied to medicinal chemistry. SIFT-MS is especially suited to exhaled breath analysis, principally because of its practical simplicity and accuracy, as is required for a worthwhile tool to support clinical diagnosis. The most recent developments have seen a large reduction in the flow rate of the sample (breath) that is needed to achieve accurate analysis, which does not disturb the helium carrier gas flow dynamics and results in a long-term stability of the SIFT-MS instrument, and improvements in sampling methodology. This allows single sequential breath exhalations to be followed over periods of many hours. So we and others have applied SIFT-MS to the promotion of the wider aspects of breath research toward the potentially valuable clinical tool that it can be.

The bulk of the research results described in this chapter originated in our Prague and Keele laboratories; details of some other SIFT-MS work mentioned in this chapter are considered further in the companion chapters by C. Turner, and B. R. Ross and A. Esarik.

4.2 DIRECT BREATH ANALYSIS

Breath analysis will take the most important step toward meaningful use in medicine when volatile biomarkers of patho-physiological disorders are recognized that can be measured to acceptable accuracy thus allowing staging of the diseases and especially recognition of the early stages of disease. Other matters, including physiological aspects such as breathing/re-breathing rates, breath hold, etc.,[34–38] which are relevant to ultimate measurement accuracy and detailed diagnostic interpretation are peripheral to this major challenge. From a practical viewpoint, direct breath analyses in the clinical setting are most desirable so that the results are immediately available to the health professional. SIFT-MS offers this tantalizing prospect, but much more research and development is needed before such is realized on a desirable scale. The developments and the preliminary results obtained in the last 5 years as reported in this section represent a step toward this goal.

4.2.1 Inter-individual and intra-individual concentration distributions of mouth-exhaled breath metabolites in the healthy population

Almost all the measurements of trace metabolites in exhaled breath have been carried out on mouth-exhaled breath, as is the case for the early SIFT-MS studies, principally because of the inherent simplicity of sampling and sample collection, but this fails to recognize that the oral cavity can be the major source of some trace gases, as is discussed later and is emphasized in the chapter by B. Ross and A. Esarik in this book. The most thorough longitudinal studies of common metabolites in the mouth-exhaled breath of significant cohorts of healthy volunteers were carried out by Turner and colleagues using SIFT-MS just a few years ago.[39–42] Such studies are vital to establish "reference ranges" for these metabolites in the healthy population if abnormal values are to be recognized in disease. The beauty of SIFT-MS analyses of breath is the very rapid, painless simplicity of data acquisition and analysis; the data flow is largely limited by the volunteer/patient recruitment rate and their willingness to participate. Thus, longitudinal studies of the common metabolites ammonia, acetone, methanol, ethanol, propanol, acetaldehyde, and isoprene in the breath of some 30 healthy volunteers over a 6-month period were carried out. The inter-individual distributions constructed from some 1500 measurement (each utilized value being the mean values of three consecutive exhalations) of each compound are shown in Figure 4.1. The concentrations given for these distributions are in *parts-per-billion by volume*, *ppbv*, as consistently given throughout this chapter. The distributions are seen to be lognormal and are quite tight with the geometrical standard deviation being typically 1.6, which is both interesting and helpful in that it is now easier to recognize abnormally high breath concentrations of these compounds that may arise. Commentaries on these data are given in the original papers[39–43] and their likely significance has been outlined in Ref. 44 and in the chapter by C. Turner in this book. We have more to say about the breath levels of individual compounds in subsequent sections of this chapter.

The data obtained in the above studies also include intra-individual variations over the 6-month sampling period, which were not reported in detail in the published papers. The spread in the concentrations for individuals is not as wide as those for the population distributions, as expected. They had been previously reported for a few individuals as obtained over several weeks of observation by Diskin et al.[45] where the coefficients of variation are seen to vary from 0.2 (tight; e.g. acetone) to 0.4 (broad; e.g. ammonia and isoprene). These variations are partly due to food intake and its influence on mouth production of compounds such as ammonia and ethanol, as discussed later.

4.2.2 Variation with age of mouth-exhaled breath concentrations of common metabolites for healthy individuals and the influence of diet

It might be expected that as the metabolic rate changes with increasing age then this will be reflected in the volatile metabolites present in exhaled breath and their concentrations, since they are considered to "mirror" the metabolic processes to some extent. The influence of diet is a contentious issue that needs to be addressed in detail in the context of breath analysis. The results described in this section relate to mouth-exhaled breath only; nose-exhaled breath analysis is described in the next section.

From the SIFT-MS studies of Turner et al.[43] described above, it is seen that there is a weak variation of breath *ammonia* concentration with age, the mean concentration of this metabolite for the restricted population of the healthy adult volunteers increasing by about a factor of 2 over a 40-year age increase. This was tentatively ascribed to the diminishing kidney function with age. However, not known at the time is that ammonia

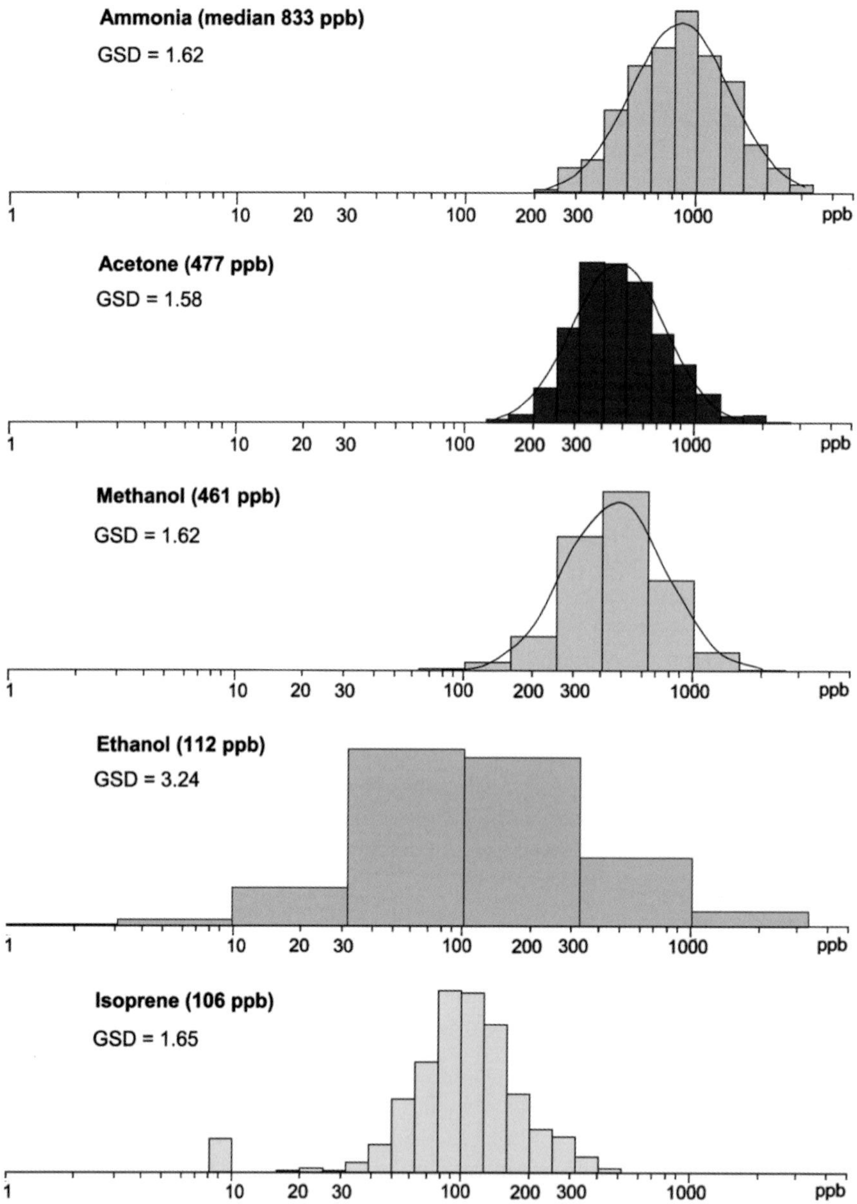

FIGURE 4.1

Distributions of the five most prominent trace metabolites present in the exhaled breath of the healthy population. The concentrations on the horizontal axes are shown by logarithmic scales in parts-per-billion by volume, ppbv. The median values are given in parentheses for each metabolite together with their geometric standard deviation, GSD. The data are taken from Refs. 39–43.

can largely originate in the oral cavity and as such must surely be dependent to some extent on oral hygiene. Since mouth-exhaled ammonia can be seriously increased by oral generation, it cannot simply reflect systemic (blood) ammonia; this is discussed again in Section 4.2.3. As shown by the SIFT-MS studies of Španěl and colleagues, the mean concentration of mouth-exhaled ammonia clearly increases with age for the young adult population (age 17/18 years; median concentration 317 ppbv) and the adult population (age 20/60; median concentration 833 ppbv), the mean mouth-exhaled breath ammonia increasing by as much as a factor of 2.[46] This increase with age is even more obvious when young children (age 4/6) are included in the variation.[47] A composite of all the SIFT-MS data is given in Figure 4.2a.

Breath *isoprene* has attracted the interest of several researchers,[48] because it is seen to be quite variable in exhaled breath of individuals, especially during exercise,[36] and is influenced by factors such as breathing rate and exercise.[49] It is present in exhaled breath at relatively low concentration (typically 100 ppbv[40,48]) and is very volatile from blood, which can partly explain its variability in exhaled breath. It is hardly

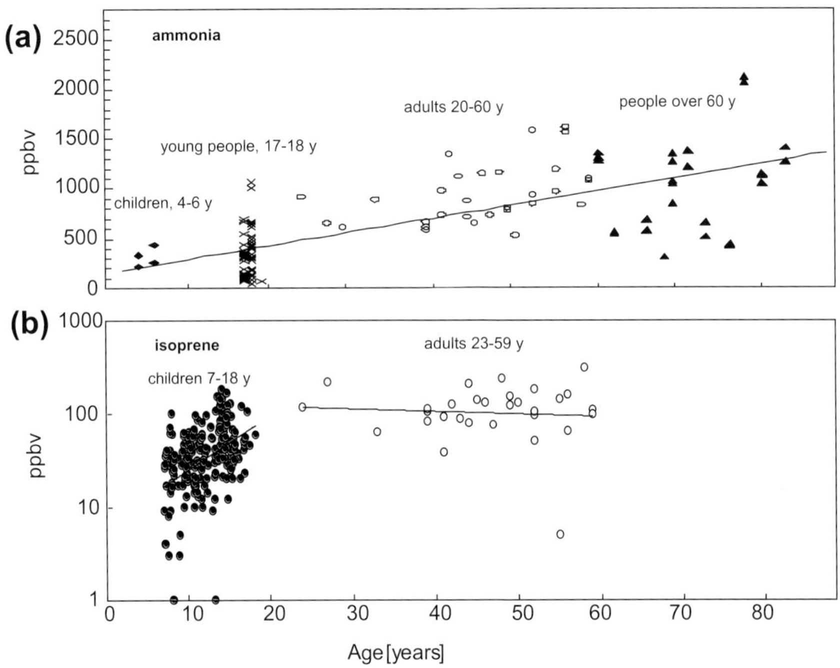

FIGURE 4.2
(a) Correlation of breath ammonia concentrations with age. Solid symbols are data on children and people over 60 years,[47] the open symbols are the data from studies of the breath of adults[43] and x are data on age-limited group of young adults.[46] (b) Variation of breath isoprene concentration with age shown on semi-logarithmic scale. Filled circles: children and young adults in the age range from 7 to 18 years.[50] Open circles: adults in the age range 24–59 years.[40]

present in the breath of the new born, but increases rapidly with age, reaching a steady value at about age 20 years, as is graphically shown by the isoprene data in Figure 4.2b, which were obtained from SIFT-MS analyses of the exhaled breath of some 200 children in the age range 7–18 years,[50,51] but also include previous data for adults.[40] What is so remarkable about these data is they were obtained in parallel with several other common breath metabolites[51] in single, direct exhalations from children as young as barely 7 years. Discernible within the data is an increase of breath isoprene during puberty, which certainly exercises the imagination (see the original paper[50,51] for more details). A study of breath isoprene (and acetone) has been carried out which combined thermal desorption to collect and concentrate the trace compounds from breath samples followed by their release into a SIFT-MS instrument for analysis. This unique experiment by Ross and colleagues is reported in.[52]

Acetone in exhaled breath has attracted much attention, principally because of its proposed association with blood glucose, diabetes, and the glycaemic state.[53,54] Yet there is no convincing experimental evidence that this association is valid, as has been discussed in a recent perspective article.[54] The situation is also complicated by biological variability in the breath acetone; while the intra-individual variations in breath acetone concentration are not great,[10] diurnal variations can be as much as a factor of two. A more serious compounding factor is diet. A recent study has shown that taking a ketogenic diet[55] can result in breath acetone increases of 3–4 times. Such has recently been seen by a SIFT-MS study in which eight healthy individuals took a ketogenic diet when the breath acetone increased several times for all eight individuals, [56] but the breath acetone concentration rapidly returns to the normal, lower level on reverting to carbohydrate-containing food. In a follow-up study, also reported in Ref. 57, it is shown that using the Atkins diet for a number of days can result in persistently high breath acetone, several times greater than that seen for the ingestion of a "normal" or regular diet, which takes about 2 days to return to the normal level on ingestion of the regular diet. While these major changes in breath acetone with diet question the value of breath acetone alone as an indicator of blood glucose, exercising dietary control by the individual may allow breath acetone, probably with other, as yet unrecognized metabolites, to be a valuable, non-invasive monitor of blood glucose. The further aspects of breath acetone vis-à-vis diabetes are discussed in more detail in Ref. 58 and in the chapter in this book by C. Turner.

4.2.3 Intra-individual studies of mouth-exhaled and nose-exhaled breath; volatile compounds deliberately generated in the oral cavity

With hindsight, it now seems obvious that mouth-exhaled breath must be at least partially contaminated by compounds generated in the oral cavity. The unfortunate condition of intra-oral halitosis is the most obvious manifestation of this, although it is important to recognise that in some 5–10% of cases the "bad breath" can also be the result of blood-borne compounds and thus can be due to systemic disease.[59] So it is self-evident that distinction must be made between trace compounds generated systemically and in the airways and those generated in the oral cavity due to bacterial and/or enzymatic activity on salivary compounds. To this end, a simple modification of the SIFT-MS sampling methodology now allows the analysis of breath exhaled via the mouth and via the nose, and indeed in the closed oral cavity.

A small vacuum pump is included in the sampling system in order to draw ambient air or exhaled breath across the entrance to the SIFT-MS sampling capillary, as depicted in Figure 4.3.[57] The flow rate of air/breath is adjusted to be much smaller than the breath exhalation rate, but greater than the flow rate through the capillary, by the in-line needle valve shown in Figure 4.3. This ensures that the normal breath exhalations from the mouth as in (a) and from the nose as in (b) are not compromised and that the static gas flow from the oral cavity with the lips closed and around the plastic sampling tube (with breath hold) do not significantly lower the pressure during the usual

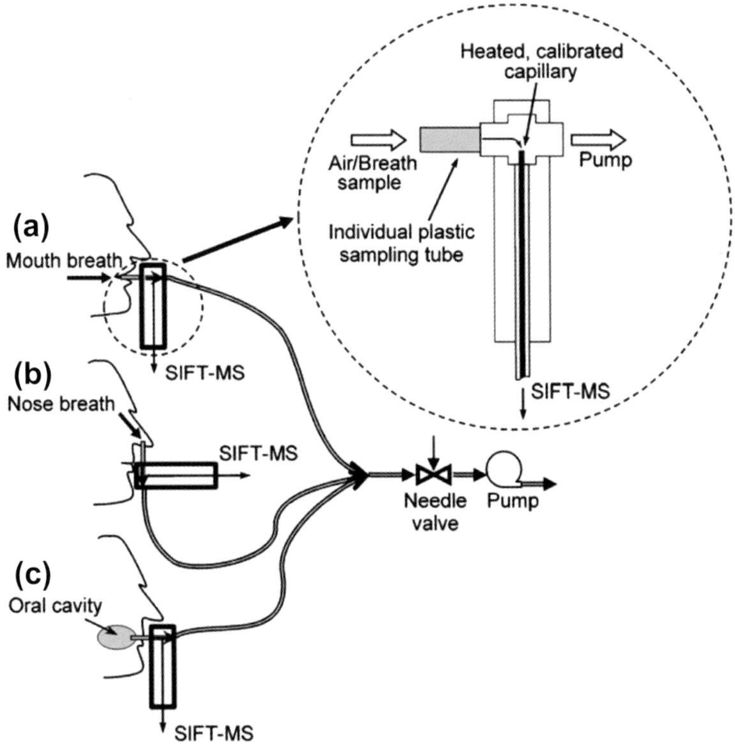

FIGURE 4.3
A representation of the sampling of (a) breath exhaled via the *mouth*, (b) breath exhaled via the *nose*, and (c) the air in the mouth *cavity*. The air/breath samples are drawn along the individual (interchangeable) plastic sampling tube and across the sampling calibrated capillary (shown in the enlargement) by the action of the small pump. The in-line needle valve is used to regulate the air/breath flow rate (reproduced with permission from IOP publishing from Ref. 57).

10 s or so of sampling time. These comparative measurements, of (a), (b), and (c), can be performed quickly and sequentially in real time for any number of volunteers by simply interchanging the personalized sampling tube illustrated in Figure 4.3.

These experiments have revealed some remarkable results, as illustrated by the data in Figure 4.4. These data show the dramatically greater *ammonia* concentration in the mouth-exhaled compared to that in nose-exhaled breath (closed mouth) and the highest concentration in the static gas in the oral cavity for a single volunteer.[57] These ammonia distributions were constructed from breath analyses obtained each day over 1 month. These data show that for this volunteer the ammonia present in mouth-exhaled breath is largely generated in the mouth, a result that has been obtained for other volunteers, as can be seen in the detailed SIFT-MS study of breath ammonia reported in Ref. 60. Now compare these ammonia data with those for *acetone* in Figure 4.4, where it can be seen that there is little difference between the acetone concentrations in the mouth-exhaled and nose-exhaled breath, which demonstrates that acetone is a truly systemic compound that largely passes into the breath from the blood at the alveolar interface. The somewhat lower concentration in the oral cavity

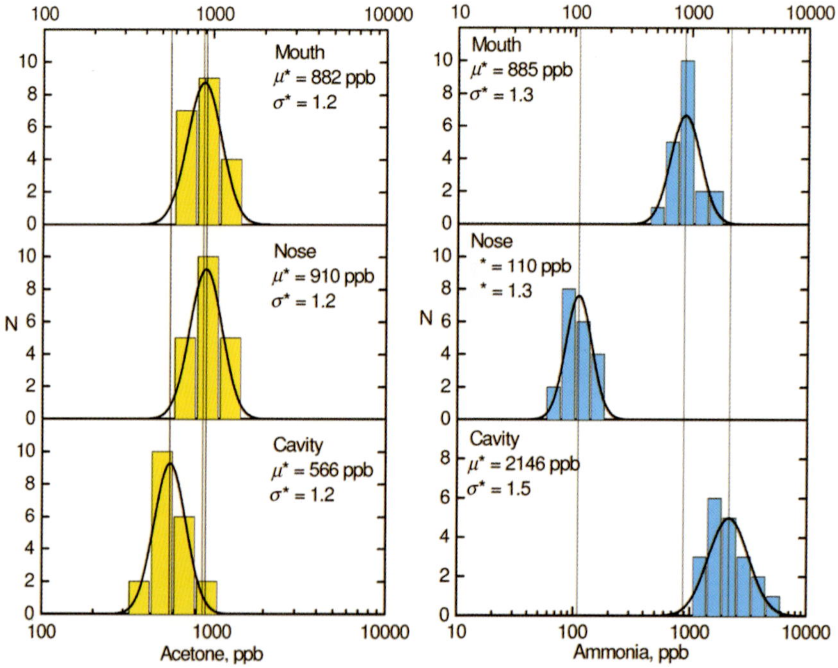

FIGURE 4.4

The concentration distributions of acetone and ammonia (in parts-per-billion, ppbv) as measured on-line by SIFT-MS in the *mouth*- and *nose*-exhaled breath and in the static gas in the oral *cavity* for one healthy male volunteer each day over a period of about one month. The μ^* are the median values of these geometric distributions and the μ^* are the geometric standard deviations.

with breath hold is a further manifestation of this. That acetone is very largely systemic has now been demonstrated unequivocally by subsequent experiments, including that in which urine headspace and breath acetone are seen to be closely equal.[7]

It has now been shown by such SIFT-MS measurements that acetone, methanol, and isoprene are very largely systemic in origin, that ammonia, ethanol, hydrogen cyanide, hydrogen sulfide, and methanol are largely mouth generated, and that the very low concentrations of propanol and acetaldehyde in exhaled breath appear to be partially systemic and partially mouth generated.[57,60] Ideally, it is desirable to perform similar experiments to identify the origin of all exhaled breath trace gas metabolites, especially those purporting to relate to particular diseased states. This is a serious challenge given that such metabolites are likely to be at the ppbv and even the parts-per-trillion, pptv, level, but it simply requires that the most commonly acquired bag and trap samples be obtained from both mouth and nose exhalations.

It is worthy of note that mouth production of ammonia is very probably the action of oral bacteria or enzymes (urease) on urea which is quite systemic and must surely be present in saliva. This mouth production can readily be demonstrated by introducing a trace of urea, solid or aqueous solution, into the mouth, when the exhaled ammonia increases dramatically.[61,62] An increase of *systemic* ammonia can be induced by the ingestion of protein[63] and even a dose of Niacin (Vitamin B3).[64] It is these observations that have cast doubt on the previous assumption that mouth-exhaled ammonia can be used to indicate the level of blood ammonia and thus be used to track the efficacy of dialysis treatment in uraemic patients suffering from end-stage renal disease, ESRD.[62] Yet evidence is now emerging indicating that blood ammonia is not significantly elevated in uraemia and so the elevated mouth-exhaled ammonia is probably due to the action of oral enzymes (urease) on salivary (systemic) urea. This now calls for detailed studies of nose-exhaled ammonia in uraemic patients.

Similarly, just a few crystals of cane sugar or a sweet drink results in a large increase of breath ethanol.[61] Thus, dietary intake control prior to breath analysis is vital, since other compounds, as yet unidentified, can surely be produced by oral bacteria. This is emphasized by the recent SIFT-MS study of oral malodor by Ross and colleagues[65] and as reviewed in the chapter by B. Ross and A. Esarik in this book. Hence, the recommendation must be to bypass the oral cavity and accept that nose-exhaled breath sampling is required to quantify systemic trace gases.

4.2.4 The special case of exhaled hydrogen cyanide

There are few, if any exhaled compounds that can be related to the presence of bacteria in the airways, but hydrogen cyanide, HCN, may be one such. The SIFT-MS contribution to this aspect of breath research began with the detection of copious amounts of HCN in the captured air above plate cultures of Pseudomonas aeruginosa, *PA*, as derived from sputum and cough swab samples provided by patients with the debilitating disease cystic fibrosis, CF.[18] The concentration of HCN was much greater above the *in vitro PA* cultures than above cultures of other bacteria in this limited study. The results of this first study suggested that HCN might be an indicator of the presence of *PA in vivo* and that breath analysis may offer a non-invasive test for *PA* colonisation of the airways.

However, it was then discovered by direct SIFT-MS breath analysis that HCN was present in the mouth-exhaled breath of all individuals, and a study of 17 healthy volunteers in the wide age range of 4–83 years[47] indicated a spread in concentrations ranging from near-zero to 62 ppbv with a median value of 10 ppbv, with no discernible variation with age in this small volunteer cohort. This work was then followed up by a parallel study of HCN (and several other volatile compounds) in mouth-exhaled breath, nose-exhaled breath and the static gas in the oral cavity,[57] using the now routine sampling technique described in the previous section. This study showed that the HCN was almost entirely produced in the oral cavity for these healthy volunteers and the mean HCN concentration in nose-exhaled breath was typically at 1 or 2 ppbv only.

The next logical step was to study the exhaled breath of children with CF to measure HCN concentrations. Thus, the mouth-exhaled breath of 40 children with CF and a similar number of children with asthma (approximately age-matched) were analyzed for both HCN and nitric oxide, NO, which is known to be elevated in the exhaled breath of asthmatics.[66] The exciting result is that the median HCN concentration in the breath of the CF patients was 14 ppbv and in that of the asthmatics just 2 ppbv, whereas the median NO concentration was 13 ppbv in the CF patients and 58 ppbv in the asthmatics. Just to say that experiments are now in train to perform nose-exhaled breath analyses of CF patients to locate the site of most HCN production.

During this time the *in vitro* studies have continued and are on-going to investigate the production of HCN (and other volatile compounds), including methylthiocyanate[67] emitted by different strains of *PA*, and preliminary results are indicating large differences between strains.[17]

4.2.5 On-line breath analysis in the operating theatre

Monitoring is essential to the management of patients within the perioperative and critical care settings. Timely recognition of pathological processes related to abnormal metabolic and inflammatory stresses is necessary to prevent organ damage. Many methods routinely used to assess patients undergoing surgery are expensive, laborious, and invasive and the required information is often delayed. Thus, there is an ongoing interest in the use of breath analysis to monitor the status of the critically ill[68] and ventilated[69] patients. So, recently, it was decided to carry out a pilot experiment to investigate if real-time breath analysis in the operating theatre could provide rapid, non-invasive, and meaningful support to monitoring and diagnosis. The close vicinity of an operating platform is a busy place with little room for relatively large analytical instrumentation. Undeterred, in the first pilot experiments of its kind, a SIFT-MS instrument was located in the operating theatre and connected by a 5 m long PEEK sampling capillary to the ventilation circuit close to the endotracheal tube in patients undergoing laparoscopic surgery. All traces of anaesthetic gases were removed from the anaesthetic circuit to ensure that these did not interfere with the breath gas analysis, and anesthesia was induced with alfentanil and propofol. In these first experiments, real-time profiles were obtained for acetone, isoprene, and propofol

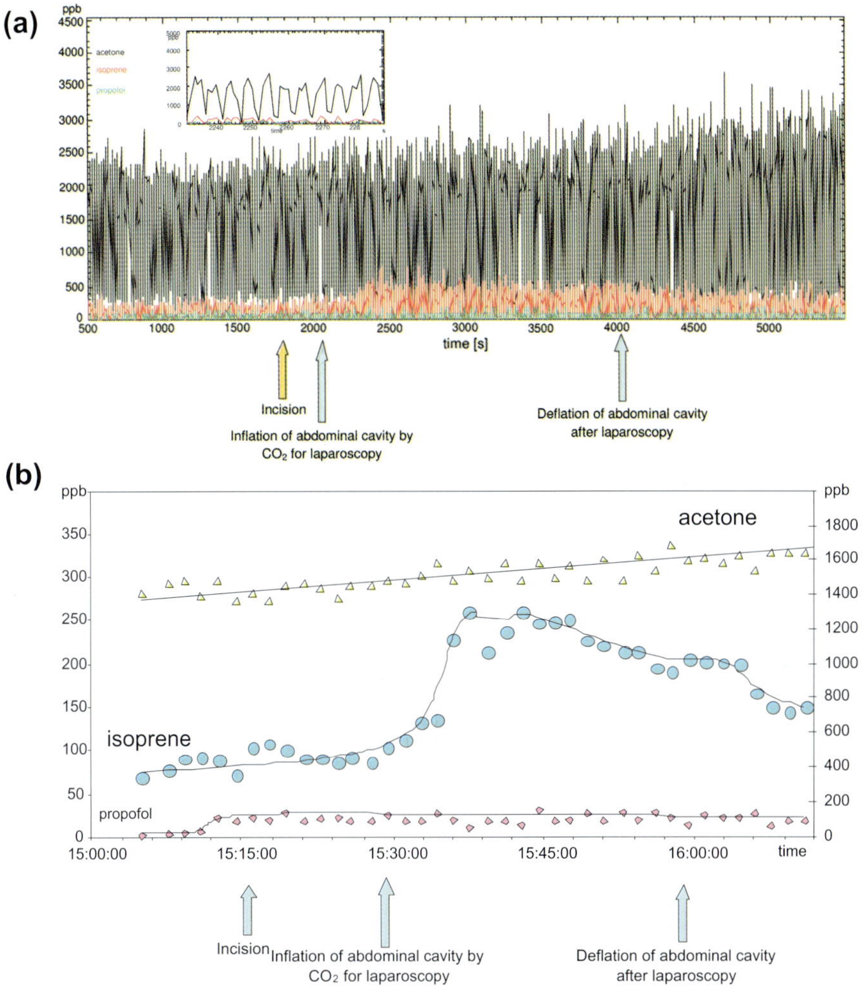

FIGURE 4.5

(a) The time dependence of acetone, isoprene, and propofol concentrations in ppbv during the full course of an operation through the initial anesthesia, incision, inflation of the abdomen by carbon dioxide and deflation over a time interval of about 80 min. The inset is an expanded one minute section of the data showing ten individual exhalation/ inhalation cycles. (b) The concentrations from the data in (a) averaged over 100 s intervals[70] (left axis; isoprene and propofol: right axis; acetone).

as well as water vapor, throughout the course of an operation from skin incision to skin closure. Breathing was induced at 10 exhalations/inhalations per minute, analysis of the above compounds being made during each inhalation/exhalation cycle. This provided several hundred analyses for each compound during each of five operations varying in duration from 20 to 80 min.

The detailed results of this study are reported in the recent paper,[70] but the general results are worth alluding to. It must first be said that the measured absolute humidity of the sampled breath after passing along the long sampling line was reduced for the 6% measured in directly exhaled breath to 3%, which is equivalent to the saturated vapor pressure of water at the operating theatre temperature. The good news is that separate experiments using both exhaled breath and standard mixtures of acetone and isoprene (together with some other common breath metabolites) indicated that the input concentrations of these compounds were not significantly lowered by passing along the sample line. This bodes well for future studies when long sampling lines are needed.

Sample SIFT-MS data for one of the operations are reproduced in Figure 4.5a and show the many breath cycles for which analyses were acquired. The acetone concentrations during the period of the operations were largely invariant, except that a small increase with time occurred due to the onset of lipolysis, this being most obvious during the longest operational period of 80 min. Of most interest are the breath isoprene results; as shown in the averaged data in Figure 4.5b, this compound actually increased following the inflation of the abdomen by carbon dioxide that is commonly used for laparoscopic abdominal operations. A possible physiological explanation that might be deduced from the observations made in a recent study,[36] which indicated that exercise leads to a temporary increase in breath isoprene due to an increasing blood supply to poorly perfused tissues, is that inflation of the body cavity similarly stimulates the release of isoprene into the blood stream from the tissues.

4.3 VOC EMISSION FROM SKIN; COMPARISON WITH VOCS IN BREATH

Little work has been carried out on the emissions of volatile organic compounds, VOCs, from skin even though, in principle, it has great utility. Encouragement is taken from a recent study involving a SPME/GC-MS survey of some VOCs, including organic fatty acids, ketones, aldehydes, esters, and alcohols, that are seen to be present in the headspace above axillary sweat samples collected from different individuals.[71] The most obvious use of skin sampling of VOCs lies in the non-invasive, real-time monitoring of blood stream volatile compounds when exhaled breath cannot be controlled as, for example, in comatose patients in intensive care units. However, before such can be realized, thorough investigations are required of skin-emitted volatile compounds and how they relate to blood/breath concentrations.

So, as a beginning to wider studies, we have carried out a pilot parallel study involving five healthy volunteers to determine changes occurring in the concentrations of several trace compounds present in exhaled breath and emitted from skin into a collection bag surrounding part of the arm, before and after ingesting 75 g of glucose in the fasting state.[72] The glucose involvement was to deliberately modify the blood sugar, which we assumed influenced the blood acetone, in order to see if skin-emitted acetone varied with blood acetone. SIFT-MS enabled real-time quantification of ammonia, methanol, ethanol, propanol, formaldehyde, acetaldehyde, isoprene, and acetone. Following the glucose ingestion, blood glucose and the trace compound concentrations were measured in the exhaled breath and the air contained in the collection bag every 30 min for 2 h. All the above compounds, except formaldehyde, were detected at the expected concentrations in exhaled breath of the five volunteers; all except isoprene, were detected in the collection bag. Ammonia, methanol, and ethanol were present at lower concentrations in the bag than in the exhaled breath. The aldehydes were present at higher concentrations in the bag than in breath. The blood glucose increased to a peak about 1 h post-ingestion, but this change was not obviously correlated with temporal changes in any of the compounds in breath or emitted by skin, except for acetone. The decrease in breath acetone was closely mirrored by skin-emitted acetone in three volunteers. Breath and skin acetone also clearly change with blood glucose and further work may ultimately enable inferences to be drawn of the blood glucose concentration from skin or breath measurements in type 1 diabetes (but see the comments on breath acetone in Section 4.2.2 and the detailed discussion in the chapter by C. Turner in this book).

These experiments, albeit preliminary, give some hope that skin VOCs may be used to monitor blood/breath VOCs; the acetone measurements seemingly being very convincing. The excess aldehydes that are apparently emitted from the skin may be due to superficial skin bacteria.

4.4 EXHALED BREATH CONDENSATE AND BRONCOALVEOLAR LAVAGE

In a similar spirit to the work reported in the previous section, we have carried out a parallel study of the volatile compounds emitted by samples of *exhaled breath condensate*, *EBC*, and those same compounds present in directly exhaled breath.[73] As far as we were aware, no data had been reported previously that would relate the concentration of volatile compounds that appear in the headspace above EBC samples to their gaseous concentrations in exhaled breath. Collection of EBC is a relatively simple non-invasive procedure, but the question to be asked is whether or not EBC carries those volatile compounds that do appear in exhaled breath. Thus, samples of EBC were collected from several subjects in a standard way and absolute concentrations of several common volatile breath metabolites (ammonia, acetone, ethanol, methanol, propanol, isoprene, hydrogen cyanide, formaldehyde, and acetaldehyde) were then quantified in the EBC headspace using SIFT-MS. The results obtained are both interesting and revealing. It is seen that the concentrations of methanol, ammonia, ethanol, and acetone in the EBC headspace correlate well with the direct breath concentrations of these compounds measured within a few minutes of EBC collection, whereas the EBC headspace concentrations of isoprene, formaldehyde, acetaldehyde, hydrogen cyanide, and propanol do not correlate with the direct breath concentrations.

These differences must, in some way, relate to the water (aerosol) solubility of these compounds; for example, isoprene is not very soluble and very readily evolves from water, so this is unlikely to be trapped in the aerosols as it passes along the bronchial tree in the exhaled breath.

Obtaining *bronchoalveolar lavage*, *BAL*, samples is very invasive and in this sense it differs from all the other sampling procedures described in this chapter. It is generally used to identify the bacteria and non-volatile compounds present in the fluid in the lungs. Our interest was to study the volatile compounds emitted from BAL samples and thus to investigate any possible diagnostic clinical value of the data obtained. Thus, in this first study,[74] a large number of BAL samples (210 in total) obtained from patients hospitalized with pneumonia in various departments of two hospitals in Prague, were analyzed for emitted volatile compounds using solid phase microextraction (SPME) coupled with gas chromatography (GC) and flame ionization detection (FID). It was seen that some 20% (9% unequivocally, 11% probably) of these samples contained volatile carboxylic acids in the homologous series from acetic acid to heptanoic acid. Importantly, the presence of these acids indicates the presence of fermenting anaerobic bacteria, which were not detected by conventional microbiological examination. Other compounds, including heptanol and cyclohexanone, were also detected in some samples by this SPME-GC-FID method. However, the cyclohexanone occurred almost exclusively in samples from patients receiving intensive care with mechanical ventilation, and so it likely originates from the plastic parts of the ventilator circuit. Representative samples of the BAL were also analyzed using gas chromatography mass spectrometry (GC-MS) of native and silylated samples and by SIFT-MS. These analyses confirmed the identities of the compounds given above and detected numerous other compounds tentatively identified as various alcohols, aldehydes, ketones, esters and hydrogen cyanide, HCN. Most of these compounds were present in the BAL headspace at very low levels and their origin and diagnostic significance remains uncertain, except, that is, for the HCN, which most probably indicates the presence of Pseudomonas aeruginosa, as discussed in Section 4.2.4.

4.5 VOCS IN URINE HEADSPACE; KETONES AND OVULATION; 3-HBA

The emission (and utility) of volatile compounds from urine have received little attention, yet this could be a valuable source of diagnostic information. The earliest SIFT-MS work was carried out several years ago when the headspace of urine donated by several patients with bladder and prostate cancer was analysed.[12,13] The results of this work were discussed in detail in the cited papers; it is sufficient to say here that formaldehyde was elevated in the headspace of the urine from the cancer patients as compared to healthy controls, and that copious amounts of nitric oxide were seen in the headspace of those acidified urine samples that were bacterially infected. Similarly, potentially important results were obtained when examining the headspace of urine donated by a healthy female volunteer across the menstrual cycle[10] when it was seen that acetone and ammonia greatly increased in the urine samples taken at the time of ovulation. These results were then reproduced for five of a group of seven ovulating women and, significantly, no such increases were detected in a control group of three other postmenopausal women.[75] It was suggested that the origin of the acetone increase during ovulation is related to an increased metabolic rate and further suggested that it might be a valuable indicator of the onset of ovulation.

The presence of VOCs in urine headspace is seemingly an under-used resource for clinical diagnosis, so we have developed SIFT-MS for their rapid and accurate analysis. It has become obvious that in addition to H_3O^+ precursor ions, the use of NO^+ precursor ions in SIFT-MS analysis is exceptionally useful, especially for the analysis of ketones. The mass spectrum shown in Figure 4.6, obtained using

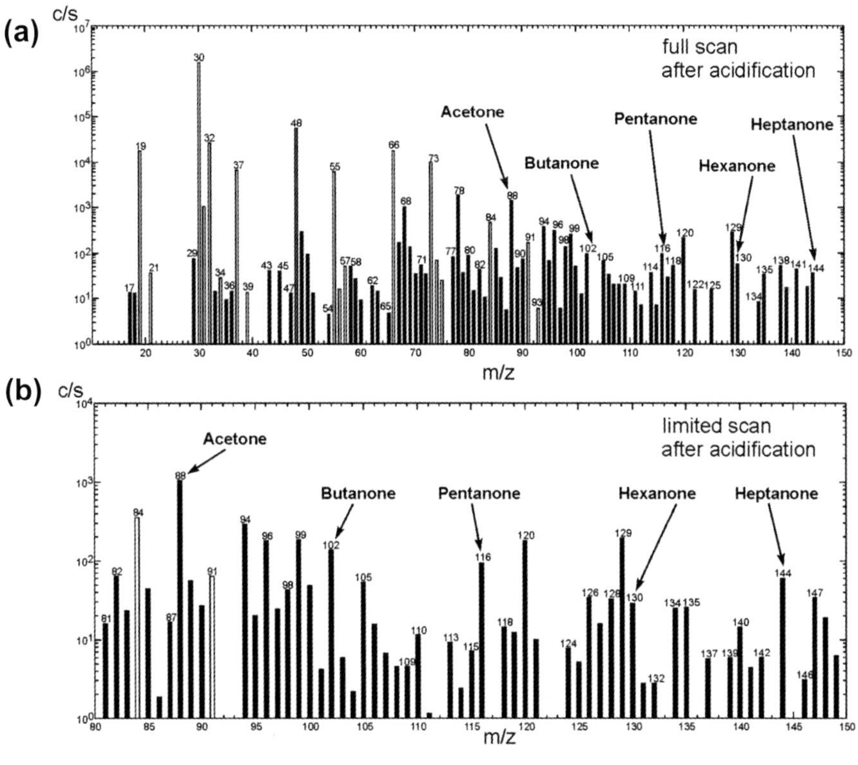

FIGURE 4.6
SIFT-MS spectra of ion count rate per second, c/s, over two ranges of mass-to-charge ratio, *m/z*, obtained using NO^+ precursor ions as the air/vapor headspace above a urine sample provided by a healthy volunteer is introduced into the helium carrier gas. The open bars indicate the NO^+ precursor ion and its hydrates. The characteristic ions of the homologous series of the ketones, *M*, are adduct ions of the form NO^+ *M*. The limited *m/z* scan range in (b) allows a greater dwell time on each ion as the scan proceeds and thus provides a better definition of the spectrum. Reproduced from Ref. 7, with permission from Wiley.

SIFT-MS in the full scan mode with NO^+ precursor ions,[7] clearly shows the homologous series of ketones ranging from acetone to heptanone. In this same study we were able to obtain kinetic data for the analysis of gas-phase 3-hydroxybutyric acid, 3-HBA which, in addition to acetone and acetoacetic acid, is one of these three so-called ketone bodies that are important in the diagnosis of ketoacidosis. A more recent study has shown how 3-HBA and the other ketones can be accurately quantified in urinary headspace.[8]

4.6 VOLATILE BIOMARKERS OF CANCER CELLS, *IN VITRO* AND *IN VIVO*

The non-invasive detection of tumors in the body, especially via the detection of volatile biomarkers in exhaled breath and those emitted by bodily fluids, is potentially of enormous value and is attracting much interest in the breath research community.[76–78] Some concerted attempts to detect cancer biomarkers in exhaled breath have been made, but with limited success so far.[79–84] This is an indication of the complexity and difficulties associated with this important aspect of breath research.

SIFT-MS research in this area has been limited and confined to the analysis of urine from cancer patients, as summarized in the previous section, and the efforts to detect biomarkers released by mammalian cells *in vitro* that we (and others[85–89]) hope will stimulate the search for breath biomarkers. A brief overview of this work is given in Ref. 90. Our earliest studies revealed that incubated CALU-1 and SKMES lung cancer cell lines release acetaldehyde molecules in amounts proportional to the number of cells in the growth medium.[16] This is an exciting and potentially very important observation that deserves much more research. Very recently, we repeated the CALU-1 studies and extended the study to include NL20 normal lung epithelial cells and 35FL121 telomerase-positive (Tel+) lung fibroblast cells. Thus, SIFT-MS has been used to quantify acetaldehyde and, additionally, carbon dioxide in the headspace of the cell cultures.[15] The results showed that acetaldehyde was generated by the CALU-1 and NL20 cell cultures in proportions to the number of cells in the medium. However, following incubation, the acetaldehyde concentrations in the headspace of the 35FL121 Tel+ cell cultures were much lower than those present in the headspace of the medium alone. The amount of carbon dioxide generated by the CALU-1 and 35FL121 Tel+ cells indicated that they were respiring normally, but much less carbon dioxide was produced by the NL20 cells, presumably indicating that normal cell metabolism was inhibited. These rather perplexing results indicate that acetaldehyde release is not specifically the action of malignant cells and so this compound alone cannot be used with confidence as a tumor diagnostic.

We have now initiated the studies of the exhaled breath of patients suffering from lung cancer, initially using bag sampling performed in the oncology unit of the local hospital in Stoke-on-Trent, UK. An initial, very preliminary yet interesting discovery is that acetic acid is elevated in the breath samples of some cancer patients compared with breath samples from healthy controls. Further analyses of the CALU-1 *in vitro* cell cultures headspace (see above), revealed that acetic acid was present when the culture was made acidic. Therefore, it seems likely that the lung tumors and/or the tissue surround them, which often has associated fluid at a low pH, are emitting acetic acid and this may be an important discreet biomarker.[90] However, much more work is needed to substantiate these preliminary results.

The collected research carried out to date indicates that volatile cancer biomarkers in breath will generally be at relatively low concentrations, perhaps at sub-ppbv levels, and so greater detection sensitivity of SIFT-MS will be required; such improvements are ongoing. Additionally, parallel SIFT-MS/GC-MS analyses of volatile compounds extracted from patients' breath are now being initiated.

4.7 ALCOHOL INGESTION AND DETECTION AND CANNABIS

Breath analysis for ethanol has a long history, for obvious reasons, and the detection methods are simple and robust, but do not provide sufficiently accurate and precise analyses to allow detailed physiological studies of ethanol dispersion and metabolism. In this regard, SIFT-MS is supreme in that direct analyses can be obtained of single breath exhalations in real time at relatively low concentrations to allow the careful tracking of breath ethanol. Our earliest investigation of this, involving only mouth-exhaled breath sampling, revealed the influence of body (liver) mass on the removal of ethanol and its metabolic conversion to acetaldehyde.[91] More recently, a study has been carried out of the decay of ethanol in mouth-exhaled and nose-exhaled breath of two healthy volunteers following the ingestion via the mouth of various doses of alcohol at different dilutions in water.[92] Continuous measurements were made over a period of about 200 min and the time interval between breath exhalations was about 20 s. Thus, the time between sampling the breath of each individual was only about 3 min, which provided many time points and well-defined breath ethanol decay curves and ethanol loss rates from the blood stream. Inspection of the mouth-exhaled and nose-exhaled breath data shows that mouth contamination of ethanol diminished to insignificant concentrations after a few minutes.

The detailed results of the analyses of nose-exhaled breath show that the peak concentrations and the decay rates of breath ethanol are dependent on the ethanol dose and the volume of ethanol/water mixture ingested. From these data, both the efficiency of the first-pass metabolism of ethanol and indications of gastric emptying rates at the various doses and ingested volumes were obtained for the two volunteers. Additionally and simultaneously, acetaldehyde, acetic acid, and acetone were measured in each single breath exhalation. Acetaldehyde was seen to track the breath ethanol. Acetic acid, a possible secondary product of this metabolism, was also detected, but this compound was shown to largely originate in the oral cavity. Breath acetone was seen to increase over the long period of measurement due to the depletion of nutrients. The details of the kinetics of ethanol metabolism are given in the original paper.[92]

A related topic that we have pursued in collaboration with local psychiatrists is that of alcoholism. Non-compliance to abstinence advice is a problem and to check for alcohol abuse the patient is expected to take an oral dose of disulfiram (Antabuse), which then results in the appearance of the marker carbon disulfide, CS_2, in exhaled breath. To investigate this phenomenon we carried out a SIFT-MS study of for CS_2 in the exhaled breath of a volunteer following the ingestion of a single dose of disulfiram and, simultaneously, for several common breath metabolites, including acetone.[93] These breath analyses were performed in single exhalations and the results were available in real time. The concentrations of breath acetone and carbon disulfide were compared with concentrations obtained from a control subject who had not ingested disulfiram. Breath carbon disulfide was seen to increase from 15 ppbv to 618 ppbv over a 28-hour period in the single individual tested following ingestion of the disulfiram, while acetone concentrations increased from 300 ppbv (normal) to over 4000 ppbv (greatly elevated). No such increases were seen in the breath of the control subject over the same period. An obvious positive correlation between breath carbon disulfide and acetone concentrations following disulfiram ingestion is seen and discussed in the original paper. The large increase in breath acetone following the disulfiram dose was unexpected and suggests interference by this compound with liver metabolism, but this needs further investigation.

Substance abuse is a major problem in society and places a great financial burden on health services. Detection of drug usage is costly and time consuming; hence the drive to find simple, non-invasive methods of detection; breath analysis is attractive in this search. Again, in collaboration with a local psychiatrist, we have begun to focus on cannabis use and to investigate if there are detectable and quantifiable volatile markers of cannabis smoking in exhaled breath. The starting point has been to carry out a SIFT-MS study of the molecular species emitted by heated "street" cannabis plant material.[94] Thus, samples of "street" cannabis leaf, held under a UK Home Office licence, were prepared by finely chopping and mixing the material. The samples were then heated in readily available commercial devices (*Blue Meanie* and *Volcano*) and the air/smoke containing the released gaseous compounds was analysed by SIFT-MS. Smoke from standard 3% National Institute on Drug Abuse (NIDA) cannabis cigarettes was also analysed. The startling result was for "street" cannabis that ammonia was present at concentrations approaching 200 parts-per-million (ppmv). This is compared with peak concentrations of 10 ppmv in the NIDA smoke samples of known provenance and of the psychoactive component tetrahydrocannabinol, THC (3%). Several other compounds were present in the smoke at lower concentrations, including acetaldehyde, methanol, acetone, acetic acid, monoterpenes, and sesquiterpenes. A sample full scan spectrum obtained when the cannabis smoke was analysed by SIFT-MS using H_3O^+ ions is shown in Figure 4.7. The richness of this spectrum indicates the presence of many, mostly toxic VOCs. Use of the above-mentioned commercial devices is claimed to be safer than direct smoking of cannabis, yet ammonia is still present in the smoke at toxic concentrations. Thus, the use of these devices to more safely smoke "street" cannabis is now open to question and further research is needed to investigate this.

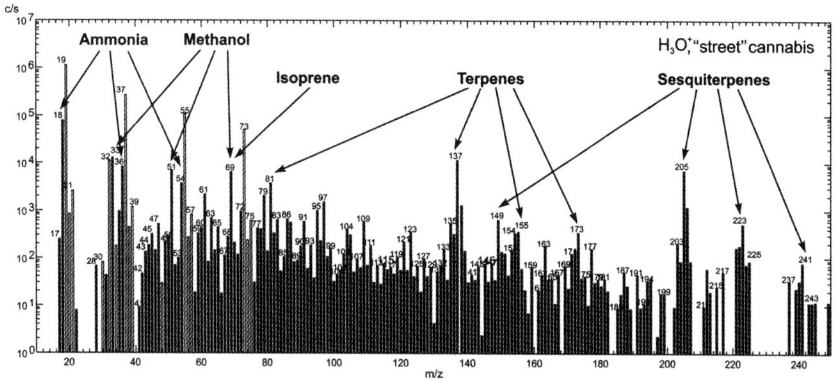

FIGURE 4.7
A SIFT-MS spectrum (counts per second, c/s, against m/z) obtained using H_3O^+ precursor ions for the smoke generated when "street" cannabis leaf is combusted at approximately 250 °C in enclosed air. Protonated trace compounds and their mono- and di-hydrates are present: ammonia (30 parts-per-million, ppmv), methanol (10 ppmv), isoprene (0.5 ppmv), monoterpenes (6 ppmv) and sesquiterpenes (3 ppmv). Reproduced from Ref. 5, with permission from Wiley.

4.8 FLOWING AFTERGLOW MASS SPECTROMETRY, FA-MS, AND TOTAL BODY WATER

Some 10 years ago we developed flowing afterglow mass spectrometry, FA-MS, by which the fractional deuterium/hydrogen content of water vapor, D/H, can be measured non-invasively and in real time in single exhalations of exhaled breath and above aqueous liquids.[95,96] Much has been written that explains the physical and ion chemical principles which underpin this analytical method. So it is sufficient to simply give a very brief outline here. The method is based on the creation of a swarm of water cluster ions, $H_3O^+(H_2O)_3$ (i.e. $H_9O_4^+$), held in an inert support gas, usually helium, in the presence of water molecules at a constant, known temperature, for time periods sufficient to allow the D, ^{17}O, and ^{18}O isotopologues of these cluster ions to reach their thermal equilibrium distribution. The abundance of D in the water vapor introduced into the support gas (the analyte) is then obtained by measuring the relative number densities of these isotopologues using quantitative mass spectrometry. These isotopologue ions have mass-to-charge ratios (m/z) of 73 ($H_9^{16}O_4^+$), 74 ($H_8D^{16}O_4^+$, and $H_9^{16}O_3^{17}O^+$) and 75 ($H_9^{16}O_3^{18}O^+$). More combinations are possible at m/z 75, including $H_8D^{16}O_3^{17}O^+$ and $H_7D_2^{16}O_4^+$, but their expected contributions are negligible (<1%) at the natural abundances of the minor isotopes D and ^{17}O and for small D enrichments. The spectrum obtained when H_3O^+ ions are introduced into the carrier gas of the FA-MS (or a SIFT-MS) instrument and water vapor is added is shown

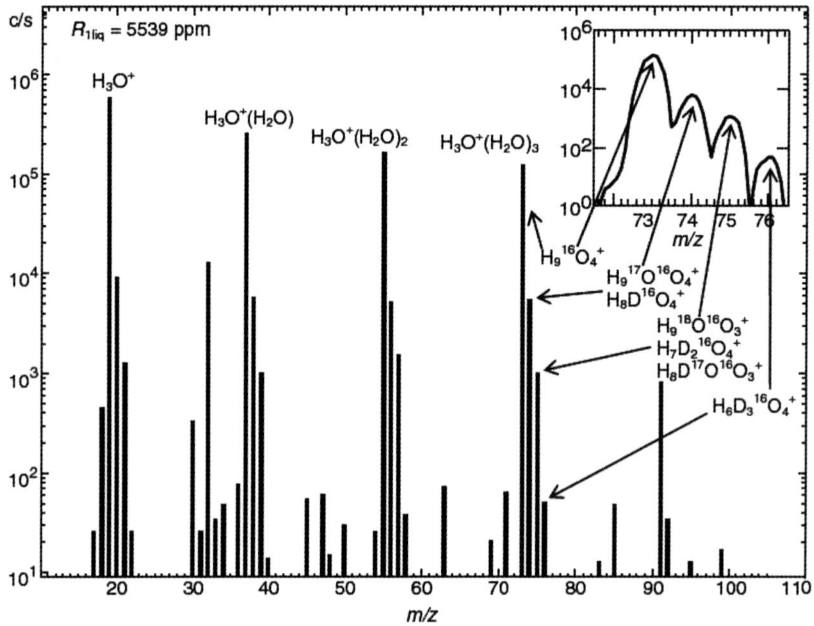

FIGURE 4.8

The SIFT-MS mass spectrum (counts per second, c/s, against ion mass-to-charge ratio, m/z) obtained as the vapor above a standard solution of D_2O in H_2O (at a D/H ratio of 5539 ppm) is introduced into the helium carrier gas. The observed isotopologues of the trihydrate of H_3O^+ are indicated in the inset. Reproduced from Ref. 97 with permission from Springer.

in Figure 4.8 (actually obtained using SIFT-MS), where some of the isotopologues mentioned above are indicated. Larger signals (count rates) of the analytical ions at m/z values of 73, 74, and 75 can be realized in FA-MS, but recently with improvements to the sensitivity of SIFT-MS this technique can now be used for D/H measurements in water vapor.[97]

The accuracy and precision of this technique has been assessed in several studies and both parameters are seen to be at the 1% level.[98] Subsequently, this FA-MS analytical method has been used by our local nephrologists under the direction of S. J. Davies to determine the total body water, TBW, in cohorts of healthy controls and patients being treated by hemodialysis, HD, and peritoneal dialysis, PD, using the principle of isotope dilution following the ingestion of an accurately known amount of D_2O.[96,99] Thus, from the temporal variations of breath deuterium, TBW can be determined non-invasively and the kinetics of water flow around the body can be tracked. The technique and some results obtained up to 2005 are reported in the previous book edited by Amann and Smith.[100]

This research work has continued apace in recent years by focusing on dialysis patients. The dispersal kinetics of HDO were determined in 12 hemodialysis patients and used to calculate the absolute TBW post-HD after full deuterium equilibration. TBW pre-HD was then determined from breath samples taken immediately prior to the next dialysis session.[101] The valuable conclusion drawn from this study is that the non-invasive FA-MS isotope dilution method for determining TBW can be applied to hemodialysis patients who show deuterium equilibration kinetics identical to normal subjects. A pre-dialysis estimation may be used to determine TBW, avoiding the necessity for the patients to remain behind after dialysis making this assessment method suitable for application in the clinical setting.

Much work has also been carried out to compare the electrical bioimpedance, BIA, method for TBW estimation (by which extracellular water ECW, and intercellular water, ICW, can be differentiated) with the deuterium dilution method as practiced using FA-MS; the results of these studies are reported in Ref. 101. It is concluded that combined near-patient measurements of absolute TBW, as determined by FA-MS and BIA, are achievable in a dialysis facility by identifying changes in body composition that are not fully appreciated by routine assessment. BIA underestimates tissue over-hydration associated with comorbidity, resulting in reduced sensitivity to longitudinal increases during a 12-month period. These FA-MS and BIA measurements of TBW couple with other important measurements in renal patients are continuing. A recent study has shown that hypoalbuminemia is an important determinant of tissue over hydration in PD patients.[102] This over hydration is not associated with an increased plasma volume, and attempts to normalize the ECW/TBW ratio in hypoalbuminemic inflamed PD patients may lead to hypovolemia and loss of residual renal function.

Pulmonary oedema is a medical condition characterized by abnormal accumulation of fluid in the extravascular space in the lung alveoli. Effective oxygenation is impaired and this leads to significant short- and long-term morbidity and mortality. The detection and monitoring of pulmonary oedema by measuring lung water volume is therefore crucial in the initiation and guidance of therapeutic intervention. The current gold-standard bedside measurement of extravascular lung water volume, EVLW, is the dilution method using various indicators, but despite the good correlation of the results with those obtained using the post-mortem gravimetric method, the invasiveness of the dilution technique limits its general application in the wider clinical setting. So we have carried out a novel experiment involving just three healthy volunteers in which the dispersal kinetics of HDO in exhaled breath following the inhalation of D_2O vapor were explored using the rapid sampling facility that is possible using FA-MS. The details of this very interesting study show how lung water might be estimated from the data. The limitations of this approach and required future work are discussed in a recent paper.[103]

The value of FA-MS for real-time measurements of TBW by breath analysis, and via blood, serum, and dialysate headspace analysis in the clinical setting is actively being explored. Also, the development of the FA-MS and SIFT-MS analytical techniques is ongoing and its extension to the measurement of high deuterium abundances up to 10,000 ppmv has recently been reported.[97]

4.9 SUMMARY REMARKS; FUTURE PROSPECTS FOR SIFT-MS AND FA-MS IN MEDICINE

The SIFT-MS and FA-MS advances summarized in this chapter largely refer to the new research results obtained since the last reports in the previous book.[32, 100, 104–108] The sensitivity of SIFT-MS and the mobility of the instrumentation have been greatly improved and now accurate on-line, real-time measurements of the concentrations of trace metabolites in exhaled breath can be made down to the ppbv level. A major desire has been to move direct breath analysis from the laboratory toward the clinical setting and this has been partially achieved with SIFT-MS (Sections 4.2.4 and 4.2.5) and and to a large extent with FA-MS (Section 4.8). A critical advance has been the routine analysis of nose-exhaled breath that minimizes oral contamination of the breath sample (Section 4.2.3). It is this and the real- time analysis feature that makes SIFT-MS superior for direct breath analysis.

Yet a greater challenge lies ahead, which is the identification and accurate quantification of new biomarkers of disease that are likely to be present in exhaled breath only at sub-ppbv levels. This calls for even greater sensitivity than is currently available to SIFT-MS. A promising approach that we are investigating is first to use off-line GC-MS, together with SPME or ATD, to identify compounds of interest in breath samples, then to research their ion chemistry and to use SIFT-MS to greater effect to monitor their concentrations in exhaled breath in real time. This will then allow much greater sample throughput, which is so desirable in the clinical setting. Surely, it is only thorough experimental work and innovation that will promote the difficult and challenging topic of breath analysis, such that it can be a valuable new diagnostic method or, at least, provide meaningful support to conventional clinical diagnostics.

Acknowledgments

We would like to express our gratitude to all our scientific, clinical and technical colleagues, especially Simon Davies, Gordon Ferns, Warren Lenney, George Hanna, Roger Bloor, Beth Enderby, Claire Turner, Andriy Pysanenko, Tianshu Wang, Kseniya Dryahina, Kristyna Sovova, Violet Shestivska, Edward Hall, and Clive Workman.

References

1. Španěl P, Smith D. Selected ion flow tube: a technique for quantitative trace gas analysis of air and breath. *Med Biol Eng Comput* 1996;**34**:409–19.
2. Smith D, Španěl P. Application of ion chemistry and the SIFT technique to the quantitative analysis of trace gases in air and on breath. *Int Rev Phys Chem* 1996;**15**:231–71.
3. Smith D, Španěl P. The novel selected-ion flow tube approach to trace gas analysis of air and breath. *Rapid Commun Mass Spectrom* 1996;**10**:1183–98.
4. Španěl P, Rolfe B, Rajan B, Smith D. The selected ion flow tube (SIFT)—a novel technique for biological monitoring. *Ann Occup Hyg* 1996;**40**:615–26.

5. Španěl P, Smith D. Progress in SIFT-MS: breath analysis and other applications. *Mass Spectrom Rev* 2011;**30**:236–67.
6. Smith D, Španěl P. Ambient analysis of trace compounds in gaseous media by SIFT-MS. *Analyst* 2011;**136**:2009–32.
7. Pysanenko A, Wang T, Španěl P, Smith D. Acetone, butanone, pentanone, hexanone and heptanone in the headspace of aqueous solution and urine studied by selected ion flow tube mass spectrometry. *Rapid Commun Mass Spectrom* 2009;**23**:1097–104.
8. Wang T, Španěl P, Smith D. Selected ion flow tube mass spectrometry of 3-hydroxybutyric acid, acetone and other ketones in the headspace of aqueous solution and urine. *Int J Mass Spectrom* 2008;**272**:78–85.
9. Španěl P, Smith D. Selected ion flow tube mass spectrometry for on-line trace gas analysis in biology and medicine. *Eur J Mass Spectrom* 2007;**13**:77–82.
10. Diskin AM, Španěl P, Smith D. Increase of acetone and ammonia in urine headspace and breath during ovulation quantified using selected ion flow tube mass spectrometry. *Physiol Meas* 2003;**24**:191–9.
11. Abbott SM, Elder JB, Španěl P, Smith D. Quantification of acetonitrile in exhaled breath and urinary headspace using selected ion flow tube mass spectrometry. *Int J Mass Spectrom* 2003;**228**:655–65.
12. Španěl P, Smith D, Holland TA, Al Singary W, Elder JB. Analysis of formaldehyde in the headspace of urine from bladder and prostate cancer patients using selected ion flow tube mass spectrometry. *Rapid Commun Mass Spectrom* 1999;**13**:1354–9.
13. Smith D, Španěl P, Holland TA, Al Singari W, Elder JB. Selected ion flow tube mass spectrometry of urine headspace. *Rapid Commun Mass Spectrom* 1999;**13**:724 –9.
14. Kato S, Burke PJ, Koch TH, Bierbaum VM. Formaldehyde in human cancer cells: detection by preconcentration-chemical ionization mass spectrometry. *Anal Chem* 2001;**73**:2992–7.
15. Sule-Suso J, Pysanenko A, Španěl P, Smith D. Quantification of acetaldehyde and carbon dioxide in the headspace of malignant and non-malignant lung cells *in vitro* by SIFT-MS. *Analyst* 2009;**134**:2419–25.
16. Smith D, Wang T, Sule-Suso J, Španěl P, El Haj A. Quantification of acetaldehyde released by lung cancer cells *in vitro* using selected ion flow tube mass spectrometry. *Rapid Commun Mass Spectrom* 2003;**17**:845–50.
17. Gilchrist FJ, Alcock A, Belcher J, Brady M, Jones A, Smith D, et al. Variation in hydrogen cyanide production between different strains of *Pseudomonas aeruginosa*. *Eur Respir J* 2011;**38**:409–14.
18. Carroll W, Lenney W, Wang T, Španěl P, Alcock A, Smith D. Detection of volatile compounds emitted by Pseudomonas aeruginosa using selected ion flow tube mass spectrometry. *Pediatr Pulmonol* 2005;**39**:452–6.
19. Španěl P, Wang TS, Smith D. Quantification of hydrogen cyanide in humid air by selected ion flow tube mass spectrometry. *Rapid Commun Mass Spectrom* 2004;**18**:1869–73.
20. Thorn RM, Reynolds DM, Greenman J. Multivariate analysis of bacterial volatile compound profiles for discrimination between selected species and strains *in vitro*. *J Microbiol Methods* 2011;**84**:258–64.
21. Allardyce RA, Langford VS, Hill AL, Murdoch DR. Detection of volatile metabolites produced by bacterial growth in blood culture media by selected ion flow tube mass spectrometry (SIFT-MS). *J Microbiol Methods* 2006;**65**:361–5.
22. Olivares A, Dryahina K, Navarro JL, Smith D, Španěl P, Flores M. SPME-GC-MS versus Selected Ion Flow Tube Mass Spectrometry (SIFT-MS) analyses for the study of volatile

compound generation and oxidation status during dry fermented sausage processing. *J Agric Food Chem* 2011;**59**:1931–8.

23. Olivares A, Dryahina K, Navarro JL, Flores M, Smith D, Španěl P. Selected ion flow tube-mass spectrometry for absolute quantification of aroma compounds in the headspace of dry fermented sausages. *Anal Chem* 2010;**82**:5819–29.
24. Smith D, Španěl P, Jones JB. Analysis of volatile emissions from porcine faeces and urine using selected ion flow tube mass spectrometry. *Bioresource Technol* 2000;**75**:27–33.
25. Dewhurst RJ, Evans RT, Mottram TT, Španěl P, Smith D. Assessment of rumen processes by selected-ion-flow-tube mass spectrometric analysis of rumen gases. *J Dairy Sci* 2001;**84**:1438–44.
26. Španěl P, Smith D. Quantification of hydrogen sulphide in humid air by selected ion flow tube mass spectrometry. *Rapid Commun Mass Spectrom* 2000;**14**:1136–40.
27. Sovova K, Ferus M, Matulkova I, Španěl P, Dryahina K, Dvorak O, et al. A study of thermal decomposition and combustion products of disposable polyethylene terephthalate (PET) plastic using high resolution Fourier transform infrared spectroscopy, selected ion flow tube mass spectrometry and gas chromatography mass spectrometry. *Mol Phys* 2008;**106**:1205–14.
28. Civis M, Civis S, Sovova K, Dryahina K, Španěl P, Kyncl M. Laser ablation of FOX-7: proposed mechanism of decomposition. *Anal Chem* 2011;**83**:1069–77.
29. Sovova K, Dryahina K, Španěl P, Kyncl M, Civis S. A study of the composition of the products of laser-induced breakdown of hexogen, octogen, pentrite and trinitrotoluene using selected ion flow tube mass spectrometry and UV-Vis spectrometry. *Analyst* 2010;**135**:1106–14.
30. Španěl P, Smith D. Selected ion flow tube—mass spectrometry: detection and real-time monitoring of flavours released by food products. *Rapid Commun Mass Spectrom* 1999;**13**:585–96.
31. Smith D, Španěl P. Selected ion flow tube mass spectrometry (SIFT-MS) for on-line trace gas analysis. *Mass Spectrom Rev* 2005;**24**:661–700.
32. Smith D, Španěl P. Selected ion flow tube mass spectrometry, SIFT-MS, for on-line trace gas analysis of breath. In: Amann A, Smith D, editors. *Breath analysis for clinical diagnosis and therapeutic monitoring*. Singapore: World Scientific; 2005. p. 3–34.
33. Ross BM. Sub-parts per billion detection of trace volatile chemicals in human breath using selected ion flow tube mass spectrometry. *BMC Res Notes* 2008;**1**:41.
34. Thekedar B, Oeh U, Szymczak W, Hoeschen C, Paretzke HG. Influences of mixed expiratory sampling parameters on exhaled volatile organic compound concentrations. *J Breath Res* 2011;**5**:016001.
35. King J, Unterkofler K, Teschl G, Teschl S, Koc H, Hinterhuber H, et al. A mathematical model for breath gas analysis of volatile organic compounds with special emphasis on acetone. *J Math Biol* 2011;**63**:959–99.
36. King J, Koc H, Unterkofler K, Mochalski P, Kupferthaler A, Teschl G, et al. Physiological modeling of isoprene dynamics in exhaled breath. *J Theor Biol* 2010;**267**:626–37.
37. King J, Unterkofler K, Teschl G, Teschl S, Mochalski P, Koc H, et al. A modeling-based evaluation of isothermal rebreathing for breath gas analyses of highly soluble volatile organic compounds. *J Breath Res* 2012;**6**:016005.
38. O'Hara ME, O'Hehir S, Green S, Mayhew CA. Development of a protocol to measure volatile organic compounds in human breath: a comparison of rebreathing and on-line single exhalations using proton transfer reaction mass spectrometry. *Physiol Meas* 2008;**29**: 309–30.

39. Smith D, Turner C, Španěl P. Volatile metabolites in the exhaled breath of healthy volunteers: their levels and distributions. *J Breath Res* 2007;**1**:014004.
40. Turner C, Španěl P, Smith D. A longitudinal study of breath isoprene in healthy volunteers using selected ion flow tube mass spectrometry (SIFT-MS). *Physiol Meas* 2006;**27**:13–22.
41. Turner C, Španěl P, Smith D. A longitudinal study of ethanol and acetaldehyde in the exhaled breath of healthy volunteers using selected-ion flow-tube mass spectrometry. *Rapid Commun Mass Spectrom* 2006;**20**:61–8.
42. Turner C, Španěl P, Smith D. A longitudinal study of methanol in the exhaled breath of 30 healthy volunteers using selected ion flow tube mass spectrometry, SIFT-MS. *Physiol Meas* 2006;**27**:637–48.
43. Turner C, Španěl P, Smith D. A longitudinal study of ammonia, acetone and propanol in the exhaled breath of 30 subjects using selected ion flow tube mass spectrometry, SIFT-MS. *Physiol Meas* 2006;**27**:321–37.
44. Smith D, Španěl P. The challenge of breath analysis for clinical diagnosis and therapeutic monitoring. *Analyst* 2007;**132**:390–6.
45. Diskin AM, Španěl P, Smith D. Time variation of ammonia, acetone, isoprene and ethanol in breath: a quantitative SIFT-MS study over 30 days. *Physiol Meas* 2003;**24**:107–19.
46. Španěl P, Dryahina K, Smith D. The concentration distributions of some metabolites in the exhaled breath of young adults. *J Breath Res* 2007;1:026001.
47. Španěl P, Dryahina K, Smith D. Acetone, ammonia and hydrogen cyanide in exhaled breath of several volunteers aged 4–83 years. *J Breath Res* 2007;**1**:011001.
48. Salerno-Kennedy R, Cashman KD. Potential applications of breath isoprene as a biomarker in modern medicine: a concise overview. *Wien Klin Wochenschr* 2005;**117**:180–6.
49. Karl T, Prazeller P, Mayr D, Jordan A, Rieder J, Fall R, et al. Human breath isoprene and its relation to blood cholesterol levels: new measurements and modeling. *J Appl Physiol* 2001;**91**:762–70.
50. Smith D, Španěl P, Enderby B, Lenney W, Turner C, Davies SJ. Isoprene levels in the exhaled breath of 200 healthy pupils within the age range 7–18 years studied using SIFT-MS. *J Breath Res* 2009;**4**:017101.
51. Enderby B, Lenney W, Brady M, Emmett C, Španěl P, Smith D. Concentrations of some metabolites in the breath of healthy children aged 7–18 years measured using selected ion flow tube mass spectrometry (SIFT-MS). *J Breath Res* 2009;**3**:036001.
52. Hryniuk A, Ross BM. Detection of acetone and isoprene in human breath using a combination of thermal desorption and selected ion flow tube mass spectrometry. *Int J Mass Spectrom* 2009;**285**:26–30.
53. Craig ME, Jones TW, Silink M, Ping YJ. Diabetes care, glycemic control, and complications in children with type 1 diabetes from Asia and the Western Pacific Region. *J Diabetes Complications* 2007;**21**:280–7.
54. Smith D, Španěl P, Fryer AA, Hanna F, Ferns GA. Can volatile compounds in exhaled breath be used to monitor control in diabetes mellitus? *J Breath Res* 2011;**5**:022001.
55. Musa-Veloso K, Likhodii SS, Cunnane SC. Breath acetone is a reliable indicator of ketosis in adults consuming ketogenic meals. *Am J Clin Nutr* 2002;**76**:65–70.
56. Španěl P, Dryahina K, Rejskova A, Chippendale T, Smith D. Breath acetone concentration; biological variability and the influence of diet. *Physiol Meas* 2011;**32**:N23–31.
57. Wang T, Pysanenko A, Dryahina K, Španěl P, Smith D. Analysis of breath, exhaled via the mouth and nose, and the air in the oral cavity. *J Breath Res* 2008;**2**:037013.

58. Turner C, Walton C, Hoashi S, Evans M. Breath acetone concentration decreases with blood glucose concentration in type I diabetes mellitus patients during hypoglycaemic clamps. *J Breath Res* 2009;**3**:046004.
59. Tangerman A, Winkel EG. Extra-oral halitosis: an overview. *J Breath Res* 2010;**4**: 017003.
60. Smith D, Wang T, Pysanenko A, Španěl P. A selected ion flow tube mass spectrometry study of ammonia in mouth- and nose-exhaled breath and in the oral cavity. *Rapid Commun Mass Spectrom* 2008;**22**:783–9.
61. Španěl P, Turner C, Wang TS, Bloor R, Smith D. Generation of volatile compounds on mouth exposure to urea and sucrose: implications for exhaled breath analysis. *Physiol Meas* 2006;**27**:N7–17.
62. Davies S, Španěl P, Smith D. Quantitative analysis of ammonia on the breath of patients in end-stage renal failure. *Kidney Int* 1997;**52**:223–8.
63. Španěl P, Davies S, Smith D. Breath ammonia dips after feeding independently of the protein content of the meat. *FASEB J* 2000;**14**:A490.
64. Smith D, Wang T, Španěl P, Bloor R. The increase of breath ammonia induced by niacin ingestion quantified by selected ion flow tube mass spectrometry. *Physiol Meas* 2006;**27**:437–44.
65. Ross BM, Dadgostar N, Bloom M, McKeown L. The analysis of oral air using selected ion flow tube mass spectrometry in persons with and without a history of oral malodor. *Int J Dent Hyg* 2009;**7**:136–43.
66. Enderby B, Smith D, Carroll W, Lenney W. Hydrogen cyanide as a biomarker for Pseudomonas aeruginosa in the breath of children with cystic fibrosis. *Pediatr Pulmonol* 2009;**44**:142–7.
67. Shestivska V, Nemec A, Drevinek P, Sovova K, Dryahina K, Španěl P. Quantification of methyl thiocyanate in the headspace of *Pseudomonas aeruginosa* cultures and in the breath of cystic fibrosis patients by selected ion flow tube mass spectrometry. *Rapid Commun Mass Spectrom* 2011;**25**:2459–67.
68. Schubert JK, Miekisch W, Geiger K, Noldge-Schomburg GF. Breath analysis in critically ill patients: potential and limitations. *Expert Rev Mol Diagn* 2004; **4**:619:29.
69. Schubert JK, Spittler KH, Braun G, Geiger K, Guttmann J. CO_2-controlled sampling of alveolar gas in mechanically ventilated patients. *J Appl Physiol* 2001;**90**:486–92.
70. Boshier PR, Cushnir JR, Mistry V, Knaggs A, Španěl P, Smith D, et al. On-line, real time monitoring of exhaled trace gases by SIFT-MS in the perioperative setting: a feasibility study. *Analyst* 2011;**136**:3233–7.
71. Curran AM, Rabin SI, Prada PA, Furton KG. Comparison of the volatile organic compounds present in human odor using SPME-GC/MS. *J Chem Ecol* 2005;**31**:1607–19.
72. Turner C, Parekh B, Walton C, Španěl P, Smith D, Evans M. An exploratory comparative study of volatile compounds in exhaled breath and emitted by skin using selected ion flow tube mass spectrometry. *Rapid Commun Mass Spectrom* 2008;**22**:526–32.
73. Cap P, Dryahina K. Pehal F, Španěl P. Selected ion flow tube mass spectrometry of exhaled breath condensate headspace. *Rapid Commun Mass Spectrom* 2008;**22**:2844–50.
74. Julak J, Stranska E, Rosova V, Geppert H, Španěl P, Smith D. Bronchoalveolar lavage examined by solid phase microextraction, gas chromatography–mass spectrometry and selected ion flow tube mass spectrometry. *J Microbiol Methods* 2006;**65**:76–86.
75. Smith D, Ismail KM, Diskin AM, Chapman G, Magnay JL, Španěl P, et al. Increase of acetone emitted by urine in relation to ovulation. *Acta Obstet Gynecol Scand* 2006;**85**: 1008–11.

76. Amann A, Corradi M, Mazzone P, Mutti A. Lung cancer biomarkers in exhaled breath. *Expert Rev Mol Diagn* 2011;**11**:207–17.
77. Szulejko JE, McCulloch M, Jackson J, McKee DL, Walker JC, Solouki T. Evidence for cancer biomarkers in exhaled breath. *IEEE Sens J* 2010;**10**: 185–210.
78. Phillips M, Cataneo RN, Ditkoff BA, Fisher P, Greenberg J, Gunawardena R, et al. Prediction of breast cancer using volatile biomarkers in the breath. *Breast Cancer Res Treat* 2006;**99**:19–21.
79. Peng G, Tisch U, Adams O, Hakim M, Shehada N, Broza YY, et al. Diagnosing lung cancer in exhaled breath using gold nanoparticles. *Nat Nanotechnol* 2009;**4**:669–73.
80. Thorpe MJ, Balslev-Clausen D, Kirchner MS, Ye J. Cavity-enhanced optical frequency comb spectroscopy: application to human breath analysis. *Opt Express* 2008;**16**:2387–97.
81. Phillips M, Cataneo RN, Condos R, Ring Erickson GA, Greenberg J, La Bombardi V, et al. Volatile biomarkers of pulmonary tuberculosis in the breath. *Tuberculosis (Edinb)* 2007;**87**:44–52.
82. Dragonieri S, Schot R, Mertens BJ, Le Cessie S, Gauw SA, Spanevello A, et al. An electronic nose in the discrimination of patients with asthma and controls. *J Allergy Clin Immunol* 2007;**120**:856–62.
83. Buszewski B, Kesy M, Ligor T, Amann A. Human exhaled air analytics: biomarkers of diseases. *Biomed Chromatogr* 2007;**21**:553–66.
84. Poli D, Carbognani P, Corradi M, Goldoni M, Acampa O, Balbi B, et al. Exhaled volatile organic compounds in patients with non-small cell lung cancer: cross sectional and nested short-term follow-up study. *Respir Res* 2005;**6**:71.
85. Brunner C, Szymczak W, Li W, Hoeschen C, Mortl S, Eckardt-Schupp F, et al. Headspace measurements of irradiated *in vitro* cultured cells using PTR-MS. *Radiat Environ Biophys* 2011;**50**:209–17.
86. Sponring A, Filipiak W, Ager C, Schubert J, Miekisch W, Amann A, et al. Analysis of volatile organic compounds (VOCs) in the headspace of NCI-H1666 lung cancer cells. *Cancer Biomark* 2010;**7**:153–61.
87. Sponring A, Filipiak W, Mikoviny T, Ager C, Schubert J, Miekisch W, et al. Release of volatile organic compounds from the lung cancer cell line NCI-H2087 *in vitro*. *Anticancer Res* 2009;**29**:419–26.
88. Filipiak W, Sponring A, Filipiak A, Ager C, Schubert J, Miekisch W, et al. TD-GC-MS analysis of volatile metabolites of human lung cancer and normal cells *in vitro*. *Cancer Epidemiol Biomarkers Prev* 2010;**19**:182–95.
89. Chen X, Xu F, Wang Y, Pan Y, Lu D, Wang P, et al. A study of the volatile organic compounds exhaled by lung cancer cells *in vitro* for breath diagnosis. *Cancer* 2007;**110**: 835–44.
90. Smith D, Španěl P, Sule-Suso J. Advantages of breath testing for the early diagnosis of lung cancer. *Expert Rev Mol Diagn* 2010;**10**:255–7.
91. Smith D, Wang T, Španěl P. On-line, simultaneous quantification of ethanol, some metabolites and water vapour in breath following the ingestion of alcohol. *Physiol Meas* 2002;**23**:477–89.
92. Smith D, Pysanenko A, Španěl P. Kinetics of ethanol decay in mouth- and nose-exhaled breath measured on-line by selected ion flow tube mass spectrometry following varying doses of alcohol. *Rapid Commun Mass Spectrom* 2010;**24**:1066–74 .
93. Bloor RN, Španěl P, Smith D. Quantification of breath carbon disulphide and acetone following a single dose of disulfiram (Antabuse) using selected ion flow tube mass spectrometry (SIFT-MS). *Addict Biol* 2006;**11**:163–9.

94. Bloor RN, Wang TS, Španěl P, Smith D. Ammonia release from heated "street" cannabis leaf and its potential toxic effects on cannabis users. *Addiction* 2008;**103**:1671–7.
95. Smith D, Španěl P. On-line determination of the deuterium abundance in breath water vapour by flowing afterglow mass spectrometry with applications to measurements of total body water. *Rapid Commun Mass Spectrom* 2001;**15**: 25–32.
96. Davies S, Španěl P, Smith D. Rapid measurement of deuterium content of breath following oral ingestion to determine body water. *Physiol Meas* 2001;**22**: 651–9.
97. Španěl P, Shestivska V, Chippendale TW, Smith D. Determination of the deuterium abundances in water from 156 to 10,000 ppm by SIFT-MS. *J Am Soc Mass Spectrom* 2011;**22**:179–86.
98. Španěl P, Smith D. Accuracy and precision of flowing afterglow mass spectrometry for the determination of the deuterium abundance in the headspace of aqueous liquids and exhaled breath water. *Rapid Commun Mass Spectrom* 2001;**15**:867–72.
99. Smith D, Engel B, Diskin AM, Španěl P, Davies SJ. Comparative measurements of total body water in healthy volunteers by online breath deuterium measurement and other near-subject methods. *Am J Clin Nutr* 2002;**76**: 1295–301.
100. Španěl P, Smith D. Flowing afterglow mass spectrometry (FA-MS) for the determination of the deuterium abundance in breath water vapour and aqueous liquid headspace. In: Amann A, Smith D, editors. *Breath analysis for clinical diagnosis and therapeutic monitoring*. Singapore: World Scientific; 2005. p. 439–55.
101. Chan C, Smith D, Španěl P, McIntyre CW, Davies SJ. A non-invasive, on-line deuterium dilution technique for the measurement of total body water in haemodialysis patients. *Nephrol Dial Transplant* 2008;**23**:2064–70.
102. John B, Tan BK, Dainty S, Španěl P, Smith D, Davies SJ. Plasma volume, albumin, and fluid status in peritoneal dialysis patients. *Clin J Am Soc Nephrol* 2010;**5**:1463–70.
103. Tan BK, Smith D, Španěl P, Davies SJ. Dispersal kinetics of deuterated water in the lungs and airways following mouth inhalation: real-time breath analysis by flowing afterglow mass spectrometry (FA-MS). *J Breath Res* 2010;**4**:017109.
104. Wang TS, Španěl P, Smith D. The combined use of SIFT-MS and FA-MS to investigate first-pass metabolism of ethanol. In: Amann A, Smith D, editors. *Breath analysis: for clinical diagnosis and therapeutic monitoring*. Singapore: World Scientific Publ. Co. Pte Ltd.; 2005. p. 479–90.
105. Villinger J, Praun S, Rossler-Graf J, Dornauer A, Neumayer H, Baumgartner E. Occupational exposure assessment through analysis of human breath and ambient air using IMR-mass spectrometry. In: Amann A, Smith D, editors. *Breath analysis for clinical diagnosis and therapeutic monitoring*. Singapore: World Scientific; 2005.
106. Turner C, Welch S, Bellingan G, Singer M, Španěl P, Smith D. Analysis of breath using SIFT-MS: a comparison of the breath composition of healthy volunteers and seriously-ill ICU patients. In: Amann A, Smith D, editors. *Breath analysis: for clinical diagnosis and therapeutic monitoring*. Singapore: World Scientific Publ. Co. Pte Ltd. 2005. p. 317–26.
107. Penault C, Španěl P, Smith D. Detection of *H. Pylori* infection by breath ammonia following urea ingestion. In: Amann A, Smith D, editors. *Breath analysis for clinical diagnosis and therapeutic monitoring*. Singapore: World Scientific; 2005. p. 393–9.
108. Bloor RN, Wang TS, Španěl P, Smith D. Applications of selected ion flow tube mass spectrometry, SIFT-MS, in addiction research. In: Amann A, Smith D, editors. *Breath analysis for clinical diagnosis and therapeutic monitoring*. Singapore: World Scientific; 2005. p. 409–20.

CHAPTER

5

The Analysis of Oral Air by Selected Ion Flow Tube Mass Spectrometry Using Indole and Methylindole as Examples

Brian M. Ross and Ashley Esarik

Northern Ontario School of Medicine and Department of Biology, Lakehead University, Thunder Bay, Ontario, Canada P7B5E1

5.1 INTRODUCTION

Most researchers engaged in breath analysis are interested in the non-invasive measurement of volatile markers of systemic, or sometimes pulmonary processes, that is, they are interested in air which has passed through the lungs.[1] When the participant is mouth breathing, however, air from the lungs must first flow through the mouth before reaching the analytical equipment. The oral cavity has been found to add important chemicals to the exhalant flow, either augmenting the concentration of the same systemically produced compound, or in some cases such as ammonia and hydrogen sulfide, being the predominant source of these compounds in exhaled air.[2–4] Therefore, it is important for breath researchers to be aware of the contributions of the oral cavity to the breath marker they are investigating. Indeed, the phenomenon of mouth derived volatiles is well known in everyday life in the form of oral malodor, a form of halitosis in which the objectionable odiferous compounds are formed in the mouth as opposed to the lungs.[5] The condition, known commonly as "bad breath," is very likely to have been encountered by all readers among their personal acquaintances and perhaps by themselves from time-to-time. In this chapter we will describe how we have used Selected Ion Flow Tube Mass Spectrometry (SIFT-MS) to investigate the trace gas chemistry of the mouth, in particular as a means to chemically analyze the causes of oral malodor.

5.2 ORAL MALODOR

Oral malodor is a very common and sometimes embarrassing condition which frequently results in visits to dental, periodontal, and dental hygienists offices,[5] usually after an acquaintance of the patient has told them about their condition. In severe cases the condition can lead to such levels of anxiety that the sufferer avoids

Volatile Biomarkers. http://dx.doi.org/10.1016/B978-0-44-462613-4.00005-2

social contact.[6,7] Such self-perceptions can be difficult to combat given that an individual's ability to ascertain the state of their own breath odor is limited.[8] The deficiency in self-assessment has lead to the need for others to rate the level of breath malodor, normally using a scale such as the organoleptic rating scale administered by a clinician. Though subjective this method of assessing the degree of malodor has become established as the main outcome measure used in oral malodor research.

The aetiology of halitosis and malodor can be systemic in nature, producing chemicals such as acetone, the breath concentration of which rises during fasting, from odiferous foodstuffs such as garlic and onions, and from bacterial degradation of foodstuffs, the latter generally being both persistent and objectionable to others given the nature of the chemicals involved.[5] Oral bacteria are most abundant upon the dorsal surface of the tongue where they catabolize foodstuffs to produce, at least in theory, a variety of odiferous compounds.[9,10] The chemical class most often associated with oral malodor is that of the sulfides, in particular hydrogen sulfide and methylmercaptan, produced in the breakdown of the sulfur containing amino acids cysteine and methionine.[9] Oral sulfide concentrations have been measured using a combination of thermal desorption and gas chromatography (GC), which allows accurate quantification of individual sulfides. Using this technique, some researchers have concluded that elevated sulfide levels are the only significant cause of oral malodor.[11] GC analysis is, however, expensive and technically demanding, as well as being slow to perform, ruling out the possibility of results being available while the patient is present in the clinic, and is, in our opinion, unsuitable for routine clinical use.

This has led to the development of solid state sulfide sensor-based devices, such as the Halimeter, which although incapable of differentiating sulfides, are inexpensive and give results in real-time, providing a routine objective measure of oral malodor.[12,13] It has been the use of such devices, however, which has challenged the view that sulfides are the only chemical cause of malodor. Specifically, oral sulfide concentrations generally correlate poorly with the organoleptic scores generated by human raters.[4,8,13] While differential sensor sensitivity to the various oral sulfides may be responsible,[8] an alternative explanation is that the production of other chemicals can also cause malodor, such that in some patients the malodor is largely nonsulfurous in nature.[4] Such non-sulfide sources may include the polyamines putrescine and cadaverine, which derive from the amino acids ornithine and lysine respectively, as well as organic acids such as acetic, butyric, and valeric acids, which are formed during the catabolism of fatty acids.[9] Furthermore, as mentioned above, systemic acetone and ammonia, and the odorants present in foodstuffs, also may contribute to malodor. The contribution of amines and short chain organic acids is biologically possible given that oral bacteria can produce them, offering a possible explanation of why subjective assessment of oral malodor is not always affirmed by sulfide measurements; indeed we have described one patient whose oral malodor is likely due to organic acid production rather than sulfides, a finding which may be due to variations in oral microflora.[4] Such conclusions remain controversial however and it is important to effectively analyze oral malodor, thereby allowing researchers to improve diagnosis and assessment, ultimately relating oral chemistry to prognosis and treatment choices.

One explanation of why the chemistry of oral malodor is not well agreed upon may derive from the methodologies used. The compounds proposed to be involved are chemically diverse making their analysis using the routinely used GC methods difficult. Moreover, GC requires that samples be introduced in a small volume, which requires the use of pre-concentration steps, including various forms of thermal desorption that can lead to the introduction of artifacts. For example, the efficient capture of the highly volatile hydrogen sulfide requires high avidity sorbents which may irreversibly bind less volatiles compounds.[14] In addition, the GC columns, chromatography conditions, and post-column detectors required for one class of compounds can be unsuitable for others.[15] While the researcher could undertake multiple analyses for different classes of compounds, this does not routinely occur. As such, erroneous conclusions may be drawn due to differential detection efficiencies, including the failure to detect some compounds even if they are present in the gas sample. A better analysis method would, arguably, minimize, or eliminate sample extraction altogether while being able to quantify the diverse chemicals which may be present in odiferous breath. One such technique is Selected Ion Flow Tube Mass Spectrometry (SIFT-MS; described in detail by Spanel and Smith[16]). Briefly, the technique ionizes trace gases using the precursor ions H_3O^+, NO^+, or O_2^+. The resulting product ions, characteristic of the gas of interest, are formed at a known rate thereby allowing absolute quantification of trace gases in real-time. Moreover, SIFT-MS can be used to quantify oral air concentrations of all the compounds postulated to be involved in oral malodor.[17–20] While our laboratory has the capability to analyze trace gases using thermal desorption gas chromatography with mass spectrometric or pulse flame photometric detection, it is for these reasons that we have predominantly utilized SIFT-MS in our studies.

5.3 ORAL AIR SAMPLING CONSIDERATIONS

Like all headspace measurements, reproducible quantification of trace gases in oral air requires that volatiles gas levels have reached equilibrium in the mouth prior to sampling. During mouth breathing there is a periodic flow of air through the mouth which moves the gas concentration away from equilibrium. As such, it is necessary to ask the subject to nose breath for a period of time prior to sampling otherwise the initial concentrations measured tend to be lower. The manufacturers of devices such as the Halimeter recommend nose breathing for 3 min (Ref. 12); however, real-time measurements using SIFT-MS suggest that a shorter time, in the order of 1 min, is all that is required to maximize gas levels for most substances. It should be noted, however, that the polyamines require longer equilibration periods, in the order of 3 min, perhaps due to these compounds being chemically adherent.

Actual oral air sampling is accomplished using negative pressure. Devices such as the Halimeter use a tube inserted into the mouth to obtain an air sample. The tube is inserted such that the end is over the tongue while not touching any parts of the mouth. The sampling rate of the Halimeter is high such that a noticeable suction is felt by the patient, requiring that they keep their lips slightly open to prevent a vacuum forming which would otherwise reduce the sampling flow rate. In addition, the subject must not blow into the tubing. Such a system is easily implemented using a SIFT-MS, which also samples using negative pressure with the advantage that the sampling rate (approximately 0.3 mL/s) is much lower than that used for the Halimeter. In our experience, this results in much less intra-sample variance, and hence more rapid measurements.

Using the Instrument Science *Profile 3* SIFT-MS in our laboratory, the normal breath sampling head can be easily adapted to analyze oral air. This unit comprises a large diameter heated stainless steel tube through which the subject exhales, with a small and constant sample of the breath flow being taken via a heated transfer line connected to the SIFT-MS instrument. This head can be adapted to sample oral air by sealing off the output of the breath sampling tube and using a tubing adapter to connect a mouth sampling tube to the intake. The tubing method of sampling requires careful consideration to avoid liquid water entering the SIFT-MS sampling head/transfer line. This can occur if the sampling tube is misplaced and starts to suction saliva, a rare occurrence in our experience if the tube is carefully placed in the mouth. In addition, water vapor can condense in the tubing since, in our laboratory, the portion which is inserted into the mouth is unheated. By keeping the portion of the sampling tube that protrudes from the mouth as short as possible, just a few centimeters, condensation has been prevented. A fully heated mouth sampling tube would avoid this issue all together but would require to be of a low diameter, much less that the sampling lines currently used by SIFT-MS instruments. An alternative design, which avoids the use of sampling tubes, has been in use in the Smith and Španěl labs. In this case the outlet of the normal heated breath sampling head is attached to a very mild vacuum to pull air from the mouth without the subject exhaling, and a portion of this flow is sampled by the SIFT-MS in the normal way (for example see Ref. 3). Careful attention must be paid to standardize and limit the oral air sampling rate when using such a method.

The reason that our laboratory does not use the method proposed by Smith, Španěl, and their colleagues was a problem we encountered when attempting to quantify the polyamines.[20] The adherent properties of putrescine and cadaverine resulted in their being lost to the metal surface of the breath sampler, as well as the glass coated copper of the SIFT-MS sampling line, such that no polyamines reached the flow tube except when sampling very high polyamine concentrations. Even when using high concentrations, the lag time between sampling introduction and detection was very long and carry-over was substantial, routinely taking many hours to return to baseline values. We therefore replaced the transfer line with heated PTFE ("Teflon") tubing, limiting the flow using a 6/1000 in. internal diameter PTFE capillary tube. We did attempt to routinely use a PTFE equivalent of the SIFT-MS breath sampling head; however, the poor heat conductance of PTFE resulted in water condensation occurring. As such, a simple PTFE transfer line contained inside a heated outer tube coupled to a PTFE mouth sampling tube was utilized. It is worth noting that the condensation problem was not limited to the mouth sampling end of the line, but also to the instrument flow tube end, at which the PTFE transfer emerges from its heated outer tube to be coupled to the short piece of stainless steel tubing which enters the instrument flow tube. Even though the exposed PTFE tubing contains gas close to the lowered flow tube pressure and is kept as short as possible, we encountered condensation when sampling high humidity oral air. This problem was solved by covering the PTFE in several millimeter of metal foil and running closely spaced loops of heating wire in direct contact with the foil.

The problem of the chemical adherence of gas phase polyamines should be noted by those using more conventional techniques such as gas chromatography with thermal desorption sampling systems. It is very likely that artifacts, including a high chance of entirely failing to detect any polyamines present, could be introduced using such multistage sampling and extraction procedures, a possibility which needs to be specifically investigated when attempting to quantify these compounds. Reports that there are no polyamines present in oral air when using GC analysis (for example see Ref. 11), which do not provide evidence that polyamines actually have traveled unimpeded from mouth to GC detector, should therefore be viewed skeptically, in our opinion.

5.4 INDOLES

We have published data concerning how the oral abundance of a variety of compounds can be altered in patients with malodor.[4] Recently, we began to include the investigation of another class of compounds that we consider may be involved in oral malodor, the indoles, the preliminary analysis of which we will describe in the remainder of this chapter. Indole (2,3-benzopyrrole) is an aromatic heterocyclic possessing a bicyclic structure in which a pyrrole ring is fused to benzene ring (Figure 5.1). Indole, and its methylated relatives, are odiferous compounds which at high concentrations have an unpleasant odor, sometimes described as fecal in nature, but also as napthene or "mothball" like.[21] Indeed, they are found in feces and comprise one of the contributors to its distinctive smell. Interestingly, elevated fecal indole levels have been reported in patients with colorectal cancer, leading to indole being proposed as a potential cancer biomarker.[22] At lower concentrations indole has a floral odor and is found in essential oils, such as that of jasmine and orange blossom,[23] as well as the headspace of some wines.[24] Plants and micro-organisms can synthesize indole from shikimic acid or anthranilate, with indole acting as a precursor of the amino acid tryptophan, being itself the metabolic source of the neurotransmitter serotonin and the hormone melatonin.[25,26] In the human gastrointestinal tract, however, indole is mainly formed in the catabolism of tryptophan by bacteria. Bacteria are also the source of the methylated indole derivative 3-methylindole (also known by the common name skatole, derived from the Greek work for dung), again formed from tryptophan albeit indirectly via the intermediate indole-3-acetic acid.[27]

As we worked with the indoles we became familiar with the distinctive aroma which we detected on the breath of a visitor to the lab. Such observations led us to attempt the measurement of breath indole levels using SIFT-MS. Two indoles had been previously characterized using this technique, but we wished to study the reactions of

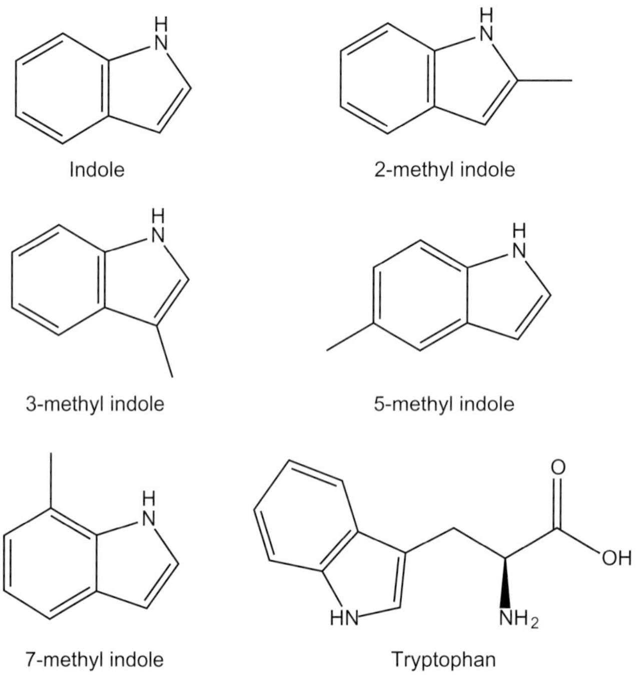

FIGURE 5.1
Structure of the indoles utilized in this study. Tryptophan, which can act as a precursor of indole and 3-methylindole, is shown for comparison.

the various available methylated indoles.[28] To accomplish this we obtained samples of >99% pure indole and 2, 3, 5, and 7-methylindole. The compounds are solids at room temperature, so we heated each to 50 °C for our experiments to increase their partial pressures. The reactions of each indole with the three precursor ions, H_3O^+, NO^+, and O_2^+ are described in Tables 5.1 and 5.2. In all cases the reaction mechanisms were straightforward, resulting in a predominant or a single product ion, in agreement with previously published details of the indole and 3-methylindole reactions.[28] Specifically, with H_3O^+ only the proton-transfer product was detected. At ambient humidity the monohydrate of the protonation product was detected for all three indoles, rising to a substantial proportion of the total product count ion count at the higher humidity of breath (6% water). The dihydrates could be detected in the presence of 6% humidity but their contribution was less than 1% in all cases and can safely be ignored for quantification purposes. For the reactions with NO^+ and O_2^+ the charge transfer products were evident, along with smaller amounts of a $(M\text{-}H)^+$ product ion detected for 2- and 3-methylindole. The reaction rate coefficients (Table 5.1) for the H_3O^+ reactions were calculated while those for NO^+ and O_2^+ were determined empirically as previously described,[29] with all compounds having

Table 5.1 Product ions generated in reaction between H_3O^+, NO^+, or O_2^+, indole and some methyl indoles.

Compound	Precursor Ion		
	H_3O^+	NO^+	O_2^+
Indole C_8H_7N MW = 117	$C_8H_7N \cdot H^+$ (100%) *m/z* 118	$C_8H_7N^+$ (100%) *m/z* 117	$C_8H_7N^+$ (100%) *m/z* 117
2-Methylindole C_9H_9N MW = 131	$C_9H_9N \cdot H^+$ (100%) *m/z* 132	$C_9H_9N^+$ (100%) *m/z* 131	$C_9H_9N^+$ (90%) *m/z* 131 $C_9H_8N^+$ (10%) *m/z* 130
3-Methylindole C_9H_9N MW = 131	$C_9H_9N \cdot H^+$ (100%) *m/z* 132	$C_9H_9N^+$ (100%) *m/z* 131	$C_9H_9N^+$ (80%) *m/z* 131 $C_9H_8N^+$ (20%) *m/z* 130
5-Methylindole C_9H_9N MW = 131	$C_9H_9N \cdot H^+$ (100%) *m/z* 132	$C_9H_9N^+$ (100%) *m/z* 131	$C_9H_9N^+$ (100%) *m/z* 131
7-Methylindole C_9H_9N MW = 131	$C_9H_9N \cdot H^+$ (100%) *m/z* 132	$C_9H_9N^+$ (100%) *m/z* 131	$C_9H_9N^+$ (100%) *m/z* 131

The relative abundance of product ion(s) is shown in parenthesis following the ion molecular structure while the mass-to-charge ratio, m/z, is shown beneath each product ion. Each compound was present in ambient air which possessed a humidity of approximately 1.5% water vapor; the flow tube pressure was approximately 1 Torr at a temperature of 300 K.

similar values. The lack of fragmentation of the methylated indoles in the H_3O^+ and NO^+ reactions makes these precursors unsuitable for differentiating between each of the methylindoles, although the minor product ion seen in the O_2^+ reaction could, in some instances, be used to investigate the presence or absence of each isomer. To our knowledge, however, only 3-methylindole is produced in biological systems, with the remainder being synthetic compounds used as precursor compounds in the chemical and pharmaceutical industries.[23]

We used these kinetics parameters to measure indole and methylindole (presumed to be 3-methylindole) in the head space of substances reported to contain these compounds, namely rat feces, the odor of decaying food matter contained in a garbage can, jasmine oil, and coal tar (Table 5.3). As expected, both indoles were detected in jasmine oil headspace containing the highest indole concentration, not unexpected since this essential oil is reported to contain approximately 2.5% indole or methylindole.[23] Importantly, the measured concentrations for each compound were similar when

Table 5.2 Rate coefficients of the reactions between H_3O^+, NO^+, and O_2^+ with indoles and some methylindoles.

Compound	Electric Dipole Polarizability (α)(10^{-24} cm^3)	Dipole Moment (μ) (D)	Rate Coefficients (10^{-9} cm^3 s^{-1}) H_3O^+	NO^+	O_2^+
Indole C_8H_7N MW = 117	14.5	2.0	[3.3]	[2.8], 2.5	[2.7], 2.6
2-Methylindole C_9H_9N MW = 131	14.5	2.0	[3.3]	[2.7], 2.5	[2.6], 2.5
3-Methylindole C_9H_9N MW = 131	14.5	2.0	[3.3]	[2.7], 2.4	[2.7], 2.3
5-Methylindole C_9H_9N MW = 131	14.5	2.0	[3.3]	[2.7], 2.4	[2.7], 2.4
7-Methylindole C_9H_9N MW = 131	14.5	2.0	[3.3]	[2.7], 2.3	[2.7], 2.5

The rate coefficients in parentheses are the calculated collision rate coefficients which, for H_3O^+, is assumed to equal the actual rate coefficient. For NO^+ and O_2^+ the rate coefficients were determined empirically relative to the H_3O^+ and are somewhat lower than the calculated collision rates.[29,30] *Each compound was present in ambient air which possessed a humidity of approximately 1.5% water vapor; the flow tube pressure was approximately 1 Torr at a temperature of 300 K.*

H_3O^+ or NO^+ were used as precursor ions, lending support to the idea that the quantified ions were indeed derived from the indoles. H_2S was measured at the same time and could be detected in significant quantities in the headspace of rat feces along with the indoles, a combination which may be expected to occur in persons with oral malodor.

We next investigated whether breath contained indoles using samples provided by eight volunteers, three having noticeable oral malodor. When rated using a subjective organoleptic 6-point scale (with 0 indicating no malodor and 5 indicating an extremely foul malodor) they possessed scores of either 3 or 4 from the mouth and 4 from the tongue, while the other subjects scored 0 or 1. Alveolar breath was found to contain very low or undetectable concentration of both indoles (data not shown). The results presented in Table 5.4 suggest that oral air contains measurable and variable quantities of indole and methylindole, with the former being the predominant compound detected. In the subjects with oral malodor, sulfide levels were elevated,

Table 5.3 Headspace indole concentration made using SIFT-MS.

Source	H_2S	Indole		Methylindole	
		H_3O^+	NO^+	H_3O^+	NO^+
Ambient air	1 ± 1	2 ± 1	1 ± 1	1 ± 0	0 ± 0
Jasmine oil	2 ± 1	413 ± 52	386 ± 50	102 ± 14	81 ± 15
Garbage can	69 ± 12	42 ± 14	34 ± 10	11 ± 5	6 ± 5
Rat feces	312 ± 14	29 ± 9	25 ± 8	17 ± 9	15 ± 3
Coal tar	28 ± 4	111 ± 12	101 ± 8	19 ± 5	21 ± 2

Values are mean ppb (by volume) ± SD of three independent measurements. For all substance except the garbage can, approximately 0.5 cm^3 *of the substance was placed in a 250 mL conical flask and sealed with aluminum foil. After a 5 min equilibration period at room temperature, the contents of the flask were sampled for approximately 30 s using a syringe needle and trace gas concentrations measured using SIFT-MS. In the case of the garbage can the headspace was sampled by placing the SIFT-MS intake under the lid of the can. Indole and methylindole levels were calculated using both the* H_3O^+ *and* NO^+ *precursor ions, which show good agreement. Hydrogen sulfide concentrations were measured using the* H_3O^+ *precursor as described previously.*[31]

as generally expected. Indole and methylindole concentrations, however, were highly elevated in one subject, slightly elevated in another, and not at all in the third. This finding suggests that indole production may contribute to some, but perhaps not all, cases of oral malodor. Such heterogeneity clearly requires further investigation using more participants, but hints that not all cases of oral malodor are qualitatively similar, and suggests the involvement either of different micro-organisms or the consumption of varying amounts of tryptophan containing foods. As with measurements of polyamines in the mouth, our results are suggestive that non-sulfide compounds do contribute to oral malodor, a subject of current controversy, but nevertheless supported by previous reports that some persons with oral malodor do not have elevated mouth sulfide concentrations. Future studies may consider whether chemical heterogeneity is predictive of prognosis or the effectiveness of various treatment strategies.

5.5 SUMMARY AND CONCLUSION

In this preliminary study we have provided evidence that SIFT-MS is suitable for the measurement of vapor phase indole concentrations. Given that these compounds are produced enzymatically by bacteria, real-time measurements using SIFT-MS may have applications in the food- and bio-technology industries. Moreover, indoles were also detected in the breath of human subjects with elevated concentrations being associated with oral malodor. Since only three cases of malodor were considered in this study a more extensive survey of oral concentrations would be required to confirm an involvement. Nevertheless, our study does illustrate the considerable power of SIFT-MS for the investigation and monitoring of oral malodor given that all actual and

Table 5.4 Oral air concentrations of sulfides and indoles measured by SIFT-MS.

Subject	Organoleptic Score		H_2S	Methylmercaptan	Indole	Methylindole
	Mouth	Tongue				
Ambient	NA	NA	1 ± 1	0 ± 0	0 ± 1	1 ± 1
S1	0	1	14 ± 2	4 ± 1	7 ± 1	3 ± 2
S2	0	1	22 ± 4	12 ± 3	5 ± 1	11 ± 2
S3	1	1	15 ± 2	4 ± 2	4 ± 2	1 ± 1
S4	1	0	17 ± 3	3 ± 2	3 ± 1	2 ± 1
S5	0	1	26 ± 4	9 ± 3	1 ± 2	8 ± 3
OM1	4	4	112 ± 20	22 ± 14	34 ± 4	18 ± 3
OM2	3	4	87 ± 10	18 ± 5	14 ± 3	17 ± 5
OM3	3	4	65 ± 18	31 ± 7	4 ± 2	3 ± 2

Oral air was sampled as described in the text from five subjects without oral malodor (S1–5) and three with significant malodor (OM1–3) as subjectively quantified using tongue and mouth organoleptic score indicating moderate to strong malodor (scores of 3 or 4). The values shown are mean oral trace gas concentrations in ppb measured using the H_3O^+ precursor ion $\pm$SD of three independent measurements.

theorized chemical causes of malodor, including, in addition to the indoles, sulfides, amines, and organic acids, can be rapidly analyzed without the need for prior sample collection. This makes the technique ideally suited for investigating the effectiveness of strategies aimed at combating this common and embarrassing condition.

References

1. Amann A, Spanel P, Smith D. Breath analysis: the approach towards clinical applications. *Mini Rev Med Chem* 2007;**7**:115–29.
2. Pysanenko A, Spanel P, Smith D. A study of sulfur-containing compounds in mouth- and nose-exhaled breath and in the oral cavity using selected ion flow tube mass spectrometry. *J Breath Res* 2008;**2**:046004.
3. Wang T, Pysanenko A, Dryahina K, Spanel P, Smith D. Analysis of breath, exhaled via the mouth and nose, and the air in the oral cavity. *J Breath Res* 2008;**2**:037013.
4. Ross BM, Dadgostar N, Bloom M, McKeown L. The analysis of oral air using selected ion flow tube mass spectrometry in persons with and without a history of oral malodour. *Int J Dent Hyg* 2009;**7**:136–43.
5. Hughes FJ, McNab R. Oral malodour—a review. *Arch Oral Biol* 2008;**53**(Suppl. 1):S1–7.
6. McKeown L. Social relations and breath odour. *Int J Dent Hyg* 2003;**1**:213–7.
7. Ng SK, Leung WK. Oral health-related quality of life and periodontal status Community. *Dent Oral Epidemiol* 2006;**34**:114–22.
8. Iwanicka-Grzegorek E, Michalik J, Kepa J, Wierzbicka M, Aleksinski M, Pierzynowska E. Subjective patients' opinion and evaluation of halitosis using halimeter and organoleptic scores. *Oral Dis* 2005;**11**(Suppl. 1):86–8.

9. Porter SR, Scully C. Oral malodour (halitosis). *BMJ* 2006;**333**:632–5.
10. van den Broek AM, Feenstra L, de Baat C. A review of the current literature on aetiology and measurement methods of halitosis. *J Dent* 2007;**35**:627–35.
11. Van den Velde S, van Steenberghe D, Van Hee P, Quirynen M. Detection of odorous compounds in breath. *J Dent Res* 2009;**88**:285–9.
12. *Halimeter instruction manual*. Chatsworth, CA, USA: Interscan Corporation; 1998.
13. Baharvand M, Maleki Z, Mohammadi S, Alavi K, Moghaddam EJ. Assessment of oral malodor: a comparison of the organoleptic method with sulfide monitoring. *J Contemp Dent Pract* 2008;**9**:76–83.
14. Kim KH. Some insights into the gas chromatographic determination of reduced sulfur compounds (RSCs) in air. *Environ Sci Technol* 2005;**39**:6765–9.
15. Grob RL. *Modern practice of gas chromatography*. 4th ed. Hoboken, NJ: John Wiley & Sons; 2004.
16. Spanel P, Smith D. Progress in SIFT-MS: breath analysis and other applications. *Mass Spectrom Rev* 2011;**30**:236–67.
17. Spanel P, Davies S, Smith D. Quantification of ammonia in human breath by the selected ion flow tube analytical method using H_3O^+ and O_2^+ precursor ions. *Rapid Commun Mass Spectrom* 1998;**12**:763–6.
18. Spanel P, Smith D. SIFT studies of the reactions of H_3O^+, NO^+ and O_2^+ with a series of volatile carboxylic acids and esters. *Int J Mass Spectrom* 1998;**172**: 137–47.
19. Spanel P, Smith D. Selected ion flow tube studies of the reactions of H_3O^+, NO^+ and O_2^+ with some organosulfur molecules. *Int J Mass Spectrom* 1998;**176**:167–76.
20. Ross BM, Babay S, Ladouceur C. The use of selected ion flow tube mass spectrometry to detect and quantify polyamines in headspace gas and oral air. *Rapid Commun Mass Spectrom* 2009;**23**:3973–82.
21. Moore JG, Jessop LD, Osborne DN. Gas-chromatographic and mass-spectrometric analysis of the odor of human feces. *Gastroenterology* 1987;**93**:1321–9.
22. Karlin DA, Mastromarino AJ, Jones RD, Stroehlein JR, Lorentz O. Fecal skatole and indole and breath methane and hydrogen in patients with large bowel polyps or cancer. *J Cancer Res Clin Oncol* 1985;**109**:135–41.
23. Van Order RB, Lindwall HG. Indole. *Chem Rev* 1942;**30**:69–96.
24. Mattivi F, Vrhovsek U, Versini G. Determination of indole-3-acetic acid, tryptophan and other indoles in must and wine by high-performance liquid chromatography with fluorescence detection. *J Chromatogr A* 1999;**855**:227–35.
25. Morgan PN, Gibson MI, Gibson F. The conversion of shikimic acid into certain aromatic compounds by cell-free extracts of aerobacter aerogenes and Escherichia coli. *Biochem J* 1963;**89**:229–39.
26. Shein HM, Wurtman RJ. Cyclic adenosine monophosphate: stimulation of melatonin and serotonin synthesis in cultured rat pineals. *Science* 1969;**166**:519–20.
27. Jensen MT, Cox RP, Jensen BB. 3-Methylindole (skatole) and indole production by mixed populations of pig fecal bacteria. *Appl Environ Microbiol* 1995;**61**:3180–84.
28. Wang TS, Spanel P, Smith D. A selected ion flow tube, SIFT, study of the reactions of H_3O^+, NO^+ and O_2^+ ions with several N- and O-containing heterocyclic compounds in support of SIFT-MS. *Int J Mass Spectrom* 2004;**237**:167–74.
29. Iachetta L, Malek L, Ross BM. The reactions of H_3O^+, NO^+ and O_2^+ with several flavourant esters studied using selected ion flow tube mass spectrometry. *Rapid Commun Mass Spectrom* 2010;**24**:815–22.

30. Su T, Chesnavich WJ. Parametrization of the ion-polar molecule collision rate-constant by trajectory calculations. *J Chem Phys* 1982;**76**:5183–5.
31. Spanel P, Smith D. Quantification of hydrogen sulphide in humid air by selected ion flow tube mass spectrometry. *Rapid Commun Mass Spectrom* 2000;**14**:1136–40.

CHAPTER

6 Smokers Breath as Seen by Proton-Transfer-Reaction Time-of-Flight Mass Spectrometry (PTR-TOF-MS)

Ingrid Kohl[*,†], **Jens Herbig**[*,†], **Jürgen Dunkl**[*,‡], **Armin Hansel**[‡], **Martin Daniaux**[§], **and Michael Hubalek**[**]

[*] *Oncoytyrol-Center for Personalized Cancer Medicine GmbH, Eduard Bodem Gasse 3, 6020 Innsbruck, Austria*

[†] *Ionimed Analytik GmbH, Eduard Bodem Gasse 3, 6020 Innsbruck, Austria*

[‡] *Institute of Ion Physics and Applied Physics, Leopold-Franzens University Innsbruck, Technikerstrasse 25, 6020 Innsbruck, Austria*

[§] *Department of Radiology, Medical University Innsbruck, Anichstrasse 35, 6020 Innsbruck, Austria*

[**] *Department of Gynecology and Obstetrics, Medical University Innsbruck, Anichstrasse 35, 6020 Innsbruck, Austria*

6.1 INTRODUCTION

Researchers are attempting to analyze exhaled breath gas in the search for chemical compounds, playing a role in disease—or more generally speaking—giving information about the status of the body. The availability of novel highly sensitive technologies has stimulated this research,[1–6] which in fact goes back to an ancient idea, as reviewed by Michael Phillips.[7]

Exhaled breath gas contains nitrogen, oxygen, water vapor, argon, and carbon dioxide as main constituents. In addition, several hundreds to thousands of volatile organic trace compounds are contained in variable concentrations as mentioned in the literature.[8,9] Some might be formed endogenously in metabolic pathways but the origin of most of these organic traces is still unknown. The composition of exhaled breath gas depends on numerous variables like life style, nutrition, activity, the inhaled air composition, and many more, as has been recently overviewed by Jonathan Beauchamp.[10] Researchers have investigated how certain behavior can influence the composition of the exhaled breath gas such as fasting,[11] exercise,[12] cigarette smoking,[13–15] alcohol intake,[16–18] intake of certain food,[19–22] the intake of a certain medicine,[23–26] the exposure to chemicals in specific environments.[10, 27–29]

When searching for characteristics in the exhaled breath of a cohort of patients, it can become necessary to collect and analyze samples over a long period of time, like weeks or even months. This requires a constant performance of the analytical

Volatile Biomarkers. http://dx.doi.org/10.1016/B978-0-44-462613-4.00006-4

instrument to extract quantitatively and qualitatively accurate results. It is therefore indispensible to determine and control the instruments performance by characterizing certain parameters, such as precision and accuracy of mass and concentration measurement.

In the presented study we report on the long-term performance of a proton-transfer-reaction time-of-flight mass spectrometer (PTR-TOF-MS)[30–37] in a breath gas analysis study over a period of 12 months. During this study, breath samples of 72 women were collected and subsequently analyzed by PTR-TOF-MS. Determined m/z values and suggested corresponding atomic compositions are summarized for the 35 most intense peaks. The data were analyzed for the differences related to cigarette smoke, and a list summarizing smoking related compounds is given.

PTR-TOF-MS is the latest development in PTR technology. Ionization by proton-transfer-reactions is an established technology, which has been proven to be useful in trace gas analysis in various fields and breath research.[38,39] PTR-TOF-MS is superior to PTR-MS (quadrupole-based systems) in view of the mass resolving power and high sampling speed due to its time-of-flight (TOF) mass discriminator.

6.2 MATERIALS AND METHODS

6.2.1 Human subjects

Breath gas samples from 72 women, recruited at the Brustgesundheitzentrum Tirol (located in Innsbruck, Austria), were collected over a period of 12 months. The women were aged 21–77 years (mean 53 years, SD 13 years). Those women, who had been shown to have breast cancer (by mammography), were excluded from the cohort and only those who considered themselves to be healthy were included. A comparison of the characteristics of this cohort with a second cohort of women affected with breast cancer will be reported elsewhere. No further information on health status, co-morbidities, routine diagnostic results, or medications has been included into the data evaluation process. The smoking status was surveyed and trace compounds expected to be present in the exhaled breath of smokers were searched for in the present study. Among the subjects were 19 self-declared cigarette smokers, aged 21–72 years (mean 50 years, SD 13 years) and 53 non-smokers, aged 25–77 years (mean 54 years, SD 13 years). Patient characteristics are summarized in Table 6.1. The self-declared smokers consumed the last cigarette more than 3 h prior to providing a breath gas sample. Their smoking rate was on an average 10 cigarettes per day (max 30, min 3).

Table 6.1 Characteristics of subjects enrolled into the study.

	Smokers	Non-Smokers	All
Number	19	53	72
Age (SD)	50 (13)	54 (13)	53 (13)

6.2.2 Statement

All experiments involving human subjects were performed according to the "Declaration of Helsinki" and in accordance with Austrian law. The study plan was approved by the ethical review board of the Medical University Innsbruck, Austria.

6.2.3 Sampling procedure

Samples were collected over a period of 12 months during 2010/2011, all in the same study room at the Department of Obstetrics and Gynecology at the Medical University Innsbruck, Austria. Patients were sent from the Brustgesundheitzentrum, a center where women attended for mammography screening.

To sample end-tidal exhaled breath gas a Breath Collecting Unit (BCU) (Ionimed Analytik GmbH, Innsbruck, Austria) was used. The test subject exhaled breath via a disposable mouthpiece carrying a re-breathing valve into a buffering tube. After completion of an exhalation, the subject withdrew from the mouthpiece until the next exhalation into the BCU was dispensed. Normal breathing was possible between the exhalations in order to prevent hyperventilation. Water condensation was avoided by heating all gas carrying surfaces of the BCU above 50 °C. For safety reasons the mouth piece remained unheated. A constant gas stream was permanently sucked from the buffering tube into the BCU toward a previously evacuated SilcoCan® (Restek Corporation, Bellefonte, Pennsylvania, USA). The SilcoCan is a leak tight stainless steel canister, equipped with an isolation valve, with all inner surfaces passivated to avoid wall loss of components and has been successfully used for breath gas sampling.[28] Although Tedlar® (Dupont) bags are widely used for breath gas storage for their comparably low price and simple handling, we refrained from using them because of their inferior storage properties.[40]

The BCU ensured that only the end-tidal fraction of exhaled air was sampled, by triggering a valve so that breath gas was sampled into the SilcoCan, in case of high CO_2 concentrations (i.e. > 20 mm Hg) were measured in the gas stream. Approximately 10 exhalations were needed to fill a 1 L SilcoCan. The total time needed to obtain a breath sample, including the interview of the volunteer and introduction to the procedure, was approximately 10 min. The sample was subsequently transported to the laboratory and analyzed by PTR-TOF-MS.

All patients were asked to evaluate the sampling procedure by assigning a grade from 1 ("no problems") to 5 (procedure was very stressful resulting in termination of the sampling). The results are summarized in Table 6.2.

Table 6.2 Grades for the sampling procedure given by the test subjects.

Grade	1	2	3	4	5	Average Grade
Number of patients	66	3	3	0	0	1.13

6.2.4 VOC sensor: Proton-transfer-reaction time-of-flight mass spectrometer

For the chemical analysis of the breath gas samples we used a proton-transfer-reaction time-of-flight mass spectrometer (PTR-TOF-MS 8000, Ionicon Analytik GmbH, Innsbruck, Austria),[30–37] which has been commercially available since 2008.

6.2.4.1 Chemical ionization

The PTR-TOF-MS combines a chemical ionization[41] source with a time-of-flight mass analyzer.[42] The chemical ionization is accomplished via a gas phase ion-molecule reaction between H_3O^+ precursor ions and (mostly) organic neutral molecules in the sample. This proton-transfer-reaction (PTR) takes place in a drift tube, operated at standard operation conditions as specified by the manufacturer, where H_3O^+ ion chemistry is dominating[43] to guarantee kinetically controlled product ion formation.[44] A proton-transfer-reaction will occur on every collision, if the proton affinity of the analyte molecule is higher than that of water (690.82 kJ mol^{-1}). As a consequence the PTR-instrument can detect most volatile organic compounds as well as some inorganic compounds at very low concentrations, with some known exceptions such as short chain alkanes C_nH_{2n+2} ($n \leqslant 7$), ethylene, and acetylene.[45] In PTR, energy transfer upon collision is generally small leaving most molecules intact, therefore directly relating a measured signal to a chemical compound. Fragmentation is, however, possible in case of weak bonds in a chemical compound and a large exothermicity of the involved proton-transfer-reaction.[45] In case of small exothermicity reverse reactions can become a significant loss term, as known for formaldehyde.[46]

6.2.4.2 Fast analysis of whole spectra

PTR-TOF-MS is capable of creating whole mass spectra in sub-microseconds, fast enough to even monitor all ions present in exhaled breath while a person exhales at a normal breathing frequency.[31] In contrast, in PTR-MS only ions of a certain unity mass-to-charge ratio, *m*/*z*, are allowed to transmit the quadrupole, while all other ions are collected onto the surfaces of the vacuum system. A complete mass spectrum is created by stepwise measuring one *m*/*z* after the other. In case of fast changing sample composition, this technical characteristic will set a limit to the possible time resolution of an experiment. In PTR-TOF-MS all product ions are periodically extracted and accelerated into the TOF mass discriminator simultaneously.

Actually, light ions travel faster through the drift region than heavy ions and hit the detector plate sooner than the heavier ones. The time span between the moment a signal is created and the preceding extraction equals the time-of-flight of an ion, which can be related to the corresponding mass by the simple equation

$$m = \frac{2Uqt^2}{l^2}, \qquad (6.1)$$

where m is the mass of an ion, U is the acceleration potential, q is the charge on the ion, t is the time-of-flight, and l is the length of the drift region an ion has to travel.

6.2.5 Data acquisition and breath sample analysis

For data acquisition, the manufacturer's acquisition software TOFDaq 1.2.86 was used. In the present study a discrete volume of all ions was extracted every 50 μs into the orthogonal field-free drift region, reflected once by an ion mirror and guided toward the multi channel detector plate where signals were generated. From each sample (human samples and calibration gas samples) a mass spectrum was created within a total integration time of 6 min. A mass range between 0 and $m/z = 800$ was scanned in all experiments. The sample container, all inlet lines and the chemical ionization region were heated >50 °C during analysis to avoid water condensation and reduce surface affinity of certain chemicals.

6.2.6 Species identification

In quadrupole PTR-MS it is important to keep in mind that more than one compound possibly contribute to one peak in a mass spectrum. PTR-MS is therefore said to be a low-selective method, requiring further complementary analysis to assist in compound identification. In PTR-MS the mass resolution is set to unity as a standard. As a consequence, chemical compounds of identical integer molecular masses (isobaric compounds with different molecular formulae) as well as isomers (compounds with the same atomic composition) create one broad peak in a mass spectrum.

The mass resolution of a PTR-TOF-MS 8000 can be tuned high enough to produce mass spectra with sufficiently resolved peaks for most isobaric compounds in a mixture of light volatile organic compounds.[30] Isomers, however, cannot be distinguished because of their identical exact mass (sum of all masses of the individual isotopes that compose a chemical compound). To overcome this limitation, researchers can take advantage of the differences in reactivity between different functional groups to distinguish aldehydes from ketones, acids from esters, and alcohols from ethers, by employing different precursor ions for the chemical ionization.[30–49]

In the here-presented work, the PTR-TOF-MS instrument was operated at a high mass resolution $R_{FWHM} = m/\Delta m$ not less than ~4000, where m is the respective mass in the mass spectrum and Δm is the spacing between two neighboring peaks, that should be separated by their full width at half maximum (FWHM). $R_{FWHM} \sim 4000$ at $m/z = 21$ increased to ~4600 at $m/z = 181$ Th in an asymptotic fashion. Similar behavior was found by Graus et al.[30] and was described by the manufacturer. As an example, at $m/z = 40$ two peaks with a mass difference of 0.01 u can be separated by their FWHM, whereas at $m/z = 180$, two peaks with 0.04 u difference in mass can be separated by their FWHM.

6.2.7 Determination of the sensitivity and LOD

The PTR-TOF-MS instruments sensitivity was determined throughout the project using a calibration gas mixture comprising 16 compounds (methanol, acetonitrile, acetaldehyde, ethanol, acrolein, acetone, isoprene, crotonaldehyde, 2-butanone, benzene, toluene, p-xylene, α-pinene, chlorobenzene, 1,2-dichlorobenzene, and 1,2,4-trichlorobenzene) diluted in pure nitrogen at concentrations of $\sim$1 ppm each (Apel Riemer Environmental Inc., Denver, Colorado, USA). The uncertainty of the calibration gas concentration was less than 7%. For these calibration experiments the calibration gas was dynamically diluted in pure air to concentrations between 45 and 285 ppbv per compound, using a dynamic gas calibration unit (GCUs, Ionimed Analytik GmbH, Innsbruck, Austria). Six minutes spectra were recorded from each dilution step and the sensitivity of the instrument for each standard gas component was calculated. The limit of detection (LOD) was determined as the 3σ uncertainty of the pure air measurement and is given for a dwell time of 2 s, which is the time allowed for integration of the signal for one single spectrum.

6.2.8 Mass scale calibration

Accurate mass measurement is needed to accurately derive the measured mass from a peak in a mass spectrum. In case of poor mass scale calibration, the addition of single spectra to sum spectra of a few minutes might cause smearing out of single peaks resulting in broad peaks, and low resolution, causing peak overlap in case of similar exact masses of single components.

It has been shown by Graus et al.[30] that PTR-TOF mass peaks can be assigned to a sum formula $C_xH_yO_z$ unambiguously in case of molecules up to *m/z* 265, based on 1 min spectra. In heavier molecules the number of possible combinations of atoms increases dramatically.

In order to guarantee high mass accuracy throughout the present breath analysis study the mass scale was continuously calibrated. This was done by following the peaks of known components, present in a spectrum at any time. These peaks were used as nodes to adjust the mass scale assignment. For mass scale calibration we chose the NO^+ peak, which was always present in the TOF spectrum as an instrumental background, the main isotope peak of acetone at *m/z* 59.0497, which was present in all breath samples, room air, and even in our purified air at low concentrations, and the product ion peak of the main isotope of 1,2,4-trichlorobenzene (Fluka, puriss. >99.0%) at *m/z* 180.9379, which was added continuously into the inlet gas stream during a measurement.

Internal calibration gas

For this purpose we prepared a low concentration calibration gas of 1,2,4-trichlorobenzene in a 6 L SilcoCan, equipped with an isolation valve. 70 μL of 1,2,4-trichlorobenzene were mixed with 4 bar pure nitrogen gas (Messer Austria GmbH, Gumpoldskirchen, Austria, nitrogen 5.0). The 1,2,4-trichlorobenzene calibration gas was streamed from the reservoir toward the PTR-TOF-MS inlet system and regulated with a steel needle valve to produce well-defined peaks in the mass spectrum, of $\sim$1000 cps (counts per second). It is important to avoid uncontrolled dilution of the breath gas samples by the internal 1,2,4-trichlorobenzene standard gas flow. This was ruled out by comparison of peak areas with and without internal 1,2,4-trichlorobenzene calibration gas flow. The mass accuracy was calculated from the calibration gas measurements, by comparing the exact mass of the calibration gas compounds with the measured mass of the corresponding signals. The mass measurement stability was determined from the calibration gas measurements, too.

The relative mass deviation is defined by the equation

$$\frac{\Delta m}{m} = \frac{m - a}{m} \tag{6.2}$$

where m is the exact mass, calculated from the atomic composition of a species, and a is the measured mass. Usually the corresponding value is multiplied by 10^6 and given in ppm. As a common rule the deviation from the real value is similar for different compounds in a similar mass range. The accuracy of the measured mass can be related to the accuracy of the time-of-flight (t) determination of an ion species by the equation

$$\frac{\Delta m}{m} = 2\frac{\Delta t}{t} \tag{6.3}$$

This relation expresses that an error in the time-of-flight will double the error in the relative mass accuracy.

In general, the determination of the time-of-flight of a light ion species is always related with a larger uncertainty than that of heavier ions,[50] a behavior that was also found in the present study. The allowed bin width of 200 picoseconds, which is the time span to trigger the input channel of the time-to-digital converter, sets a theoretical limit to the uncertainty in time and therefore in mass, of at least ±0.5 bins ($\Delta t = 100$ ps). As an example, the mass accuracy will be 15 ppm for an ion at m/z 59 (time-of-flight = 13.578 μs) and 8 ppm in the case of an ion at m/z 181 (time-of-flight = 23.784 μs).

6.2.9 Raw data analysis

PTR-TOF-MS produces a large amount of information. A quick look at a PTR-TOF-MS spectrum of ambient air or exhaled breath gas will reveal that on virtually every integer mass two or three peaks can be seen at minimum. Several studies have tried to advance the data evaluation methodologies for the Ionicon PTR-TOF-MS, in order to make use of the full potential of this technology. However, no automated universally applicable procedure has been accepted up to now, but solutions to specific problems have been developed.[30,35–37,51,52] Customized tools can be found in the Internet[†] or have been presented in the doctoral thesis of Thorsten Titzmann.[50]

In the present study, the goal of raw data analysis was to extract a summary about the peaks contained in a spectrum, their corresponding m/z and peak intensity. A certain degree of automation in raw data analysis was desired due to the large number of samples in our study, and the large number of peaks contained in one spectrum.

† http://www.tomkarl.com/toftools.htm, https://sites.google.com/site/ptrtof/

This peak summary was used for further statistical data analysis, e.g. the comparison of two subgroups, like smokers with non-smokers in the present case. This comparison leaded to statistically significant differences between these two subgroups, pointing to chemical compounds, which might originate from cigarette smoke inhalation.

Comparison of breath gas data from non-smokers and smokers can be used as a benchmark test for the data quality, by judging the results of such a statistical analysis against the current knowledge on VOC in tobacco smoke. In such an approach it is an important prerequisite that those signals arising from the same chemical compound are compared by statistical analysis. In contrast to PTR-TOF-MS, the non-selective quadrupole PTR-MS does not fulfill this condition, but broad peaks are followed and evaluated. Chemical noise due to different chemical components of the same nominal mass might mask certain differences between two subgroups. Still, in other cases, differences between nominal peaks are still high enough, as the important compound is the main contributor to its intensity.

Even though mass resolution is high in PTR-TOF-MS, both single peaks as well as overlapping peaks can be found in a TOF spectrum. It is not clear in every case whether a peak should be treated as a single peak originating from a single ion species or not. A criterion to decide whether multi-peak analysis has to be performed is difficult to define for the whole range of TOF. Reasons are manifold, for instance, the peak broadness (R_{FWHM}) is not constant but increases with increasing time-of-flight, due to statistical uncertainty of the absolute time-of-flight determination.[53] Another difficulty comprises peaks originating from the detection system that can be found as spectral artifacts, especially when count rates are high. These artifacts must be distinguished from real peaks of ionic species.

The following approach was applied to the data presented in this chapter. In a first step, all measured breath gas TOF spectra were added to one sum spectrum. By this simple procedure, random peaks at low signal-to-noise-ratio were removed and non-random peaks became sharper. The definition of peaks that partly overlapped was improved in terms of the center (the maximum of the peak) and the full width. Based on this sum spectrum, a base set of 790 peaks was identified in the mass range of m/z 15–800. Each PTR-TOF spectrum was searched for this base set subsequently. Signal intensities and corresponding m/z were determined from Gaussian fits using Matlab (MathWorks version 7.8.0.347). Statistical significances of the differences between the means/medians of smoker's samples and non-smoker's samples were investigated using the Student's t-test and the Mann-Whitney U-test ($p < 0.05$ was considered to be statistically significant).

6.3 RESULTS AND DISCUSSION

6.3.1 Sensitivity and LOD

During the course of the study we performed 15 calibration experiments. The sensitivity of the PTR-TOF-MS for 15 compounds is listed in Table 6.3. In contrary to quadrupole PTR-MS, in PTR-TOF-MS the sensitivities for lower masses were similar

Table 6.3 Sensitivity, standard deviation, and detection limit for 15 organic compounds. No value can be given for 1,2,4-trichlorobenzene due to varying background signals as a result of internal calibration gas use.

Compounds Sum Formula	Compound	Calculated Mass of the Ion Species	Sensitivity (ncps/ppbv)[b]	Standard Deviation (%)	LOD[a] (ppb)
CH_4O	Methanol	33.0340	12	5	0.54
C_2H_3N	Acetonitrile	42.0344	26	10	0.14
C_2H_4O	Acetaldehyde	45.0340	24	11	0.35[c]
C_2H_6O	Ethanol	47.0497	2	17	0.61
C_3H_4O	Acrolein	57.0340	22	13	0.12
C_3H_6O	Acetone	59.0497	27	11	0.21
C_5H_8	Isoprene	69.0704	12	13	0.26
C_4H_6O	Crotonaldehyde	71.0497	34	11	0.07
C_4H_8O	2-Butanone	73.0653	30	12	0.10
C_6H_6	Benzene	79.0548	20	10	0.09
C_7H_8	Toluene	93.0704	24	10	0.07
C_8H_{10}	*o*-Xylene	107.0861	28	11	0.05
C_6H_5Cl	Chlorobenzene	113.0158	27	8	0.08
$C_{10}H_{16}$	α-Pinene	137.1330	15	14	0.09
$C_6H_4Cl_2$	1,2-Dichlorobenzene	146.9768	38	10	0.04

[a] *Value based on 2 s single spectra.*
[b] *Sensitivities are given in ncps/ppbv, normalized counts per second per volume mixing ratio; the normalization process has become a standard practice in PTR-MS applications, where the measured count rates are normalized to the number of primary ions.*
[c] *The LOD of acetaldehyde at m/z 45.0340 is influenced by an overlap with the ethanol product of the H abstraction reaction $O_2^+ + H_3C - CH_3 - OH$. The given value is corrected for this overlap.*

to those of higher masses, which was due to the fact that not only light compounds, but also heavy compounds were transmitted well through the TOF-MS. The sensitivity for ethanol is generally known to be small in the PTR-MS detection.[54] The standard deviation of concentration measurements in the calibration gas addition experiments was $\leqslant 17\%$, based on a 6 min integration time. Within this standard deviation there was no trend in sensitivity over the period of 12 months, which was an important prerequisite for a long-term breath gas analysis project. The sensitivities were within the same order-of-magnitude found by Ruuskanen et al.[35]; however, the deviation in sensitivities between their calibrations was only 2.2% during an 8 days field campaign.

6.3.2 Mass accuracy

The stability and high quality of mass measurement is an important prerequisite to the determination of the atomic compositions of the product ions and meaningful comparisons of measured data gained at different times. Mass accuracy relates to the closeness of the agreement between the calculated exact mass of a species and its measured mass. The exact mass of an ion species can be calculated from its atomic composition using the atomic masses of the main isotopes. This value comprises the true value to which the measured masses are compared. The mass accuracy is a result of the error associated with the PTR-TOF measurement, together with the calibration procedure and errors in the data evaluation procedures.

In the present study the PTR-TOF-MS was operated over a period of 12 months under realistic conditions. Constraints were limited man power and room temperature changes as large as 10°, between hot and cold weather. The temperature changes during a single measurement of one sample were negligible. High mass accuracy was accomplished by re-calibrating the assignment of the mass scale to the measured time-of-flight once a minute. The mass scale was assigned to the three known peaks of NO^+, acetone, and 1,2,4-trichlorobenzene, which were present in all mass spectra at all times. The mass accuracy and mass measurement stability was determined from the calibration gas measurements. Known VOCs were used as test masses.

The mass differences between calculated and measured m/z were (except the isoprene fragment which showed a greater uncertainty) less than 0.0016 (or less than 29 ppm of the ion mass). The standard deviation of the measured m/z was less than 0.0025 and the coefficient of variation was less than 29 ppm.

As predicted by Eq. (6.3), the relative uncertainty decreased with increasing mass numbers in the present study. The relative mass difference, as listed in Table 6.4, lied within the measured standard deviation of the measured ion species mass. A clear exception was the isoprene fragment (m/z 41.0381). Its peak area was small compared to the peaks of the other VOCs contained in the calibration gas mixture, leading to a weak definition of the measured peak and therefore larger uncertainty in mass determination.

The results in this present study were less accurate than those presented by Graus et al.,[30] who undertook calibration experiments of a similar instrument over a period of 6 weeks. Taking into account that TOF measurements of the present study were done with a bin width of 200 s, our resulting values were within the expected range.[50]

6.3.3 The exhaled breath composition of the women volunteers

The most prevalent constituents of exhaled breath gas of 53 non-cigarette smoking women were m/z 59.05 ($C_3H_7O^+$, acetone) [‡], followed by m/z 33.034 (CH_5O^+, methanol), m/z 69.07 ($C_5H_9^+$, isoprene), m/z 47.05 ($C_2H_7O^+$, ethanol), m/z 45.034 ($C_2H_5O^+$, acetaldehyde), m/z 61.0299 ($C_2H_5O_2^+$, acetic acid), m/z 75.0446 ($C_3H_7O_2^+$, hydroxypropanone) and m/z 57.034 ($C_3H_5O^+$, propenal). The large signal at m/z 43.02 ($C_2H_3O^+$, ketene) was at least partly caused by fragmentation of higher m/z oxygen-containing VOCs. We found m/z 51.046, 65.06, 77.06, 79.042, that can be explained by water clusters of methanol, ethanol, acetone, and acetic acid.

[‡]For the description of the results we first give the measured m/z of the ion plus its atomic composition and one suggestion of a possible chemical compound in brackets.

Table 6.4 Summary of the highest signals in breath gas samples of non-smoking women. Please note that the compounds are suggestions and not a definite result.

Peak Information					Non-Smokers	
Base Set Mass	**Atomic Composition of the Ion**	**Possible Compound**	**Theoretical Mass**	**Base Set Mass Minus Theoretical Mass**	**Mean (ppbv)**	**Standard Deviation (ppbv)**
33.034	CH_5O^+	Methanol	33.0340	0.0000	312	159
41.04	$C_3H_5^+$	Propadiene, Isoprene fragment	41.0391	0.0009	56	21
42.035	$C_2H_4N^+$	Acetonitrile	42.0344	0.0006	9	7
43.02	$C_2H_3O^+$	Fragments of oxy-VOCs	43.0184	0.0016	93	59
43.055	$C_3H_7^+$	Cyclopropane, Propanol fragment	43.0548	0.0002	59	43
45.034	$C_2H_5O^+$	Acetaldehyde	45.0340	0.0000	89	134
47.013	$CH_3O_2^+$	Formic acid, fragments of oxy-VOCs	47.0133	−0.0003	21	45
47.05	$C_2H_7O^+$	Ethanol	47.0497	0.0003	130	213
51.046	$CH_4O{\cdot}H_3O^+$	Methanol water	51.0446	0.0014	82	50
56.047	$C_3H_6N^+$	Propanenitrile	56.0500	−0.0030	6	3
57.034	$C_3H_5O^+$	Propenal	57.0340	0.0000	32	64
57.07	$C_4H_9^+$	1-Butene	57.0704	−0.0004	12	8
59.05	$C_3H_7O^+$	Acetone	59.0497	0.0003	1802	984
61.0299	$C_2H_5O_2^+$	Acetic acid	61.0290	0.0009	68	64
61.0528	$C_3H_9O^+$	Propanol	61.0653	−0.0125	10	5
63.027	$C_2H_7S^+$	Dimethyl sulfide	63.0268	0.0002	17	10
63.045	$C_2H_7O_2^+$	1,2-Ethanediol	63.0446	0.0004	8	13
65.06	$C_2H_6O{\cdot}H_3O^+$	Ethanol water	65.0603	−0.0003	20	51

(Continued)

Table 6.4 (Continued)

Peak Information					Non-Smokers	
Base Set Mass	Atomic Composition of the Ion	Possible Compound	Theoretical Mass	Base Set Mass Minus Theoretical Mass	Mean (ppbv)	Standard Deviation (ppbv)
69.034	$C_4H_5O^+$	Furan	69.0340	0.0000	5	6
69.07	$C_5H_9^+$	Isoprene	69.0704	−0.0004	280	143
71.051	$C_4H_7O^+$	Crotonealdehyde	71.0497	0.0013	10	6
73.065	$C_4H_9O^+$	Butanal	73.0653	−0.0003	9	5
75.0446	$C_3H_7O_2^+$	1-Hydroxy-2-propanone	75.0446	0.0000	55	113
77.06	$C_3H_6O{\cdot}H_3O^+$	Acetone water	77.0603	−0.0003	21	14
79.042	$C_2H_4O_2{\cdot}H_3O^+$	Acetic acid water	79.0395	0.0025	5	4
81.07	$C_6H_9^+$	Monoterpene fragment	81.0704	−0.0004	12	17
83.089	$C_6H_{11}^+$	1,5-Hexadiene	83.0961	−0.0071	9	23
87.045	$C_4H_7O_2^+$	2,3-Butanedione	87.0466	−0.0016	5	5
89.058	$C_4H_9O_2^+$	2-Hydroxy-butanal	89.0603	−0.0023	11	8
93.038	$C_6H_5\ O^+$	3-Ethynylfuran	93.0340	0.0040	18	20
93.0705	$C_7H_9^+$	Toluene	93.0704	0.0001	12	31
95.05	$C_6H_7O^+$	Phenol	95.0497	0.0003	21	14
99.083	$C_6H_{11}O^+$	2-Hexenal	99.0810	0.0020	5	9
137.135	$C_{10}H_{17}^+$	Monoterpene	137.1330	0.0020	13	19
151.115	$C_6H_{15}O_4^+$	1,2,4,5-Hexanetetrol	151.0970	0.0180	21	144

In Table 6.4, we summarized 35 *m/z* that had 5 or more ppbv on average, together with propositions of possible atomic compositions and possible compounds. The difference between the predefined base set mass (see Section 6.2.9) and the theoretical mass corresponding to the suggested sum formula were also listed for comparison. This deviation gave a clue on how reliable the assignment could be. The mean concentrations were given in ppbv together with the corresponding standard deviation.[**] Only peaks genuinely arising from the breath gas samples were included in Table 6.4. Isotopic peaks and instrumental background peaks were left out.

Note that the attributions of measured *m/z* and derived atomic compositions to possible chemical compounds are tentative and were done empirically in accordance to the literature on breath gas analysis.[1,6,8,55,56] Please keep in mind, that formation of fragments is possible in PTR-TOF-MS and might account for some of the found peaks too.

6.3.4 Breath compounds enhanced by cigarette smoking

The breath of 19 self-declared smokers was analyzed and compared to the breath composition of the non-smoking women. The continuous exposure to various chemicals during frequent inhalation of cigarette smoke instead of normal air is reflected by high concentrations of such chemicals in the exhaled breath. The sum of all reported concentrations of those *m/z* listed in Table 6.4 amounts to 3336 ppbv in non-smoking women compared to 4408 ppbv in smokers.

Mühlberger and coauthors[57] made a similar observation using single photon ionization time-of-flight mass spectrometry for the investigation of exhaled breath of smokers and non-smokers with a LOD down to 50 ppb. Their experiments found acetone, isoprene, and small amounts of phenol in the breath of non-smokers. In contrast, the breath of smokers contained a larger variety of compounds, such as well known cigarette smoking markers 2,5-dimethylfuran, butadiene, and benzene.[15,39,55] Ions derived from hydrocarbons such as butane, cyclopentadiene, pentene, toluene, and xylene occurred solely in the spectrum of smoker's breath. The single photon ionization time-of-flight mass spectrometry used by Mühlberger et al. is not capable of resolving isobaric species. Therefore, Mühlberger based his assignment on the literature on cigarette smoke constituents. In PTR-TOF-MS butadiene (*m/z* 55.0548) might be masked by the large $H_3O^+ \cdot (H_2O)_2$ signal (*m/z* 55.0395).

It has been known for a long time that smokers can be identified unambiguously by a high fraction of acetonitrile and benzene in exhaled breath gas.[58] This is evident in the present data set.

Smokers showed the largest signal increase compared to non-smokers in *m/z* 42.035, highly likely reflecting acetonitrile (C_2H_3N), *m/z* 79.055 highly likely reflecting benzene (C_6H_6), *m/z* 83.05 highly likely reflecting methylfuran (C_5H_6O), and *m/z* 97.065 likely caused by 2,5-dimethylfuran (C_6H_8O). This is demonstrated in a box plot representation in Figure 6.1. In some cases the outliers marked by crosses

[**] Concentrations (Volume Mixing Ratios) were calculated in parts-per-billion per volume (ppbv) using a reaction rate constant of $2 \cdot 10^{-9}$ cm^3 s^{-1}.

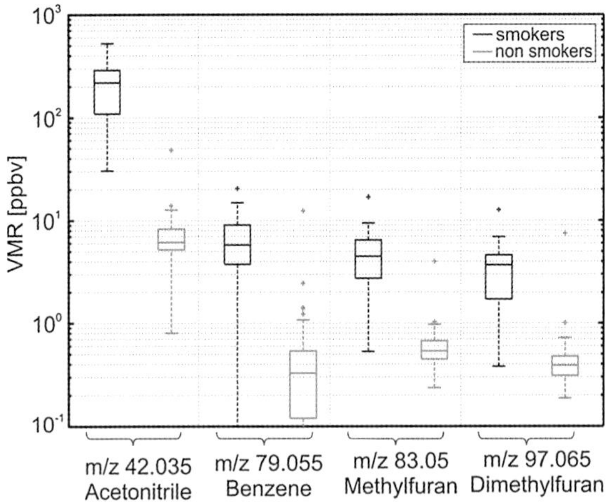

FIGURE 6.1

Box plot representation of the volume mixing ratios (VMR) of four compounds considerably increased in the group of smokers, when compared to self declared non-smokers. The boxes contain 50% of the data distributed around the median value, upper and lower whiskers enclose the lower and upper quartile of values. Outliers are depicted with crosses.

might be caused by smokers who declared themselves as non-smokers. Please note that the group size of non-smokers is larger by a factor 2.7 than the group size of smokers. There is a large distribution of the measured values within the group of smokers representing the difference in exposure to those chemicals, inhaled with cigarette smoke. The large distribution might also be a result of the different removal time of such chemicals from the body, which is known to be extremely short for benzene, and as long as a few days for acetonitrile.[58]

Kischkel et al.[14] found acetonitrile, benzene, 2,5-dimethylfuran, acetone, toluene, butanal, and hexanal significantly elevated in the breath of smokers using SPME-GC-MS analysis. Our study did not reproduce significant differences in acetone levels, which Kischkel attributed to a response to nicotine deprivation in smokers, who were requested to stop smoking at least 1 h before the breath sample was collected. Toluene showed a large spread in found concentrations within the group of non-smokers, which might be an indication that smokers did not declare themselves correctly, or a result of an exposure from other sources than smoking.

Our study found a statistically significant difference between smokers and non-smokers for the signals of *m/z* 73.0650 attributed to butanal, and for the small signal of *m/z* 101.0960, attributed to hexanal.

As true for our data set, Bajtarevic et al.[39] confirmed acetonitrile, as well as toluene, benzene, 2-methylfuran, 2,5-dimethylfuran, furan, 1,3-cyclohexadiene, 1,3-cyclopentadiene, 2-methyl-1-butene, and 1,4-pentadiene to be smoking-related compounds contained in exhaled breath by GC-MS. In mass spectrometry however,

1,4-pentadiene is indistinguishable from isoprene (C_5H_8). In the present data set there was no statistically significant difference between smokers and non-smokers signal at *m/z* 69.07 (C_5H_8). 1,3-cyclohexadiene (protonated *m/z* 81.0704) might contribute to the signal that was found at *m/z* 81.07 in the present data set, which was attributed to the monoterpene fragment in Tables 6.4 and 6.5.

Table 6.5 represents a summary on smoking related *m/z* that were identified in the present data set.

Aromatic compounds such as benzene, toluene, xylene, furan, and its derivatives are frequently mentioned in the literature to be connected with cigarette smoke.[56,57,59,60]

Our study found significantly elevated signals that are highly likely caused by aromatic compounds such as *m/z* 93.038 ($C_6H_5O^+$, 2-ethynyl-furan), *m/z* 93.0705 ($C_7H_9^+$, toluene), *m/z* 79.055 ($C_6H_7^+$, benzene), *m/z* 69.034 ($C_4H_4O^+$, furan), *m/z* 83.05 ($C_5H_7O^+$, 2-methylfuran), *m/z* 99.045 ($C_5H_7O_2^+$, 3-furanmethanol), *m/z* 97.065 ($C_6H_9O^+$, 2,5-dimethylfuran), *m/z* 107.086 ($C_8H_{11}^+$, xylene), *m/z* 68.063 ($C_4H_6N^+$, pyrrole), *m/z* 111.05 ($C_6H_7O_2^+$, catechol), and *m/z* 80.05 (C_5H_5N, pyridine). Furan is mentioned in the literature as a smoking compound.[61] Its molecular weight is very similar to the weight of isoprene, which always causes a large peak in PTR-based mass spectrometers in breath gas analysis. There is a chance, that the furan peak (*m/z* 69.0340) is masked by the large isoprene peak (*m/z* 69.0704 Th) in some cases.

Aliphatic VOCs are frequently mentioned in the literature as smoking compounds.[39,56,57,59,60]

Compared to the non-smokers, the smokers breath samples collected during the present study exhibited significant increases in oxygenated VOCs such as *m/z* 43.02 ($C_2H_3O^+$, ketene), *m/z* 61.0299 ($C_2H_5O_2^+$, acetic acid) plus *m/z* 79.042 ($C_2H_4O_2 \cdot H_3O^+$, acetic acid water), *m/z* 75.0446 ($C_3H_7O_2^+$, 1-hydroxy-2-propanone), *m/z* 89.058 ($C_4H_9O_2^+$, 3-hydroxybutanal), *m/z* 71.051 ($C_4H_7O^+$, crotonealdehyde), *m/z* 73.065 ($C_4H_9O^+$, butanal), *m/z* 61.053 ($C_3H_9O^+$, propanol) plus *m/z* 43.055 ($C_3H_7^+$, propanol fragment), *m/z* 87.045 ($C_4H_7O_2^+$, 2,3-butanedione), *m/z* 101.061 ($C_5H_9O_2^+$, 2,4-pentanedione), *m/z* 87.081 ($C_5H_{11}O^+$, pentanal), *m/z* 73.03 ($C_3H_5O_2^+$, methylglyoxal), *m/z*;115.078 ($C_6H_{11}O_2^+$, 2,3-hexanedione), *m/z* 103.078 ($C_5H_{11}O_2^+$, 1-methyl-1-propanol), *m/z* 85.065 ($C_5H_9O^+$, 2-methyl-2-butenal), *m/z* 117.092 ($C_6H_{13}O_2^+$, 4-hydroxy-4-methylpentan-2-one), and *m/z* 129.093 ($C_7H_{13}O_2^+$, cyclohexane carboxylic acid).

It is a contradiction that water clusters of methanol and acetone showed significant differences between the groups of smokers and non-smokers while their H^+ adducts did not. We cannot exclude that other components contributed to the signals of the water cluster *m/z*s.

Carbohydrates such as *m/z* 57.07 ($C_4H_9^+$, 1-butene), *m/z* 83.089 ($C_6H_{11}^+$, 1,5-hexadiene), *m/z* 137.135 ($C_{10}H_{17}^+$, monoterpene) plus *m/z* 81.07 ($C_6H_9^+$, monoterpene fragment), *m/z* 67.055 ($C_5H_7^+$, 1,3-cyclopentadiene), *m/z* 95.085 ($C_7H_{11}^+$, 2-norbornene), *m/z* 71.086 ($C_5H_{11}^+$, cyclopentan), *m/z* 97.107 ($C_7H_{13}^+$, norbornane), *m/z* 139.15 ($C_{10}H_{19}^+$, 3,7-dimethyl-1,6-octadiene), *m/z*;109.102 ($C_8H_{13}^+$,

Table 6.5 Summary of those signals significantly elevated in the exhaled breath gas samples of cigarette smokers; Signals are ordered according to their mean concentration in smokers samples.

Peak Information					Smokers (19 Samples)		Non-Smokers (53 Samples)		Student's *t*-Test	Ratio	Whitney Mann *U*-Test
Base Set Mass	Atomic Composition of the Ion	Possible Compound	Theoretical Mass	Base Set Mass Minus Theoretical Mass	Mean/ppbv	SD/ppbv	Mean/ppbv	SD/ppbv	*p*-Value	Smoker/Non-Smoker	*p*-Value
42.035	$C_2H_4N^+$	Acetonitrile	42.0344	0.0006	265	160	9	7	0.0%	29.1	2.8%
43.02	$C_2H_3O^+$	Fragments of oxy-VOCs	43.0184	0.0016	213	137	93	59	0.0%	2.3	46.6%
51.046	$CH_7O_2^+$	Methanol water	51.0446	0.0014	163	160	82	50	0.2%	2.0	0.1%
61.0299	$C_2H_5O_2^+$	Acetic acid	61.0290	0.0009	157	122	68	64	0.0%	2.3	55.7%
75.0446	$C_3H_7O_2^+$	1-Hydroxy-2-propanone	75.0446	0.0000	108	107	55	113	7.8%	2.0	2.6%
43.055	$C_3H_7^+$	Cyclopropane, Propanol fragment	43.0548	0.0002	86	48	59	43	2.7%	1.5	5.4%
41.04	$C_3H_5^+$	Propadiene, Isoprene fragment	41.0391	0.0009	70	29	56	21	3.3%	1.2	2.6%
77.06	$C_3H_6O{\cdot}H_3O^+$	Acetone water	77.0603	−0.0003	44	45	21	14	0.1%	2.1	0.6%

(*Continued*)

Table 6.5 (Continued)

Peak Information					Smokers (19 Samples)		Non-Smokers (53 Samples)		Student's *t*-Test	Ratio	Whitney Mann *U*-Test
Base Set Mass	Atomic Composition of the Ion	Possible Compound	Theoretical Mass	Base Set Mass Minus Theoretical Mass	Mean/ppbv	SD/ppbv	Mean/ppbv	SD/ppbv	*p*-Value	Smoker/Non-Smoker	*p*-Value
89.058	$C_4H_9O_2^+$	2-Hydroxybutanal	89.0603	−0.0023	39	44	11	8	0.0	3.5	13.5
57.07	$C_4H_9^+$	1-Butene	57.0704	−0.0004	27	12	12	8	0.0	2.2	4.2
71.051	$C_4H_7O^+$	Crotonealdehyde	71.0497	0.0013	27	26	10	6	0.0	2.6	58.3
93.038	$C_6H_5O^+$	2-Ethynylfuran	93.0340	0.0040	26	27	18	20	16.6	1.5	0.0
81.07	$C_6H_9^+$	Monoterpene fragment, 1,3-Cyclohexadiene	81.0704	−0.0004	25	14	12	17	0.4	2.1	5.9
83.089	$C_6H_{11}^+$	1,5-Hexadiene	83.0831	0.0059	25	37	9	23	3.3	2.8	0.5
93.0705	$C_7H_9^+$	Toluene	93.0704	0.0001	24	13	12	31	12.0	1.9	0.1
137.135	$C_{10}H_{17}^+$	Monoterpene	137.1330	0.0020	18	15	13	19	28.4	1.4	0.0
73.065	$C_4H_9O^+$	Butanal	73.0653	−0.0003	18	10	9	5	0.0	2.1	0.0
61.053	$C_3H_9O^+$	Propanol	61.0653	−0.0123	17	12	10	5	0.0	1.7	7.6
79.042	$C_2H_4O_2 \cdot H_3O^+$	Acetic acid water	79.0422	−0.0002	16	16	5	4	0.0	3.5	0.7

(*Continued*)

Table 6.5 (Continued)

Peak Information					Smokers (19 Samples)		Non-Smokers (53 Samples)		Student's *t*-Test	Ratio	Whitney Mann *U*-Test
Base Set Mass	Atomic Composition of the Ion	Possible Compound	Theoretical Mass	Base Set Mass Minus Theoretical Mass	Mean/ppbv	SD/ppbv	Mean/ppbv	SD/ppbv	*p*-Value	Smoker/Non-Smoker	*p*-Value
87.045	$C_4H_7O_2^+$	2,3-Butanedione	87.0446	0.0004	13	15	5	5	0.2	2.4	23.0
79.055	$C_6H_7^+$	Benzene	79.0548	0.0002	12	9	1	3	0.0	10.7	2.2
67.055	$C_5H_7^+$	1,3-Cyclopentadiene	67.0548	0.0002	12	7	4	2	0.0	2.8	1.0
69.034	$C_4H_4O^+$	Furan	69.0340	0.0000	11	5	5	6	0.0	2.1	0.0
89.073	$C_3H_9N_2O^+$	*N,N'*-Dimethylurea	89.0715	0.0015	10	14	2	2	0.0	4.9	0.1
56.047	$C_3H_6N^+$	Propanenitrile	56.0500	−0.0030	10	7	6	3	0.1	1.8	1.0
95.085	$C_7H_{11}^+$	2-Norbornene	95.0861	−0.0011	10	6	3	3	0.0	3.4	0.1
83.05	$C_5H_7O^+$	2-Methylfuran	83.0497	0.0003	9	7	1	1	0.0	8.1	0.3
71.086	$C_5H_{11}^+$	Cyclopentan, 2-Methyl-1-butene	71.0861	−0.0001	8	4	3	2	0.0	2.4	0.0
101.061	$C_5H_9O_2^+$	2,4-Pentanedione	101.0603	0.0007	8	11	4	2	0.8	2.2	0.4
99.045	$C_5H_7O_2^+$	3-Furanmethanol	99.0446	0.0004	7	12	2	3	0.5	3.3	0.0
97.065	$C_6H_9O^+$	2,5-Dimethylfuran	97.0653	−0.0003	7	5	1	2	0.0	7.0	0.0

(*Continued*)

Table 6.5 (Continued)

Peak Information					Smokers (19 Samples)		Non-Smokers (53 Samples)		Student's *t*-Test	Ratio	Whitney Mann *U*-Test
Base Set Mass	Atomic Composition of the Ion	Possible Compound	Theoretical Mass	Base Set Mass Minus Theoretical Mass	Mean/ppbv	SD/ppbv	Mean/ppbv	SD/ppbv	*p*-Value	Smoker/Non-Smoker	*p*-Value
87.081	$C_5H_{11}O^+$	Pentanal	87.0810	0.0000	7	3	4	2	0.0	1.6	0.0
107.086	$C_8H_{11}^+$	Xylene	107.0861	−0.0001	7	4	2	2	0.0	3.6	0.2
73.03	$C_3H_5O_2^+$	Methylglyoxal	73.0290	0.0010	7	7	4	2	1.3	1.8	0.0
97.107	$C_7H_{13}^+$	Norbornane	97.1017	0.0053	6	4	2	3	0.0	2.7	0.0
113.048	$C_6H_9S^+$	2,4-Dimethylthiophene	113.0425	0.0055	6	7	2	2	0.3	2.5	0.0
62.03	$CH_4NO_2^+$	Nitromethane	62.0242	0.0058	6	3	4	2	0.4	1.5	3.2
68.063	$C_4H_6N^+$	Pyrrole	68.0500	0.0130	5	2	4	2	0.6	1.4	0.2
111.05	$C_6H_7O_2^+$	Catechol	111.0446	0.0054	5	8	2	1	0.6	2.4	1.8
139.15	$C_{10}H_{19}^+$	3,7-Dimethyl-1,6-octadiene	139.1487	0.0013	5	10	2	8	24.8	2.2	0.6
115.078	$C_6H_{11}O_2^+$	2,3-Hexanedione	115.0759	0.0021	5	8	1	2	0.5	3.4	0.6
109.102	$C_8H_{13}^+$	1-Vinylcyclohexene	109.1017	0.0003	5	4	2	8	24.3	2.1	1.7
85.104	$C_6H_{13}^+$	2,3-Dimethyl-2-butene	85.1017	0.0023	3	2	1	1	0.0	2.8	0.7

(*Continued*)

Table 6.5 (Continued)

Peak Information					Smokers (19 Samples)		Non-Smokers (53 Samples)		Student's *t*-Test	Ratio	Whitney Mann *U*-Test
Base Set Mass	Atomic Composition of the Ion	Possible Compound	Theoretical Mass	Base Set Mass Minus Theoretical Mass	Mean/ppbv	SD/ppbv	Mean/ppbv	SD/ppbv	*p*-Value	Smoker/Non-Smoker	*p*-Value
103.078	$C_5H_{11}O_2^+$	2-Methyl-1-propanol	103.0759	0.0021	3	3	2	1	0.2	2.2	0.2
85.065	$C_5H_9O^+$	2-Methyl-2-butenal	85.0653	−0.0003	3	3	1	1	0.0	2.0	0.0
80.05	$C_5H_6N^+$	Pyridine	80.0500	0.0000	3	2	1	1	0.0	2.5	1.5
117.092	$C_6H_{13}O_2^+$	4-Hydroxy-4-methylpentan-2-one	117.0916	0.0004	3	3	1	1	0.0	2.6	2.6
123.12	$C_9H_{15}^+$	1,5-Dimethyl-3-methylenecyclohexene	123.1174	0.0026	3	2	1	2	0.0	3.5	0.0
129.093	$C_7H_{13}O_2^+$	Cyclohexanecarboxylic acid	129.0916	0.0014	3	4	1	1	0.4	3.9	0.0
101.096	$C_6H_{13}O^+$	Hexanal	101.0966	−0.0006	2	2	1	1	0.0	2.3	0.0

Peaks according to the basis set m/z are listed. Measured and proposed theoretical m/z are given together with the corresponding atomic composition and a suggestion of a possible compound. Statistically significant differences between the group of non-smokers and smokers were evaluated calculating the p-value using the Student's t-test and the Whitney Mann U-test. $p < 0.05$ was considered to be statistically significant.

1-vinylcyclohexene), and *m/z* 123.12 ($C_9H_{15}^+$, 1,5-dimethyl-3-methylenecyclohexene) were found significantly elevated due to smoking, as well as nitrogen-containing compounds such as *m/z* 42.035 ($C_2H_4N^+$, acetonitrile), *m/z* 89.073 ($C_3H_9N_2O^+$, *N,N'*-dimethylurea), *m/z* 56.047 ($C_3H_6N^+$, propanenitrile), and *m/z* 62.03 ($CH_4NO_2^+$, nitromethane) and sulfur-containing compounds such as *m/z* 113.048 Th ($C_6H_9S^+$, 2,4-dimethylthiophene).

It is important to note that the mentioned compounds represent possible suggestions that fit the found sum formulas. We consider it is possible that the order according to the peak intensity listed in Table 6.5 is not absolute but might change if more volunteers were included in the study. It cannot be excluded that passive smoking and incorrect allocation to the groups of smokers and non-smokers possibly biased the results.

In contrast to Kushch et al.,[56] who found an effect of smoking on the concentration of seven masses using conventional quadrupole PTR-MS, we could find several more significantly higher signals in the range from 33 *m/z* to 150 *m/z* using high mass resolving PTR-TOF-MS. While Kushch et al. sampled mixed breath gas into partly permeable Tedlar bags,[40] we sampled end-tidal exhaled breath gas via CO_2 control into leak tight SilcoCans, excluding uncontrolled dilution of end tidal-breath gas.

6.3.5 Structure of mass spectral peaks

Breath gas samples contain a large number of volatile organic compounds. Using PTR-TOF-MS in breath gas analysis, it became evident that more than one compound often underlies one broad PTR-MS mass peak of unity resolution. Figures 6.2–6.5

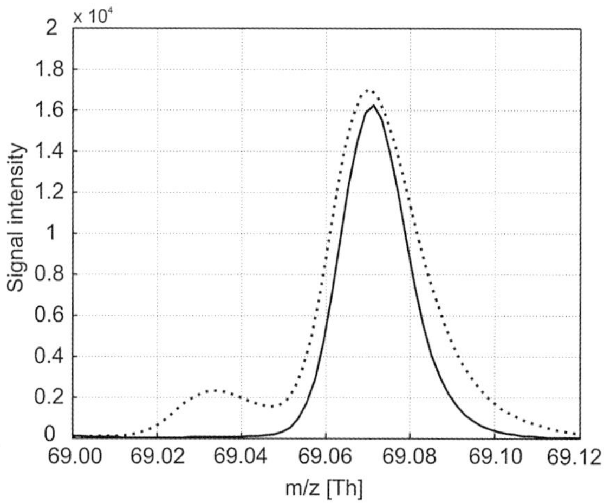

FIGURE 6.2
Detail of a TOF mass spectrum at *m/z* 69. The left peak was attributed to furan (*m/z* 69.0340, C_4H_3O), the right peak was attributed to isoprene (*m/z* 69.0704, C_5H_8).

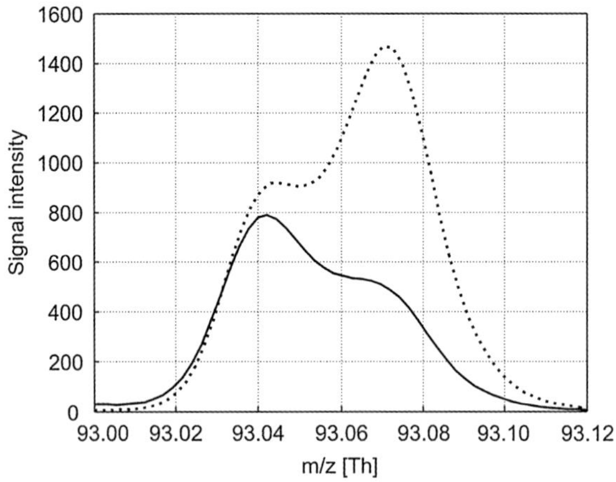

FIGURE 6.3

Detail of a TOF mass spectrum at *m/z* 93. The left peak at *m/z* 93.0380 was explained by an atomic composition of C_6H_4O which is the sum formula (besides other compounds) of 2-ethynylfuran. The right peak at *m/z* 93.0705 was explained by toluene (C_7H_8).

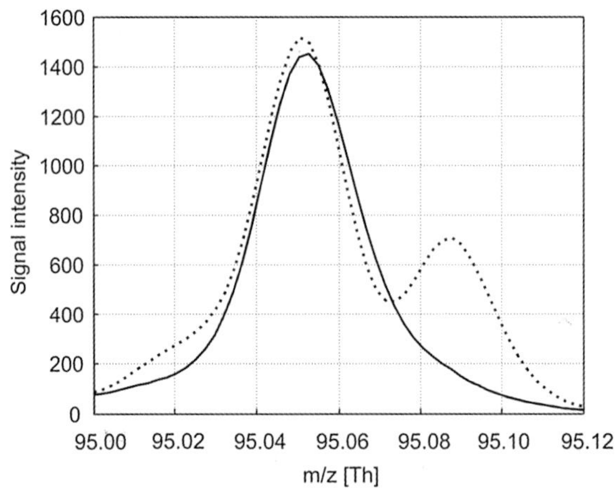

FIGURE 6.4

Detail of a TOF mass spectrum at *m/z* 95. The signal at *m/z* 95.0497 was highly likely caused by phenol (C_6H_6O). The compound causing the signal at *m/z* 95.085 was explained by an atomic composition of C_7H_{10} and might represent 2-norbornene.

represent details of four spectral regions, *m/z* 69, 93, 95, and 107. Baseline corrected smoothed traces of different study subjects are shown. Smoothing procedures did not affect peak areas. In each figure the trace of a cigarette smoking woman (dotted trace) is compared with a non-smoking woman (continuous line).

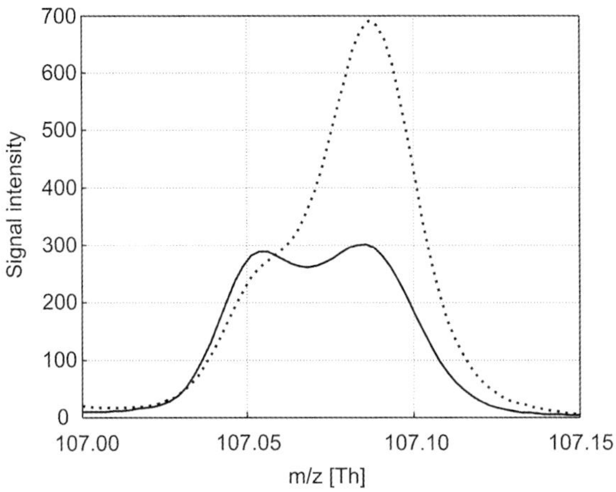

FIGURE 6.5
Detail of a TOF mass spectrum at *m/z* 107. The signal at *m/z* 107.0497 was highly likely caused by benzaldehyde (C_7H_6O), the compound causing the signal at *m/z* 107.0861 was highly likely xylene (C_8H_{10}).

Figure 6.2 shows the spectral range of *m/z* 69, where the large peak of isoprene (C_5H_8) is expected at *m/z* 69.0704. The average level of isoprene in the present cohort of women was ~280 ppbv, which was within the range of isoprene concentrations observed in other studies on healthy volunteers using PTR-MS.[62,63] Furan (C_4H_3O) is of similar molecular mass and a corresponding peak is expected at *m/z* 69.0340 in the TOF spectrum. Indeed, a peak at *m/z* 69.0340 could be found in our PTR-TOF-MS analysis of smoking women, as shown in Figure 6.2.

In PTR-TOF-MS spectra two peaks could be found in the spectral range of *m/z* 93 (Figure 6.3). Both peaks were found to be statistically significantly elevated in cigarette smokers compared to non-smokers, with the right peak increased more than 1.9-fold in smokers. The left peak at *m/z* 93.0380 might have been caused by 2-ethynylfuran (C_6H_4O). The right peak at *m/z* 93.0705 might have been caused by toluene (C_7H_8), a known smoking compound.

At *m/z* 95 (Figure 6.4) we found two peaks: The left peak at *m/z* 95.0497 was highly likely caused by phenol (C_6H_6O), which did not discriminate smokers from non-smokers in the present data set. This was in accordance with the results of Mühlberger et al.[57] The compound causing the peak at *m/z* 95.085 could be explained by an atomic composition of C_7 H_{10} representing 2-norbornene or another unsaturated aliphatic compound. The corresponding peak was statistically significantly elevated in the group of smokers compared to the group of non-smokers by a factor of 3.4.

Figure 6.5 demonstrates that the mass resolving power of PTR-TOF-MS was high enough to distinguish a peak at *m/z* 107.0497, which was highly likely caused by benzaldehyde (C_7H_6O), from the peak at *m/z* 107.0861, which was highly likely caused by xylene (C_8H_{10}), a known cigarette smoking compound. The potential benzaldehyde peak was only seen in 13 out of 19 cigarette smokers. The potential xylene peak was 3.6-fold elevated in the group of smokers compared to the non-smokers.

6.4 CONCLUSION

PTR-TOF-MS has recently become available for trace gas analysis. It provides powerful features for VOC determination in breath gas research:

(1) High mass resolving power enables VOCs of different atomic composition but same nominal mass (isobaric compounds) to be distinguished. This is an important pre-requisite to correctly determine the concentrations of certain compounds. (2) PTR-TOF-MS provides full spectral information in each measurement cycle. Gases can now be fully characterized in real time, even when changes in composition and concentration are fast. As demonstrated in the present study, PTR-TOF-MS is capable of providing more detailed information on the complex composition of exhaled breath gas samples.

In the present study, we accurately measured the *m/z* of the mass spectral peaks obtained in the analysis of exhaled breath. 35 mass spectral peaks were found in the majority of spectra at VMR higher than 5 ppbv and the corresponding chemical compositions of these measured *m/z* and tentative suggestions were made on the origin of the neutral compounds. Taking into account current knowledge, low weight VOCs could be identified in several cases. These high quality data could be used for further statistical data analysis and the identification of chemical compounds in exhaled breath related to cigarette smoking. Several of the smoking compounds found were in agreement with the current knowledge.

Acknowledgments

This work was supported by the COMET Center ONCOTYROL and funded by the Federal Ministry for Transport, Innovation and Technology (BMVIT) and the Federal Ministry of Economics and Labour/the Federal Ministry of Economy, Family and Youth (BMWA/BMWFJ), the Tiroler Zukunftsstiftung (TZS) and the State of Styria represented by the Styrian Business Promotion Agency (SFG) and supported by the Innsbruck Medical University, the University for Health Sciences, Medical Informatics and Technology, Verein Sanitätsbetrieb der Autonomen Provinz Bozen and TILAK - Tiroler Landeskrankenanstalten GmbH.

References

1. Lindinger W, Hansel A, Jordan A. Proton-transfer-reaction mass spectrometry (PTR-MS): on-line monitoring of volatile organic compounds at pptv levels. *Chem Soc Rev* 1998;**27**:347–54.

2. Smith D, Spanel P. The novel selected-ion flow tube approach to trace gas analysis of air and breath. *Rapid Commun Mass Sp* 1996;**10**:1183–98.
3. Pauling L, Robinson AB, Teranishi R, Cary P. Quantitative analysis of urine vapor and breath by gas-liquid partition chromatography. *Proc Natl Acad Sci U S A* 1971;**68**: 2374–6.
4. Teranishi R, Mon TR, Robinson AB, Cary P, Pauling L. Gas chromatography of volatiles from breath and urine. *Anal Chem* 1972;**44**:18–20.
5. Baumbach JI. Ion mobility spectrometry coupled with multi-capillary columns for metabolic profiling of human breath. *J Breath Res* 2009;**3**:034001.
6. Lindinger W, Hansel A, Jordan A. On-line monitoring of volatile organic compounds at pptv levels by means of proton-transfer-reaction mass spectrometry (PTR-MS) - Medical applications, food control and environmental research. *Int J Mass Spectrom* 1998;**173**: 191–241.
7. Phillips M. Breath tests in medicine. *Sci Am* 1992;**267**:74–9.
8. Phillips M, Herrera J, Krishnan S, Zain M, Greenberg J, Cataneo RN. Variation in volatile organic compounds in the breath of normal humans. *J Chromatogr B Biomed Sci Appl* 1999;**729**:75–88.
9. Miekisch W, Schubert JK, Noeldge-Schomburg GF. Diagnostic potential of breath analysis-focus on volatile organic compounds. *Clin Chim Acta* 2004;**347**:25–39.
10. Beauchamp J. Inhaled today, not gone tomorrow: pharmacokinetics and environmental exposure of volatiles in exhaled breath. *J Breath Res* 2011;**5**:037103.
11. Statheropoulos M, Agapiou A, Georgiadou A. Analysis of expired air of fasting male monks at Mount Athos. *J Chromatogr B Analyt Technol Biomed Life Sci* 2006;**832**:274–9.
12. Schwoebel H, Schubert R, Sklorz M, Kischkel S, Zimmermann R, Schubert JK, et al. Phase-resolved real-time breath analysis during exercise by means of smart processing of PTR-MS data. *Anal Bioanal Chem* 2011;**401**:2079–91.
13. Prazeller P, Karl T, Jordan A, Holzinger R, Hansel A, Lindinger W. Quantification of passive smoking using proton-transfer-reaction mass spectrometry. *Int J Mass Spectrom* 1998;**178**:L1–4.
14. Kischkel S, Miekisch W, Sawacki A, Straker EM, Trefz P, Amann A, et al. Breath biomarkers for lung cancer detection and assessment of smoking related effects - confounding variables, influence of normalization and statistical algorithms. *Clin Chim Acta* 2010;**411**:1637–44.
15. Gordon SM, Wallace LA, Brinkman MC, Callahan PJ, Kenny DV. Volatile organic compounds as breath biomarkers for active and passive smoking. *Environ Health Perspect* 2002;**110**:689–98.
16. Smith D, Wang T, Spanel P. On-line, simultaneous quantification of ethanol, some metabolites and water vapor in breath following the ingestion of alcohol. *Physiol Meas* 2002;**23**:477–89.
17. Jones AW, Andersson L. Comparison of ethanol concentrations in venous blood and end-expired breath during a controlled drinking study. *Forensic Sci Int* 2003;**132**:18–25.
18. Lindberg L, Brauer S, Wollmer P, Goldberg L, Jones AW, Olsson SG. Breath alcohol concentration determined with a new analyzer using free exhalation predicts almost precisely the arterial blood alcohol concentration. *Forensic Sci Int* 2007;**168**:200–7.
19. Lindinger W, Taucher J, Jordan A, Hansel A, Vogel W. Endogenous production of methanol after the consumption of fruit alcohol. *Clin Exp Res* 1997;**21**:939–43.

20. Taucher J, Hansel A, Jordan A, Lindinger W. Analysis of compounds in human breath after ingestion of garlic using proton-transfer-reaction mass spectrometry. *J Agr Food Chem* 1996;**44**:3778–82.
21. Hansanugrum A, Barringer SA. Effect of milk on the deodorization of malodorous breath after garlic ingestion. *J Food Sci* 2010;**75**:C549–58.
22. Friedman MI, Preti G, Deems RO, Friedman LS, Munoz SJ, Maddrey WC. Limonene in expired lung air of patients with liver disease. *Dig Dis Sci* 1994;**39**:1672–6.
23. Beauchamp J, Kirsch F, Buettner A. Real-time breath gas analysis for pharmacokinetics: monitoring exhaled breath by on-line proton-transfer-reaction mass spectrometry after ingestion of eucalyptol-containing capsules. *J Breath Res* 2010;**4**:026006.
24. Harrison GR, Critchley AD, Mayhew CA, Thompson JM. Real-time breath monitoring of propofol and its volatile metabolites during surgery using a novel mass spectrometric technique: a feasibility study. *Br J Anaesth* 2003;**91**:797–9.
25. Takita A, Masui K, Kazama T. On-line monitoring of end-tidal propofol concentration in anesthetized patients. *Anesthesiology* 2007;**106**:659–64.
26. Mattison LK, Fourie J, Hirao Y, Koga T, Desmond RA, King JR, et al. The uracil breath test in the assessment of dihydropyrimidine dehydrogenase activity: pharmacokinetic relationship between expired $^{13}CO_2$ and plasma [2-^{13}C]dihydrouracil. *Clin Cancer Res* 2006;**12**:549–55.
27. Schubert JK, Miekisch W, Birken T, Geiger K, Noldge-Schomburg GF. Impact of inspired substance concentrations on the results of breath analysis in mechanically ventilated patients. *Biomarkers* 2005;**10**:138–52.
28. Pleil JD, Smith LB, Zelnick SD. Personal exposure to JP-8 jet fuel vapors and exhaust at air force bases. *Environ Health Perspect* 2000;**108**:183–92.
29. Wisthaler A, Weschler CJ. Reactions of ozone with human skin lipids: sources of carbonyls, dicarbonyls, and hydroxycarbonyls in indoor air. *Proc Natl Acad Sci U S A* 2010;**107**: 6568–75.
30. Graus M, Muller M, Hansel A. High resolution PTR-TOF: quantification and formula confirmation of VOC in real time. *J Am Soc Mass Spectr* 2010;**21**:1037–44.
31. Herbig J, Muller M, Schallhart S, Titzmann T, Graus M, Hansel A. On-line breath analysis with PTR-TOF. *J Breath Res* 2009;**3**:027004.
32. Jordan A, Haidacher S, Hanel G, Hartungen E, Mark L, Seehauser H, et al. A high resolution and high sensitivity proton-transfer-reaction time-of-flight mass spectrometer (PTR-TOF-MS). *Int J Mass Spectrom* 2009;**286**:122–8.
33. Mayhew CA, Sulzer P, Petersson F, Haidacher S, Jordan A, Mark L, et al. Applications of proton transfer reaction time-of-flight mass spectrometry for the sensitive and rapid real-time detection of solid high explosives. *Int J Mass Spectrom* 2010;**289**:58–63.
34. Petersson F, Sulzer P, Mayhew CA, Watts P, Jordan A, Mark L, et al. Real-time trace detection and identification of chemical warfare agent simulants using recent advances in proton transfer reaction time-of-flight mass spectrometry. *Rapid Commun Mass Sp* 2009;**23**: 3875–80.
35. Ruuskanen TM, Mueller M, Schnitzhofer R, Karl T, Graus M, Bamberger I, et al. Eddy covariance VOC emission and deposition fluxes above grassland using PTR-TOF. *Atmos Chem Phys* 2011;**11**:611–25.
36. Titzmann T, Graus M, Muller M, Hansel A, Ostermann A. Improved peak analysis of signals based on counting systems: Illustrated for proton-transfer-reaction time-of-flight mass spectrometry. *Int J Mass Spectrom* 2010;**295**:72–7.

37. Cappellin L, Biasioli F, Granitto PM, Schuhfried E, Soukoulis C, Costa F, et al. On data analysis in PTR-TOF-MS: From raw spectra to data mining Sensor. *Actuat B-Chem* 2011;**155**:183–90.
38. Blake RS, Whyte C, Hughes CO, Ellis AM, Monks PS. Demonstration of proton-transfer reaction time-of-flight mass spectrometry for real-time analysis of trace volatile organic compounds. *Anal Chem* 2004;**76**:3841–5.
39. Bajtarevic A, Ager C, Pienz M, Klieber M, Schwarz K, Ligor M, et al. Noninvasive detection of lung cancer by analysis of exhaled breath BMC. *Cancer* 2009;**9**:348.
40. Beauchamp J, Herbig J, Gutmann R, Hansel A. On the use of Tedlar® bags for breath-gas sampling and analysis. *J Breath Res* 2008;**2**:046001.
41. Munson MSB, Field FH. Chemical Ionization Mass Spectrometry I. General Introduction. *J Am Chem Soc* 1966;**88**:2621–30.
42. Gross JH. *Mass spectrometry*. Berlin: Springer-Verlag; 2004.
43. de Gouw J, Warneke C. Measurements of volatile organic compounds in the earths atmosphere using proton-transfer-reaction mass spectrometry. *Mass Spectrom Rev* 2007;**26**:223–57.
44. Hansel A, Jordan A, Holzinger R, Prazeller P, Vogel W, Lindinger W. Proton-transfer reaction mass-spectrometry - online trace gas-analysis at the ppb level. *Int J Mass Spectrom* 1995;**149**:609–19.
45. Spanel P, Smith D. Selected ion flow tube studies of the reactions of H_3O^+, NO^+, and O_2^+ with several aromatic and aliphatic hydrocarbons. *Int J Mass Spectrom* 1998;**181**:1–10.
46. Wisthaler A, Apel EC, Bossmeyer J, Hansel A, Junkermann W, Koppmann R, et al. Technical Note: Intercomparison of formaldehyde measurements at the atmosphere simulation chamber SAPHIR. *Atmos Chem Phys* 2008;**8**:2189–200.
47. Jordan A, Haidacher S, Hanel G, Hartungen E, Herbig J, Mark L, et al. An online ultra-high sensitivity proton-transfer-reaction mass-spectrometer combined with switchable reagent ion capability (PTR+SRI-MS). *Int J Mass Spectrom* 2009;**286**:32–8.
48. Smith D, Chippendale TWE, Spanel P. Selected ion flow tube, SIFT, studies of the reactions of H_3O^+, NO^+, and O_2^+ with some biologically active isobaric compounds in preparation for SIFT-MS analyses. *Int J Mass Spectrom* 2011;**303**:81–9.
49. Pysanenko A, Spanel P, Smith D. Analysis of the isobaric compounds propanol, acetic acid and methyl formate in humid air and breath by selected ion flow tube mass spectrometry SIFT-MS. *Int J Mass Spectrom* 2009;**285**:42–8.
50. Titzmann T. Automatisierte Methoden zur quantitativen und qualitativen analyse von PTR-TOF-Massenspektren. In *Department for mathematics, informatics and physics*. University of Innsbruck: Innsbruck, Austria; 2010.
51. Cappellin L, Biasioli F, Schuhfried E, Soukoulis C, Mark TD, Gasperi F. Extending the dynamic range of proton transfer reaction time-of-flight mass spectrometers by a novel dead time correction. *Rapid Commun Mass Sp* 2011;**25**:179–83.
52. Cappellin L, Biasioli F, Fabris A, Schuhfried E, Soukoulis C, Mark TD, et al. Improved mass accuracy in PTR-TOF-MS: Another step towards better compound identification in PTR-MS. *Int J Mass Spectrom* 2010;**290**:60–3.
53. Coles JN, Guilhaus M. Resolution limitations from detector pulse-width and jitter in a linear orthogonal-acceleration time-of-flight mass-spectrometer. *J Am Soc Mass Spectr* 1994;**5**:772–8.
54. Inomata S, Tanimoto H. A deuterium-labeling study on the reproduction of hydronium ions in the PTR-MS detection of ethanol. *Int J Mass Spectrom* 2009;**285**:95–9.

55. Libardoni M, Stevens PT, Waite JH, Sacks R. Analysis of human breath samples with a multi-bed sorption trap and comprehensive two-dimensional gas chromatography (GC×GC). *J Chromatogr B Analyt Technol Biomed Life Sci* 2006;**842**:13–21.
56. Kushch I, Schwarz K, Schwentner L, Baumann B, Dzien A, Schmid A, et al. Compounds enhanced in a mass spectrometric profile of smokers' exhaled breath versus non-smokers as determined in a pilot study using PTR-MS. *J Breath Res* 2008;**2**:026002.
57. Muhlberger F, Streibel T, Wieser J, Ulrich A, Zimmermann R. Single photon ionization time-of-flight mass spectrometry with a pulsed electron beam pumped excimer VUV lamp for on-line gas analysis: Setup and first results on cigarette smoke and human breath. *Anal Chem* 2005;**77**:7408–14.
58. Jordan A, Hansel A, Holzinger R, Lindinger W. Acetonitrile and benzene in the breath of smokers and nonsmokers investigated by proton-transfer reaction mass-spectrometry (PTR-MS). *Int J Mass Spectrom* 1995;**148**:L1–3.
59. Talhout R, Schulz T, Florek E, van Benthem J, Wester P, Opperhuizen A. Hazardous compounds in tobacco smoke. *Int J Env Res Pub He* 2011;**8**:613–28.
60. Van Berkel JJBN, Dalling JW, Moeller GM, Godschalk RWL, Moonen E, Wouters EFM, et al. Development of accurate classification method based on the analysis of volatile organic compounds from human exhaled air. *J Chromatogr B* 2008;**861**:101–7.
61. Hoffmann D, Hoffmann I, El-Bayoumy K. The less harmful cigarette: A controversial issue A tribute to Ernst L. Wynder. *Chem Res Toxicol* 2001;**14**:767–90.
62. Kushch I, Arendacka B, Stolc S, Mochalski P, Filipiak W, Schwarz K, et al. Breath isoprene-aspects of normal physiology related to age, gender and cholesterol profile as determined in a proton transfer reaction mass spectrometry study. *Clin Chem Lab Med* 2008;**46**:1011–8.
63. King J, Mochalski P, Kupferthaler A, Unterkofler K, Koc H, Filipiak W, et al. Dynamic profiles of volatile organic compounds in exhaled breath as determined by a coupled PTR-MS/GC-MS study. *Physiol Meas* 2010;**31**:1169–84.

CHAPTER

7 Exhaled Breath Analysis in Occupational Medicine

Massimo Corradi and Antonio Mutti

Department of Clinical and Experimental Medicine, University of Parma, Via Gramsci 14, I-43126 Parma, Italy

7.1 RATIONALE FOR THE USE OF EXHALED BREATH ANALYSIS IN OCCUPATIONAL MEDICINE

Respiratory diseases are among the main causes of worldwide mortality and their incidence is increasing: the World Health Organization calculates that in 2020 these pathologies will represent the third most common cause of death in the world.[1] Cigarette smoke represents the most common worldwide cause of disease and every year provokes 5.4 million of deaths in the world,[2] but occupational exposure to pneumotoxic substances, such as gas or vapor, aerosols, smoke, and dust, represents the second most relevant risk factor.[3] Studies in several European countries (including Italy), the United States, and Canada have demonstrated that the percentage of the adult population at risk of respiratory diseases due to occupational causes ranges from 10% to 25%.[4] Bronchial asthma approaches a prevalence of 36.2% in the field of intermediation, chartering, and in other activities, and 31% in agriculture.[4] Occupational diseases of the respiratory tract as reported to the Italian National Institute for insurance against industrial injury in 2007 numbered 1461 in industry and services and 94 in agriculture (as compared to only 65 cases in 2003).[5]

Within the health protection of workers, international laws attribute a key role to occupational health physicians in both primary and secondary prevention of respiratory injury induced by airborne pollutants, such a role being exploited through health surveillance and biological monitoring.[6] Health surveillance classically comprehends the strategies and methods to detect and systematically assess adverse effects of work on the health of workers.[7] Medical examination performed during health surveillance of workers at risk of occupational respiratory diseases represents an extraordinary and often unique opportunity to collect health information. In fact, the majority of occupational respiratory diseases show a slow and asymptomatic onset; thus the worker does not feel the need to refer to a doctor in order to estimate his/her respiratory clinical condition.[8]

Currently, functional tests (e.g. the spirometry) represent the only diagnostic approach to assess pulmonary status of workers exposed to airborne pollutants during health surveillance.[9] However, besides the need for subject collaboration and competent operator performance, spirometry can detect only late, irreversible alterations,[10]

Volatile Biomarkers. http://dx.doi.org/10.1016/B978-0-44-462613-4.00007-6

and its role as a diagnostic tool to identify early signs of pulmonary damage is contentious.[11] The key aspect of secondary prevention is, indeed, the identification of functional or structural changes at early and still reversible stages, in order to develop appropriate ambient or personal preventive strategies (such as personal safety procedures, change of duty). Furthermore, spirometry is neither sufficiently informative about lung pathobiology nor able to distinguish possible damages induced by exposure to occupational pollutants.[12] This causes strong practical and interpretative problems, also with the employer. The occurrence of false positive tests in healthy subjects may lead to further difficulties in data interpretation.[13] Whereas the validity of spirometry as a tool for health surveillance in subjects at risk of occupational respiratory pathologies has not been sufficiently demonstrated, its low sensitivity has been demonstrated. For example, Brant et al. showed that in bakers, health surveillance performed relying on medical history and spirometry leads to a remarkable underestimation of the actual number of subjects with work-related respiratory problems.[11]

The second task, often carried out in parallel with health surveillance, is biological monitoring, which is the quantification of potentially toxic substances or their metabolites in accessible tissues and fluids in order to assess levels of exposure and to identify the possible need for action, e.g. removal from exposure, adoption of protective devices or hygienic intervention.[7] Traditionally, biological monitoring is performed using blood and urine samples, which are representative of internal dose and effects at the systemic level. Therefore, they cannot reflect pathogenetic events at the respiratory level. This is mainly true for those substances that, because of their strong reactivity or poor solubility, act at the site of entry, in this case the respiratory apparatus.[14] The quantification of target tissue dose in association with respiratory biomarkers of effect and susceptibility could add important and more specific information in comparison with traditional biological monitoring.[11] In this way, it would be also possible to overcome the limits derived from functional tests.[15] The use of biological markers sampled at target tissue would therefore represent a remarkable step ahead in health surveillance and biological monitoring of workers at risk of occupational respiratory diseases.

Many approaches to directly evaluation of lung inflammation in diseases have been proposed, including bronchial biopsy, bronchial lavage, or bronchoalveolar lavage.[16] These methods are quite invasive and are mainly performed in clinical settings. In the last few years, less invasive techniques such as induced sputum analysis and exhaled breath analysis[17,18] have been successfully introduced in the evaluation of lung diseases. Induced sputum, which is obtained through the inhalation of hypertonic solution, can give direct information on the kind and degree of bronchial inflammation and has brought a better understanding of the role of inflammatory mediators found in the airways of asthmatic patients. A good correlation between inflammatory mediator levels and clinical and functional findings was found in asthmatic subjects.[19,20] The evaluation of cells and mediators in induced sputum can help to better characterize asthmatic patients, in addition to clinicald and functional findings.[17]

Exhaled NO measurement is another technique widely adopted in monitoring asthmatic inflammation. NO is a biological transmitter involved in airways and blood vascular tone regulation; it is produced by many cells within airways, both resident and recruited during the inflammatory process. Its physiopathology is still unclear, but it has been demonstrated that NO levels in exhaled air are higher in asthmatic than in healthy subjects.[21] Exhaled NO levels have been shown to increase during spontaneous or induced asthma exacerbation, and decrease after anti-inflammatory treatment.[19] There are very few published data on the use of exhaled NO in health surveillance programs.[20]

Exhaled air can also be used to provide additional information. In particular, it is now possible to analyze the composition of the exhaled breath condensate (EBC), which is a matrix obtained by cooling exhaled breath under normal breathing. This represents a very promising approach to obtaining information on the status of the airways. In fact, it is possible to identify markers of exposure, effect, and susceptibility in EBC and therefore to monitor pulmonary status as a consequence of the exposure to chemical compounds, such as metallic elements contained in cigarette smoke and polluted environments.[22] EBC collection is completely non-invasive and easily carried out at the workplace and at home, and repeated measurements can be done under controlled conditions. Several molecules have been found in the EBC, such as hydrogen peroxide, nitrite, nitrate, prostaglandins, leukotrienes, aldehydes, metals, etc.[23]

This chapter focuses on the main applications of exhale breath analysis in occupational medicine.

7.2 EXHALED NITRIC OXIDE

Nitric oxide (NO) is an endogenous, soluble gas whose highly reactive molecules are involved in many biological and pathophysiological organic processes. It is generated from the oxidation of L-arginine to L-citrulline by nitric oxide synthase (NOS).[24] The type II, inductible NOS (iNOS), is released by many cells in the lung (macrophages, fibroblasts, neutrophils, and epithelial cells).[25] Expression of iNOS is upregulated by interferon γ, tumor necrosis factor α, and interleukin 1β, all cytokines known to be active in airway inflammation. These cytokines are suspected to be source of airway NO in subjects with asthma.[26] In the clinical setting, measurements of fractional exhaled nitric oxide (F_ENO) are suggested as a sensitive measurement of airway inflammation[19] and as a diagnostic tool for identification of asthma.[27] Several studies have demonstrated that levels of F_ENO were significantly elevated in asthma compared with nonasthmatic subjects.[26,27] With more than 900 publications, F_ENO is widely accepted as a non-invasive marker of airway inflammation for research. American Thoracic Society (ATS) standardized techniques for measuring F_ENO; the technique of measurement is the following: the subject inhales to total lung capacity and then exhales at a constant flow rate of 50 mL/s. A constant flow is essential, as F_ENO levels are flow dependent. After a few seconds, a steady NO concentration plateau is reached and this is taken as the F_ENO value. Repeated exhalations are performed to obtain two or three reproducible values.[28] The advantage of this technique is its non-invasiveness, simplicity, possibility of equipment transport to the home, which allows measurements with sufficient accuracy and reproducibility, and easy diagnostic procedure also in children.

7.2.1 Exhaled NO as diagnostic tool in occupational respiratory diseases

F_ENO is generally accepted as a diagnostic tool for asthma, but to date, there are only limited number of studies that examined the role of F_ENO in occupational asthma.[20] The use of F_ENO can be a useful source of information in the interpretation of occupational challenge tests in cases with borderline functional changes.[29] The study of Piipari et al.[30] used F_ENO measurements to assess the asthmatic airway inflammation during specific bronchial challenge tests using occupational agents in patients with suspected occupational asthma. In patients with normal or slightly increased basal NO level, a significant increase in F_ENO was seen, but patients with high basal NO level and significant bronchoconstriction did not show a significant NO elevation. The results indicate that F_ENO measurements give useful additive information for the interpretation of occupational challenge tests in cases with borderline functional changes, increasing the validity of the tests. The investigation of Menzies et al.[31] that examined the markers of inflammation using F_ENO in bar workers before and after a legislative ban on smoking in public places has demonstrated that asthmatic bar workers had less airway inflammation, with reduction in F_ENO after the ban. The study of Olin et al.[32] investigated whether exposure to high peaks of ozone resulted in an increased prevalence of asthma or respiratory symptoms among bleacher workers and whether NO was elevated in the exhaled air of these workers. They had a higher median concentration of F_ENO in comparison with controls, and the results of this study confirmed that workers who have been repeatedly exposed to gassings have an increased prevalence of adult-onset asthma, although these authors did not find significant difference in NO between exposed subjects and controls in the previous study.[33] Lund et al.[34] have found an increased NO in exhaled air in non-smoking aluminum pot room workers in comparison with controls employed at the same plant but working outside of the pot room atmosphere. The authors conclude that NO can be used as an early marker of asthma in aluminum pot room workers. Other recently published studies, applied in occupational settings, have demonstrated an increase in F_ENO in work-related isocyanate exposure,[35] in latex allergen exposure in symptomatic healthcare workers,[36] in exposure to the organic solvents toluene, xylene, and methyl ethyl ketone in shoe and leather workers.[36] The study of Lehtonen et al.[37] showed a significant increase of F_ENO levels in patients with asbestosis. On the other hand, the investigation of Tan et al.[38] reported results that demonstrate a decrease of F_ENO levels in occupational latex exposure in healthcare workers. F_ENO was measured before and after a controlled latex challenge at the beginning and end of the working week, F_ENO did not show a clear correlation with routine workplace exposure. A decrease of F_ENO was also reported in the study of Akpinar-Elci et al.[39] who found significantly lower F_ENO levels in a high-exposure group of workers in a microwave popcorn production plant.

In conclusion, many investigators have found that F_ENO rises and lung function deteriorates after specific occupational challenge tests are valuable parameters to differentiate between occupational and non-occupational airway disorders. Naturally, F_ENO is influenced by many non-occupational agents. However, the percentage of F_ENO increase was found to be a more appropriate diagnostic tool than absolute F_ENO values.

7.3 EXHALED BREATH CONDENSATE (EBC)

EBC allows the assessment of the lung pathobiology, because there is strong evidence that abnormality in EBC composition may reflect biochemical changes of airway lining fluid.[40–42] The EBC is essentially formed from water vapor and the group of elements that are expired in the form of bio-aerosols. All the regulation mechanisms that form EBC are not yet known; however, there is an accepted hypothesis that liquid drops from the respiratory airways surfaces are released and transported by the exhaled air flow and consequently can be collected in the condenser. The air presented in the airway has a constant temperature of about 37 °C, relatively independent of the temperature of inhaled air. By contact with a superficies of a low temperature the exhaled water vapor condensate passes from the liquid phase; exploiting this physical phenomenon it is possible to collect the substances present in the exhaled air.[42] The systems reported in the literature generally collect 1 mL of EBC in 5–15 min.[42] The factors that determine the number of formalized aerosol have not yet been studied in detail, even if there is a considerable probability that equilibrium between the surface tension, the velocity, and the humidity of inhaled air, and changes of exhaled turbulence play an important role.[42] Although predominantly derived from water vapor, EBC has dissolved within it non-volatile compounds, including cytokines, lipids, surfactants, ions, oxidation products, adenosine, histamine, acetylcholine, and serotonin. In addition, ECB traps potentially volatile water-soluble compounds, including ammonia, hydrogen peroxide, nitrogen oxides (NOx), ethanol, and other volatile organic compounds. EBC has a readily measurable pH.[43] EBC collection is suitable to assess biomarkers of local inflammation and oxidative stress, which may be sensitive endpoints for identifying early biochemical changes in the airways that occur in exposed subjects. The assays have provided evidence of airway and lung redox deviation, acid-base status, and the degree and type of inflammation in acute and chronic asthma, chronic obstructive pulmonary disease, adult respiratory distress syndrome, occupational diseases, and cystic fibrosis.[42]

The American Thoracic Society (ATS)/European Respiratory Society (ERS) has developed the guidelines for EBC collection and measurement of exhaled biomarkers, and has collaborated on publishing recommendations, which summarize current knowledge, including optimal collection techniques.[22] The general recommendations for oral EBC collection are available: collection should occur during tidal breathing in the sitting position using a nose clip and a saliva trap, at a defined cooling temperature and collection time (temperature and time can vary depending on the study objective); surfaces contacting the EBC should be inert to the compounds of interest in a given study; inclusion of expiratory flow resistance of filters is not required. EBC samples should be immediately frozen after collection and stored at −70 °C.[8] As ECB mainly consists of water that is practically free of potentially interfering solutes, it is an ideal biological fluid for elemental determinations based on relatively common techniques, such as electrothermal atomic absorption spectroscopy (ETAAS), or less frequently available reference techniques such as inductively coupled plasma-mass spectrometry (ICP-MS).[44]

Collection of EBC is a completely non-invasive method of sampling the respiratory tract that can be repeated several times with short intervals between sampling. Collection devices can be portable and can be used in a variety of settings, including intensive care units, outpatient clinics, workplaces, and at home.[45] The interest derives in part from the promise that EBC will provide a substrate by which inflammatory and biochemical components of lung disease can be non-invasively assessed and that it therefore could substantially improve on spirometry as a non-invasive tool for diagnosis and management.[46] Through use of appropriate equipment, samples have been collected at worksites, schools, hospitals, and clinics as well as in homes.

To date, there are only a few head-to-head comparisons of EBC assays with other methods of assessing airway biochemistry and inflammation, such as exhaled NO, BAL, IS, and biopsy. In general, there does appear to be correlation of EBC assays with other assays of lung inflammation. It must be kept in mind that the ionic composition of the human airway lining fluid in health is not known, to say nothing of disease, so there is as yet no available gold standard by which to make comparisons.[42]

7.3.1 EBC in occupational settings

Most publications related to the use of the EBC as a diagnostic tool have targeted non-occupational diseases, mostly asthma and COPD.[42] These studies have provided clear evidence that breath analysis may have an important place in clinical management of asthma and COPD.

Considering that EBC is mostly water practically free of interfering solutes, it represents an ideal biological matrix for elemental characterization. Published data show that several toxic metals and trace elements are detectable in EBC, raising the possibility of using this medium to quantify the lung tissue dose of pneumotoxic substances occurring in occupational settings. This novel approach may represent a significant advance over the analysis of alternative media (blood, serum, urine, hair), which are not as reliable and reflect systemic rather than lung (target tissue) levels of both toxic metals and essential trace elements. Several recently published studies demonstrate that EBC can be used assessing the target tissue dose of pneumotoxic substances from polluted workplaces. Goldoni et al.[44] have showed that cobalt (Co) and tungsten (W) can be measured in the EBC of occupational exposed workers and thus suggest the potential use of this matrix as a novel approach to monitor target tissue dose and effects occurring in the respiratory tract following exposure to pneumotoxic substances. The concentration of malondialdehyde (MDA) in EBC has been used as a biomarker of pulmonary oxidative stress. MDA levels increased depending on Co concentration and were enhanced by co-exposure to W. Such correlation between EBC MDA and both Co and W levels was not observed in urinary concentration of either element.

Another study of Goldoni et al.[45] followed the kinetics of Cr(VI) in chrome plating workers using EBC samples. Kinetic data showed that airborne Cr(VI) was reduced by 50% in airway lining fluid sampled at the end of exposure and that there was a further 50% reduction after 15 h. The persistence of Cr(VI) in EBC supports the use of EBC in assessing target tissue levels of Cr(VI). Caglieri et al.[46] investigated Cr levels in EBC of workers exposed to Cr(VI) and assessed their relationship with biochemical changes in the airways by analyzing EBC biomarkers of oxidative stress, namely, hydrogen peroxide (H_2O_2) and MDA. Cr-EBC levels increased from the beginning to the end of Friday but were considerably lower on Monday morning. A similar trend was observed for $H_2O_2^-$ EBC levels and MDA-EBC levels. All inscribed levels correlated with each other as well as with urinary Cr levels. The results of this study demonstrate that EBC is a suitable matrix that can be used to investigate both Cr levels and biomarkers of free radical production because it may represent the epithelial-lining fluid of workers exposed to Cr(VI). Another study used EBC as a tool for a detection of toluene concentration in healthy subjects after a short period of exposure (20 min). EBC samples were collected before and at the end of the exposure, while the environmental concentration of toluene was continuously monitored. Before the exposure toluene concentrations in the EBC were below the detectable limit in all subjects, while after the exposure toluene was detectable in all EBC samples. A significant correlation was found between the environmental toluene levels and toluene in the EBC. Furthermore, a significant relationship between increased EBC toluene levels and urinary hippuric acid after the exposure was found. The authors of this study conclude that EBC is a promising matrix for toluene assessment, although its application in humans requires further investigations.[36] The review of Chapman et al.[47] highlights the use of non-invasive methods, such as analysis of exhaled breath biomarkers, which could potentially be extremely useful in patients with asbestos-related lung diseases.

7.4 EXHALED VOLATILE ORGANIC COMPOUNDS

The term volatile organic compounds (VOCs) refers to organic compounds that are in the vapor phase at room temperature.[48] Many VOCs are the chemicals used to manufacture and prepare many building materials, interior furnishings, textiles, office equipment, cleaners, personal care supplies, and pesticides. That is why they are a source of pollution and of concern for indoor air. Some VOCs are present in exhaled breath and so breath analysis has been used to assess VOC exposure.[48,49]

There has been substantial progress in sampling techniques over the last decades. In 1980, an apparatus composed by the Rahn-Otis valve connected to a gasometer and to a pump equipped with activated charcoal tubes was used to investigate respiratory absorption and excretion of hexane and cyclohexane in shoe factory workers.[50] Such an apparatus represented substantial progress over techniques relying on poorly standardized methods based on the use of glass tubes and manual sampling procedures, but was difficult to assemble, cumbersome, and not applicable to routine measurements in occupational settings. Moreover, air collection required at least a few minutes (the flow rate of the pump could not exceed 200 mL/min), thus confining its use to research based on carefully standardized protocols. It is worth mentioning that the rapid phase of excretion of most tested solvents ranges from 5 to 10 min, which makes sampling procedures critical. Just a few years ago, a simple device (BioVoc®) was developed to address the same biomonitoring techniques at the UK Health and Safety Executive (HSE),[51] thus making collection of alveolar air at the workplace feasible and meaningful. An improvement to the HSE technique was proposed in our laboratory by Poli et al.[52] who used solid-phase microextraction (SPME) instead of the cartridges proposed by HSE, which are potentially contaminated by the pollutants present in ambient air.

7.4.1 Exhaled volatile organic compounds in occupational medicine

Breath sampling and analysis is superficially an attractive technique for biological monitoring. It has been used widely over the past 20 years in studies in the occupational and environmental area primarily as a research tool.[53] However, it has not yet found its way into routine biological monitoring programs, possibly because there are no commercially available breath samplers or because the data generated are difficult to interpret. Recent technical developments[54,55] would tend to indicate the imminent use of routine breath analysis for biological monitoring of workers exposed to solvents. This view is further strengthened by the need to move away from blood sampling toward non-invasive biological monitoring procedures and the decision by some authorities to set biological monitoring guidance values.

Breath analysis will be readily accepted for routine screening when it is reliable, robust, and inexpensive. Such is the case with measurements of alcohol in breath when they are used for evidential purposes. In the case of alcohol, we are dealing with a single substance at a relatively high concentration and for which there are suitable analytical methods. Also, there are well-validated data sets which assist with the interpretation of results. In the occupational and environmental setting the position is not so clear. There are numerous pollutants on people's breath and the concentrations are orders-of-magnitude less than those of alcohol. Although the relevant toxicokinetic data of many volatile compounds are well known, the relations between the breath concentration, the time after exposure, and the exposure conditions are not always well established and the significance of individual variability has not yet been studied sufficiently.

Instruments for direct and on-line breath analysis based on mass spectrometry are now available,[56] but they are bulky and relatively expensive. Further developments of microsensor arrays or immunoassay technology may possibly result in cheaper direct reading options.

References

1. *Health for All Policy Framework for the WHO European Region: 2005 Update.* Copenhagen: WHO Regional Office for Europe; 2006.
2. *Global Initiative for Chronic Obstructive Lung Disease. Global strategy for the diagnosis, management, and prevention of chronic obstructive pulmonary disease*; 2009. http://www.goldcopd.org; [accessed 28.02.2012].
3. Balmes J, Becklake M, Blanc P, Henneberger P, Kreiss K, Mapp C, et al. American thoracic society statement: occupational contribution to the burden of airway disease. *Am J Respir Crit Care Med* 2003;**167**:787–97.
4. Kogevinas M, Anto JM, Sunyer J, Tobias A, Kromhout H, Burney P. Occupational asthma in Europe and other industrialised areas: a population-based study. European community respiratory health survey study group. *Lancet* 1999;**353**:1750–4.
5. Inail: Istituto Nazionale Per L'asscurazione Contro Gli Infortuni Sul Lavoro. http://www.inail.it; [accessed 28.02.2012].
6. Marinaccio A, Di Marzio D, Binazzi A, Napoli G, Businelli A, Crema M, et al. The surveillance of occupational risks. Problems and perspectives of the recent legislative framework after the national preliminary survey. *Med Lav* 2010;**101**:252–61.
7. Ayres JG, Boyd R, Cowie H, Hurley JF. Costs of occupational asthma in the UK. *Thorax* 2011;**66**:128–33.
8. Erdogan MS, Martin CJ. Miscellaneous occupational lung diseases. *Arch Environ Occup Health* 2008;**63**:93–6.
9. Eisner MD, Iribarren C, Yelin EH, Sidney S, Katz PP, Ackerson L, et al. Pulmonary function and the risk of functional limitation in chronic obstructive pulmonary disease. *Am J Epidemiol* 2008;**167**:1090–101.
10. Derom E, van Weel C, Liistro G, Buffels J, Schermer T, Lammers E, et al. Primary care spirometry. *Eur Respir J* 2008;**31**:197–203.
11. Brant A, Nightingale S, Berriman J, Sharp C, Welch J, Newman Taylor AJ, et al. Supermarket baker's asthma: how accurate is routine health surveillance? *Occup Environ Med* 2005;**62**:395–9.
12. Rushton L. Occupational causes of chronic obstructive pulmonary disease. *Rev Environ Health* 2007;**22**:195–212.
13. Vedal S, Crapo RO. False positive rates of multiple pulmonary function tests in healthy subjects. *Bull Eur Physiopathol Respir* 1983;**19**:263–6.
14. Mutti A, Corradi M. Recent developments in human biomonitoring: non-invasive assessment of target tissue dose and effects of pneumotoxic metals. *Med Lav* 2006;**97**:199–206.
15. Do R, Bartlett KH, Dimich-Ward H, Chu W, Kennedy SM. Biomarkers of airway acidity and oxidative stress in exhaled breath condensate from grain workers. *Am J Respir Crit Care Med* 2008;**178**:1048–54.
16. Zhang JY, Wenzel SE. Tissue and BAL based biomarkers in asthma. *Immunol Allergy Clin North Am* 2007;**27**:623–32.
17. Fireman E. Induced sputum as a diagnostic tactic in pulmonary diseases. *Isr Med Assoc J* 2003;**5**:524–7.
18. Gerritsen WB, Asin J, Zanen P, van den Bosch JM, Haas FJ. Markers of inflammation and oxidative stress in exacerbated chronic obstructive pulmonary disease patients. *Respir Med* 2005;**99**:84–90.

19. Ricciardolo FL, Sterk PJ, Gaston B, Folkerts G. Nitric oxide in health and disease of the respiratory system. *Physiol Rev* 2004;**84**:731–65.
20. Lemiere C. Induced sputum and exhaled nitric oxide as noninvasive markers of airway inflammation from work exposures. *Curr Opin Allergy Clin Immunol* 2007;**7**:133–7.
21. Corradi M, Montuschi P, Donnelly LE, Pesci A, Kharitonov SA, Barnes PJ. Increased nitrosothiols in exhaled breath condensate in inflammatory airway diseases. *Am J Respir Crit Care Med* 2001;**163**:854–8.
22. Horvath I, Hunt J, Barnes PJ, Alving K, Antczak A, Baraldi E, et al. Exhaled breath condensate: methodological recommendations and unresolved questions. *Eur Respir J* 2005;**26**:523–48.
23. Malerba M, Ragnoli B, Corradi M. Non-invasive methods to assess biomarkers of exposure and early stage of pulmonary disease in smoking subjects. *Monaldi Arch Chest Dis* 2008;**69**:128–33.
24. Redington AE, Meng QH, Springall DR, Evans TJ, Creminon C, Maclouf J, et al. Increased expression of inducible nitric oxide synthase and cyclo-oxygenase-2 in the airway epithelium of asthmatic subjects and regulation by corticosteroid treatment. *Thorax* 2001;**56**: 351–7.
25. Ricciardolo FL, Sterk PJ, Gaston B, Folkerts G. Nitric oxide in health and disease of the respiratory system. *Physiol Rev* 2004;**84**:731–65.
26. Yates DH. Role of exhaled nitric oxide in asthma. *Immunol Cell Biol* 2001;**79**:178–90.
27. Berkman N, Avital A, Breuer R, Bardach E, Springer C, Godfrey S. Exhaled nitric oxide in the diagnosis of asthma: comparison with bronchial provocation tests. *Thorax* 2005;**60**:383–8.
28. Silkoff P. Exhaled nitric oxide as a diagnostic tool. *Am J Respir Crit Care Med* 2003;**167**:665–6; author reply 665–6.
29. Swierczynska-Machura D, Krakowiak A, Wiszniewska M, Palczynski C. Diagnostics of occupational asthma: measurements of exhaled nitric oxide. *Med Pr* 2005;**56**:485–90.
30. Piipari R, Piirila P, Keskinen H, Tuppurainen M, Sovijarvi A, Nordman H. Exhaled nitric oxide in specific challenge tests to assess occupational asthma. *Eur Respir J* 2002;**20**: 1532–7.
31. Menzies D, Nair A, Williamson PA, Schembri S, Al-Khairalla MZ, Barnes M, et al. Respiratory symptoms, pulmonary function, and markers of inflammation among bar workers before and after a legislative ban on smoking in public places. *JAMA* 2006;**296**:1742–8.
32. Olin AC, Andersson E, Andersson M, Granung G, Hagberg S, Toren K. Prevalence of asthma and exhaled nitric oxide are increased in bleachery workers exposed to ozone. *Eur Respir J* 2004;**23**:87–92.
33. Olin AC, Ljungkvist G, Bake B, Hagberg S, Henriksson L, Toren K. Exhaled nitric oxide among pulpmill workers reporting gassing incidents involving ozone and chlorine dioxide. *Eur Respir J* 1999;**14**:828–31.
34. Lund MB, Oksne PI, Hamre R, Kongerud J. Increased nitric oxide in exhaled air: an early marker of asthma in non-smoking aluminium potroom workers? *Occup Environ Med* 2000;**57**:274–8.
35. Barbinova L, Baur X. Increase in exhaled nitric oxide (eNO) after work-related isocyanate exposure. *Int Arch Occup Environ Health* 2006;**79**:387–95.
36. Maniscalco M, De Laurentiis G, Pentella C, Mormile M, Sanduzzi A, Carratu P, et al. Exhaled breath condensate as matrix for toluene detection: a preliminary study. *Biomarkers* 2006;**11**:233–40.

37. Lehtonen H, Oksa P, Lehtimaki L, Sepponen A, Nieminen R, Kankaanranta H, et al. Increased alveolar nitric oxide concentration and high levels of leukotriene B(4) and 8-isoprostane in exhaled breath condensate in patients with asbestosis. *Thorax* 2007;**62**: 602–7.
38. Tan K, Bruce C, Birkhead A, Thomas PS. Nasal and exhaled nitric oxide in response to occupational latex exposure. *Allergy* 2001;**56**:627–32.
39. Akpinar-Elci M, Stemple KJ, Elci OC, Dweik RA, Kreiss K, Enright PL. Exhaled nitric oxide measurement in workers in a microwave popcorn production plant. *Int J Occup Environ Health* 2006;**12**:106–10.
40. Corradi M, Zinelli C, Caffarelli C. Exhaled breath biomarkers in asthmatic children. *Inflamm Allergy Drug Targets* 2007;**6**:150–9.
41. Mutti A, Corradi M, Goldoni M, Vettori MV, Bernard A, Apostoli P. Exhaled metallic elements and serum pneumoproteins in asymptomatic smokers and patients with COPD or asthma. *Chest* 2006;**129**:1288–97.
42. Hunt J. Exhaled breath condensate: an evolving tool for noninvasive evaluation of lung disease. *J Allergy Clin Immunol* 2002;**110**:28–34.
43. Montuschi P, Kharitonov SA, Ciabattoni G, Barnes PJ. Exhaled leukotrienes and prostaglandins in COPD. *Thorax* 2003;**58**:585–8.
44. Goldoni M, Catalani S, De Palma G, Manini P, Acampa O, Corradi M, et al. Exhaled breath condensate as a suitable matrix to assess lung dose and effects in workers exposed to cobalt and tungsten. *Environ Health Perspect* 2004;**112**:1293–8.
45. Goldoni M, Caglieri A, Poli D, Vettori MV, Corradi M, Apostoli P, et al. Determination of hexavalent chromium in exhaled breath condensate and environmental air among chrome plating workers. *Anal Chim Acta* 2006;**562**:229–35.
46. Caglieri A, Goldoni M, Acampa O, Andreoli R, Vettori MV, Corradi M, et al. The effect of inhaled chromium on different exhaled breath condensate biomarkers among chrome-plating workers. *Environ Health Perspect* 2006;**114**:542–6.
47. Chapman EA, Thomas PS, Yates DH. Breath analysis in asbestos-related disorders: a review of the literature and potential future applications. *J Breath Res* 2010;4:034001.
48. Gordon SM, Szidon JP, Krotoszynski BK, Gibbons RD, O'Neill HJ. Volatile organic compounds in exhaled air from patients with lung cancer. *Clin Chem* 1985;**31**:1278–82.
49. Delfino RJ, Gong H, Linn WS, Hu Y, Pellizzari ED. Respiratory symptoms and peak expiratory flow in children with asthma in relation to volatile organic compounds in exhaled breath and ambient air. *J Expo Anal Environ Epidemiol* 2003;**13**:348–63.
50. Mutti A, Falzoi M, Lucertini S, Cavatorta A, Franchini I, Pedroni C. Absorption and alveolar excretion of cyclohexane in workers in a shoe factory. *J Appl Toxicol* 1981;**1**:220–3.
51. *Health and Safety Executive (HSE). Occupational exposure limits 1996: Summary Criteria. EH40 ? 96*. 1996, Sudbury: HSE Books.
52. Poli D, Carbognani P, Corradi M, Goldoni M, Acampa O, Balbi B, et al. Exhaled volatile organic compounds in patients with non-small cell lung cancer: cross sectional and nested short-term follow-up study. *Respir Res* 2005;**6**:71.
53. Wilson HK, Monster AC. New technologies in the use of exhaled breath analysis for biological monitoring. *Occup Environ Med* 1999;**56**:753–7 .
54. Silva LI, Freitas AC, Rocha-Santos TA, Pereira ME, Duarte AC. Breath analysis by optical fiber sensor for the determination of exhaled organic compounds with a view to diagnostics. *Talanta* 2011;**83**:1586–94.

55. Wu Y, Wen F, Liu D, Kong H, Zhang C, Zhang S. Analysis of 2-propanol in exhaled breath using in situ enrichment and cataluminescence detection. *Luminescence* 2011;**26**:125–9.
56. Wang CJ, Yang NH, Liou SH, Lee HL. Fast quantification of the exhaled breath condensate of oxidative stress 8-iso-prostaglandin F2alpha using on-line solid-phase extraction coupled with liquid chromatography/electrospray ionization mass spectrometry. *Talanta* 2010;**82**:1434–8.

CHAPTER

8 Volatile Organic Compounds in Human Breath: Biogenic Origin and Point-of-Care Analysis Approaches

Alexander A. Aksenov[*], Michael Schivo[†,‡], Hamzeh Bardaweel[*], Yuriy Zrodnikov[*], Alice M. Kwan[*], Konstantin Zamuruyev[*], William H.K. Cheung[*], Daniel J. Peirano[*], and Cristina E. Davis[*,‡]

[*]*Department of Mechanical and Aerospace Engineering, One Shields Avenue, University of California, Davis, CA 95616, USA*

[†]*Center for Comparative Respiratory Biology and Medicine, 451 Health Sciences Drive, #6517, Davis, CA 95616, USA*

[‡]*Clinical and Translational Sciences Center (CTSC), UC Davis Medical Center, Sacramento, CA 95817, USA*

8.1 BIOGENIC ORIGIN OF VOLATILE COMPOUNDS IN HUMAN EXHALED BREATH

Exhaled human breath is a complex mixture of volatile and non-volatile compounds that are produced when inhaled ambient air is expelled out of the lungs through bronchi, trachea, and nasal and oral cavities. The composition of ambient air will change when it is inhaled as it becomes enriched in carbon dioxide and depleted of oxygen by ~5%, and also by trapping droplets from oral/nasal mucosa of the epithelial lining. Other than air, the main components of exhaled breath are water vapor, hydrogen, ammonia, carbon monoxide, methane—with the latter three in the parts-per-million (ppm) range, e.g. methane concentration can be as high as 30 ppm.[1] Exhaled breath also contains trace amounts of volatile organic compounds (VOCs) such as of acetone, methanol, and ethanol, typically 100s of ppb and many other small metabolite compounds—mainly in the parts-per-billion (ppb) or parts-per-trillion (ppt) range. To understand the complex field of VOC analysis of human breath, it is necessary to identify the relevant lung structures which may contribute to a VOC mixture or metabolite signature of specific disorders. The human lung contains 1800 mL total tissue/cellular volume which can contain another 2400 mL of gas, and the relative contributions of the different compartments are: alveolar walls 250–300 mL, epithelium 60–80 mL, endothelium 50–70 mL, interstitium 100–185 mL, alveolar

Volatile Biomarkers. http://dx.doi.org/10.1016/B978-0-44-462613-4.00008-8

macrophages 50 mL, supportive tissues 250 mL, and the blood capillary system 400 mL. Surrounding tissues comprise about 900 mL of volume.[2] By necessity, VOCs must enter the airway compartment to be sampled in exhaled breath. Where, exactly, VOCs enter this compartment is an active area of research at this time. Some trace compounds can be attributed to oral origin,[3] e.g. ammonia that is largely generated in the oral cavity by bacterial action on salivary urea.[4]

The alveolar interface is at the distal end of the lung where poorly oxygenated and CO_2-rich blood enters the pulmonary arterial capillaries surrounding an alveolus. The interface between the capillary endothelium and type-I alveolar cells is extraordinarily thin (0.1 μm), thus facilitating gas transfer. Theoretically, systemic VOC production in distant parts of the body results in VOCs entering the blood stream at a distant organ where it is transported through the body. It is then speculated that the VOCs transfer from circulating blood to the alveolus and then exhaled in the breath, though evidence for this chain of events has often been indirect. A study assessing 54 women with biopsy-proven breast cancer compared to 204 control subjects found significant and reproducible discernment between groups based on breath VOCs.[5] Another group was able to distinguish between subjects with breast, lung, colon cancer, and healthy controls using single-array nanosensors applied to detect these types of compounds in exhaled breath.[6] Though these studies did not identify the specific chemical structures or origin of the VOCs, it is plausible that they were introduced into the alveolar space from the peripheral vasculature. The alveolus itself is comprised of type I and type II pneumocytes; the latter demonstrate plasticity and, hence, probable VOC production in times of stress and repair.[7,8] Continuing along the lung structure, beyond the alveoli are the small airways called bronchioles. These are comprised of several cell types including bronchial epithelial cells, Clara cells, and occasionally populations of subepithelial lymphocytes which may expand in response to infection or inflammation.[9] Damage to airway epithelial cells occurs in lung diseases such as asthma, chronic obstructive pulmonary disease, and bronchiectasis, and certain volatiles may be present as a result. One of the most widely studied and now best established volatile compound that is routinely analyzed is exhaled nitric oxide (eNO).[10] Nitric oxide is a potent vaso- and bronchodilator, and airway epithelial inflammation that is observed during an acute asthma exacerbation has been linked to increased eNO production through inducible NO synthase expression.[11] Cell culture studies have demonstrated increased NO production in small airway epithelial cells exposed to common inflammatory cytokines (IL1-beta, TNF-alpha, and IFN-gamma), and the response is more robust than in proximal cultured human tracheal cells.[12] Fractional eNO (F_ENO) levels measured in exhaled breath correlate well with several clinical and laboratory measurements in asthma including lung function, airway hyper-responsiveness, bronchodilator response, asthma symptoms, sputum eosinophil count, serum immunoglobulin levels, and allergen assessment.[13,14] There are now standardizations in F_ENO measurement and interpretation,[10,15] and devices such as the NIOX-MINO® have excellent reproducibility and are now used clinically on a frequent basis.[16] Although eNO is believed to be derived from the distal epithelial cells in asthma, other tissues also produce nitric oxide including endothelium, neuronal tissue, and nasal mucosa.[17,18] Efforts to compartmentalize eNO attempt to localize the regional production of NO into larger and smaller airways including the alveoli (Calv(NO)).[19] Though this approach is rational, Calv(NO) identification is not routinely used and requires further investigation. This fact confounds precise localization of F_ENO to the distal airway epithelium in many studies and clinical practice. Another volatile compound derived from the airway epithelium is exhaled carbon monoxide (eCO). Carbon monoxide is produced through several mechanisms involving oxidative stress, and it functions in inflammation, cell death, cell proliferation, and cell differentiation.[20–22] Studies to date have focused on asthma management using eCO levels.[23] eCO is easily measured with appropriate sensors; however, several confounders such as cigarette smoking and air pollution limit its clinical niche as a method to assess airway disorders.

One recent hypothesis is that VOCs observed in exhaled breath could originate directly from epithelial cells lining the lung/bronchial surface or be introduced from other tissues. The exchange of gases in lungs, specifically adsorption of oxygen and release of carbon dioxide, occurs in the pulmonary alveoli by passive diffusion of gases between the alveolus and the blood in lung capillaries. Certain volatile chemicals that are dissolved in blood can become volatilized in the exhaled air and are likely to contribute to the total chemical composition of exhaled breath. The blood is an extremely complex mixture that is unique to each individual and its composition varies with genetic makeup, environment, diet, and many other potentially confounding factors. Therefore, the variety of minor volatiles in breath can be staggering, and the variability of breath composition make chemical analysis very complicated. The VOCs detected in breath are, essentially, reflective of the entire organism's metabolism. Many cells emit volatile organic compounds[24–26] that may be of use in non-invasive disease detection. *In vivo* human studies have used exhaled VOCs to separate disease states,[27–30] though many of these studies have not always identified the exact compounds which function in disease pathology. For example, Phillips et al. looked at oxidative stress products in the exhaled breath of subjects pre- and post-influenza vaccine.[31] They found a significant increase in both VOC patterns and alkane derivatives associated with vaccination. These studies are important to understanding real-world changes in breath VOC composition, though they do not identify from where within the respiratory tract the VOCs emanate.

8.1.1 VOC production by neoplastic cells

The validity and credibility of a reported breath biomarker could be greatly increased if there is a clear understanding of the biomarker's origin, i.e. the biochemical process which led to production of that particular chemical and how variations in the biomarker presence or amount or chemical structure are indicative of certain biological states. Unfortunately, it is often very difficult to extricate the details of mechanistic processes occurring at the cellular level, and these studies are inherently challenging. A very useful approach that can be taken is to study production of volatiles that originate from single cell lines. This way it may be possible to eliminate a multitude of confounding factors associated with an entire organism. The cell lines can be tightly controlled with regard to their genotype, environment, and developmental stage. The ultimate goal is to establish a connection between intricate cellular mechanisms in biological/physiological processes, and the end-result, off-gassed volatile compounds. Upon establishing mechanisms of VOCs production, transfer, and expiration, the science of breath biomarkers will certainly make further advances. This approach will also aid in differentiating VOC production in oral and nasal cavities by bacteria rather than an organism's cells.

At the present time, the field of volatiles profiling in single cells is still nascent. Clinical studies utilizing exhaled breath are abundant, but many currently lack the ability to link VOCs to specific cell types.[32] Only a limited number of studies have looked at *in vitro* cell cultures to measure VOCs from specific cells. Examples include headspace analysis of VOCs to separate cancer cells from bronchial epithelial cells (BECs)[33–36] and nitric oxide chemoluminescent analysis of small airway BECs in asthma.[12] In some cases, *in vitro* headspace analysis determined specific VOCs which were modified between cell lines. Acetaldehyde,[4,33,37] other aldehydes, and tert-butyl ether[35] are some differentially expressed VOCs identified. The single cell line studies to date are mainly focused on neoplastic cell lines with the major clinical effort focused on early malignancy detection. The chemicals that are specifically released by malignant cell lines could be used in future screens as biomarkers for the early detection of lung cancer, or possibly other cancers. However, the use of VOCs for cancer screening and disease monitoring is currently limited by a lack of validated cancer-derived metabolites in the breath. In the study of melanoma-released volatile compounds, it has been demonstrated that it might be possible to detect tumors *in vivo* by analyzing VOCs emitted from different body compartments, such as breath and skin.[38] In that study, the authors measured VOCs mixtures released by well-characterized tumor cells derived from human malignancies by using an array of broadly selective chemical sensors. The patterns of VOCs emitted by two different melanoma tumorigenic cell lines transplanted ad hoc into a murine animal model produced during the exponential phase of tumor growth were shown to be distinguishable, as well as tumor-specific. The existence of volatile organic compounds specifically released by the cancer cell line was also established for the lung cancer cell line A549 as well as human bronchial epithelial cells (HBEpC) and human fibroblasts (hFB) cells.[34] VOCs were detected in the headspace of cell cultures for both cancer cell line and media control using gas chromatography/mass spectrometry (GC/MS). VOC differences and similarities between normal and transformed cells have also been identified. Compounds such as 2-pentanone and 2, 4-dimethyl-1-heptene were found to increase in the headspace of A549, HBEpC, and hFB cell cultures. Some compounds, e.g. branched hydrocarbons and alcohols, were more abundant in untransformed than in A549 cells. A variety of aldehydes and other compounds were found at lower abundances in the headspace of all cell lines compared to medium controls. These results may indicate that certain groups of VOCs may be preferentially associated with the transformed phenotype. Similar studies were carried out for the lung cancer cell lines NCI-H2087[39] and CALU-1[40] *in vitro*. In these studies, the cells were incubated in a sealed fermenter for several hours and the headspace of the cultures was sampled and analyzed by GC/MS. It was found that abundance of certain compounds such as 2-ethyl-1-hexanol and 2-methylpentane in the headspace of NCI-H2087 cells was elevated compared with medium controls. 2-methylpentane was also reported to be found in exhaled breath of lung cancer patients in contrast to that from healthy volunteers. These findings may demonstrate that certain volatile compounds can be cancer-cell derived and thus indicative of the presence of a tumor. Abundance of some compounds, e.g. acetaldehyde, 3-methylbutanal, butyl acetate, acetonitrile, acrolein, methacrolein, 2-methylpropanal, 2-butanone, 2-methoxy-2-methylpropane, 2-ethoxy-2-methylpropane, and hexanal for CALU-1 cells was found to decrease compared to media control, which suggests that these compounds are not released but rather consumed by CALU-1 cells. In the study using proton transfer reaction mass spectrometry, it was possible to clearly separate the cancerous and non-cancerous cell lines from each other[33] by applying multivariate statistical analysis using 42 selected VOCs biomarkers. These biomarkers were detected in the headspace over the cell cultures from pairs of cancerous and non-cancerous human cell lines from lung epithelium (A549), retinal pigment epithelium (hTERT-RPE1), and compared to squamous lung carcinoma cells (EPLC) and immortalized human bronchial epithelial cells (BEAS2B) VOCs, respectively. Surprisingly, the acetaldehyde was shown to be consumed by the cancerous cell lines, but not by the non-cancerous cells.[37] Similarly, in the study of non small cell lung cancer cell line NCI-H1666, derived from a bronchoalveolar carcinoma, it was found that the NCI-H1666 cells consumed several aldehydes, such as butyl acetate and the tert-butyl but showed no unequivocal release of VOCs specific to that cancer cell type.[35] The underlying causes of this peculiar phenomenon of VOC consumption rather than release are currently unknown. The human non-small cell lung carcinoma A549 cell line was also studied with for volatile biomarkers associated with cisplatin-induced apoptosis and necrosis.[41] It is presumed that these compounds are indicative of drug-induced cell death pathways and can serve as new biomarkers of these processes.

The studies of VOCs produced by human non-cancerous single cell lines are sparse. The cell lines that may be of particular interest due to their possible VOCs contribution to expired breath are epithelial cells, neutrophils, and macrophages. In a normal distal bronchus, the epithelium lines the airway and a basement membrane, elastic fibers, and smooth muscle enclose the bronchus. In disease states such as pulmonary fibrosis and asthma, fibrocytes may migrate into the system and deposit collagen aberrantly.[42,43] Considerable interest exists in defining volatile components associated with abnormal fibrocyte activity in these lung disorders. Several culture models of fibrocytes have been developed which ultimately may facilitate VOC detection.[44–46] These studies have shown that circulating fibrocytes express high levels of the CXCR4 chemokine when exposed to hypoxic environments *in-vitro*,[44] and that co-cultured airway smooth muscle cells in asthma induce fibrocyte movement.[46] Utilizing these models, headspace analysis might provide correlating VOCs seen in activated fibrocytes. This concept is further supported by the findings of mesenchymal cells (of which fibrocytes are a type) in the airway lavage fluid of patients with pulmonary fibrosis and asthma.[47] This implies that fibrocytes recruited to subepithelial bronchial tissue may eventually spill into the airway lumen, a location prime for direct VOC detection.

8.1.2 VOC production on cellular level by non-neoplastic cells

Neutrophils are white blood cells which function in the innate immune system; these cells respond to chemoattractant factors induced by local and systemic inflammation.[48,49] They are present in the airways in acute exacerbations of severe/refractory asthma,[50] and they are often found during pulmonary infections in patients with asthma, COPD, sickle cell disease, and several vasculitides. As neutrophils release abundant digestive enzymes designed to kill bacteria,[49] they appear to be good candidates for volatile production. Early cell culture work in this area demonstrated that human neutrophilic promyelocyte leukemia cells (HL60), released aldehydes in trace quantities.[51] Other studies have identified reactive oxidation species emanating from neutrophils *in-vitro*. These include a study showing increased nitric oxide release from neutrophils treated with advanced glycation end-products[52] and a study demonstrating a loss of pulmonary surfactant-D function via myeloperoxidase-induced damage from neutrophils cultured from a murine model of bacterial pneumonia.[53] This latter study did not directly measure VOCs, though the potential clearly exists. A clinical study of 14 ozone-responsive subjects exposed to ozone and room air assessed total VOC breath prints using an electronic nose (eNose) technology.[54] Though no distinct VOCs were identified, breath print patterns correlated with blood neutrophil counts indicating that neutrophilic inflammation could be identified in exhaled breath. Researchers have performed abundant work on VOC detection in cystic fibrosis (CF) subjects with recurrent *Pseudomonas aeruginosa* infections. Though CF airways are often sites of intense inflammation and cellular degradation, *P. aeuruginosa* is known to cause suppurative lung infections with copious neutrophils. VOC studies in CF have identified 2-aminoacetophenone as a potential biomarker of infection,[55,56] discriminated between CF subjects with and without *P. aeruginosa* infection,[57,58] and identified methyl thiocyanate as a potential pathogenic mechanism.[59] In some strains of *P. aeuruginosa*, hydrogen cyanide was noted to be produced in significant quantities.[60]

Another cell type of potential interest are macrophages which are present in high quantities in the distal human lung. They function as antigen presenting cells and comprise a part of the adaptive immune system. Macrophages phagocytize and control foreign elements such as dust and bacteria (including *Mycobacterium tuberculosis*). To date, few studies have assessed VOCs emitted specifically from macrophages. A small study looking at goats chronically infected with a *Mycobacterium sp* compared to healthy goats was able to discriminate between the groups based on exhaled breath and fecal VOCs.[61] A larger study in humans assessed breath VOCs in 226 subjects at high risk for pulmonary tuberculosis (MTb), and VOC results were compared to sputum smear, culture, and chest radiography.[62] The VOC patterns had an accuracy of 85% with a sensitivity of 84% and a specificity of 64.7% when the smear, culture, and radiography were all positive or all negative for MTb. Specific VOCs isolated were alkane derivatives, cyclohexane, and benzene derivatives. Though these studies do not address macrophages specifically, they do so indirectly as macrophages are a predominant cell type activated in Mycobacterial lung disease. Future studies in the field may focus on macrophage cell culture to assess which VOCs are elaborated from cells exposed to MTb antigens. Other ways to study macrophages may be to investigate other lung diseases associated with macrophage-predominance. These include respiratory bronchiolitis-associated interstitial lung disease, certain forms of amiodarone toxicity, and pulmonary alveolar proteinosis; most of these diseases, however, are rare.

An *in vitro* study of human B-cells was carried out: the human leukocyte antigen super locus C1R HLA class I A and B locus negative cell line was stably transfected with specific A and B alleles.[63] It was shown that human B-cells produce a set of volatile organic compounds measurable by GC/MS and that a specific "odor fingerprints" of mammalian B-cells can be observed. The results of that study suggested that the HLA-related volatile organic compounds are produced directly on the cellular level and that the presence of a cell-specific HLA allele influences production of certain VOCs resulting in an allele-specific odor fingerprint. It was hypothesized that the observed VOCs appeared to provide a series of "snapshots" into a metabolic pathway controlled by particular alleles. Given that the unique "odor fingerprints" depend on expression of specific HLA sequences, these findings may pertain to fields of human scent studies,[64] as well as breath analysis.[65,66] In both of these research areas, the VOCs in human emanations are elucidated to better understand underlying physiological processes. Apart from the relative small number of confirmed breath biomarkers, one of the main challenges in using breath analysis as a diagnostic tool is significant subject-to-subject variability.[67,68] If individuals with discrete allele sets produce unique chemical compounds that may be released into the blood stream and, in turn, be exhaled, a great chemical complexity of the exhaled mixture and great inter-subject variability can be expected, given that each individual person has a unique HLA gene set. Consequently, the understanding of human breath composition as pertains to certain immune related conditions may need to entail understanding individual differences in HLA-related VOC chemicals, or at least understanding how to account for this person-to-person source of variation.

8.1.3 Viral and bacterial single cell VOCs

In addition to healthy cell lines, studies of VOC profile changes upon viral infection will have been conducted. The C1R B lymphoblastoid cells were studied using GC/MS with respect to VOC profile changes upon infection with avian H_9N_2 and H_6N_2 and human H_1N_1 influenza viruses.[69] The removal of confounding factors associated with complexity of an entire organism allowed for determining of a number of VOC biomarkers associated with viral infection. Some of the biomarkers were found to be upregulated or downregulated differently for different virus strains, and there was a time-dependence of response post-infection. The results suggest that the viral infection of a cell influences production of VOCs and release of branched hydrocarbons and oxygenated compounds such as esters can be attributed to the viral infection.[31]

The observation of the former is in agreement with clinical studies using live attenuated influenza vaccine, while observation of the latter is consistent within increased oxidative stress[70] associated with (influenza) infection. Indeed, it has been reported that the compounds such as aldehydes, ketones, and organic acids when found in human breath[71,72] and body fluids may signal the presence of radical oxygen species caused by disease and/or bacterial infections, as well as metabolic disorders.[73] Establishing a putative link between observed VOCs and biological pathways associated with viral infection may lead to better understanding of the disease and provide novel diagnostics and treatment options. To date, very few studies have assessed respiratory cellular changes associated with viral infection as a step toward improving diagnostics. One study looked at upregulated gene expression in peripheral blood associated with human rhinovirus (HRV), respiratory syncytial virus (RSV), and influenza infection.[74] Though the results are promising, the technology is invasive in its plasma requirement and does not reflect changes to respiratory cell-produced VOCs. Another study assessed changes in the cytokine profile in COPD subjects infected with HRV.[75] This study showed an association HRV-infection (specifically HRV proteinase 2A) with skewed T helper cells (TH1 and TH2) balance, and the authors suggest that this imbalance may lead to COPD exacerbations. Two studies have assessed cytokine profiles in the supernatant over virus-infected human bronchial epithelial cells (BEC) cultures with an air-liquid interface design. The authors found increased levels of IL-6, IL-8, and RANTES (Regulated on Activation, Normal T Expressed, and Secreted) protein in RSV–compared to adenovirus-infected cells,[76] and there was an increased expression of IL-1-alpha, IL-8, IP-10, and RANTES in HRV-infected BECs.[77] No direct correlations of presence of certain VOCs to the infection states that are used in clinical setting have been established to date.

In addition to studies of eukaryotic cells, a number of studies have been conducted on prokaryotic cells (specifically, bacteria) with regard to their VOC emission fingerprints. These studies are very relevant to the field of breath analysis since associating VOC biomarkers to the presence of certain bacteria may open a new avenue in rapid infection diagnosis. A large number of breath VOC result from bacterial activity in oral cavity. Bacteria may be more amenable to the single-cell line VOC profiling as these are stand-along organisms and are often easier to culture than specific tissue cell lines. Intricate studies linking production of VOCs and certain metabolic pathway may also be easier to accomplish. An example of such research is the mechanistic study of structural and regulatory genes required to make the gas dimethyl sulfide in bacteria.[78] It was demonstrated that the bacterial gene dddD is required for the process of bacterial catabolism of dimethylsulfoniopropionate, a secondary metabolite made by marine algae, to produce dimethyl sulfide, a key compound in global sulfur and carbon cycle. Volatile compounds produced by growing and proliferating bacteria have been profiled and used for speciation, as demonstrated by *in vitro* study of bacterial headspace using differential mobility spectrometry (DMS).[79] In general, more than 200 compounds were identified as microbial VOCs, i.e. chemical compounds that are formed in the metabolism of fungi and bacteria.[80] Unfortunately, none of these compounds can be regarded as exclusively of microbial origin or as specific for certain microbial species.[80] Nevertheless, identification of biological VOC components and the interpretation of biological VOC characteristics are constantly improving. The combination of suitable sampling and bio-information discrimination techniques carries a great promise for the future.[81]

All of these examples demonstrate the tremendous potential of single cell-line studies for the breath analysis research field and beyond. The removal of confounding factors associated with the inherent complexity of an entire organism may allow for discovery of biomarkers that may be otherwise obscured in an initial pilot clinical trial. Even in cases where such biomarkers do not provide immediate diagnostic usefulness, it may be beneficial to use such information to elucidate underlying biochemical processes that produce the observed VOCs. Correlation between these single-cell and clinical studies may also serve as a cross-validation and reduce possibilities of reporting false biomarkers.

8.2 MINIATURE MOBILE POINT-OF-CARE DIAGNOSTIC SYSTEMS FOR VOC BREATH BIOMARKERS: TRENDS AND FUTURE REQUIREMENTS

Volatile organic compounds (VOCs) in exhaled human breath provide a promising platform for the diagnosis of various diseases including cancer, asthma, and respiratory infections. Different sensor technologies have been proposed for VOC measurement. Some of the proposed technologies have the potential to be applied for exhaled breath analysis. The novel devices typically use materials with favorable properties, such as enhanced absorptivity, for VOC pre-concentration and sensing. To date, these materials are mostly polymers,[83] metal oxides,[84,85] or a combination of the two,[84] as well as a few optical methods.[86] Polymer-based VOCs detection has been used in many VOCs sensors.[87] In the presence of a specific VOC, different polymers exhibit measurable responses such as changes in conductivity.[83] Because these devices need to be able to measure different VOCs, many applications have combined various sensors and materials into a single array.[84,87–90] This has been referred to as an "electronic nose." One embodiment of the device is shown in Figure 8.1. Analysis of the measured signals from these sensors is typically combined with a statistical algorithm to interpret the hardware output, such as neural networks or principle components, which can link sensor response combinations to identify and estimate the concentration of the specific VOC present in a clinical sample. Electronic nose uses sensor micro-array devices made from a base material such as SnO_2 coupled with pattern recognition analysis for identification of VOCs.[89] The technique relies on differences in selectivity toward different VOCs or complex mixtures of VOCs. One embodiment is a chip having four devices consisting of a SnO_2 gas sensitive layer with different doping concentrations of platinum, palladium, and gold on each of the devices.[89] Underneath that layer are gold finger-electrodes used as contact pads to record sensor output signals. Because operation of an array type of device such as this has been shown to be highly dependent on temperature effects, a heating element is also

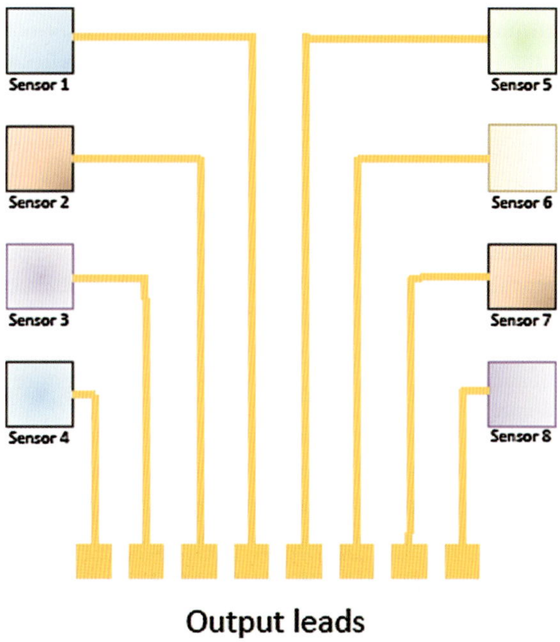

FIGURE 8.1
A Schematic representation of an electronic nose array chip, cf. also Ref. 82.

added on the back side of the array to maintain a steady temperature. For this specific SnO_2 array, 350°C was the optimum temperature for maximum sensitivity.[89] As gas samples are introduced to the sensors, changes in resistance are measured at the contact pads. Steady and transient responses vary from one sensor to another, creating a complex signal which will vary greatly depending on the VOC being tested. Identification of VOCs is done by initially measuring known training samples to create a neural network algorithm which defines characteristics used to differentiate between samples. Testing samples are then measured on the device and the VOCs are identified based on those characteristics.

Conductive polymers have been implemented directly as sensors for breath analysis.[83] Nano-scale copolymers show complex changes in conductivity in response to different VOC analytes.[83] One benefit of using such polymers is that they can be easily and repeatably deposited using an inkjet system. Other features of this method include the operation at room temperature, making it more robust compared to other detection methods. Another device reported in the literature takes advantage of polymer electrical permittivity characteristics to monitor changes in capacitance values in the presence of different VOC analytes.[87] The chemi-capacitative sensor features a capacitor having the polymer as a dielectric to absorb the VOC analyte. When an analyte is absorbed it alters the permittivity of the dielectric, and thus the capacitance changes as well. Analytes have different absorption rates in different polymers. The use of both solid and liquid polymers as the dielectric material on an array of multiple capacitors has been assessed.[87] Polymers can be deposited as liquid droplets using a piezoceramic inkjet head onto a micro-manufactured capacitor with a gap size in the range of a few micrometers. Typical limits of detection for polymer-based devices are 100 ppm for industrial solvents and 1 ppm for explosive materials.[87] The sensitivity of such devices is insufficient for breath analysis, in their current configurations. Metal oxide gas sensors have also been implemented in VOCs detection. Voltage measured at leads of the device rises in the presence of VOCs. Wolfrum et al. successfully used pattern recognition to distinguish between different VOCs by developing calibration models.[84] However, these types of devices frequently require high temperatures in the range of approximately 400°C for proper operation. Current studies are looking at ways of implementing micro-structured hot plates capable of using ultra-low power, in the mW range per sensor, to provide heating.[85]

Surface acoustic wave (SAW) devices are another category of miniature VOC sensors that operate on the principle of applying an acoustic wave along the surface of a substrate and measuring properties of this wave.[91] Typically a Rayleigh wave is created by applying a time-varying electric field to a piezoelectric material in contact with a substrate. As the piezoelectric material cyclically deforms under the electric field, and the effects are transferred to the contacted substrate causing the generation of the wave, adsorbed matter on the surface will alter wave properties. Devices such as this can be coated with a layer having properties conducive to the adsorption of VOCs. An array of eight devices was manufactured using a ZnO thick film, with deposited aluminum interdigitated electrodes used to apply the electrical field and a layer of different polymer film on each device.[88] Frequency changes in the wave are recorded when VOC mixtures are present. Principal components analysis (PCA) and neural network methods have been used to measure and differentiate gas mixtures at the ppm level based on characteristics of a training data set. Single and multiwall carbon nanotubes (SWCNT and MWCNT) have also been implemented as the VOC absorptive selective layer for SAW devices.[92]

Optical devices are unique compared to other methods of VOC detection because they can be used to remotely measure VOCs and are not highly dependent on relative humidity. Materials known as vapochromic complexes change color or refractive index in the presence of the analyte.[86] Specificity of response to a certain VOC is determined by the structure of such materials, as different structures will respond differently. One method making use of these properties also features an array for sensing. Vapochromic complexes are deposited by the electrostatic self-assembly (ESA) method. A LED source induces a color change in the material which changes the absorbance spectra. It has been shown that arrays can be used to selectively identify a VOC by analyzing the optical properties from all the devices of the array together using principle component analysis (PCA) analysis.[90] Another optical VOC sensor has used various chemoresponsive dyes printed on solid materials creating a disposable and relatively inexpensive set-up.[93] The method showed differences in responses at the ppb level.

8.2.1 Point-of-care devices: requirements and challenges

Several limitations on the currently available VOCs detection devices arise. Among these limitations are the poor selectivity between VOCs, poor long-term stability which leads to a potentially high cost-of-ownership, and very large sensor-to-sensor variations. The currently available VOC detection devices are not capable of selectively detecting sub-ppm levels of certain VOCs. Specificity, selectivity, robustness in operation, reproducible manufacturing uniformity, and long-life stability are a combination of features that is not offered by current sensors at an acceptable cost level.

There is an urgent need to develop miniature chemical sensors which can be used to detect VOC breath biomarkers and offer point-of-care diagnoses. The miniature point-of-care chemical sensors would provide immediate, near-real-time diagnoses of the patient health status that can assist with important medical care decisions. Current technologies require sending samples from the patient to a medical center core laboratory facility where the samples are analyzed over time, and the results are then returned.

The evolution of microelectromechanical systems (MEMS) and microfabrication methods has resulted in several new miniature mobile technologies that show great promise in the area of breath diagnostics. Examples of currently developed technologies include miniaturized total analysis systems (μTAS)[94–97] and micro-fluidics systems. The aforementioned technologies are able to perform complex tasks including fluid pumping, pressure sensing, motion detection, fluid separation, fluid transportation, and bio-agent and chemical identification. Thus, the potential of MEMS and microfabrication technologies in miniature chemical sensing for future mobile point-of-care diagnostic systems is extremely promising. The next generation of VOC detection MEMS-based chemical sensors will need to overcome a variety of challenges and meet a number of consumer requirements in order to reliably provide point-of-care VOCs detection results to users. The reliability and accuracy of the chemical sensor are essential, as well as fast response and minimal delay in operation.

One of the essential requirements necessary for the successful introduction of breath VOCs detection devices in the market is the demand for portability. This significantly enhances the ease of use for the consumer and allows VOC detection tests to be performed in a variety of environments, if necessary. However, the need for portability is closely followed by the need for miniaturization of the sensing mechanism. Recent developments in MEMS fabrication techniques have allowed the realization of such demands. Research in the area of miniaturized total chemical analysis systems has shown that these small MEMS devices can dramatically shorten chemical analysis time. However, a potential drawback of miniaturizing MEMS sensing is that it could reduce the sample quantity processing capability. The latter could be also an advantage, allowing for users to collect fewer samples and possibly reducing sampling time for breath diagnostics. Another potential limitation of MEMS sensing mechanisms may be the ability to accurately report the presence of a compound if the concentration in the sample is lower than the detection limit of the device. One way of combating this problem is to incorporate high performance micro pre-concentrators into VOC detection devices. Pre-concentrators are designed to eliminate interfering chemicals and increase the concentration of the target compound in the sample. Some examples of existing micro pre-concentrators solutions include coated planar surfaces, micro-channels, and packed compartments.[98,99] Due to the small size of the micro pre-concentrators, wall-coated varieties are at a disadvantage due to the small adsorption surface area.[98] As the need to detect more dilute samples arises, pre-concentrators will likely be a critical element that allows future VOC sensors to accurately extract data. The detection of VOCs in different environments and from different patients could also create potential technological obstacles. Extra effort must be made to establish a method to effectively remove residue in between repeated samples to avoid the risk of cross contamination and inaccurate readings. With advancing capabilities in manufacturing, many MEMS devices can be mass produced cost effectively, introducing the option of disposable VOC detection devices.

Another consideration is the connectivity and immediate data transfer that are required for future VOCs detection devices. Current available technologies, such as Bluetooth and WiFi, can be used for online data transfer. Since the mobile point-of-care chemical sensor will be used by patients and inexperienced users, the chemical sensor should be user-friendly and easy to use. Moreover, many methods of measuring VOCs involve using materials, particularly metal oxides, which require temperatures as high as 500°C.[83] This affects the robustness of some sensor types since a heating element and additional power is required. In such arrangements, other components of the device must also be capable of handling such temperature ranges. Other requirements include affordability, the ability to detect new VOCs as new biomarkers are identified, and the ability to measure multiple constituents with sensitivity at trace levels.[100]

8.3 ADVANCES IN DATA ANALYSIS REQUIRED FOR BREATH SENSOR TECHNOLOGIES

Modern lab-based instrumentations used for the analysis and characterization of biological samples are ever increasing in their analytical capability and sophistication. They are able to detect and measure large numbers of variables simultaneously at any given moment (e.g. mass-to-charge ratio m/z in mass spectrometry or wave numbers in spectroscopy). Such systems, when coupled to chromatographic separation devices such as gas/liquid chromatography, offer an additional level of chemical information, and the subsequent generated data set increase in complexity and dimensionality. The sheer volume of data (in terms of the number of common variables and total number of observations made) represent a real and significant challenge in terms of digital storage, curation, and interpretation of information. Conventional univariate statistical approaches may not be practical or feasible in analyzing complex data sets from breath analysis sensors accurately, objectively, or within a reasonable time frame. Chemometrics, or multivariate analysis (MVA), is in itself an entire discipline within metabolomics, and has exciting applications in the breath analysis area. Here, only a brief overview of the methodologies and rationales commonly applied in the analysis of multivariate data sets is given. For further detailed explanations readers are strongly suggested to refer to the references given within this section. MVA can be separated into the three main categories of data preprocessing, unsupervised data analysis, and supervised data analysis.

8.3.1 Data preprocessing

Data preprocessing is an important, but often underappreciated step prior to actual MVA, as inappropriate understanding and application of data preprocessing can severely bias the data analysis obtained giving widely inaccurate and misleading results. Data preprocessing as applied to spectrometric and spectroscopic data sets generally encompasses the following steps: alignment, baseline removal, data normalization, and scaling. Additional steps may be incorporated depending on the structure of the data and experiment design implemented. Some of the most commonly applied preprocessing steps are briefly described below.

8.3.1.1 Alignment

In hyphenated instrumentation such as a gas chromatography differential mobility spectrometry (GC/DMS), gas chromatography mass spectrometry (GC/MS), and high performance liquid chromatography mass spectrometry (HPLC/MS), slight variations in sample flow rate will induce instrumental drift over time. Such variations are ultimately inevitable. In univariate targeted analysis, where only part of the analytical space is utilized for quantitative analysis, the effect of such misalignment is minimal and can often be ignored. However, in non-targeted profiling where the entire analytical space is utilized for analysis, it is essential that the chromatograph or spectra are properly aligned so that they have the same start and stop times, where each corresponding variable within different samples are matched accordingly. Correlation optimized warping (COW) is a well-known and widely used alignment algorithm.[101] It has been successfully applied in both aligning spectroscopic and spectrometric data sets. PCA is applied to misaligned data sets and the resulting trends observed are likely due to either differences in alignment or drifting between samples rather than to actual biological differences.

8.3.1.2 Baseline correction

Differences in background intensity due to samples being analyzed at different points in time often made a significant contribution toward the clustering observed. This effect can be largely reduced by adjusting the background intensity to be equal between all samples prior to MVA. One form of background correction is based on the asymmetric least squared (ALS) adaptive baseline estimation algorithm, where the shape and intensity differences are taken into account and subtracted from the chromatograph/spectra.[102] If the data are not properly baseline-corrected prior to

PCA, the effect would be similar to misaligned data sets where the subsequent clustering observed would likely be due to the changing baseline intensity rather than any biological differences.

8.3.1.3 Scaling

Scaling in MVA is applied to adjust for large scale differences in variables and to account for heteroscedasticity (differences in the variances between variables) within the data set. By adjusting variance to be equal across all variables within the data set, this would enable PCA model to capture as much of the systematic variation as possible. This is due to every variable having an equal probability to enter the PCA model. There are two types of scaling relevant to two way data, row scaling and column scaling, where individual variables are scaled by the inverse of its standard deviation. In conjunction with mean centering across the row vector, this scaling is often referred to as autoscaling (scaled to unit variance or column standardization).[103]

8.3.1.4 Mean centering

Mean centering is applied to remove any common offset and to adjust the distribution of the data points to be more symmetrical. Mean centering is achieved by subtracting the mean of each variable from itself. The process also has the effect of reducing the Rank of the model and improving the overall fit of the PCA model. This operation is automatically implemented prior to unsupervised MVA[103] (PCA).

8.3.1.5 Normalization

Normalization is applied to adjust for variations from sample to sample, such as min-max normalization where each variable is divided by the absolute differences between the maximum and minimum intensities. The data set is scaled so that the influence of each sample is between a predefined range with respect to each other (i.e. between 0 and 1). This is also implemented to avoid differences between large values within the data set.[104]

8.3.2 Unsupervised learning methodology

PCA is a well-known form of unsupervised learning methodology,[105,106] and has many useful properties as an exploratory data analysis (EDA) tool in MVA. It is a variance capturing method that seeks to describe the most significant variations within the data set through the use of a few latent variables called principal components (PC), via a linear combination of the original features. Given a two way data set (X), PCA can be written as follows:

$$(X) = T \cdot P + E$$

The data set is decomposed into a set of loadings (P—defined as variations in variables) and scores (T—variations in samples). E is the residual error. The first PC captures the largest total explained variance (TEV) within the data set with subsequent PCs having monotonically lower contributions. In most cases, the variations induced by instrumental noise are much smaller than those caused by biological factors. Therefore, by disregarding the latter PCs, where little or no relevant information is contained, random instrument noise variations can be removed from the entire data set while preserving any major trends present. When applied in such a manner, PCA can be considered a form of dimensionality reduction and data compression in which the major trends of the data set are captured by the first few PCs, containing the majority of the TEV.

PCA is often employed to visualize dominant trends within the data set and can be used as an indirect form of outlier detection. The spatial separation between data points within the data set can be used to relate samples to one another. Data points which cluster close together will have a similar multivariate profile while objects which are far apart will be more dissimilar. By assessing the compactness (shape) and purity (degree of overlap) between clustering within the scores plots, it is possible to directly evaluate the analytical reproducibility of replicate analysis.

In order to visualize the overall relationships of the samples within the data set, the first few PCs in the T (scores plot) are plotted against each other (i.e. PC1 versus PC2, PC1 versus PC3, and PC2 versus PC3). This facilitates identification of any major trends, groupings and outliers within the data set. Analogous to the scores plot, the loadings plot, which plots the first few PCs in P against each other, displays the relative weight, influence of each individual variable and its contribution toward the separation observed within the scores plot. The greater the magnitude of a variable within the loading vector, the greater contribution it has toward the trend observed in the corresponding scores plot.

An important feature of PCA is that the directions of the scores plot correspond to the directions in the loadings plot. This enables the identification and correlation of variables which are responsible for the separation observed between different groups in the scores plot.[107] By examining the cumulative TEV, with respect to the number of PCs extracted, it is possible to assess the proportionality of the data that has been successfully modeled through PCA. The closer it is to 100%, the more accurate the data have been modeled.

8.3.3 Supervised data analysis

In supervised learning, the primary goal is to generate a model which can correctly associate the response of the instrument with the target by minimizing the error between them. Supervised data analysis methodology in MVA involves the use of a classifier or discriminating function to generate a rule of classification based on a sample set with a known class membership (training set). The rule is then tested by applying it to a data set that was not used in training in order to assess and evaluate the classifier's predictive ability. This rule is generated through many iterations of training and testing steps which slowly correct the classifier in its knowledge of the rule.

There are numerous classifiers available in supervised learning. The most commonly encountered are artificial neural networks (ANNs) and support vector machines (SVMs). The following sections give brief explanations of the classifiers mentioned and conceptually how they operate.

8.3.3.1 Artificial neural networks (ANNs)

Neurons are replaced with a simple processing unit (called a node) which can take a numerical input and transform it into an output. This typically can be achieved by the combination of the inputs into a sigmoid function. These processing units are connected and organized in a manner similar to that of the human brain and they Artificial neural networks (ANNs). These have been extensively applied in numerous different scientific disciplines for non-linear classification and calibration applications. Extensive and detailed descriptions of ANNs are readily available in the literature.[108,109] ANNs are biologically inspired classifiers that consist of processing sub units (neurons) that function similarly to that of the biological counterpart. They receive a signal or stimuli, process it, and provide a response that is passed on to other cells. In ANNs, they exhibit similar characteristics; they can learn classification behaviors from previous examples and apply it to new ones.

The structure of MLP consists of the input layer connected to the output layer via the hidden layer(s). Each node of the input layer is connected to the node of the first hidden layer through an abstract connection called the synapse. Each synapse has an associative weight which scales the input passing through it, along with a bias term with a modifiable weight. This bias term allows for corrected outputs that range from 0 to 1. The nodes in the hidden layer(s) take the input signals from all nodes of the previous layer (Figure 8.2). The combination of these values with the bias term are input into a sigmoid function and fed to the output node through output synapses. The output node(s) use the same technique to provide a value for evaluation of accuracy of the entire network.[109]

For classification purposes a sigmoid function is generally used, while for calibration a summation function is employed instead. Training of MLP typically uses a standard back propagation (BP) algorithm[110] in which the input is applied to the network, an output is produced, and the correction is "pushed" backwards through the network, correcting the weights of each synapse and the bias term at each node. The process is iteratively repeated over multiple samples until a suitable level of error is obtained. Training an MLP network is a highly inefficient process as all connections are adjusted simultaneously and the number of nodes and layers are specified prior to training. This process is very time consuming and computationally intensive to optimize. Additionally, ANNs are prone to over-training in which the ANN becomes so accustomed to the training data that it is inapplicable to any other samples resulting in decreased predictive ability.

However, MLP is widely employed because an MLP network consisting of only a single hidden layer containing a large number of nodes is able to learn and apply any arbitrary non-linear solution to a suitable degree of accuracy.[111–113] While it may

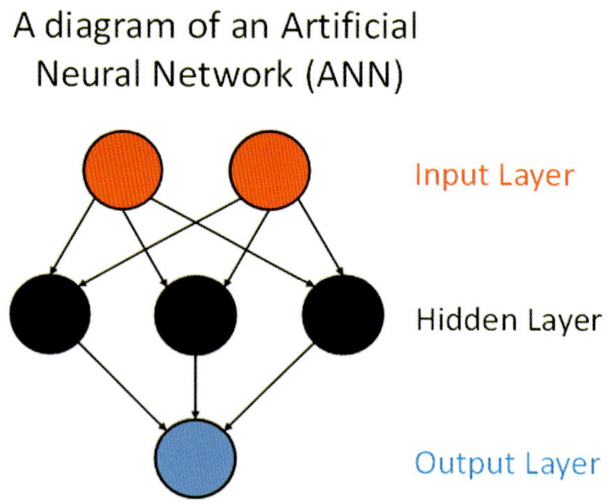

FIGURE 8.2
A three-layer artificial neural network. The nodes of the input layer communicate with the hidden layer through synapses. Inside each node, a sigmoid function is being calculated based on the sum of the inputs and the bias term to create an out.

be unable to output a rule, in the proper training environment, an MLP can find a solution to any solvable problem.

8.3.3.2 Support vector machines (SVMs)

SVM is a form of kernel method originally developed in the mid-1990s[114,115] and has since become widely accepted and applied in different applications for analyzing non-linearly separable classification problems. Many of the drawbacks associated with non-linear classifiers (like ANNs) have largely been resolved by the introduction of SVMs. SVMs require fewer parameters for optimization, are less prone to over-training and noisy data, and have a dramatic improvement in computational efficiency. SVMs work by transforming a non-linearly separable problem into a linearly separable one by the use of a suitable mapping function (kernel function). After transformation a linear solution can be found within the feature space defined by the mapping function.

The use of a mapping function enables the determination of a non-linear boundary to be found in a hyper-planar boundary which is able to separate the classes in a higher dimension feature space. In addition, the optimal hyper-planar boundary should be equidistant from the outer edges of each respective class which in turn maximizes the margin between them. SVMs therefore focus on the data points which are closest to the separation boundary (support vectors) and ignore those that are already well separated. Thus the separation boundary is based on a subset of samples rather than the whole data set.

SVM can be divided into two categories, hard margin and soft margin. Hard margin SVMs assume two classes to be perfectly separable and the goal is to determine the optimal boundary that exactly separates the two classes with a maximum possible margin between them. This is not always the case. By contrast, soft margin SVMs are able to tolerate a certain degree of misclassifications and are designed to find a compromise between the complexity of the model and generalization performance of the model. The lower the tolerance for the rate of classification error, the more complex the hyper-planar boundary will become. This, in turn, may lower the generalization performance of the model.[104,116]

Acknowledgments

This work was generously and partially supported by several funding agencies. The content of this work is solely the responsibility of the authors and does not necessarily represent the official view of these agencies. Partial support is acknowledged from: UL1 RR024146 from the National Center for Research Resources (NCRR), a component of the National Institutes of Health (NIH), and NIH Roadmap for Medical Research [CED]; Gilead Sciences, Inc. [CED]; the Defense Advanced Research Projects Agency (DARPA) [CED]. The Hartwell Foundation [CED], the United States Air Force [CED], the Office of Naval Research [CED], fellowship support from the American Association of University Women [AMK]; fellowship support from P42ES004699 from the National Institute of Environmental Health Sciences [KZ], and fellowship support from the National Heart, Lung and Blood Institute T32-HL007013 and #T32-ES007059 [MS] and a UC Davis Clinical and Translational Science Award (CTSA) NCRR UL1 RR024146 [MS]; UC Davis School of Medicine and NIH #8KL2TR000134-07 K12 mentored training award [MS]; US Department of Veteran Affairs, Post-9/11 GI-Bill [DJP].

References

1. Dryahina K, Smith D, Spanel P. Quantification of methane in humid air and exhaled breath using selected ion flow tube mass spectrometry. *Rapid Commun Mass Spectrom* 2010;**24**:1296–304.
2. Albertine KH. Anatomy of the lungs. In: Mason RJ, Broaddus VC, Martin T, King Jr T, Schraufnagel D, Murray JF, et al., editors. *Murray and Nadel's textbook of respiratory medicine*. Philadelphia: Saunders Elsevier; 2010.
3. Wang T, Pysanenko A, Dryahina K, Spanel P, Smith D. Analysis of breath, exhaled via the mouth and nose, and the air in the oral cavity. *J Breath Res* 2008;**2**:037013.
4. Smith D, Wang T, Sule-Suso J, Spanel P, El Haj A. Quantification of acetaldehyde released by lung cancer cells *in vitro* using selected ion flow tube mass spectrometry. *Rapid Commun Mass Spectrom* 2003;**17**:845–50.
5. Phillips M, Cataneo RN, Saunders C, Hope P, Schmitt P, Wai J. Volatile biomarkers in the breath of women with breast cancer. *J Breath Res* 2010;**4**:026003.

6. Peng G, Hakim M, Broza YY, Billan S, Abdah-Bortnyak R, Kuten A, et al. Detection of lung, breast, colorectal, and prostate cancers from exhaled breath using a single array of nanosensors. *Br J Cancer* 2010;**103**:542–51.
7. Dobbs LG, Johnson MD, Vanderbilt J, Allen L, Gonzalez R. The great big alveolar TI cell: evolving concepts and paradigms. *Cell Physiol Biochem* 2010;**25**:55–62.
8. Signorelli S, Jennings P, Leonard MO, Pfaller W. Differential effects of hypoxic stress in alveolar epithelial cells and microvascular endothelial cells. *Cell Physiol Biochem* 2010;**25**:135–44.
9. Randall TD. Bronchus-associated lymphoid tissue (BALT) structure and function. *Adv Immunol* 2010;**107**:187–241.
10. Dweik RA, Boggs PB, Erzurum SC, Irvin CG, Leigh MW, Lundberg JO, et al. An official ATS clinical practice guideline: interpretation of exhaled nitric oxide levels (F_ENO) for clinical applications. *Am J Respir Crit Care Med* 2011;**184**:602–15.
11. Eiserich JP, Patel RP, O'Donnell VB. Pathophysiology of nitric oxide and related species: free radical reactions and modification of biomolecules. *Mol Aspects Med* 1998;**19**:221–357.
12. Jiang J, Malavia N, Suresh V, George SC. Nitric oxide gas phase release in human small airway epithelial cells. *Respir Res* 2009;**10**:3.
13. Covar RA, Szefler SJ, Martin RJ, Sundstrom DA, Silkoff PE, Murphy J, et al. Relations between exhaled nitric oxide and measures of disease activity among children with mild-to-moderate asthma. *J Pediatr* 2003;**142**:469–75.
14. Kovesi T, Kulka R, Dales R. Exhaled nitric oxide concentration is affected by age, height, and race in healthy 9- to 12-year-old children. *Chest* 2008;**133**:169–75.
15. Smith AD, Cowan JO, Brassett KP, Herbison GP, Taylor DR. Use of exhaled nitric oxide measurements to guide treatment in chronic asthma. *N Engl J Med* 2005;**352**:2163–73.
16. Kapande KM, McConaghy LA, Douglas I, McKenna S, Hughes JL, McCance DR, et al. Comparative repeatability of two handheld fractional exhaled nitric oxide monitors. *Pediatr Pulmonol* 2011.
17. Nakagawa T, Johnson RJ. Endothelial nitric oxide synthase. *Contrib Nephrol* 2011;**170**:93–101.
18. Hesslinger C, Strub A, Boer R, Ulrich WR, Lehner MD, Braun C. Inhibition of inducible nitric oxide synthase in respiratory diseases. *Biochem Soc Trans* 2009;**37**:886–91.
19. Garcia-Rio F, Casitas R, Romero D. Utility of two-compartment models of exhaled nitric oxide in patients with asthma. *J Asthma* 2011;**48**:329–34.
20. Ryter SW, Choi AM. Heme oxygenase-1/carbon monoxide: from metabolism to molecular therapy. *Am J Respir Cell Mol Biol* 2009;**41**:251–60.
21. Horvath I, Donnelly LE, Kiss A, Paredi P, Kharitonov SA, Barnes PJ. Raised levels of exhaled carbon monoxide are associated with an increased expression of heme oxygenase-1 in airway macrophages in asthma: a new marker of oxidative stress. *Thorax* 1998;**53**:668–72.
22. Donnelly LE, Barnes PJ. Expression of heme oxygenase in human airway epithelial cells. *Am J Respir Cell Mol Biol* 2001;**24**:295–303.
23. Zhang J, Yao X, Yu R, Bai J, Sun Y, Huang M. Exhaled carbon monoxide in asthmatics: a meta-analysis. *Respir Res* 2010;**11**:50.
24. Labows JN, McGinley KJ, Webster GF, Leyden JJ. Headspace analysis of volatile metabolites of Pseudomonas aeruginosa and related species by gas chromatography-mass spectrometry. *J Clin Microbiol* 1980;**12**:521–6.

25. Chou S, Chedore P, Kasatiya S. Use of gas chromatographic fatty acid and mycolic acid cleavage product determination to differentiate among *Mycobacterium genavense*, *Mycobacterium fortuitum*, *Mycobacterium simiae*, and *Mycobacterium tuberculosis*. *J Clin Microbiol* 1998;**36**:577–9.
26. Acevedo CA, Sanchez EY, Reyes JG, Young ME. Volatile profiles of human skin cell cultures in different degrees of senescence. *J Chromatogr B: Analyt Technol Biomed Life Sci* 2010;**878**:449–55.
27. Fens N, Zwinderman AH, van der Schee MP, de Nijs SB, Dijkers E, Roldaan AC, et al. Exhaled breath profiling enables discrimination of chronic obstructive pulmonary disease and asthma. *Am J Respir Crit Care Med* 2009;**180**:1076–82.
28. Dragonieri S, Annema JT, Schot R, van der Schee MP, Spanevello A, Carratu P, et al. An electronic nose in the discrimination of patients with non-small cell lung cancer and COPD. *Lung Cancer* 2009;**64**:166–70.
29. Dragonieri S, Schot R, Mertens BJ, Le Cessie S, Gauw SA, Spanevello A, et al. An electronic nose in the discrimination of patients with asthma and controls. *J Allergy Clin Immunol* 2007;**120**:856–62.
30. Basanta M, Jarvis RM, Xu Y, Blackburn G, Tal-Singer R, Woodcock A, et al. Non-invasive metabolomic analysis of breath using differential mobility spectrometry in patients with chronic obstructive pulmonary disease and healthy smokers. *Analyst* 2010;**135**:315–20.
31. Phillips M, Cataneo RN, Chaturvedi A, Danaher PJ, Devadiga A, Legendre DA, et al. Effect of influenza vaccination on oxidative stress products in breath. *J Breath Res* 2010;**4**:026001.
32. Ibrahim B, Basanta M, Cadden P, Singh D, Douce D, Woodcock A, et al. Non-invasive phenotyping using exhaled volatile organic compounds in asthma. *Thorax* 2011; **66**:804–9.
33. Brunner C, Szymczak W, Hollriegl V, Mortl S, Oelmez H, Bergner A, et al. Discrimination of cancerous and non-cancerous cell lines by headspace-analysis with PTR-MS. *Anal Bioanal Chem* 2010;**397**:2315–24.
34. Filipiak W, Sponring A, Filipiak A, Ager C, Schubert J, Miekisch M, et al. TD-GC-MS analysis of volatile metabolites of human lung cancer and normal cells *in vitro*. *Cancer Epidemiol Biomarkers Prev* 2010;**19**:182–95.
35. Sponring A, Filipiak W, Ager C, Schubert J, Miekisch W, Amann A, et al. Analysis of volatile organic compounds (VOCs) in the headspace of NCI-H1666 lung cancer cells. *Cancer Biomark* 2010;**7**:153–61.
36. Chen X, Xu F, Wang Y, Pan Y, Lu D, Wang P, et al. A study of the volatile organic compounds exhaled by lung cancer cells *in vitro* for breath diagnosis. *Cancer* 2007; **110**:835–44.
37. Sule-Suso J, Pysanenko A, Spanel P, Smith D. Quantification of acetaldehyde and carbon dioxide in the headspace of malignant and non-malignant lung cells *in vitro* by SIFT-MS. *Analyst* 2009;**134**:2419–25.
38. Pennazza G, Santonico M, Martinelli E, Paolesse R, Tamburrelli V, Cristina S, et al. Monitoring of melanoma released volatile compounds by a gas sensors array: from *in vitro* to *in vivo* experiments. *Sensor Actuat B—Chem* 2011;**154**:288–94.
39. Sponring A, Filipiak W, Mikoviny T, Ager C, Schubert J, Miekisch W, et al. Release of volatile organic compounds from the lung cancer cell line NCI-H2087 *in vitro*. *Anticancer Res* 2009;**29**:419–26.

40. Filipiak W, Sponring A, Mikoviny T, Ager C, Schubert J, Miekisch W, et al. Release of volatile organic compounds (VOCs) from the lung cancer cell line CALU-1 *in vitro*. *Cancer Cell Int* 2008;**8**:s17.
41. Pyo JS, Ju HK, Park JH, Kwon SW. Determination of volatile biomarkers for apoptosis and necrosis by solid-phase microextraction-gas chromatography/mass spectrometry: a pharmacometabolomic approach to cisplatin's cytotoxicity to human lung cancer cell lines. *J Chromatogr B: Analyt Technol Biomed Life Sci* 2008;**876**:170–4.
42. Singh SR, Hall IP. Airway myofibroblasts and their relationship with airway myocytes and fibroblasts. *Proc Am Thorac Soc* 2008;**5**:127–32.
43. Andersson-Sjoland A, Nihlberg K, Eriksson L, Bjermer L, Westergren-Thorsson G. Fibrocytes and the tissue niche in lung repair. *Respir Res* 2011;**12**:76.
44. Mehrad B, Burdick MD, Strieter RM. Fibrocyte CXCR4 regulation as a therapeutic target in pulmonary fibrosis. *Int J Biochem Cell Biol* 2009;**41**:1708–18.
45. Hong KM, Belperio JA, Keane MP, Burdick MD, Strieter RM. Differentiation of human circulating fibrocytes as mediated by transforming growth factor-beta and peroxisome proliferator-activated receptor gamma. *J Biol Chem* 2007;**282**:22910–20.
46. Saunders R, Siddiqui S, Kaur D, Doe C, Sutcliffe A, Hollins F, et al. Fibrocyte localization to the airway smooth muscle is a feature of asthma. *J Allergy Clin Immunol* 2009; **123**:376–84.
47. Nihlberg K, Larsen K, Hultgardh-Nilsson A, Malmstrom A, Bjermer L, Westergren-Thorsson G. Tissue fibrocytes in patients with mild asthma: a possible link to thickness of reticular basement membrane? *Respir Res* 2006;**7**:50.
48. Phillipson M, Kubes P. The neutrophil in vascular inflammation. *Nat Med* 2011; **17**:1381–90.
49. Parker D, Prince A. Innate immunity in the respiratory epithelium. *Am J Respir Cell Mol Biol* 2011;**45**:189–201.
50. Bhakta NR, Woodruff PG. Human asthma phenotypes: from the clinic, to cytokines, and back again. *Immunol Rev* 2011;**242**:220–32.
51. Shin HW, Umber BJ, Meinardi S, Leu SY, Zaldivar F, Blake DR, et al. Acetaldehyde and hexanaldehyde from cultured white cells. *J Transl Med* 2009;**7**:31.
52. Bansal S, Siddarth M, Chawla D, Banerjee BD, Madhu SV, Tripathi AK. Advanced glycation end products enhance reactive oxygen and nitrogen species generation in neutrophils *in vitro*. *Mol Cell Biochem* 2012;**361**:289–96.
53. Crouch EC, Hirche TO, Shao B, Boxio R, Wartelle J, Benabid R, et al. Myeloperoxidase-dependent inactivation of surfactant protein D *in vitro* and *in vivo*. *J Biol Chem* 2010;**285**:16757–70.
54. Biller H, Holz O, Windt H, Koch W, Muller M, Jorres RA, et al. Breath profiles by electronic nose correlate with systemic markers but not ozone response. *Respir Med* 2011;**105**:1352–63.
55. Scott-Thomas AJ, Syhre M, Pattemore PK, Epton M, Laing R, Pearson J, et al. 2-Aminoacetophenone as a potential breath biomarker for *Pseudomonas aeruginosa* in the cystic fibrosis lung. *BMC Pulm Med* 2010;**10**:56.
56. Scott-Thomas A, Pearson J, Chambers S. Potential sources of 2-aminoacetophenone to confound the *Pseudomonas aeruginosa* breath test, including analysis of a food challenge study. *J Breath Res* 2011;**5**:046002.

57. Robroeks CM, van Berkel JJ, Dallinga JW, Jobsis Q, Zimmermann LJ, Hendriks HJ, et al. Metabolomics of volatile organic compounds in cystic fibrosis patients and controls. *Pediatr Res* 2010;**68**:75–80.
58. Savelev SU, Perry JD, Bourke SJ, Jary H, Taylor R, Fisher AJ, et al. Volatile biomarkers of *Pseudomonas aeruginosa* in cystic fibrosis and noncystic fibrosis bronchiectasis. *Lett Appl Microbiol* 2011;**52**:610–3.
59. Shestivska V, Nemec A, Drevinek P, Sovova K, Dryahina K, Spanel P. Quantification of methyl thiocyanate in the headspace of *Pseudomonas aeruginosa* cultures and in the breath of cystic fibrosis patients by selected ion flow tube mass spectrometry. *Rapid Commun Mass Spectrom* 2011;**25**:2459–67.
60. Carroll W, Lenney W, Wang T, Spanel P, Alcock A, Smith D. Detection of volatile compounds emitted by *Pseudomonas aeruginosa* using selected ion flow tube mass spectrometry. *Pediatr Pulmonol* 2005;**39**:452–6.
61. Purkhart R, Kohler H, Liebler-Tenorio E, Meyer M, Becher G, Kikowatz A, et al. Chronic intestinal Mycobacteria infection: discrimination via VOC analysis in exhaled breath and headspace of feces using differential ion mobility spectrometry. *J Breath Res* 2011;**5**:027103.
62. Phillips M, Basa-Dalay V, Bothamley G, Cataneo RN, Lam PK, Natividad MP, et al. Breath biomarkers of active pulmonary tuberculosis. *Tuberculosis (Edinb)* 2010;**90**:145–51.
63. Aksenov AA, Gojova A, Zhao W, Morgan JT, Sankaran S, Sandrock CE, Davis CE. Characterization of volatile organic compounds in human leukocyte antigen heterologous expression systems: a cell's "chemical odor fingerprint". *ChemBioChem* 2012;**13**(7): 1053–59.
64. Gallagher M, Wysocki CJ, Leyden JJ, Spielman AI, Sun X, Preti G. Analyses of volatile organic compounds from human skin. *Br J Dermatol* 2008;**159**:780–91.
65. Grob NM, Aytekin M, Dweik RA. Biomarkers in exhaled breath condensate: a review of collection, processing and analysis. *J Breath Res* 2008;**2**:037004.
66. Risby TH. Volatile organic compounds as markers in normal and diseased states. In: Marczin N, Yacoub M, editors. *Disease markers in exhaled breath: basic mechanisms and clinical applications*. Washington DC: IOS Press; 2002. p. 418.
67. Phillips M, Herrera J, Krishnan S, Zain M, Greenberg J, Cataneo RN. Variation in volatile organic compounds in the breath of normal humans. *J Chromatogr B: Biomed Sci Appl* 1999;**729**:75–88.
68. Pleil JD. Influence of systems biology response and environmental exposure level on between-subject variability in breath and blood biomarkers. *Biomarkers* 2009;**14**:560–71.
69. Aksenov A, Zhao W, Sandrock C, Davis CE, Production of odor by cell infected with influenza viruses, in press.
70. Schwarz KB. Oxidative stress during viral infection: a review. *Free Radic Biol Med* 1996;**21**:641–9.
71. Dalton P, Gelperin A, Preti G. Volatile metabolic monitoring of glycemic status in diabetes using electronic olfaction. *Diabetes Technol Ther* 2004;**6**:534–44.
72. Preti G, Clark L, Cowart BJ, Feldman RS, Lowry LD, Weber E, et al. Non-oral etiologies of oral malodor and altered chemosensation. *J Periodontol* 1992;**63**:790–6.
73. Castro L, Freeman B. Reactive oxygen species in human health and disease. *Nutrition* 2001;**17**:161–5.

74. Zaas AK, Chen M, Varkey J, Veldman T, Hero 3rd AO, et al. Gene expression signatures diagnose influenza and other symptomatic respiratory viral infections in humans. *Cell Host Microbe* 2009;**6**:207–17.
75. Singh M, Lee SH, Porter P, Xu C, Ohno A, Atmar RL, et al. Human rhinovirus proteinase 2A induces TH1 and TH2 immunity in patients with chronic obstructive pulmonary disease. *J Allergy Clin Immunol* 2010;**125**:1369–78 [e2].
76. Yoon JS, Kim HH, Lee Y, Lee JS. Cytokine induction by respiratory syncytial virus and adenovirus in bronchial epithelial cells. *Pediatr Pulmonol* 2007;**42**:277–82.
77. Lopez-Souza N, Favoreto S, Wong H, Ward T, Yagi S, Schnurr D, et al. *In vitro* susceptibility to rhinovirus infection is greater for bronchial than for nasal airway epithelial cells in human subjects. *J Allergy Clin Immunol* 2009;**123**:1384–90 [e2].
78. Todd JD, Rogers R, Li YG, Wexler M, Bond PL, Sun L, et al. Structural and regulatory genes required to make the gas dimethyl sulfide in bacteria. *Science* 2007;**315**:666–9.
79. Shnayderman M, Mansfield B, Yip P, Clark HA, Krebs MD, Cohen SJ, et al. Species-specific bacteria identification using differential mobility spectrometry and bioinformatics pattern recognition. *Anal Chem* 2005;**77**:5930–7.
80. Korpi A, Jarnberg J, Pasanen AL. Microbial volatile organic compounds. *Crit Rev Toxicol* 2009;**39**:139–93.
81. Zhang ZM, Li GK. A review of advances and new developments in the analysis of biological volatile organic compounds. *Microchem J* 2010;**95**:127–39.
82. http://www.mse.umd.edu/images/nanosensors-lg.jpg.
83. Li B, Sauve G, Iovu MC, Jeffries-El M, Zhang R, Cooper J, et al. Volatile organic compound detection using nanostructured copolymers. *Nano Lett* 2006;**6**:1598–602.
84. Wolfrum EJ, Meglen RM, Peterson D, Sluiter J. Metal oxide sensor arrays for the detection, differentiation, and quantification of volatile organic compounds at sub-parts-per-million concentration levels. *Sensor Actuat B—Chem* 2006;**115**:322–9.
85. Elmi I, Zampolli S, Cozzani E, Mancarella F, Cardinali GC. Development of ultra-low-power consumption MOX sensors with ppb-level VOC detection capabilities for emerging applications. *Sensor Actuat B—Chem* 2008;**135**:342–51.
86. Elosua C, Bariain C, Matias IR, Arregui FJ, Vergara E, Laguna M. Optical fiber sensing devices based on organic vapor indicators towards sensor array implementation. *Sensor Actuat B—Chem* 2009;**137**:139–46.
87. Patel SV, Mlsna TE, Fruhberger B, Klaassen E, Cemalovic S, Baselt DR. Chemicapacitive microsensors for volatile organic compound detection. *Sensor Actuat B—Chem* 2003;**96**:541–53.
88. Fernandez MJ, Fontecha JL, Sayago I, Aleixandre M, Lozano J, Gutierrez J, et al. Discrimination of volatile compounds through an electronic nose based on ZnOSAW sensors. *Sensor Actuat B—Chem* 2007;**127**:277–83 .
89. Srivastava AK. Detection of volatile organic compounds (VOCs) using SnO_2 gas-sensor array and artificial neural network. *Sensor Actuat B—Chem* 2003;**96**:24–37.
90. Taurino AM, Distante C, Siciliano P, Vasanelli L. Quantitative and qualitative analysis of VOCs mixtures by means of a microsensors array and different evaluation methods. *Sensor Actuat B—Chem* 2003;**93**:117–25.
91. Wohltjen H, Dessy R. Surface acoustic-wave probe for chemical-analysis. 1. Introduction and instrument description. *Anal Chem* 1979;**51**:1458–64.
92. Penza M, Antolini F, Antisari MV. Carbon nanotubes as SAW chemical sensors materials. *Sensor Actuat B—Chem* 2004;**100**:47–59.

93. Suslick KS, Rakow NA, Sen A. Colorimetric sensor arrays for molecular recognition. *Tetrahedron* 2004;**60**:11133–8.
94. Kricka LJ. Microchips, microarrays, biochips and nanochips: personal laboratories for the 21st century. *Clin Chim Acta* 2001;**307**:219–23.
95. Leung YF, Tam PO, Tong WC, Baum L, Choy KW, Lam DS, et al. High-throughput conformation-sensitive gel electrophoresis for discovery of SNPs. *Biotechniques* 2001;**30**:334–5, 338–40.
96. Vo-Dinh T, Cullum BM, Stokes DL. Nanosensors and biochips: frontiers in biomolecular diagnostics. *Sensor Actuat B—Chem* 2001;**74**:2 11.
97. Manz A, Graber N, Widmer HM. Miniaturized total chemical-analysis systems—a novel concept for chemical sensing. *Sensor Actuat B—Chem* 1990;**1**:244–8.
98. Alfeeli B, Agah M. MEMS-based selective preconcentration of trace level breath analytes. *IEEE Sens J* 2009;**9**:1068–75.
99. Strand N, Bhushan A, Schivo M, Kenyon NJ, Davis CE. Chemically polymerized polypyrrole for on-chip concentration of volatile breath metabolites. *Sensor Actuat B—Chem* 2010;**143**:516–23.
100. Davis CE, Frank M, Mizaikoff B, Oser H. The future of sensors and instrumentation for human breath analysis. *IEEE Sens J* 2010;**10**:3–6.
101. Tomasi G, van den Berg F, Andersson C. Correlation optimized warping and dynamic time warping as preprocessing methods for chromatographic data. *J Chemometr* 2004; **18**:231–41.
102. Peng JT, Peng SL, Jiang A, Wei JP, Li CW, Tan J. Asymmetric least squares for multiple spectra baseline correction. *Anal Chim Acta* 2010;**683**:63–8.
103. Bro R, Smilde AK. Centering and scaling in component analysis. *J Chemometr* 2003;**17**:16–33.
104. Xu Y, Zomer S, Brereton RG. Support vector machines: a recent method for classification in chemometrics. *Crit Rev Anal Chem* 2006;**36**:177–88.
105. Pearson K. On lines and planes of closest fit to systems of points in space. *Philos Mag* 1901;**2**:559–72.
106. Wold S. Multivariate data analysis: converting chemical data tables to plots. In: Brandt J, Ugi II, editors. *Computer applications in chemical research and education*. Heidelberg: Hüthig-Verlag; 1989. p. 101–28.
107. Lindon JC, Holmes E, Nicholson JK. So whats the deal with metabonomics? Metabonomics measures the fingerprint of biochemical perturbations caused by disease, drugs, and toxins. *Anal Chem* 2003;**75**:384a–91a.
108. Smits JRM, Melssen WJ, Buydens LMC, Kateman G. Using artificial neural networks for solving chemical problems. 1. Multilayer feedforward networks. *Chemometr Intell Lab* 1994;**22**:165–89.
109. Beavis RC, Colby SM, Goodacre R, Harrington PB, Reilly JP, Sokolow S, et al. Artificial intelligence and expert systems in mass spectrometry. In: Meyers RA, editor. *Encyclopedia of analytical chemistry*. Chichester: John Wiley & Sons; 2000. p. 11558–97.
110. Werbos PJ. *The roots of back-propagation: from ordered derivatives to neural networks and political forecasting*. Chichester: Wiley; 1994.
111. Funahashi K. On the approximate realization of continuous-mappings by neural networks. *Neural Networks* 1989;**2**:183–92.

112. Hornik K, Stinchcombe M, White H. Universal approximation of an unknown mapping and its derivatives using multilayer feedforward networks. *Neural Networks* 1990; **3**:551–60.
113. White H. Connectionist nonparametric regression—multilayer feedforward networks can learn arbitrary mappings. *Neural Networks* 1990;**3**:535–49.
114. Boser BE, Guyon IM, Vapnik VN. A training algorithm for optimal margin classifiers. In: Haussler D, editor. *Fifth annual workshop on computational learning theory*. Pittsburgh, PA: ACM Press; 1992.
115. Vapnik VN. *The nature of statistical learning theory*. New York: Springer; 1995.
116. Brereton RG. *Chemometrics: data analysis for the laboratory and chemical plant*. Chichester: Wiley; 2003.

CHAPTER

9 Breath Analysis in Critically Ill Patients—Potential and Limitations

Jochen K. Schubert and Wolfram Miekisch

University of Rostock, Department of Anaesthesia and Intensive Care, Schillingallee 35, D-18057 Rostock, Germany

9.1 INTRODUCTION

Breath analysis represents a new diagnostic approach in medicine and holds potential as a novel non-invasive diagnostic tool. In contrast to blood or tissue analysis, sampling of breath is completely non-invasive and testing can be done repeatedly and frequently without any burden to the patient and without any risk to the staff collecting the samples. Analysis of volatile substances in expired air opens a non-invasive window onto biochemical processes in the lung, in the airways, and in the whole body. Requirements for clinical application of this innovative technique include time-resolved, on-site, and point-of-care applicability. Modern analytical methods like sensor technology can meet these requirements. However, many sensors and analytical methods are not fast enough to realize breath-to-breath resolved sampling. Therefore, additional effort is necessary to provide well-defined and reproducible breath samples that can be fed into the measuring devices. Some substances in exhaled breath are blood borne and originate from the alveoli, others, such as NO and CO, are generated in the airways, other substances represent contaminants from ambient air and some other compounds may be generated in the oral cavity. Hence, thoroughly controlled sampling is a key requirement for reliable analysis of breath biomarkers.

In order to introduce sensor technology into clinical breath analysis, automatic, controlled, and adjustable sampling devices have to be at hand. These devices have to be easy to handle, non-expensive, and must be adaptable to any sensor in terms of sample volume, sampling flow, and sample origin (spontaneously breathing or mechanically ventilated individuals). For reasons of standardization, relevant data like sampling flow rate, total volume, or CO_2 partial pressures during sampling have to be recorded continuously.

Breath tests are attractive because they are non-invasive and can be repeated frequently without any risk to the patient. This is especially important in the dynamically changing state of critically ill patients. Despite a number of very promising results revealing interesting diagnostic properties of different markers, analysis of volatile compounds in the breath of the critically ill has not yet been introduced into clinical practice. The main obstacles are technical problems such as sampling,

Volatile Biomarkers. http://dx.doi.org/10.1016/B978-0-44-462613-4.00009-X

pre-concentration, and analysis as well as basic methodological issues such as normalization, expression of data, and the problem of background contamination. Apart from NO, generally accepted standards of sampling, pre-concentration, and analysis do not exist. For this reason, considerable effort has been expended in establishing reproducible and reliable analytical methods and in understanding how various compounds end up in the breath. Three broad questions challenge researchers studying volatile markers in the breath of critically ill patients. How should breath samples be collected? Given the low concentrations of VOCs in breath, how can they be detected and quantified in a fast and reliable way? And finally, is there a clear link between certain volatile substances and particular diseases?

This chapter is intended to describe the diagnostic potential of endogenous organic volatile substances in the breath of critically ill patients. Since many of these patients need ventilator support, aspects of breath analysis under mechanical ventilation will be addressed. Analytical procedures, problems concerning the physiological meaning of breath markers, and future developments will be discussed.

9.2 TECHNICAL ASPECTS OF BREATH ANALYSIS IN CRITICALLY ILL PATIENTS

The collection of gaseous samples represents a crucial issue in breath analysis.[1,2] Dilution and contamination of samples and loss of analytes during the sampling procedures have resulted in a huge variation of substance concentrations reported in different studies. These effects are even more pronounced in the clinical environment and under mechanical ventilation because of positive airway pressures.[3,4] For each substance, it must be determined if its source was actually the patient or if it came from ambient air or previous exposure. As substance concentrations found in exhaled air fall in the range from μmol/L to fmol/L (ppmv to pptv), analytical methods have to be highly sensitive. Identification of the respiratory phases as well as substance identification has to be unequivocal and reproducible.

9.2.1 Sampling

Since no general standards of sampling procedures exist for VOC sampling numerous studies still apply mixed expiratory sampling or sampling techniques use fixed parameters like time after beginning of expiration instead of physiological parameters.[3] Hence, reproducibility and reliability of sampling methods and analytical procedures are of critical importance.

As substance concentrations found in exhaled air fall in the range from μmol/L to fmol/L (ppmv to pptv), analytical results are easily affected by impurities from inspired air or by dilution with dead space air, especially when positive airway pressure occurs during mechanical ventilation.[4] Typically, a time resolution <100 ms is required for the parameter used for recognition of the respiratory phases. This holds true for direct (real-time) analysis[5] as well as for controlled (alveolar) sampling.

Sampling without controlled identification of respiratory phases bears the risks of dilution by dead space gas in the case of mixed expiratory sampling or shows large variations in the case of time controlled sampling.[3] The distinction between alveolar (end tidal) and non-alveolar (airways) can easily be achieved by means of expired CO_2 concentrations.[6] (Figure 9.1) Furthermore a differentiation can be made between a mixed expiration phase, containing dead space gas and alveolar gas, and pure alveolar gas.

In principle, there are two basic approaches to breath collection:

- Mixed expiratory sampling means that total breath including dead space air is collected.
- Alveolar sampling means that pure alveolar gas is collected.

Mixed expiratory sampling has been used most often since this method is easy to perform in spontaneously breathing subjects and requires no additional equipment. Concentrations of endogenous volatile substances in alveolar air are 2–3 times higher than those found in mixed expiratory samples because there is no dilution by dead space gas.[6] Moreover, alveolar breath samples have the lowest concentration of contaminants.[6,7]

Sampling may be done for a single breath or for a certain period of time. If single breath sampling is done one has to be sure, that this single breath is representative of all subsequent breaths. Especially for spontaneously breathing subjects, averaging is necessary because the breath-to-breath concentrations may vary considerably.

In most studies, exhaled breath was collected at the bedside into some sort of recipient and the gaseous samples were pre-concentrated and analyzed at the laboratory. Inert Tedlar bags or stainless steel electro-polished canisters[8] were used for the sampling. In other cases, sampling was done directly at the bedside by means of adsorption onto special adsorption materials[9–11] or by direct cryofocussation.[12]

9.2.2 Pre-concentration and desorption

Pre-concentration is necessary since most substance concentrations in exhaled breath fall in the nmol/L–pmol/L (ppbv–pptv) range. Pre-concentration can be achieved by adsorption onto sorbent traps, coated fibers (SPME),[13–15] or by direct cryofocussation.[12]

Due to the different boiling-points of the VOCs, adsorbents in sorbent traps have to be selected carefully to avoid breakthrough as well as memory effects. Organic polymers (e.g. Tenax TA),[16] activated charcoal,[11] different types of graphitized carbon (e.g. Carbopack X), and carbon molecular sieves (e.g. Carboxen 1021) have been used.[17] Since breath samples are water saturated and, for example, in ventilated patients additional humidification is applied, the relatively high content of water (increasing with increasing sample volume) has to be taken into account. Organic polymers are least affected by high water contents in the samples, but these have low breakthrough volumes, especially for small hydrocarbons. By contrast, carbon molecular sieves and graphitized carbon have high breakthrough volumes for these

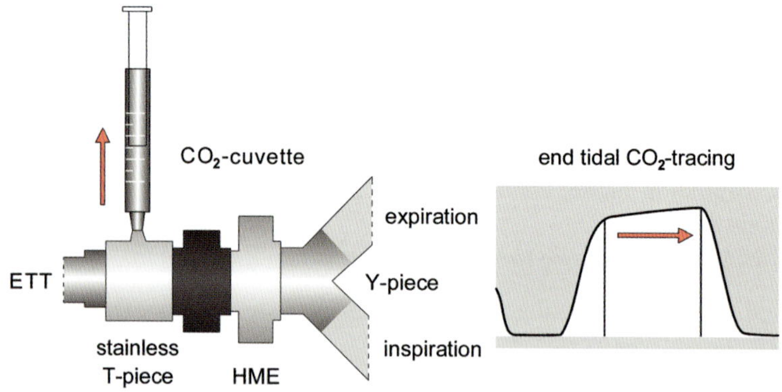

FIGURE 9.1

Schematic drawing of CO_2 controlled alveolar sampling. ETT, endotracheal tube, HME, heat and moisture exchanger. Exhaled air is withdrawn automatically or manually (syringe) during the alveolar phase of expiration.

compounds (e.g. ethane). These advantages are offset by possible memory effects when these compounds are used as the only adsorption material. The problem can be solved by using multibed sorbent traps.[17,18] Memory and breakthrough effects are minimized in that the strong sorbents such as graphitized carbon or molecular sieves are protected by a layer of a weaker adsorbent consisting, for instance, of an organic polymer such as Tenax.[18] For further analysis, the volatile substances have to be released from the adsorption material. This can be achieved by heating the trap (thermal desorption) or by means of microwave energy. Thermodesorption can be done automatically by means of commercially available devices. A disadvantage of these techniques are the relatively high sampling volumes (typically 100–3000 mL) that are needed for reliable and sensitive analysis that result in long sampling times, and potential problems related to the high water content in the samples.

Breath volatiles can also be pre-concentrated by means of SPME.[13] Due to the physical properties of the available fibers, relatively small volumes (typically 10–20 mL) can be used but the number of substances that can be adsorbed is limited. Desorption of volatile compounds from these coated fibers is done by direct heating in the gas chromatographic (GC) inlay. This adsorption technique is not affected by high water contents of the samples and SPME may be automated for adsorption as well as for desorption. As SPME is basically a distribution-based technique, amounts of the target analytes (and, therefore, detection limits) cannot be increased by further increasing the sample volume. Recent developments combine the advantages of both techniques (low sample volumes, good stability, lower detection limits by increase of sample volume) by applying miniaturized micro extraction techniques such as needle trap micro extraction (NTME).[19] These techniques have successfully been applied

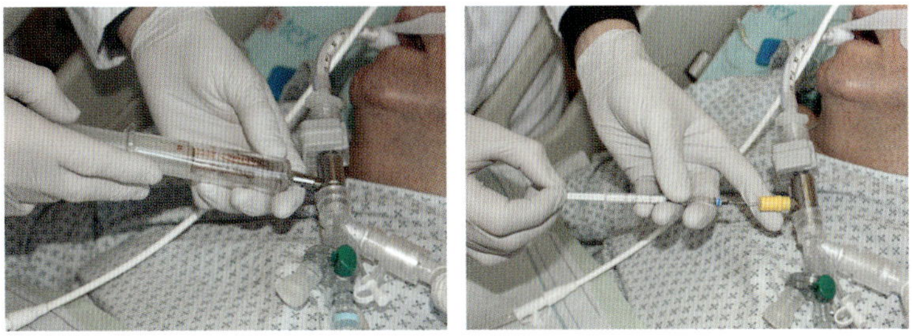

FIGURE 9.2
Alveolar sampling from mechanically ventilated patients by means of gas tight syringes for SPME analysis and by means of needle trap devices for needle trap micro extraction (NTME).

in the clinical environment and in combination with mechanical ventilation.[9,20] (see Figure 9.2)

9.2.3 Analysis and detection

Separation of volatile compounds has been done mostly by means of gas-liquid chromatography, but also gas–solid chromatography has been used.[21] Hydrocarbons in the nmol/L–pmol/L (ppbv–pptv) range, such as ethane, pentane, or isoprene, usually are analyzed using GC coupled to flame ionization detection (FID)[11,21,22] or mass-selective detection.[23,24] The choice of the separation column is crucial. In former studies the chromatographic separation of pentane and isoprene was not satisfactory and the sum of isoprene and pentane was determined as pentane.[11,22,25] Also, other substances like isopentane may co-elute with pentane.[26] As a consequence, mass spectrometric detection has to be applied for the identification of unknown substances. Only if a substance had been identified unequivocally by mass spectrometry and only if separation from all possible contaminants was perfect, easier and more economic detection systems such as FID can be used.

A critical issue is the high water content of breath samples, which may affect pre-concentration, separation, and detection of single compounds. This is especially true for mechanically ventilated patients if active humidifiers are used in the respiratory circuit. Other confounding sources include high concentrations of anesthetics, by-products of plastic materials, and disinfectants. To account for these problems, hyphenated separation techniques such as heart cut GC-GC or comprehensive GCxGC in combination with mass spectrometry represent a promising choice for breath analysis in critically ill patients.[9]

Real-time continuous measurements of defined compounds offer an interesting alternative for the monitoring of fast changes in volatile marker profiles. Effects of physiological parameters onto VOC exhalation may also be assessed in this way. Single compounds can be monitored in real-time by means of direct MS techniques such as PTR-MS.[5] Quantification of alveolar concentrations of VOC's is possible when time resolutions <200 ms can be realized or when adequate sampling systems are used.[27,28] In general, the method of separation as well as the kind of detection has to be adapted to the problems and substances under investigation.

9.2.4 Technical development—trends in breath analysis

Breath sampling, pre-concentration, and analysis still require sophisticated equipment and excellent skills. Recent developments and technical progress, however, show a trend toward improved and simplified analytical techniques which could be employed for routine use in the near future.

Sampling can be standardized by automatic devices being able to separate alveolar air from dead space and inspiratory gas.[6,29] Pre-concentration of some volatile substances has been simplified considerably by the introduction of micro extraction techniques such as SPME[13] and NTME[19,20] facilitating automatic pre-concentration as well as automatic desorption. Promising developments of real-time determination of volatile substances are going on. Ethane can already be detected in the pmol/L (pptv) range via laser spectroscopy,[30] other compounds have been detected and identified without relevant delay by means of selected ion flow tube mass spectrometry (SIFT-MS)[31] or PTR-MS.[5,27] Combination of direct mass spectrometry with TOF analyzers, when mass resolutions of 5000 and more can be realized, bears enormous potential for the application in critically ill patients (see Figure 9.3). However, before

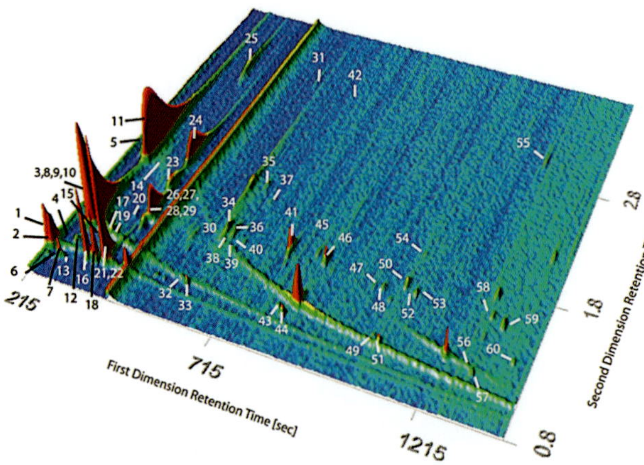

FIGURE 9.3

Detail of a two-dimensional chromatogram showing exhaled substances (pentane to *n*-decane) from a ventilated patient after induction of anesthesia. Peaks can be identified according to mass spectra, relative retention times, and retention index (alkanes). Reprinted with permission from Ref. 9, Copyright 2010 American Chemical Society.

these devices can be used in the clinical environment, approval for their use in the clinical environment has to be obtained.

Non-specific sensor array technology (electronic noses)[32,33] may gain importance for diagnostic purposes when specific patterns of disease markers have been defined and interference factors can be excluded. Specific sensors for single substances hold promise if occurrence or the time course of these substances is in close relation with disease.

9.3 METHODOLOGICAL ASPECTS

9.3.1 Normalization; expression of data

Crucial problems hampering the use of exhaled VOCs in clinical practice are the lack of standardization of analytical methods and the huge variation of results between different studies. Due to substantially different technical approaches, data cannot be compared to each other if there is no normalization to some generally accepted standard substance or standard condition. This issue is still a matter of debate and the impact of physiological variables such as cardiac output, breathing pattern,[34] and ventilation/perfusion ratio of the lung[35] on the results of breath analysis have not been explored in sufficient detail. Under normal conditions (healthy volunteers) these effects may be negligible. In patients with severe hemodynamic or pulmonary dysfunction, however, data should be corrected or normalized by means of end-tidal and/or arterial PCO_2 measurement.

9.3.2 Inspired concentrations

Other parameters that can affect results are inspired substance concentrations. Addition of inhaled concentrations to exhaled concentrations may blur the concentration differences present in the blood. Expired samples are diluted or contaminated by inspired and/or dead space gas depending on the ratio of alveolar and dead space ventilation,[4] which itself depends on the breathing pattern.[34] Alveolar concentration gradients are altered through inspired substances and substance elimination is affected accordingly, i.e. less substance is eliminated from the blood into the alveoli than venous concentrations would induce without the presence of inspired substances. The extent of this influence is controlled by the ventilation/perfusion ratio in the lung.[35,36] In addition, background concentrations of volatile compounds in inspiratory air have to be taken into account in order to distinguish between endogenous substances and exogenous contaminants.

Different approaches to overcome this problem have been adopted. Substance concentrations in ambient or inspiratory air have been measured and exhaled concentrations corrected by subtracting inspiratory from expiratory concentrations or by calculating "alveolar gradients" (expired-inspired concentration normalized to an internal (GC) standard[24]). Others tried to eliminate ambient concentrations by having patients or volunteers breathe pure air for a certain time before measurement.[17]

This approach is the most effective but cumbersome and time consuming, and will not be acceptable for routine clinical purposes.

Calculating alveolar gradients or any other subtraction methods are easy to perform, but these methods do not take into account the complexity of pulmonary adsorption and exhalation of volatile substances. Excretion and intake of volatile substances depend on the ventilation/perfusion ratio in the lung and on the alveolar concentration gradients of the substances.[35] Due to the complexity of pulmonary adsorption and exhalation of volatile substances,[35,36] these effects are not linear. These complex inter-dependencies cannot be accounted for by simply subtracting inspired from expired substance concentrations. The more a patient's lung is diseased the more these problems will gain importance.

This reasoning is in contrast to the findings of other researchers who use subtraction methods to account for inspired substance concentrations. The recognition pattern Phillips et al. uses to detect lung cancer,[37] for instance, is based on the calculation of alveolar gradients (expired-inspired concentration normalized to an internal standard[24]) regardless of the relative amount of inspired concentrations. The patients these workers investigated substantially differed from critically ill patients in that they were spontaneously breathing. Hence, the likelihood of severe deviation from normal breathing patterns and the presence of ventilation/perfusion mismatch were much lower. The non-linear effects of inspired substance concentrations most probably are linked to the occurrence of dead space ventilation and shunt perfusion, which is a common and frequent problem in mechanically ventilated patients.[38,39] This may explain why Phillips et al. were able to obtain clinically relevant results by using a subtraction method. In addition, no information on the ratio of substance concentrations in blood and breath in their patients is available, since Phillips et al. did not determine marker concentrations in blood. This bears the risk that results might change if for any reason patients' breathing patterns are modified.[34]

9.3.3 Data interpretation

Data interpretation is a crucial step in breath analysis and, therefore, every effort should be made to implement data cross-testing and statistical validation during this process to ensure that the measurements made and the marker patterns observed are actually valuable and representative for the intended purpose.

To avoid artifacts in breath testing, every effort should be made to implement method validation, data cross-testing, and statistical validation during this procedure. The most crucial data analysis related problems can be classified into three groups: confounding variables (CV), which have a real correlation with both the disease state and a breath marker, but lead to the erroneous conclusion that disease and breath are in a causal relationship, voodoo correlations (VC), which can be understood as statistically true correlations, which arise coincidentally in the vast number of measured variables, and statistical misconceptions in the study design (SMSD).

- *Confounding variables (CV):* CV have a real statistical correlation with both the disease state and a breath marker. This may lead to the erroneous conclusion that disease and breath are in a causal relationship. Typical confounding variables are environmental and medical history, host factors such as gender, age, weight, etc. and parameters that could affect the quality of breath data such as subject breathing mode, effects of breath sampling, and effects of the analytical technique itself.[3,4]

- *Voodoo correlations (VC):* The number of measured variables quickly overwhelms the number of samples that can feasibly be taken. As a consequence, the chances of finding coincidental "voodoo" correlations grow proportionally. Voodoo correlations can typically be expected in the following scenarios: insufficient number of patients, (too) many measurement variables, the use of advanced statistical data-mining methods, and the use of non-independent data for validation.[40,41]
- *Statistical misconceptions in the study design (SMSD):* Non-prospective, non-blinded and non-randomized trials, *a priori* biased study populations or group selection with unrealistically high disease prevalence.

A typical example of CV during measurements in critically ill patients is the effect of contaminations from the clinical environment. If the control group is not well matched (e.g. if diseased patients are compared to healthy volunteers from a different environment), then volatile substances (such as disinfectants) originating from the clinical environment will be always present in higher concentrations in the diseased group even though there is no correlation to the disease. Indoor air in hospitals also contains many different compounds such as isopropanol, ethanol, isoflurane,[42] sevoflurane,[43] or *p*-xylene. Compounds contained in cleaning agents, such as limonene, may also be stored in the human body and subsequently be released into the breath. These effects must be taken into account and be addressed during data interpretation, especially in a clinical setting. A considerable influence on the concentrations of volatile compounds in exhaled breath may also be generated by other confounding variables such as the sampling procedure itself,[3] the effect of storage and effects of physiological parameters like cardiac output and ventilatory flow.

Voodoo correlations can be understood as statistically true correlations, which arise coincidentally in the vast number of measured variables. For the first time, Vul et al. critically challenged many publications in the field of social neuroscience, which were reporting "mysteriously high correlations." His article[44] has triggered a vivid and still ongoing discussion.

When a large number of variables is measured, a (random) variable can coincidentally correlate with, for example, the disease state. This statistically unlikely scenario becomes dramatically more manifest when the number of variables (m) is large compared to the number of independent measurements (n). As a general rule, n should be larger than five times m. The ability of sophisticated analytical instruments to measure a vast number of compounds (m, e.g. $m \sim 1000$ for GCxGCxTOF) in breath thus quickly overwhelms the number of samples that can feasibly be taken and the number of distinct patients that can be recruited. As such, the chances of finding coincidental correlations become real.

The n/m ratio determines the susceptibility for voodoo correlations, which rises, when the number of measurements is decreased, e.g. by dividing the study population into subgroups. Vice Versa, when *m* is increased, e.g. by calculating *additional* variables such as any "corrected" values, the likeliness of VC also rises, falsely giving better results. In this light, the dramatic effect that sophisticated data-mining tools play for voodoo correlations can also be understood. Using a method that can combine two (or more) variables to enhance results (pattern analysis), m^*m ($^*m, \ldots$) new variables can be created and multiply the problem.

If reliability, sensitivity, specificity, and relevance of clinical tests are to be assessed, prospective, blinded, and randomized trials have to be undertaken. Retrospective or observational studies do not have great value in this context.[45] Disease prevalence within the study groups represents another problem that is often underestimated. Frequently, there is a disease prevalence of near 100% in the patient group as only diseased subjects are included; controls are chosen from the normal population. Any data concerning reliability and clinical relevance of the test derived from such a study must not be transferred to a setting where the disease prevalence is low, as it is the case when (screening) tests are used in the general population. The classification of study participants as healthy or diseased should to be confirmed through solid clinical data.

9.3.4 Physiological meaning, biochemical pathways

If volatile substances found in the breath are to be used as markers of disease or health their physiological meaning and their pathway of generation should be known. As substances may be generated within the body and may simultaneously be taken in from the environment, mechanisms of storage and metabolism of these substances can gain importance.

9.3.5 Study design

In order to build breath tests that can be used in practice, their clinical relevance in terms of sensitivity, specificity, positive, and negative predictive values has to be established. For that purpose, studies have to be powered sufficiently and adequate statistical methods have to be applied. In addition, possible effects of contaminations from the laboratory or from the specific environment have to be taken into account, since qualitative and quantitative composition of background air may vary considerably from one place to the other.

9.4 CLINICAL ASPECTS OF BREATH ANALYSIS IN CRITICALLY ILL PATIENTS

9.4.1 Oxidative stress; inflammation

Reactive oxygen species (ROS), such as the superoxide anion, O_2^-, or the hydroxyl radical, OH, physiologically act as defense mechanisms against microbial attack. ROS are generated by activated granulocytes and can potentially damage any cellular structure. For that reason, there are a number of protective mechanisms to prevent the organism from being destroyed by its own weapons. Important antioxidant (enzyme) systems are glutathione, superoxide dismutase, catalase, and vitamins C and E. Under healthy conditions, ROS activity is restricted to limited regions of external attack or inflammation and is well balanced by antioxidant protection of the body itself. In some diseased states, however, the balance between ROS activity and protection may be impaired when antioxidant systems are overwhelmed or exhausted. This typically happens during systemic inflammatory reaction syndrome (SIRS) or sepsis when inflammatory reactions are triggered throughout the body by infectious (sepsis) or non-infectious (SIRS) stimuli. Ischemia/reperfusion also leads to generation of ROS that may locally or systematically exhaust antioxidant defense mechanisms. Whenever ROS activity takes place in an uncontrolled way the organism itself will be damaged by what is called "oxidative stress." Multi organ dysfunction (MODS) or multi organ failure (MOF) that are among the leading reasons of morbidity and mortality in critical care are caused in this way. For that reason, any measure or device facilitating diagnostics or therapy of oxidative stress can improve the outcome of critically ill patients.

ω-3 and ω-6 fatty acids are basic components of cell membranes. *In vitro* studies have shown that ethane and pentane are generated when cell cultures are exposed to ROS.[46] Animal and clinical studies have demonstrated a close correlation between clinical conditions with high peroxidative activity and the exhalation of ethane and pentane.[22] Hence, these aliphatic hydrocarbons are regarded as *in vitro* and *in vivo* markers of lipid peroxidation. Furthermore, levels of exhaled pentane and ethane correlate well with other lipid peroxidation markers such as malondialdehyde (MDA), thiobarbituric acid reactive substances (TBARS), or glutathione.[47] Breath markers often proved to be more sensitive than the serum markers MDA and TBARS.[48] Although there are other potential sources of hydrocarbons in the body, such as protein oxidation and colonic bacterial metabolism, these are apparently of limited importance and do not interfere with the interpretation of the hydrocarbon breath test for ethane and pentane.[49]

Due to the physiological ratio of ω-3 to ω-6 fatty acids, four times more pentane than ethane is generated through lipid peroxidation. Pentane, but not ethane, may easily be metabolized by hepatic cytochrome P450 enzymes.[50] As a consequence, exhaled pentane concentrations should be interpreted cautiously whenever liver function varies between patients or during the sampling period.

Propane and butane are mainly derived from protein oxidation and fecal flora and their role as markers of lipid peroxidation is doubtful. Methylated hydrocarbons have also been described as lipid peroxidation markers.[51,52] However, the biochemical pathways of generation and the physiological meaning of these compounds have not yet been elucidated in sufficient depth.[53] Hydrocarbons as stable end products of lipid peroxidation show only low solubility in blood and are therefore excreted into breath within minutes of their formation in tissues. Hence, exhaled concentrations of ethane and *n*-pentane can be used to monitor the degree of oxidative damage in the body.[17,54]

Elevated concentrations of exhaled *n*-alkanes were found following reperfusion of ischemic liver[55] following abdominal ischemia[56] in ischemic heart disease,[57] in myocardial infarction,[58] following cardio pulmonary bypass[59] and in acute rejection after heart transplant.[60] Exhaled pentane concentrations were different in patients with systemic inflammatory reaction syndrome (SIRS) and sepsis.[48] The differentiation between these pathological inflammatory conditions could have important consequences. SIRS is best treated with anti-inflammatory agents whereas sepsis demands consequent antibiotic use.

The data obtained during hemodialysis are not unequivocal. Spittle et al.[61] found increased concentration of exhaled ethane during hemodialysis, whereas Capodicasa et al. did not find any change in the exhaled concentrations of *n*-alkanes under these conditions.[62] In addition, a very large variation of exhaled ethane concentrations has been observed.[63] From these results it is not possible to evaluate the role of lipid peroxidation during hemodialysis.

In order to control lipid peroxidation, numerous studies have been undertaken where the influence of vitamin supplements or special diets were examined.[47,64–68] Supplementation of vitamins E and C, as well as a low-fat-high-vegetable diet, significantly decreased exhaled ethane and/or pentane concentrations. The mechanism proposed by the authors of these studies is an increase in radical scavenging capacities by means of vitamin or vegetable supplementation. Since inflammation and oxidative stress are among the most important reasons of disease and organ damage in critically ill patients, breath tests involving lipid peroxidation markers, such as ethane and pentane, may soon play an important role in diagnosis and therapy control in the intensive care unit.

9.4.2 Lung diseases

The lung is among those organs that are frequently affected in the critically ill. Pneumonia and acute respiratory distress syndrome (ARDS) are well known and dreaded complications in critical care. Chronic lung diseases such as asthma, chronic obstructive pulmonary disease, COPD, or pulmonary hypertension can seriously affect patients' outcomes.

Many lung diseases, including COPD, asthma, bronchiectasis, cystic fibrosis, interstitial lung disease, and ARDS are linked to chronic or acute inflammation. Since peroxidation and other reactions of ROS are basic mechanisms of inflammatory processes, organic lipid peroxidation markers should be elevated under these conditions. In fact, pentane and ethane levels are increased in patients with asthma,[69,70] COPD,[71] obstructive sleep apnea,[72] pneumonia,[73] and ARDS.[2,14]

Another substance linked to airway inflammation is nitric oxide, NO. Exhaled NO concentrations have been extensively studied in asthma and other inflammatory lung diseases and will not be discussed here. Since NO and VOCs such as ethane and pentane are generated in different regions of the lung, breath markers may give information on the local distribution of inflammatory processes.

Isoprene (2-methyl-1,3-butadiene) is always present in human breath, and is thought to be formed along the mevalonic pathway of cholesterol synthesis,[74,75] but small fraction of exhaled isoprene may be of bacterial origin.[76] There is experimental evidence

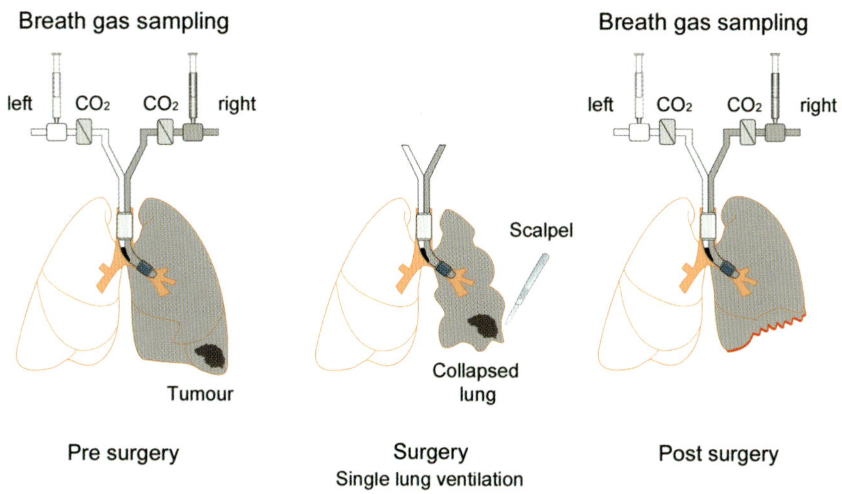

FIGURE 9.4
Schematic drawing of parallel breath gas sampling from non-diseased and diseased lungs pre and post-surgery. T: stainless steel T-pieces with Luer Lock adapter and gas tight syringe CO_2: Fast CO_2-mainstream monitoring. Modified from Ref. 87.

that isoprene exhalation may be related to oxidative damage to the fluid lining of the lung[77] and the body.[78] Exhaled isoprene concentrations have been found to be significantly lower in ARDS patients when compared to any other patient group.[2,73] Human breath isoprene concentrations show a circadian rhythm with a maximum at around 6 a.m. and a minimum at about 6 p.m. Concentrations of breath isoprene seem also to be age dependent. Isoprene concentrations are significantly lower in children.[79,80]

Breath volatiles are linked to acute lung diseases such as pneumonia and ARDS and to chronic affections such as asthma or COPD and exhibit, therefore, a considerable potential for diagnostics and surveillance of the critically ill. Since many bacteria produce volatile substances when they are growing[81–86] even recognition of infectious agents from exhaled air seems possible.

Other interesting aspects arise from the determination of potential lung disease markers in the perioperative setting, e.g. in patients undergoing lung surgery for tumor resection (see Figure 9.4).

Inter-individual variations can be minimized and potential markers as well as confounders can easily be identified when exhaled air from healthy and diseased lungs is analyzed in the same patient before, during and after tumor resection.[87]

9.4.3 Metabolism

Metabolic disorders and starvation play an important role in critical care medicine. Protein catabolism and anaerobic metabolic pathways cause additional morbidity in the critically ill. Easy to measure and reliable markers for these pathological states are lacking.

Acetone (dimethyl ketone) is one of the most abundant compounds in human breath. Acetone is formed by decarboxylation of acetoacetate which derives from lipolysis or lipid peroxidation. Ketone bodies like acetone are oxidized via the Krebs cycle in peripheral tissue. Ketone bodies in blood (including acetoacetate and β-hydroxybutyrate) are increased in ketonemic subjects, in times of fasting or starving or during diet. Breath acetone concentrations are increased in patients with (uncontrolled) diabetes mellitus.[88] As acetone is produced by spontaneous decarboxylation of acetoacetate it is impossible to quantify the fraction that arises from lipid peroxidation.

Acetaldehyde is probably produced by oxidation of endogenous ethanol. As a consequence, acetaldehyde concentrations in breath are always lower than the corresponding ethanol concentrations, which in the exhaled breath of human subjects are normally very much lower than the levels found in human breath after alcohol ingestion. The potential source of endogenous ethanol is the intestinal bacterial flora.[89] 2-Propanol is postulated to be a product of an enzyme mediated reduction of acetone. As for acetaldehyde, concentrations of 2-propanol in humans are always lower than acetone concentrations.[90] The origin of breath methanol may also be the intestinal bacterial flora.

Breath volatiles can provide valuable information on metabolic states in the critically ill. Since measurements are non-invasive and can be repeated frequently a continuous metabolic surveillance seems possible.

9.4.4 Organ dysfunction

Sulfur-containing compounds like ethyl mercaptan, dimethylsulfide or dimethyldisulfide are responsible for the characteristic odor in the breath of cirrhotic patients.[91] Sulfur-containing compounds are generated in humans by incomplete metabolism of methionine in the transamination pathway.[92] Mercaptans are easily oxidized to their respective sulfides. Under normal conditions, concentrations of sulfur-containing compounds in human blood and breath are very low. Impairment of liver function increases the level of sulfur-containing compounds.

The characteristic odor of uremic breath due to elevated levels of dimethylamine and trimethylamine has been known for a long time. In 1977 amines were conclusively identified and quantified in human breath.[93] Significant levels of ammonia will appear in the blood if the removal of ammonia through conversion to urea is limited due to an impairment of liver function. Ammonia has also been identified in the breath of healthy volunteers and in higher concentrations in uraemic patients.[94]

9.4.5 Transplantation

A growing number of transplant recipients have to be treated in the intensive care unit. Most often, the reason is chronic or acute rejection of the transplanted organ and infectious complications due to immune-suppression. A crucial problem in these patients is the distinction between organ dysfunction due to infection or any other complication and organ dysfunction due to organ rejection. Up to now, invasive procedures such as biopsy (heart, liver, and kidney) or bronchoscopy (lung) are necessary for that purpose.

Allograft rejection following organ transplantation has been linked to exhaled alkane concentrations. Pentane concentrations were found to be elevated in acute rejection of transplanted hearts.[95] One has to remember, however, that pentane as a lipid peroxidation marker is linked to inflammation rather than being a specific marker for allograft rejection. A volatile marker not found in the exhaled breath of healthy individuals is carbon disulfide. This substance seems to be generated as a by-product of methionine metabolism and may act as an acute marker for organ rejection after lung transplantation.[96]

9.4.6 Drug monitoring

Monitoring of intravenous drugs represents another interesting aspect of breath analysis with a high clinical relevance. Breath analysis could offer a non-invasive means of intravenous drug monitoring if robust correlations between drug concentrations in breath and blood can be established. A decade ago, detection of the intravenous anesthetic propofol (1,3-diisopropylphenol) in the breath has been reported by Harrison et al.[97] Correlations between blood and breath concentrations of propofol in ventilated patients have been found.[27,98–103] However physiological parameters such as changes in the ventilation/perfusion ratio or cardiac output led to a deterioration of the relationship between breath and blood propofol concentrations and have to be taken into account before levels of drug blood concentrations can be estimated by breath measurements.

9.5 CONCLUSIONS

Volatile organic compounds are mainly blood borne and therefore enable monitoring of different processes and different organ systems in the body. Single substances or sets and patterns of exhaled markers can be correlated to clinically relevant conditions such as oxidative stress, lung diseases, organ dysfunction, or allograft rejection. Exhaled ethane and pentane concentrations are elevated during oxidative stress and in inflammatory diseases. Acetone is linked to dextrose metabolism and lipolysis. Exhaled isoprene concentrations show correlations with pneumonia and ARDS. Exhaled levels of sulfur-containing compounds are elevated in liver failure and allograft rejection. Estimation of blood drug levels by means of continuous breath analysis seems possible, if physiological limitations are taken into account.

Despite the considerable progress achieved in sampling and analytical procedures, up to now breath analysis has not yet been introduced into clinical practice. This is mainly due to the lack of standardization and normalization of sampling procedures and the absence of generally accepted evaluation criteria for data. In addition, biochemical pathways and exhalation kinetics have been investigated only for a small number of volatile substances found in human breath. Many critically ill patients need ventilator support because of pneumonia or ARDS. Hence, special attention has to be paid to pathological conditions of the lung as changes of ventilation and perfusion may seriously affect exhalation mechanisms.

Breath analysis can offer important advantages when applied in the critically ill. In the intensive care environment, where situations are changing dynamically and sometimes rapidly, breath analysis seems attractive since it is non-invasive and can be repeated almost indefinitely. If the diagnostic potential of this method is to be used in clinical practice, standardization has to be promoted and more has to be learned about origins, physiological meaning, and exhalation kinetics of the volatile substances. Further study is required as well in the clinical environment as in the controllable setting of animal models in order to improve sensitivity and specificity of the exhaled biomarkers.

Regarding the technical and scientific progress of recent years, one may expect that analysis of VOCs in exhaled air of critically ill patients can be introduced into clinical practice in the near future. Ongoing development of real-time analysis will further expand the potential of breath analysis as non-invasive and fast instrument of diagnosis and therapy surveillance.

Acknowledgments

The authors like to thank R. Fuchs for his valuable help in designing figures and diagrams.

References

1. Miekisch W, Schubert JK, Noeldge-Schomburg GF. Diagnostic potential of breath analysis-focus on volatile organic compounds. *Clin Chim Acta* 2004;**347**:25–39.
2. Schubert JK, Miekisch W, Geiger K, Noldge-Schomburg GFE. Breath analysis in critically ill patients: potential and limitations. *Expert Rev Mol Diagn* 2004;**4**:619–29.
3. Miekisch W, Kischkel S, Sawacki A, Liebau T, Mieth M, Schubert JK. Impact of sampling procedures on the results of breath analysis. *J Breath Res* 2008;**2**:026007.
4. Schubert JK, Miekisch W, Birken T, Geiger K, Noldge-Schomburg GF. Impact of inspired substance concentrations on the results of breath analysis in mechanically ventilated patients. *Biomarkers* 2005;**10**:138–52.
5. Schwoebel H, Schubert R, Sklorz M, Kischkel S, Zimmermann R, Schubert JK, et al. Phase-resolved real-time breath analysis during exercise by means of smart processing of PTR-MS data. *Anal Bioanal Chem* 2011;**401**:2079–91.
6. Schubert JK, Spittler KH, Braun G, Geiger K, Guttmann J. CO(2)-controlled sampling of alveolar gas in mechanically ventilated patients. *J Appl Physiol* 2001;**90**:486–92.
7. Schubert JK, Esteban-Loos I, Geiger K, Guttmann J. *In vivo* evaluation of a new method for chemical analysis of volatile components in the respiratory gas of mechanically ventilated patients. *Technol Health Care* 1999;**7**:29–37.

8. Pleil JD, Lindstrom AB. Measurement of volatile organic compounds in exhaled breath as collected in evacuated electropolished canisters. *J Chromatogr B: Biomed Appl* 1995;**665**:271–9.
9. Mieth M, Schubert JK, Groger T, Sabel B, Kischkel S, Fuchs P, et al. Automated needle trap heart-cut GC/MS and needle trap comprehensive two-dimensional GC/TOF-MS for breath gas analysis in the clinical environment. *Anal Chem* 2010;**82**:2541–51.
10. Phillips M, Greenberg J. Ion-trap detection of volatile organic compounds in alveolar breath. *Clin Chem* 1992;**38**:60–5.
11. Mueller W, Schubert J, Benzing A, Geiger K. Method for analysis of exhaled air by microwave energy desorption coupled with gas chromatography-flame ionization detection-mass spectrometry. *J Chromatogr B: Biomed Sci Appl* 1998;**716**:27–38.
12. Knutson MD, Viteri FE. Concentrating breath samples using liquid nitrogen: a reliable method for the simultaneous determination of ethane and pentane. *Anal Biochem* 1996;**242**:129–35.
13. Grote C, Pawliszyn J. Solid-phase microextraction for the analysis of human breath. *Anal Chem* 1997;*69*:587–596.
14. Miekisch W, Schubert JK, Vagts DA, Geiger K. Analysis of volatile disease markers in blood. *Clin Chem* 2001;**47**:1053–60.
15. Sabel B, Schubert J, Miekisch W, Fuchs P, Kischkel S, Mieth M. Comparison of different preconcentration techniques in breath analysis under real life clinical conditions. *Eur J Med Res* 2010;**15**:1.
16. Tangerman A, Meuwese-Arends MT, van Tongeren JH. New methods for the release of volatile sulfur compounds from human serum: its determination by Tenax trapping and gas chromatography and its application in liver diseases. *J Lab Clin Med* 1985;**106**:175–82.
17. Risby TH, Sehnert SS. Clinical application of breath biomarkers of oxidative stress status. *Free Radic Biol Med* 1999;**27**:1182–92.
18. Larstad M, Loh C, Ljungkvist G, Olin AC, Toren K. Determination of ethane, pentane and isoprene in exhaled air using a multi-bed adsorbent and end-cut gas-solid chromatography. *Analyst* 2002;**127**:1440–5.
19. Trefz P, Kischkel S, Hein D, James ES, Schubert JK, Miekisch W. Needle trap micro-extraction for VOC analysis: effects of packing materials and desorption parameters. *J Chromatogr A* 2012;**1219**:29–38.
20. Mieth M, Kischkel S, Schubert JK, Hein D, Miekisch W. Multibed needle trap devices for on site sampling and preconcentration of volatile breath biomarkers. *Anal Chem* 2009;**81**:5851–7.
21. Lemoyne M, Van Gossum A, Kurian R, Jeejeebhoy KN. Plasma vitamin E and selenium and breath pentane in home parenteral nutrition patients. *Am J Clin Nutr* 1988;**48**:1310–5.
22. Aghdassi E, Allard JP. Breath alkanes as a marker of oxidative stress in different clinical conditions. *Free Radic Biol Med* 2000;**28**:880–6.
23. Jansson BO, Larsson BT. Analysis of organic compounds in human breath by gas chromatography-mass spectrometry. *J Lab Clin Med* 1969;**74**:961–6.
24. Phillips M. Method for the collection and assay of volatile organic compounds in breath. *Anal Biochem* 1997;**247**:272–8.
25. Springfield JR, Levitt MD. Pitfalls in the use of breath pentane measurements to assess lipid peroxidation. *J Lipid Res* 1994;**35**:1497–504.

26. Mitsui T, Naitoh K, Tsuda T, Hirabayashi T, Kondo T. Is endogenous isoprene the only coeluting compound in the measurement of breath pentane? *Clin Chim Acta* 2000;299: 193–8.
27. Kamysek S, Fuchs P, Schwoebel H, Roesner JP, Kischkel S, Wolter K, et al. Drug detection in breath: effects of pulmonary blood flow and cardiac output on propofol exhalation. *Anal Bioanal Chem* 2011;**401**:2093–102.
28. Boshier PR, Cushnir JR, Mistry V, Knaggs A, Spanel P, Smith D, et al. On-line, real time monitoring of exhaled trace gases by SIFT-MS in the perioperative setting: a feasibility study. *Analyst* 2011;136:3233–7.
29. Miekisch W, Hengstenberg A, Kischkel S, Beckmann U, Mieth M, Schubert JK. Construction and evaluation of a versatile CO(2) controlled breath collection device. *IEEE Sens J* 2010;**10**:211–5.
30. von Basum G, Dahnke H, Halmer D, Hering P, Murtz M. Online recording of ethane traces in human breath via infrared laser spectroscopy. *J Appl Physiol* 2003;**95**:2583–90.
31. Spanel P, Smith D. Selected ion flow tube: a technique for quantitative trace gas analysis of air and breath. *Med Biol Eng Comput* 1996;**34**:409–19.
32. Mohamed EI, Bruno E, Linder R, Alessandrini M, Di Girolamo A, Poppl SJ, et al. A novel method for diagnosing chronic rhinosinusitis based on an electronic nose. *An Otorrinolaringol Ibero Am* 2003;**30**:447–57.
33. Di Natale C, Macagnano A, Martinelli E, Paolesse R, D'Arcangelo G, Roscioni C, et al. Lung cancer identification by the analysis of breath by means of an array of non-selective gas sensors. *Biosens Bioelectron* 2003;18:1209–18.
34. Cope KA, Watson MT, Foster WM, Sehnert SS, Risby TH. Effects of ventilation on the collection of exhaled breath in humans. *J Appl Physiol* 2004;**96**:1371–9.
35. Wagner PD, Saltzman HA, West JB. Measurement of continuous distributions of ventilation-perfusion ratios: theory. *J Appl Physiol* 1974;**36**:588–99.
36. West JB, Wagner PD. Pulmonary gas exchange. *Am J Respir Crit Care Med* 1998;157: S82–7.
37. Phillips M, Cataneo RN, Cummin AR, Gagliardi AJ, Gleeson K, Greenberg J, et al. Detection of lung cancer with volatile markers in the breath. *Chest* 2003;**123**:2115–23.
38. Dembinski R, Max M, Bensberg R, Bickenbach J, Kuhlen R, Rossaint R. High-frequency oscillatory ventilation in experimental lung injury: effects on gas exchange. *Intensive Care Med* 2002;**28**:768–74.
39. Feihl F, Eckert P, Brimioulle S, Jacobs O, Schaller MD, Melot C, et al. Permissive hypercapnia impairs pulmonary gas exchange in the acute respiratory distress syndrome. *Am J Respir Crit Care Med* 2000;**162**:209–15.
40. Fiedler K. Voodoo correlations are everywhere-not only in neuroscience. *Perspect Psychol Sci* 2011;**6**:163–71.
41. Poldrack RA, Mumford JA. Independence in ROI analysis: where is the voodoo? *Soc Cogn Affect Neurosci* 2009;**4**:208–13.
42. Cope KA, Merritt WT, Krenzischek DA, Schaefer J, Bukowski J, Foster WM, et al. Phase II collaborative pilot study: preliminary analysis of central neural effects from exposure to volatile anesthetics in the PACU. *J Perianesth Nurs* 2002;**17**:240–50.
43. Rieder J, Lirk P, Summer G, Gunkel A, Riccabona U, Innerhofer P, et al. Exposure to sevoflurane in otorhinolaryngologic operations. *Can J Anaesth* 2001;**48**:934.
44. Vul E, Harris C, Winkielman P, Pashler H. Puzzlingly high correlations in fMRI studies of emotion, personality, and social cognition. *Perspect Psychol Sci* 2009;**4**:274–90.

45. Higgins JP, Altman DG, Gotzsche PC, Juni P, Moher D, Oxman AD, et al. The Cochrane Collaboration's tool for assessing risk of bias in randomised trials. *BMJ* 2011;**343**:d5928.
46. Pitkanen OM, Hallman M, Andersson SM. Determination of ethane and pentane in free oxygen radical-induced lipid peroxidation. *Lipids* 1989;**24**:157–9.
47. Aghdassi E, Wendland BE, Steinhart AH, Wolman SL, Jeejeebhoy K, Allard JP. Antioxidant vitamin supplementation in Crohn's disease decreases oxidative stress. A randomized controlled trial. *Am J Gastroenterol* 2003;**98**:348–53.
48. Scholpp J, Schubert JK, Miekisch W, Geiger K. Breath markers and soluble lipid peroxidation markers in critically ill patients. *Clin Chem Lab Med* 2002;**40**: 587–94.
49. Kneepkens CM, Lepage G, Roy CC. The potential of the hydrocarbon breath test as a measure of lipid peroxidation. *Free Radic Biol Med* 1994;**17**:127–60.
50. Allerheiligen SR, Ludden TM, Burk RF. The pharmacokinetics of pentane, a by-product of lipid peroxidation. *Drug Metab Dispos* 1987;**15**:794–800.
51. Phillips M, Cataneo RN, Greenberg J, Grodman R, Gunawardena R, Naidu A. Effect of oxygen on breath markers of oxidative stress. *Eur Respir J* 2003;**21**: 48–51.
52. Phillips M, Cataneo RN, Greenberg J, Gunawardena R, Rahbari-Oskoui F. Increased oxidative stress in younger as well as in older humans. *Clin Chim Acta* 2003;**328**:83–86.
53. Mitsui T, Kondo T. Inadequacy of theoretical basis of breath methylated alkane contour for assessing oxidative stress. *Clin Chim Acta* 2003;**333**:91; author reply 93–4.
54. Kohlmuller D, Kochen W. Is n-pentane really an index of lipid peroxidation in humans and animals? A methodological reevaluation. *Anal Biochem* 1993;**210**:268–76.
55. Kazui M, Andreoni KA, Norris EJ, Klein AS, Burdick JF, Beattie C, et al. Breath ethane: a specific indicator of free-radical-mediated lipid peroxidation following reperfusion of the ischemic liver. *Free Radic Biol Med* 1992;**13**:509–15.
56. Kazui M, Andreoni KA, Williams GM, Perler BA, Bulkley GB, Beattie C, et al. Visceral lipid peroxidation occurs at reperfusion after supraceliac aortic cross-clamping. *J Vasc Surg* 1994;**19**:473–7.
57. Mendis S, Sobotka PA, Leja FL, Euler DE. Breath pentane and plasma lipid peroxides in ischemic heart disease. *Free Radic Biol Med* 1995;**19**:679–84.
58. Weitz ZW, Birnbaum AJ, Sobotka PA, Zarling EJ, Skosey JL. High breath pentane concentrations during acute myocardial infarction. *Lancet* 1991;**337**:933–5.
59. Andreoni KA, Kazui M, Cameron DE, Nyhan D, Sehnert SS, Rohde CA, et al. Ethane: a marker of lipid peroxidation during cardiopulmonary bypass in humans. *Free Radic Biol Med* 1999;**26**:439–45.
60. Sobotka PA, Gupta DK, Lansky DM, Costanzo MR, Zarling EJ. Breath pentane is a marker of acute cardiac allograft rejection. *J Heart Lung Transplant* 1994;**13**:224–9.
61. Spittle MA, Hoenich NA, Handelman GJ, Adhikarla R, Homel P, Levin NW. Oxidative stress and inflammation in hemodialysis patients. *Am J Kidney Dis* 2001;38:1408–13.
62. Capodicasa E, Trovarelli G, De Medio GE, Pelli MA, Lippi G, Verdura C, et al. Volatile alkanes and increased concentrations of isoprene in exhaled air during hemodialysis. *Nephron* 1999;**82**:331–7.
63. Handelman GJ, Rosales LM, Barbato D, Luscher J, Adhikarla R, Nicolosi RJ, et al. Breath ethane in dialysis patients and control subjects. *Free Radic Biol Med* 2003;**35**:17–23.

64. Van Gossum A, Shariff R, Lemoyne M, Kurian R, Jeejeebhoy K. Increased lipid peroxidation after lipid infusion as measured by breath pentane output. *Am J Clin Nutr* 1988;**48**:1394–9.
65. Sobotka PA, Brottman MD, Weitz Z, Birnbaum AJ, Skosey JL, Zarling EJ. Elevated breath pentane in heart failure reduced by free radical scavenger. *Free Radic Biol Med* 1993;**14**:643–7.
66. Miller 3rd ER, Appel LJ, Risby TH. Effect of dietary patterns on measures of lipid peroxidation: results from a randomized clinical trial. *Circulation* 1998;**98**:2390–5.
67. Smith D, Spanel P, Davies S. Trace gases in breath of healthy volunteers when fasting and after a protein-calorie meal: a preliminary study. *J Appl Physiol* 1999;**87**:1584–8.
68. Urso ML, Clarkson PM. Oxidative stress, exercise, and antioxidant supplementation. *Toxicology* 2003;**189**:41–54.
69. Paredi P, Kharitonov SA, Barnes PJ. Elevation of exhaled ethane concentration in asthma. *Am J Respir Crit Care Med* 2000;**162**:1450–4.
70. Olopade CO, Zakkar M, Swedler WI, Rubinstein I. Exhaled pentane levels in acute asthma. *Chest* 1997;**111**:862–5.
71. Paredi P, Kharitonov SA, Leak D, Ward S, Cramer D, Barnes PJ. Exhaled ethane, a marker of lipid peroxidation, is elevated in chronic obstructive pulmonary disease. *Am J Respir Crit Care Med* 2000;**162**:369–73.
72. Olopade CO, Christon JA, Zakkar M, Hua C, Swedler WI, Scheff PA, et al. Exhaled pentane and nitric oxide levels in patients with obstructive sleep apnea. *Chest* 1997;**111**:1500–4.
73. Schubert JK, Muller WP, Benzing A, Geiger K. Application of a new method for analysis of exhaled gas in critically ill patients. *Intensive Care Med* 1998;24:415–21.
74. Deneris ES, Stein RA, Mead JF. *In vitro* biosynthesis of isoprene from mevalonate utilizing a rat liver cytosolic fraction. *Biochem Biophys Res Commun* 1984;**123**:691–696.
75. Deneris ES, Stein RA, Mead JF, Acid-catalyzed formation of isoprene from a mevalonate-derived product using a rat liver cytosolic fraction. *J Biol Chem* 1985;260:1382–5.
76. Kuzma J, Nemecek-Marshall M, Pollock WH, Fall R. Bacteria produce the volatile hydrocarbon isoprene. *Curr Microbiol* 1995;**30**:97–103.
77. Foster WM, Jiang L, Stetkiewicz PT, Risby TH. Breath isoprene: temporal changes in respiratory output after exposure to ozone. *J Appl Physiol* 1996;**80**:706–10.
78. Mendis S, Sobotka PA, Euler DE. Expired hydrocarbons in patients with acute myocardial infarction. *Free Radic Res* 1995;**23**:117–22.
79. Nelson N, Lagesson V, Nosratabadi AR, Ludvigsson J, Tagesson C. Exhaled isoprene and acetone in newborn infants and in children with diabetes mellitus. *Pediatr Res* 1998;**44**:363–7.
80. Smith D, Spanel P, Enderby B, Lenney W, Turner C, Davies SJ. Isoprene levels in the exhaled breath of 200 healthy pupils within the age range 7–18 years studied using SIFT-MS. *J Breath Res* 2009;**4**:017101.
81. Bean HD, Zhu J, Hill JE. Characterizing bacterial volatiles using secondary electrospray ionization mass spectrometry (SESI-MS). *J Vis Exp* 2011.
82. Ferreira JA, Dias E, Rocha SM, Coimbra MA. Process for detecting *Helicobacter pylori* using aliphatic amides. *Anal Bioanal Chem* 2011;**401**:1889–98.

83. Phillips M, Basa-Dalay V, Bothamley G, Cataneo RN, Lam PK, Natividad MP, et al. Breath biomarkers of active pulmonary tuberculosis. *Tuberculosis (Edinb)* 2010;**90**: 145–51.
84. Purkhart R, Kohler H, Liebler-Tenorio E, Meyer M, Becher G, Kikowatz A, et al. Chronic intestinal Mycobacteria infection: discrimination via VOC analysis in exhaled breath and headspace of feces using differential ion mobility spectrometry. *J Breath Res* 2011;**5**:027103.
85. Shestivska V, Nemec A, Drevinek P, Sovova K, Dryahina K, Spanel P. Quantification of methyl thiocyanate in the headspace of *Pseudomonas aeruginosa* cultures and in the breath of cystic fibrosis patients by selected ion flow tube mass spectrometry. *Rapid Commun Mass Spectrom* 2011;**25**:2459–67.
86. Syhre M, Scotter JM, Chambers ST. Investigation into the production of 2-pentylfuran by *Aspergillus fumigatus* and other respiratory pathogens *in vitro* and human breath samples. *Med Mycol* 2008;**46**:209–15.
87. Kischkel S, Miekisch W, Fuchs P, Schubert JK. Breath analysis during one-lung ventilation in cancer patients. *Eur Respir J* 2012.
88. Lebovitz HE. Diabetic ketoacidosis. *Lancet* 1995;**345**:767–72.
89. Cope K, Risby T, Diehl AM. Increased gastrointestinal ethanol production in obese mice: implications for fatty liver disease pathogenesis. *Gastroenterology* 2000;**119**:1340–7.
90. Davis PL, Dal Cortivo LA, Maturo J. Endogenous isopropanol: forensic and biochemical implications. *J Anal Toxicol* 1984;**8**:209–12.
91. Chen S, Zieve L, Mahadevan V. Mercaptans and dimethyl sulfide in the breath of patients with cirrhosis of the liver. Effect of feeding methionine. *J Lab Clin Med* 1970;**75**:628–35.
92. Scislowski PW, Pickard K. The regulation of transaminative flux of methionine in rat liver mitochondria. *Arch Biochem Biophys* 1994;**314**:412–6.
93. Simenhoff ML, Burke JF, Saukkonen JJ, Ordinario AT, Doty R. Biochemical profile or uremic breath. *N Engl J Med* 1977;**297**:132–5.
94. Davies S, Spanel P, Smith D. Quantitative analysis of ammonia on the breath of patients in end-stage renal failure. *Kidney Int* 1997;**52**:223–8.
95. Holt DW, Johnston A, Ramsey JD. Breath pentane and heart rejection. *J Heart Lung Transplant* 1994;**13**:1147–8.
96. Studer SM, Orens JB, Rosas I, Krishnan JA, Cope KA, Yang S, et al. Patterns and significance of exhaled-breath biomarkers in lung transplant recipients with acute allograft rejection. *J Heart Lung Transplant* 2001;**20**:1158–66.
97. Harrison GR, Critchley AD, Mayhew CA, Thompson JM. Real-time breath monitoring of propofol and its volatile metabolites during surgery using a novel mass spectrometric technique: a feasibility study. *Br J Anaesth* 2003;**91**:797–9.
98. Miekisch W, Fuchs P, Kamysek S, Neumann C, Schubert JK. Assessment of propofol concentrations in human breath and blood by means of HS-SPME-GC-MS. *Clin Chim Acta* 2008;**395**:32–37.
99. Grossherr M, Hengstenberg A, Meier T, Dibbelt L, Gerlach K, Gehring H. Discontinuous monitoring of propofol concentrations in expired alveolar gas and in arterial and venous plasma during artificial ventilation. *Anesthesiology* 2006;**104**:786–90.
100. Grossherr M, Hengstenberg A, Meier T, Dibbelt L, Igl BW, Ziegler A, et al. Propofol concentration in exhaled air and arterial plasma in mechanically ventilated patients undergoing cardiac surgery. *Br J Anaesth* 2009;**102**:608–613.

101. Grossherr M, Hengstenberg A, Papenberg H, Dibbelt L, Wollenberg B, Schmucker P, et al. Propofol in bronchoalveolar lavage during anaesthesia. *Clin Chim Acta* 2011;**412**:190–3.
102. Grossherr M, Spies E, Scheel A, Hengstenberg A, Gehring H, Dibbelt L. Differences of propofol concentrations in mammalian whole blood and in corresponding plasma samples analyzed by high performance liquid chromatography. *Clin Lab* 2007;**53**:315–9.
103. Grossherr M, Varadarajan B, Dibbelt L, Schmucker P, Gehring H, Hengstenberg A. Time course of ethanol and propofol exhalation after bolus injection using ion molecule reaction-mass spectrometry. *Anal Bioanal Chem* 2011;**401**:2063–7.

CHAPTER

10 Analysis of Cancer Biomarkers in Exhaled Breath and Comparison with Sensory Indications by Dogs

Bogusław Buszewski*, **Tomasz Ligor***, **Joanna Rudnicka***, **Tadeusz Jezierski**†, **Marta Walczak**†, **and Anna Wenda-Piesik**‡

**Department of Environmental Chemistry and Bioanalytics, Faculty of Chemistry, Nicolaus Copernicus University, 7 Gagarin St., 87-100 Toruń, Poland*

†Institute of Genetics and Animal Breeding of Polish, Academy of Sciences, Department of Animal Behaviour, Jastrzębiec, 05-552 Wólka Kosowska, Poland

‡Department of Plant Growth Principles and Experimental Methodology, University of Technology and Life Sciences, 20 Kordeckiego St., 85-225 Bydgoszcz, Poland

10.1 INTRODUCTION

Lung cancer is one of the most aggressive tumors, being a major cause of death for both men and women in industrialized countries. Predominant factors in the development of lung cancer are active and passive smoking, because cigarette smoke contains more than 200 carcinogenic and mutagenic substances. Other common initiators of lung cancer are exposure to radon, cadmium, arsenic, beryllium, and asbestos.[1–3] Lung cancer is classified into two broad groups: small cell lung carcinoma (SCLC) (20–25% frequency of occurrence) and non-small cell lung carcinoma (NSCLC) (70–75%). The latter category includes adenocarcinomas (25–30%), squamous cell (30–35%), and large cell carcinomas (10–15%). This classification is largely dependent on histology, and establishes the method of treatment and is indicative of the prognosis. If at the time of diagnosis of SCLC metastasis has occurred then it cannot be cured by surgery and treatment can only be by chemotherapy and radiotherapy.[4]

In recent years there has been a growing interest in and a search for a non-invasive, painless technique that is also agreeable to patients, which would facilitate diagnosis of early-stage lung cancer obviating invasive medical routines. Volatile organic compounds (VOCs) in exhaled breath can provide valuable information concerning the health of individuals. Different analytical techniques have been used for breath analysis such as sensor arrays, PTR-MS, SIFT-MS, TDLS.[5–16]

A relatively new kind of mass analyzer is a time-of-flight mass spectrometer, TOF-MS. The GC/TOF-MS instrument provides a fast data acquisition rate. In addition, the ion ratios in a mass spectrum do not change across the chromatographic

Volatile Biomarkers. http://dx.doi.org/10.1016/B978-0-44-462613-4.00010-6

elution peak. Therefore, software can be used to deconvolute complex chromatographic co-elutions and extract the electron ionization mass spectrum of each compound.[17]

It is well known that dogs are able to detect human body odor and the odor of substances like explosives or narcotics. Williams and Church presented the first case report that an untrained pet dog evidently made its owner aware of a lesion on the owner's thigh that was subsequently diagnosed as malignant melanoma.[2,18] These authors hypothesized for the first time that dogs are able to detect cancer in humans on the base of the odor signature. The dogs demonstrated spontaneously a constant interest in a lesion on the owner's legs by sniffing, licking, and trying to bite it off even through clothing. After the carcinoma lesions had been excised the dogs showed no further interest in the site. In the last 7 years, several papers have discussed the possibility of using specially trained sniffer dogs to detect different kinds of neoplastic diseases in humans, such as melanoma, lung, breast, bladder, prostate, and ovary cancers, on the base of the odor of the carcinoma tissue, exhaled breath, or urine.[19–22] Balseiro and Correia hypothesized that the VOCs produced by tumors, and detected by dogs, are products of the major histocompatibility complex (MHC) genes.[23]

Figure 10.1 represents the canine sniffing process. Olfactory receptors are located in the back of the nasal cavity. In the mucous membrane of this part of cavity, which is called the neighborhood olfactory nose, supports the nursing cells and the stem cells of the olfactory receptors. Between these cells there are from several to hundreds of olfactory cells. Their fibers reach the olfactory bulb within which there are neurons of the olfactory nerve, which transfers information to the "smell brain."[24]

The ability to explore odor stimuli depends on the size of the olfactory epithelium and olfactory cells. A German Shepherd dog has more than 200 million olfactory cells on an area of about 170 cm^2, while a human being has about 5 million cells on

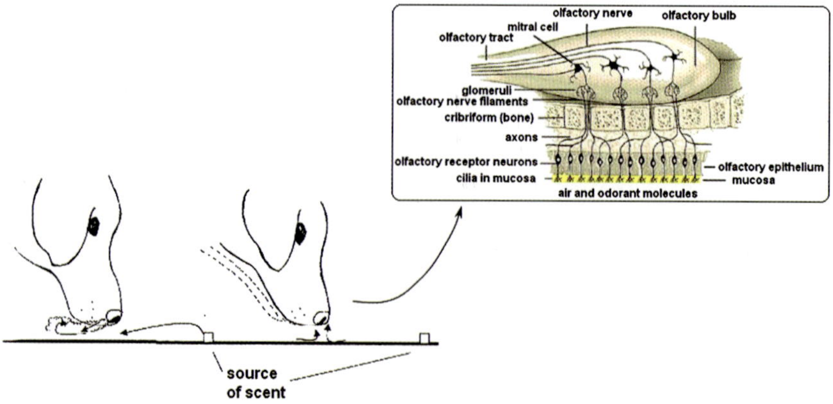

FIGURE 10.1
Canine sniffing.

about 5 cm^2 of olfactory epithelium. Olfactory receptors are irritated only by gaseous substances in the air, which, by diffusion, reach the mucous membrane and dissolve in the thin layer covering it. Most odoriferous substances are volatile organic compounds containing a few to several carbon atoms in the molecule. Applying GC-MS to cancer screening on the basis of odors emitted by the human body, in practice may be problematic, not only because of the sampling procedure, but also because of the difficulty in interpreting the results.

More than 30 VOCs have been identified in the breath of cancer patients and healthy controls, and the VOCs may be present in different combinations and quantities. Therefore, employing a single VOC as a cancer biomarker is currently unrealistic.[25–27] Since multivariate analysis of several biomarkers in combinations can probably better indicate disease, sophisticated methods as fuzzy logic have been used to analyze GC-MS data. On the other hand, the canine detection of VOC emissions in a scent lineup, are of a binary character, i.e. in the form of clear-cut (yes/no) response, which makes the interpretation of results much simpler. In this chapter, our own experiments on the training and the use of dogs to detect odor markers of lung cancer in breath samples will be described. The results obtained by dogs will be compared with those obtained by GC/TOF-MS measurements.

10.2 EXPERIMENTAL

10.2.1 GC-TOF/MS experiments

The analyses reported in this chapter were performed on a 7890A gas chromatograph (Agilent, Waldbronn, Germany) equipped with CP-PoraBOND-Q 25 m × 0.25 m × 3 μm column (Varian Inc., Middelburg, The Netherlands) coupled with a TruTOF mass spectrometer (Leco, St. Joseph, MI, USA). The oven temperature programme was as follows: initial temperature 40 °C held for 2 min, then ramped at 10 °C/min to 140 °C and then ramped at 5 °C/min to 270 °C and held for 5 min. The temperature of the split-splitless injector was 200 °C. Mass spectra were acquired over a mass-to-charge, m/z, range from 30 to 300 at a rate of 50 spectra/s using an electron ionization (EI) energy of 70 eV. Both the ion source and line transfer temperatures were set to 200 °C. The acquisition of chromatographic data was achieved using ChromaTOF software (Leco). A manual solid phase microextraction, SPME, holder, and Carboxen/Polydimethylsiloxane (CAR/PDMS) (75 μm) coated fiber (Supelco, Bellefonte, USA) was used for trace gas collection from breath samples. All chromatographic standards (aldehydes, alcohols, hydrocarbons, and ketones) used for calibration were purchased from Sigma-Aldrich (Steinheim, Germany).

10.2.2 Breath collection for GC/MS analysis

Alveolar breath samples were collected by means of a CO_2 controlled sampler (Department of Anesthesiology and General Intensive Care, Innsbruck Medical University, Innsbruck, Austria) in 1 L Tedlar bags. Then a 200 mL sample was transferred to another bag using a glass syringe and the SPME fiber was introduced into the bag through a septum and exposed to the sample held at a room temperature of 25 °C. After a 15 min extraction time the SPME fiber was placed into the hot GC injector port for 2 min at 200 °C where the adsorbed compounds were desorbed. Ambient air samples were also taken in the same way to provide baseline levels of ambient compounds.

10.2.3 Calibration

Gaseous standard was prepared by injecting 1 or 3 μL of each liquid compound into a 1 L glass bulb where it immediately evaporated. Afterwards, the mixture was subsequently diluted in Tedlar bags filled with nitrogen to obtain concentration in the range 0.5–200 parts-per-billion by volume, ppbv.

10.2.4 Experiments with dogs

Two male dogs (German Shepherd mixed breed) that had successfully undergone a three-phase training in scent lineups were used for the lung cancer detection trials. The dogs were 20–22 months old and were naïve in scent detection and identification at the beginning of the training. The dogs underwent a basic obedience training prior to the training in the lineup and showed interest in sniffing and retrieving objects both in rooms and outdoors. The experimental procedure and the housing conditions for the dogs were approved by the Local Ethical Commission for Animal Experimentation in Warsaw.

10.2.5 Odor samples

The breath odor samples for the canine training and detection were taken by exhaling 2–3 times through disposable polypropylene sampling tubes (Defencetek, Pretoria, South Africa) 15 cm long and 3 cm diameter (Figure 10.2) within which were

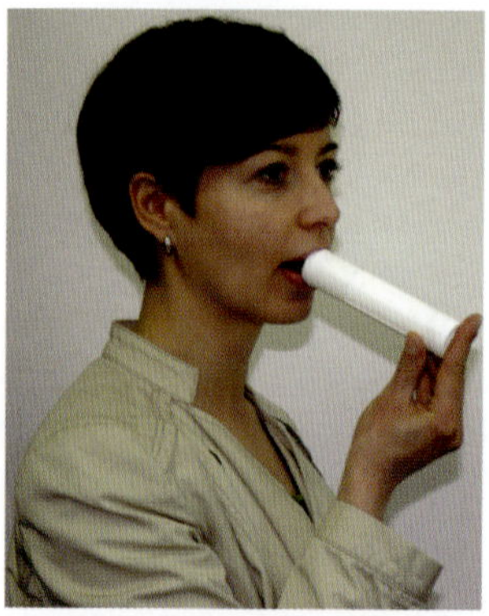

FIGURE 10.2
Taking breath samples for canine training or testing.

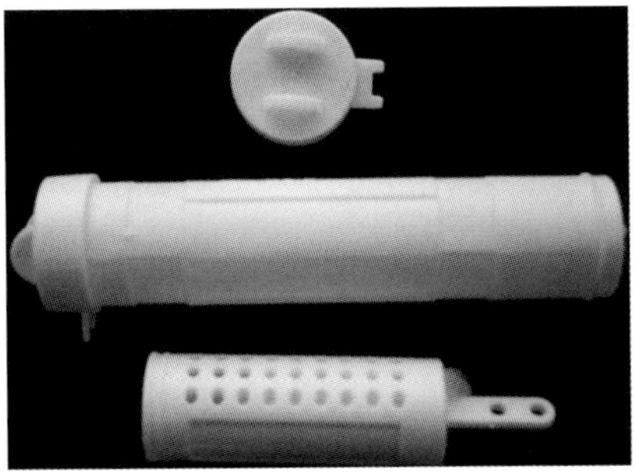

FIGURE 10.3
Sampling tube with removable insertions filled with cotton absorbent.

removable insertions filled with cotton absorbent (Figure 10.3). To provide testing samples for the dogs, the removable insertions were taken out of the tubes and put in sterile polypropylene boxes covered with lids with several holes in them.

10.2.6 Training procedure

Training of the dogs was conducted at the Institute of Genetics and Animal Breeding, Polish Academy of Science, Jastrzębiec, Poland. Each dog made 8–10 trials during the day depending on the interest and motivation it showed for work, its stamina, or how quickly it became bored, as judged subjectively by the dog handler. The training was divided into a preliminary phase and three training phases. The aim of the preliminary phase was to teach the dogs to visit and sniff all stands in the lineup.

During training phase I, the dogs were trained to indicate one target stand that contained a (cancer) pattern sample coupled with a food scent (a piece of food) by sitting down in front of the target sample. The remaining stands contained four blank (sterile) tubes. For each trial the location of the pattern sample was changed randomly. Initially, the location of the pattern sample was known to the handler who was to give the dog commands in order to teach it to indicate the pattern sample. A clicker sound was given by the experimenter immediately after a correct indication of the pattern sample and the dog was praised and rewarded by the handler with a food snack. No clicker and no reward followed a false alert toward a control sample and the dog was rebuked with "no" spoken by the handler. Correct indication, false alerts (FA), misses (MI), and the dog's hesitations were recorded during each trial. Most of the trials were "active" (single blind). In order to check the dogs' tendency to make false alerts randomly, the "zero" trials (without a pattern sample) were conducted, but not more than once on a training day. The criterion for a dog to pass to the next training phase was that it performed a minimum of 40% correct indications, without false indications in at least 100 trials.

In phase II of the training, the breath cancer sample was no longer included with a piece of food. Otherwise, the training procedure was the same as in the previous training phase I.

During the last training phase III, the cancer pattern sample (without food scent) was placed randomly among the four control samples, and the previous training procedure was the same as in the previous training phase. After the dogs fulfilled the criterion of 40% correct trials in training phase III, the working phase began. Phases I–III of the training took a total of 31 weeks.

During the working phase, pattern samples taken from patients with lung cancer were tested, i.e. one pattern sample was randomly placed amidst the four control samples, and conducted similar to the training phase III. After the first false alert during a trial, the dog was given a chance to indicate correctly; however, the trial was classified as FA (false alert). If the dog made two FAs within a trial or checked all olfactory samples in the lineup three times without indicating any of them, it was recalled by the handler to the starting position. In the latter case the trial was classified as a miss (MI).

10.2.7 Breath samples

Samples of breath were collected from 44 healthy volunteers and 29 patients with lung cancer, including 18 persons with SCLC and 11 with NSCLC. The samples from patients with lung cancer were collected in the Department of Lung Diseases of Collegium Medicum NCU. The study was approved by the Local Ethic Commission. For each person, a questionnaire was filled in that included age, gender, coexistent diseases, prescribed drugs, smoking habits, and last the meal and for the patients, a note concerning the cancer and its stage.

10.2.8 Statistical evaluation

All statistical analyses were computed using "`Statistica 8.0`" and "`Stat Soft`®." For data classification Shapiro-Wilks *W* test, ANOVA, Pearson's correlations, chi-square (χ^2) test, PCA (principal component analysis), and FA (factor analysis) have been applied. They allowed to determine the individual compounds for groups of patients with cancer diagnosis and healthy person and indicate correlation between dog indications and chemical analyses.

10.3 RESULTS AND DISCUSSION

10.3.1 Calibration and methods

The linearity, precision, and detection limits for VOCs determination in human breath are shown in Table 10.1. The relative standard deviation (RSD) is within the range from 3.3% to 9.5% for hydrocarbons, alcohols, aldehydes, ketones, and aromatic compounds. The calibration curves are linear within the following ranges in parts-per-billion by volume, ppbv: aliphatic hydrocarbons, 0.9–150.0; alcohols, 1.6–163.5;

Table 10.1 Validation parameters for volatile organic compounds by the current GC-MS method. LOQ is the limit of quantification.

Compounds	Linearity (ppb)	R^2	RSD (%)	LOD (ppb)	LOQ (ppb)
Pentane	1.5–150.0	0.998	5.2	0.49	1.47
Hexane	1.5–150.0	0.994	3.4	0.48	1.44
2-Methylpentane	0.9–92.6	0.995	9.5	0.31	0.93
3-Methylpentane	0.9–94.1	0.988	9.3	0.32	0.94
1-Propanol	1.6–163.5	0.995	5.1	0.53	1.58
2-Propanol	1.6–159.6	0.998	9.4	0.52	1.57
Butanal	1.3–135.6	0.997	8.4	0.44	1.32
Propanal	1.7–170.4	0.998	7.2	0.52	1.56
2-Methylpropanal	1.3–133.9	0.996	6.2	0.44	1.32
Acetone	1.6–166.4	0.991	8.9	0.54	1.62
2-Butanone	1.3–136.5	0.997	3.3	0.45	1.35
2-Pentanone	1.3–103.7	0.996	6.2	0.44	1.32
Benzene	1.3–136.7	0.995	4.6	0.43	1.29
Toluene	1.1–114.7	0.991	5.9	0.37	1.11
Ethylbenzene	1.0–99.8	0.992	4.5	0.32	0.96
Ethyl acetate	0.9–87.8	0.992	4.5	0.32	0.96
o-Xylene	1.0–100.1	0.994	4.8	0.33	0.99
Furan	1.6–165.6	0.999	4.9	0.51	1.53
Acetonitrile	2.3–234.0	0.996	3.4	0.75	2.25

aldehydes, 1.3–170.4; and ketones, 1.3–166.4. The linear regression correlation curve gave R^2 as at least 0.99. The lowest values for the limit of detection, LOD, obtained for hydrocarbons varied from 0.31 to 0.49 ppbv.

10.3.2 Exhaled air

All compounds detected in human breath were compared with their concentrations in ambient air and only those compounds showing concentrations at least 10% higher than ambient air are considered. Figure 10.4 presents a chromatogram obtained for a breath sample from a patient with lung cancer.

The compounds identified are listed in Table 10.2. Methanol, acetaldehyde, acetone, and isoprene were found in all the breath samples. Acetonitrile, benzene, toluene, furan, 2-methylfuran, 3-methylfuran, 2,5-dimethylfuran were found only in the breath of the smokers.

The concentration of pentane, which is regarded as a oxidative stress biomarker, ranged from 6.8 to 14.3 ppbv for healthy person and from 0.7 to 17.5 ppbv for the cancer patients (Table 10.3). Furan and its derivative are considered as markers of smoking. 1-Propanol was observed in cancer patients at concentrations in the range 4.37–13.15 ppbv. This compound has been postulated as potential cancer marker in breath.[28] Moreover, 2-propanol was found in the breath of healthy and ill persons.

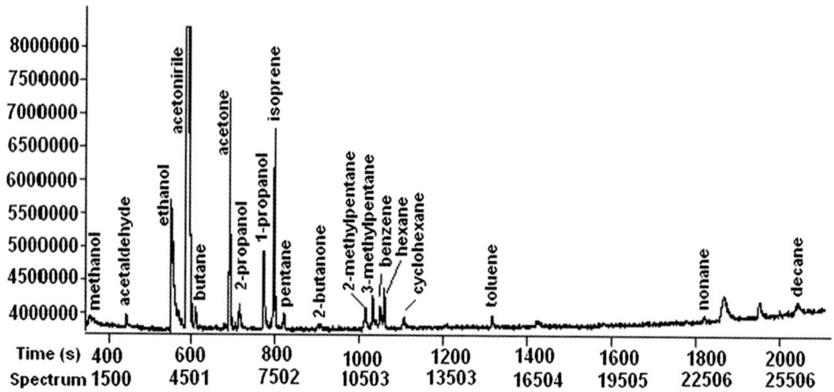

FIGURE 10.4
A GC-TOF/MS chromatogram of the exhaled breath from a patient with lung cancer.

However, its concentration was on a similar level in both exhaled breath and ambient air.

The original data of VOC concentrations were log transformed and tested for normality by the Shapiro-Wilks *W* test ($p < 0.05$). ANOVA was used to explore variations between control groups versus patients. The effect of origin (groups) on

Table 10.2 Compounds detected in the exhaled breath of 44 healthy volunteers and 29 patients with lung cancer; smoking and non-smoking volunteers and patients are included (T_R—retention time).

Compounds	CAS Number	T_R (s)	Healthy Person	Patients
Methyl alcohol	67-56-1	356.0	44	29
Propyne	74-99-7	361.9	2	0
Propane	74-98-6	364.3	10	21
Acetaldehyde	75-07-0	442.2	44	29
Ethyl alcohol	64-17-5	556.0	34	23
1-Butene-3-yne	689-97-4	562.08	2	0
Isobutane	75-28-5	569.5	14	23
1-Butene	106-98-9	576.3	30	2
1,3-Butadiene	106-99-0	576.8	3	3
Acetonitrile	75-05-8	591.5	11	18
Butane	106-97-8	609.1	33	22
2-Propenal	107-02-8	655.5	35	23
Furan	110-00-9	677.3	6	2
Propanal	123-38-6	682.8	36	2
Acetone	67-64-1	688.5	44	29

Table 10.2 (Continued)

Compounds	CAS Number	T_R (s)	Healthy Person	Patients
Carbon disulfide	75-15-0	705.0	24	2
Isopropyl alcohol	67-63-0	712.7	30	27
Dimethyl sulfide	75-18-3	718.4	24	2
1-Hydroxy-2-propanone	116-09-6	748.9	24	2
1-Propanol	71-23-8	771.4	3	23
2-Pentene	109-68-2	791.7	11	0
Isoprene	78-79-5	796.6	44	29
Pentane	109-66-0	820.2	28	2
Cyclopentane	287-92-3	828.4	10	1
1,3-Pentadiene	504-60-9	835.3	5	1
Methacrolein	78-85-3	842.2	24	4
2-Methylpropanal	78-84-2	859.1	29	2
Methyl vinyl ketone	78-94-4	875.0	27	2
2,3-Butanedione	431-03-8	900.3	6	2
2-Methylfuran	534-22-5	901.3	8	2
Butanal	123-72-8	902.1	6	2
2-Butanone	78-93-3	908.0	3	21
3-Methylfuran	930-27-8	917.5	6	0
Ethyl acetate	141-78-6	951.5	24	26
2-Methylpentane	107-83-5	1014.9	20	2
3-Methylpentane	96-14-0	1031.8	20	2
Benzene	71-43-2	1040.4	22	20
Methylcyclopentane	96-37-7	1050.8	19	2
Hexane	110-54-3	1059.6	23	2
Cyclohexane	110-82-7	1107.2	12	1
2-Pentanone	107-87-9	1148.7	24	24
2,5-Dimethylfuran	625-86-5	1151.7	5	0
Pentanal	110-62-3	1164.1	6	2
Toluene	108-88-3	1317.5	35	21
2,4,4-Trimethyl-1-pentene	107-39-1	1424.5	30	2
Hexanal	66-25-1	1432.6	6	1
Butyrolactone	96-48-0	1445.5	21	2
Ethylbenzene	100-41-4	1571.0	18	16
o-Xylene	95-47-6	1616.8	22	23
Benzaldehyde	100-52-7	1683.2	2	5
Nonane	111-84-2	1821.1	19	2
6-Methyl-5-hepten-2-one	110-93-0	1870.2	18	2
Acetophenone	98-86-2	1932.8	13	1
Decane	124-18-5	2044.7	3	2
Styrene	100-42-5	2364.5	11	2

Table 10.3 Concentration ranges for compounds detected in exhaled breath of healthy volunteers and cancer patients.

Compounds	Concentration Range for Healthy Person (ppb)	Concentration Range for Lung Cancer Patients (ppb)
Pentane	7–14	0.7–18
Hexane	2–6	0.8–2
2-Methylpentane	2–11	0.3–4
3-Methylpentane	1–9	0.6–9
1-Propanol	–	4–13
2-Propanol	3–4	3–7
Ethyl acetate	1–8	4–23
Butanal	1.5–2	0.8–3
Propanal	1.6–3	0.7–4
2-Methylpropanal	5–10	7–94
Acetone	44–531	35–391
2-Butanone	0.5–3	0.5–3
Benzene	1.3–15	0.8–4
Toluene	1.1–37	2–17
Ethylbenzene	2–18	1–3
o-Xylene	2–7	2–8
2-Pentanone	2–4	3–9
Furan	1.6–3	1–3
Acetonitrile	6–29	11–24

VOCs content in breath air is presented in Table 10.4, which indicates significantly variations between the groups ($p < 0.05$) for 17 trace compounds; the most distinctive effect was detected for ethanol, butane, 1-propanol, 2-propanol, hexane, and xylene ($p < 0.0001$). Means were separated by unequal N HSD test (Tukey's honestly significant difference), at $p = 0.05$. Statistically greater contents of 2-propanol, 1-propanol, isoprene, hexanal, propanal, and butane appeared in the exhaled breath of the cancer patients than the breath of healthy controls.

10.3.3 Data from the experiments with dogs

Detection sensitivity and specificity were calculated according to the scheme given in Table 10.5. A yes/no response toward each sniffed sample in the lineup indicated at 50% probability of a correct response by chance. To evaluate differences in the dogs' indications between pattern and control samples and between dogs the chi-square (χ^2) test was applied.

In active trials the dogs indicated correctly the pattern samples of the breath from lung cancer patients in 82.2% of cases (the detection specificity). The false positive indications toward healthy controls amounted to 17.6% of the trials. The difference in the dogs' indications of pattern samples versus controls in single blind trials were

Table 10.4 ANOVA results (data log transformed).

ID	Substance	SS	df	MS	SS_E	df_E	MS_E	*F*	*p*
1	Ethanol	75.3	3	25.1	100.6	69	1.5	17.2	0.000
2	Acetonitrile	25.2	3	8.4	153.3	63	2.4	3.5	0.022
3	Butane	92.8	3	30.9	59.7	58	1.0	30.0	0.000
4	Ethyl ether	11.1	3	3.7	60.4	49	1.2	3.0	0.039
5	1-Propanol	106.7	3	35.6	51.0	58	0.9	40.4	0.000
6	2-Methylpentane	22.0	3	7.3	140.1	56	2.5	2.9	0.042
7	Isoprene	7.6	3	2.5	40.6	69	0.6	4.3	0.008
8	Pentane	22.3	3	7.4	143.2	68	2.1	3.5	0.019
9	2-Propanol	241.5	3	80.5	47.8	61	0.8	102.7	0.000
10	2-Butanone	8.9	3	3.0	25.8	49	0.5	5.7	0.002
11	Hexanal	1.7	2	0.8	0.7	10	0.1	12.1	0.002
12	Benzene	16.1	3	5.4	40.7	50	0.8	6.6	0.001
13	Hexane	47.0	3	15.7	100.1	66	1.5	10.3	0.000
14	Heptane	10.1	2	5.1	10.1	12	0.8	6.0	0.016
15	Toluene	10.6	3	3.5	51.1	70	0.7	4.8	0.004
16	*o*-Xylene	59.8	3	19.9	5.6	13	0.4	46.6	0.000
17	Propanal	3.0	2	1.5	7.0	26	0.3	5.6	0.009

ANOVA with grouping variable. All effects are significant at $p < 0.05$.

Table 10.5 Scheme for the calculation of the detection sensitivity and specificity.

		True Presence of Cancer Disease	
		Cancer odor present (cancer pattern sample)	No cancer odor (control sample)
Dog's reaction	Indication (sitting in front of a sample)	(a) Correct positive	(b) False positive (false alert)
	No indication	(c) False negative (miss)	(d) Correct negative

(a), (b), (c), (d) = number of particular reactions.
Detection sensitivity = a/(a + c).
Detection specificity = d/(b + d).

highly significant ($\chi^2 = 1056.5$, df = 1, $p < 0.000$). There were significant differences between dogs in detection sensitivity ($\chi^2 = 25.17$, df = 1, $p < 0.001$, respectively) but no significant differences in detection specificity.

10.3.4 Correlation between dog indications and chemical analyses

The percent of the dog's indications when investigating the patients correlated with the VOC content of the exhaled breath of the patient. Two data sets for the healthy controls ($N = 49$) and for the patients ($N = 29$) contained the % of the dog's indications and the chosen compounds. For the patient group, a positive Pearson's correlations between the dog's indications and the VOCs in exhaled breath reached a significant power with positive or negative tendency. Table 10.6 presents these indications for several specific compounds. For example, ethyl acetate and 2-pentanone correlated with the dog's indication positively ($r = 0.85$ and $r = 0.97$, respectively), while the concentrations for acetonitrile, propanal, and 1-propanol were negatively correlated with dog's indications ($r = -0.78$, $r = -0.87$, and $r = -0.98$, respectively).

Two multivariate methods of analysis were applied to the data, i.e. FA (factor analysis) and PCA (principal component analysis), separately for the control and patient group with 19 dimensions presenting the % age of the dog's indications and 18 for the VOCs (Table 10.7). Data of each dimension were standardized within the individuals and to obtain a meaningful structure of principal components. The number of factors has been finally limited to 2. Factor analysis with varimax rotation was carried out as a first approach to the data sets. For the patient group two factors were essential for classification having Eigen values greater than 1. These two factors explained over 80%

Table 10.6 Pearson's correlation coefficients *r*, between several compounds in breath and the dog's indications.

% Indications Versus	*r*
Ethanol	0.27
2-Methylpropane	0.29
Acetonitrile	−0.78
Butane	0.26
Propanal	−0.87
Acetone	−0.77
Carbon disulfide	−0.67
2-Propanol	−0.60
1-Propanol	−0.98
Isoprene	0.07
Pentane	0.47
2-Butanone	−0.51
Ethyl acetate	0.85
2-Methylpentane	0.14
Benzene	0.11
Hexane	0.50
2-Pentanone	0.97
Toluene	0.30

Table 10.7 Correlations between breath VOCs and dog's indications.

Variable	Factor 1	Factor 2
% indications	0.22	0.97
Ethanol	0.97	0.09
2-Methylpropane	0.74	0.04
Acetonitrile	0.16	−0.91
Butane	0.92	0.00
Propanal	0.15	−0.86
Acetone	0.32	−0.79
Carbon disulfide	−0.73	−0.45
2-Propanol	0.58	−0.78
1-Propanol	−0.28	−0.91
Isoprene	0.96	−0.11
Pentane	0.76	0.39
2-Butanone	−0.95	−0.30
Ethyl acetate	0.35	0.86
2-Methylpentane	0.59	0.12
Benzene	0.86	−0.15
Hexane	0.37	0.55
2-Pentanone	0.45	0.89
Toluene	−0.86	0.49
Explained variance	8.19	7.14
%	43.00	38.00

of the total variation. Factor 1 gave positive signs of loading coefficients for ethanol, 2-methylpropane, butane, isoprene, pentane, and benzene, while negative signs were apparent for carbon disulfide, 2-butanone, and toluene. Factor 2 was associated with the dog's indication (positive loading) and those compounds that were not classified by factor 1.

The sign of loading explain the role of the VOCs in the dog's indications in the same way as simple correlations. Hence, the first factor could be related to indifference by the dog while the second factor is concerned with the dog's indications. PCA was also used to show the relations between VOCs and dog's indications in two components.

The wheel-plot classification in Figure 10.5 indicates which of the compound is correlated to the first factor and which to the second factor. For the healthy controls two first factors were essential for classification but explained only 51% of the variance. The interpretation of such results is uncertain at this time.

10.4 CONCLUSIONS

These preliminary investigations on the applications of GC-TOF/MS and canine scent for the detection of volatile cancer biomarkers show promising possibilities. It is interesting to note that some biomarker candidates appear to be detected by dogs.

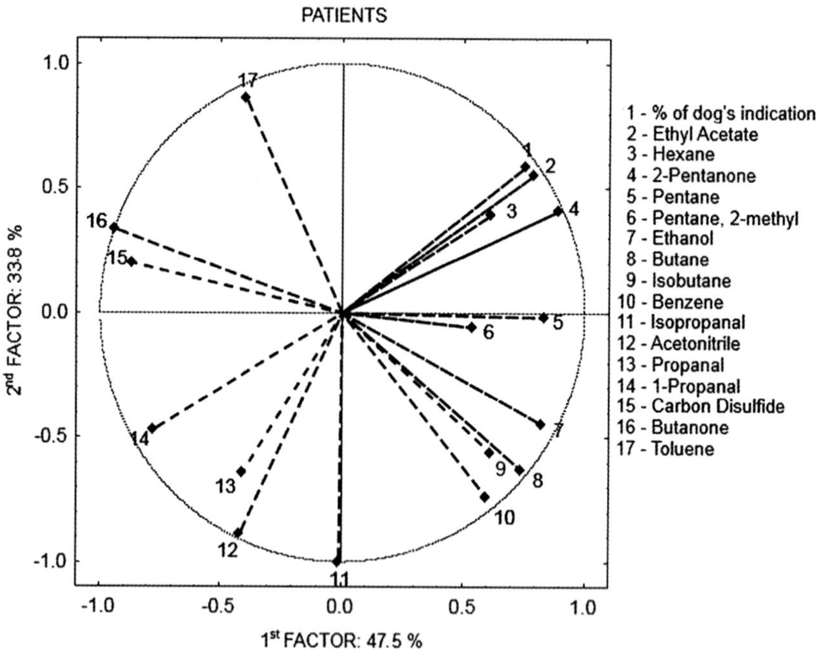

FIGURE 10.5
Wheel-plot classification.

From the present results it can be concluded that trained dogs can distinguish breath samples of patients with diagnosed cancer disease from those of healthy people, at a "better than by chance" rate. Additionally, the canine method has the following advantages: the dog training and testing is relatively simple and inexpensive in comparison to analytical equipment (GC/MS) and relatively high detection sensitivity and specificity in relation to pattern samples is indicated. However, it is not known which chemical compound(s) the dogs detect or for what odor or other stimuli the dogs respond.

Acknowledgments

This research work was supported by Grant N N 204 026238 (2010–2013).

References

1. Baker RR. Smoke generation inside a burning cigarette: modifying combustion to develop cigarettes that may be less hazardous to health. *Prog Energ Combust* 2006;**32**:373–85.
2. Williams H, Pembroke A. Sniffer dogs in the melanoma clinic? *Lancet* 1989;**1**:734.

3. Yao S, DellaVentura G, Petibois C. Analytical characterization of cell-asbestos fiber interactions in lung pathogenesis. *Anal Bioanal Chem* 2010;**397**:2079–89.
4. Edelman MJ, Gandara DR. Lung cancer. In: Casciato DE, Territo MC, editors. *Manual of clinical oncology*. Philadelphia: Lippincott Williams & Wilkins; 2008.
5. Abbott SM, Elder JB, Spanel P, Smith D. Quantification of acetonitrile in exhaled breath and urinary headspace using selected ion flow tube mass spectrometry. *Int J Mass Spectrom* 2003;**228**:655–65.
6. Buszewski B, Kesy M, Ligor T, Amann A. Human exhaled air analytics: biomarkers of diseases. *Biomed Chromatogr* 2007;**21**:553–66.
7. Diskin AM, Spanel P, Smith D. Increase of acetone and ammonia in urine headspace and breath during ovulation quantified using selected ion flow tube mass spectrometry. *Physiol Meas* 2003;**24**:191–9.
8. Gibson G, Monk SD, Padgett M. A field-portable, laser-diode spectrometer for the ultrasensitive detection of hydrocarbon gases. *J Mod Optic* 2002;**49**:769–76.
9. Machado RF, Laskowski D, Deffenderfer O, Burch T, Zheng S, Mazzone PJ, et al. Detection of lung cancer by sensor array analyses of exhaled breath. *Am J Respir Crit Care Med* 2005;**171**:1286–91.
10. Mazzone PJ, Hammel J, Dweik R, Na J, Czich C, Laskowski D, et al. Diagnosis of lung cancer by the analysis of exhaled breath with a colorimetric sensor array. *Thorax* 2007;**62**:565–8.
11. Miekisch W, Schubert JK. From highly sophisticated analytical techniques to life-saving diagnostics: technical developments in breath analysis. *TrAC Trend Anal Chem* 2006;**25**:665–73.
12. Schwarz K, Filipiak W, Amann A. Determining concentration patterns of volatile compounds in exhaled breath by PTR-MS. *J Breath Res* 2009;**3**:027002.
13. Skeldon KD, McMillan LC, Wyse CA, Monk SD, Gibson G, Patterson C, et al. Application of laser spectroscopy for measurement of exhaled ethane in patients with lung cancer. *Respir Med* 2006;**100**:300–6.
14. Turner C, Spanel P, Smith D. A longitudinal study of ammonia, acetone and propanol in the exhaled breath of 30 subjects using selected ion flow tube mass spectrometry, SIFT-MS. *Physiol Meas* 2006;**27**:321–37.
15. Wehinger A, Schmid A, Mechtcheriakov S, Ledochowski M, Grabmer C, Gastl GA, et al. Lung cancer detection by proton transfer reaction mass-spectrometric analysis of human breath gas. *Int J Mass Spectrom* 2007;**265**:49–59.
16. Bajtarevic A, Ager C, Pienz M, Klieber M, Schwarz K, Ligor M, et al. Noninvasive detection of lung cancer by analysis of exhaled breath. *BMC Cancer* 2009;**9**:348.
17. Mieth M, Schubert JK, Groger T, Sabel B, Kischkel S, Fuchs P, et al. Automated needle trap heart-cut GC/MS and needle trap comprehensive two-dimensional GC/TOF-MS for breath gas analysis in the clinical environment. *Anal Chem* 2010;**82**:2541–51.
18. Church J, Williams H. Another sniffer dog for the clinic? *Lancet* 2001;**358**:930.
19. Horvath G, Jarverud GA, Jarverud S, Horvath I. Human ovarian carcinomas detected by specific odor. *Integr Cancer Ther* 2008;**7**:76–80.
20. McCulloch M, Jezierski T, Broffman M, Hubbard A, Turner K, Janecki T. Diagnostic accuracy of canine scent detection in early- and late-stage lung and breast cancers. *Integr Cancer Ther* 2006;**5**:30–9.
21. Pickel D, Manucy GP, Walker DB, Hall SB, Walker JC. Evidence for canine olfactory detection of melanoma. *Appl Anim Behav Sci* 2004;**89**:107–16.

22. Willis CM, Church SM, Guest CM, Cook WA, McCarthy N, Bransbury AJ, et al. Olfactory detection of human bladder cancer by dogs: proof of principle study. *BMJ* 2004;**329**:712.
23. Balseiro SC, Correia HR. Is olfactory detection of human cancer by dogs based on major histocompatibility complex-dependent odor components? A possible cure and a precocious diagnosis of cancer. *Med Hypotheses* 2006;**66**:270–2.
24. Jezierski T, Walczak M, Glanc D, Górecka A, Dziubiska N. *Sniffing sense of dogs and its practical application*. Jastrzbiec: Institute of Genetics and Animal Breeding, Polish Academy of Sciences; 2008.
25. Ligor T, Ligor M, Amann A, Ager C, Bachler M, Dzien A, et al. The analysis of healthy volunteers' exhaled breath by the use of solid-phase microextraction and GC-MS. *J Breath Res* 2008;**2**:046006.
26. Phillips M, Altorki N, Austin JH, Cameron RB, Cataneo RN, Greenberg J, et al. Prediction of lung cancer using volatile biomarkers in breath. *Cancer Biomark* 2007;**3**:95–109.
27. Phillips M, Gleeson K, Hughes JM, Greenberg J, Cataneo RN, Baker L, et al. Volatile organic compounds in breath as markers of lung cancer: a cross-sectional study. *Lancet* 1999;**353**:1930–3.
28. Ligor M, Ligor T, Bajtarevic A, Ager C, Pienz M, Klieber M, et al. Determination of volatile organic compounds in exhaled breath of patients with lung cancer using solid phase microextraction and gas chromatography mass spectrometry. *Clin Chem Lab Med* 2009;**47**:550–60.

PART

D

Nitric Oxide, NO, and Carbon Monoxide, CO

CHAPTER

11 Added Value with Extended NO Analysis

Marieann Högman[*,†] **and Pekka Meriläinen**[*]

[*]*Department of Medical Sciences, Respiratory Medicine and Allergology, Uppsala University, SE 75105 Uppsala, Sweden*

[†]*Centre for Research and Development, Uppsala University/County Council of Gävleborg, SE 801 88 Gävle, Sweden*

11.1 BACKGROUND

NO is involved in many processes in the respiratory system in health[1,2] and can be altered during disease.[3] Individuals with atopic asthma have higher than normal NO concentrations[4] and this is due to increased expression of the inducible NO synthase (iNOS) and availability of the substrate L-arginine in airway epithelial cells of asthmatics.[5] There is an abundant expression in the airways of iNOS[6] due to continuous transcriptional activation.[5] Asthma is an obstructive inflammatory disease that effects the airways and NO has been said to be an inflammatory marker in asthma. Lately it has been put forward that asthma can also be a systemic disease.[7] Since the disease is characterized by recurrent attacks and varies in frequency and severity from person to person and is due to inflammation, the need for monitoring is obvious. However, it was recently concluded in a Cochrane review that, at present, defining the dose of inhaled corticosteroids based on exhaled nitric oxide cannot be routinely advocated.[8] In severe asthma the peripheral airways are also affected and treatment given by inhalation may not reach these airways. Therefore, it is desirable to have insight into how widespread the inflammation is. One way of doing this is to model NO in the respiratory system and thereby get NO values from the alveolar region as well as from the airways. This has been done by Lehtimäki et al. and they have found in more severe asthma and in nocturnal asthma that NO from the alveolar region is increased.[9,10]

The discovery of the flow dependency of exhaled NO was made with the new generation of fast responding NO analyzers and led to the opportunity to model NO production. This discovery shed light on the different results published in healthy controls as in disease and was done by two research groups independently.[11,12] A European task force was formed in 1997 and the latest published guidelines is an ERS and ATS joint statement which was to set the flow rate at 50 mL/s ($F_ENO_{0.05}$), but other flow rates could be used but must be stated.[13] As can be seen in Figure 11.1, middle panel; higher flow rates results in lower NO values. $F_ENO_{0.05}$ was chosen because it gives a good reading of the analyzer and it is easy for the patient

Volatile Biomarkers. http://dx.doi.org/10.1016/B978-0-44-462613-4.00011-8

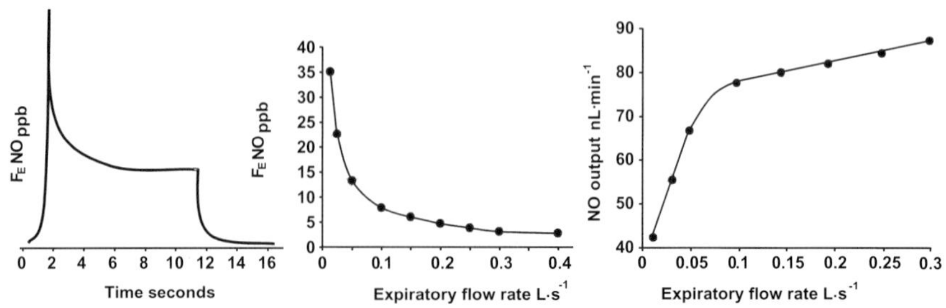

FIGURE 11.1

The characteristics of an exhaled breath at a flow rate of 0.050 L/s ($F_ENO_{0.05}$, left panel). The peak represents the inhaled NO not taken up and with a steady exhalation flow the NO levels out and with the inhalation of low concentration of NO the curve rapidly goes toward environment concentrations. The middle panel shows the flow dependency and each plateau value and plateau values at the different exhalation flow rates are plotted. The right panel shows the output of NO for each flow rate and this is the plot that is used for both the linear and the non-linear models.

to perform. The studies included in the previously mentioned Cochrane review are all done with one exhalation flow rate.

Exhalation at different flow rates gives different NO concentrations in the exhaled gas. Contrary to the CO_2 measurement, the concentration of which peaks at the end exhalation, NO concentrations peaked at the beginning of the exhalation; see Figure 11.1 left panel. Therefore, less NO must be present in the alveoli so the gradient in NO number density allows the application of Fick's first law of diffusion.[14] For simplicity, low concentration enters a tube and travels along the tube while molecules diffuse into the tube depending on the transfer rate across the tube and the concentration surrounding the tube. Hence, during an inhalation air is drawn down to the lung periphery by pressure difference (convection), the alveoli expand, and gas exchange takes place driven by a concentration gradient (diffusion). With the movement of air and the addition of NO from conducting airways into the alveolar region, equilibrium for the different gases is soon reached and the exhalation begins. In health, the concentration of NO in the alveolar gas phase (C_ANO) is most likely in the same order as for the alveolar tissue ($C_{alv}NO$), i.e. close to zero, since blood is a sink for NO. A bolus of near-zero NO gas starts the transport upwards to be exhaled. During the passage in the airway the bolus picks up NO from the small and large airways and the increase of NO in the bolus depends on the time of contact. From a system theory perspective, the conducting airways can be modeled as a black box adding NO to the gas flow from the lungs. The experimentally obtained mutual relationships between NO output and NO input for a set of exhalation flows can be used to construct a transfer function able to predict the system NO output for any other flow. However, instead of a purely mathematical approach it is important to be able to link all the information defining the system to physical/physiological properties of the human airways parameters. In Figure 11.1, right panel, the NO output from the lung is illustrated and is used for both linear and non-linear NO models of the lung.

11.2 A TWO COMPARTMENT MODEL

A simplification of the lung can be made based on a two compartment model (2CM), consisting of one rigid compartment representing the airways (the tube), and one expansile compartment representing the gas exchange areas,[15] see Figure 11.2. Fick's first law of diffusion can then be applied to the mechanism controlling the NO transfer to the airway lumen. Measurements of exhaled NO at three or more flow rates can be exploited to determine three different parameters defining the NO production. The abbreviations have been agreed upon[16] and are given in Table 11.1.

The expansile compartment of the lung holds the fraction of NO in the gas phase of the alveolar region, C_ANO in ppb. The rigid conducting airway system lumped into a single tube is characterized by the airway tissue concentration of NO (wall concentration, $C_{aw}NO$ in ppb) and the total airway compartment diffusing capacity, transfer factor, or conductance for radial mass transfer of NO from the airway wall to the gas stream ($D_{aw}NO$ in mL/s). The flux of NO from the wall to the lumen depends on the distance z along the airway and it has a total value $J_{aw}NO = D_{aw}NO \cdot (C_{aw}NO - C_ANO)$ at flows high enough to keep the local lumen NO value close to C_ANO all the way along the airway. The highest potential total

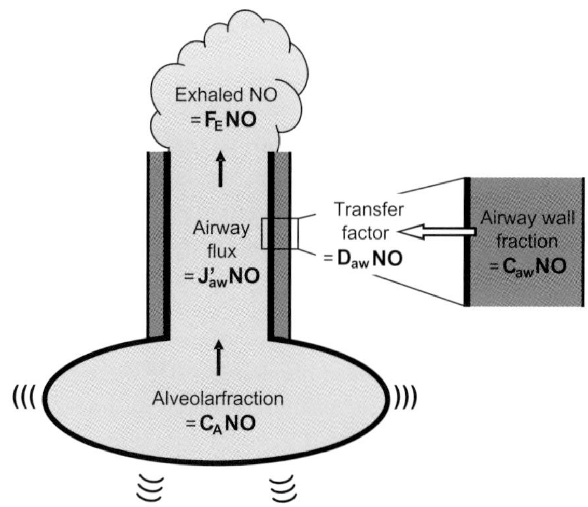

FIGURE 11.2
A simplified model of the NO production in the lung. It consists of an expansile compartment representing the gas exchange areas, and a rigid compartment representing the airways.

Table 11.1 Abbreviations for the NO parameters of the lung.

F_ENO	Fractional concentration of exhaled NO in the gas phase (ppb) The exhalation flow rate is determined in L/s and is given as a subscript; for example, at the recommended flow rate of 50 mL/s the abbreviation becomes $F_ENO_{0.05}$ (ppb)
C_ANO	Calculated fractional concentration of NO in the gas phase of the alveolar region (ppb)
$C_{alv}NO$	Calculated alveolar tissue concentration of NO (ppb)
$C_{aw}NO$	Calculated tissue concentration of NO of the airway wall (ppb)
$D_{aw}NO$	Calculated airway compartment diffusing capacity, transfer factor, or conductance for radial mass transfer of NO from the airway wall to the gas stream (mL/s)
$J'_{aw}NO$	Calculated total maximum flux (rate of radial transport) of NO in the airway compartment (nL/s), which occurs when the gas phase concentration in the alveolar compartment is close to zero, and is equal to the product $C_{aw}NO \cdot D_{aw}NO$
$J_{aw}NO$	Calculated flux of NO in the airway compartment (nL/s). This flux is dependent on the value of C_ANO. If C_ANO is close to zero the flux is equal to the total flux
$\dot{V}_{NO}$	Elimination rate of NO from the breath during exhalation (nL/min)
$\dot{V}_E$	Exhalation flow rate (mL/s)

flux is then $J'_{aw}NO = D_{aw}NO \cdot C_{aw}NO$, and is obtained when C_ANO is zero. The definitions and physical meanings of the two NO fluxes, $J_{aw}NO$ and $J'_{aw}NO$, have been inadequately described and understood in several publications on exhaled NO. This is related to the attempts to solve the three independent parameters from measurements at only two flows, which is possible only by assuming C_ANO to be zero, making $J_{aw}NO$ equal to $J'_{aw}NO$.

As seen in Figure 11.1, right panel, the curve shows two phases where the steeper curve represents the airways portion and the gentle slope the alveolar portion. These two approximately linear curves meet around 50–60 mL/s, just at the recommended flow rate, irrespective of a high $F_ENO_{0.05}$ value or low (unpublished observations). C_ANO shows no correlations with $F_ENO_{0.05}$ but has a good correlation with $J_{aw}NO$, which also strengthen the two compartment model. Other findings supporting the two compartment model are from Verbanck et al. who have shown that a position change from sitting to prone or supine position only changes the C_ANO value and not the NO produced in the airway.[17] It is well known that the alveolar volume changes with body position and that the airways are not affected by this change.

11.3 DIFFERENT NO MODELS

Different approaches have been used to calculate the NO parameters and most of the models were presented during 1998–2000.[15,18–21] Comparisons of the different models have been published and the non-linear model has been found to provide a better fit to experimental[22] data and is, therefore, suggested to be the best method.[23] The advantage of the multiple flow analysis relies on the simplest calculations and may therefore be the most robust.[24]

11.3.1 The Tsoukias and George two compartment model (2CM)

The Tsoukias and George two compartment model (2CM) is a two parameter model giving $J_{aw}NO$ and C_ANO,[15] where NO concentrations at multiple constant exhalation flow rates between 100 and 500 mL/s are obtained[25] and $\dot{V}_{NO}$ is plotted versus the flow rates; see Figure 11.3. The slope and the intercept of the resulting linear regression line provide estimates of C_ANO and $J_{aw}NO$ according to Eq. (11.1) and the *r*-value of the line gives the quality of the fit.

$$\dot{V}_{NO} = C_ANO \cdot \dot{V}_E + J_{aw}NO \tag{11.1}$$

11.3.2 The Silkoff technique (ST)

The Silkoff technique (ST) utilizes a non-linear regression technique using Eq. (11.2) which is the basic equation of the 2CM for presenting the dependence of exhaled

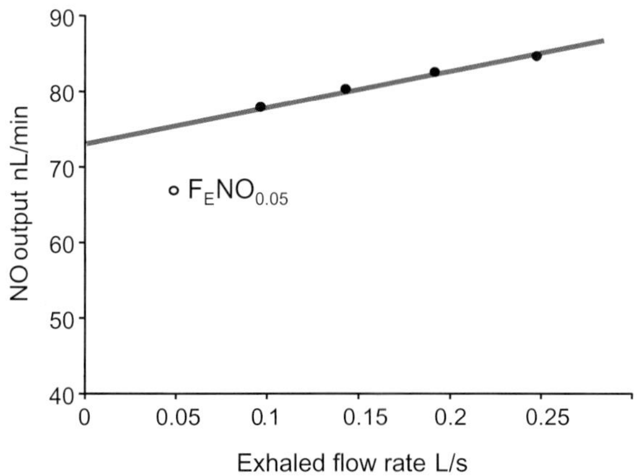

FIGURE 11.3
2CM model by Tsoukias and George. The slope of the regression line will give the C_ANO and the intercept on the *Y*-axis gives the NO-flux, $J_{aw}NO$. $F_ENO_{0.05}$ is not one of these flow rates.

NO values as a function of flow.[21] Silkoff et al. used nine flow rates between 4 and 1550 mL/s to estimate the three NO parameters.

$$F_E NO = C_{aw}NO + (C_A NO - C_{aw}NO) \cdot \exp(-D_{aw}NO/\dot{V}_E) \tag{11.2}$$

Consequently, the NO output is obtained from Eq. (11.2) by multiplying both sides by $\dot{V}_E$:

$$\begin{aligned} \dot{V}_{NO} &= \dot{V}_E \cdot F_E NO \\ &= \dot{V}_E \cdot (C_{aw}NO + (C_A NO - C_{aw}NO) \cdot \exp(-D_{aw}NO/\dot{V}_E)) \end{aligned} \tag{11.3}$$

For all the flows $\dot{V}_E \gg D_{aw}NO$, the exponential in Eq. (11.3) can be linearized and replaced by $(1 - D_{aw}NO/\dot{V}_E)$; then Eq. (11.3) becomes:

$$\begin{aligned} \dot{V}_{NO} &= \dot{V}_E \cdot F_E NO = C_A NO \cdot \dot{V}_E + (C_{aw}NO - C_A NO) \cdot D_{aw}NO \\ &= C_A NO \cdot \dot{V}_E + J_{aw}NO \cdot D_{aw}NO \end{aligned} \tag{11.4}$$

Silkoff et al. also suggests that $D_{aw}NO$ and $J'_{aw}NO$ can be estimated from two low flow rates (<50 mL/s) by using the slope ($-D_{aw}NO$) and the intercept $(J'_{aw}NO/D_{aw}NO)$ of a plot of $\dot{V}_{NO}$ versus $F_E NO$. However, this is only a semi-quantitative approximation which can be shown by obtaining $\dot{V}_E$ as a function of $F_E NO$ using Eq. (11.2) and substituting the result in Eq. (11.3).

11.3.3 The Pietropaoli technique (PT)

The Pietropaoli technique (PT) includes six constant-exhalation vital capacity maneuvers with flow rates between 6 and 1355 mL/s.[20] $C_A NO$ and $J'_{aw}NO$ are estimated together with the estimation of $D_{aw}NO$. The theory of this technique was first published by Hyde et al.[26] A plot of $F_E NO$ versus $1/\dot{V}_E$ for $\dot{V}_E > 200$ mL/s renders the intercept as an estimation of $C_A NO$ according to Eq. (11.5).

$$F_E NO = C_A NO + J'_{aw}NO \cdot 1/\dot{V}_E \tag{11.5}$$

When comparing Eq. (11.5) to Eq. (11.4), they become identical if $J'_{aw}NO$ is replaced by $J_{aw}NO$, which is the accurate expression for non-zero $C_A NO$ values. $C_A NO$ and $J_{aw}NO$ can be solved from two data points if flow values high enough to allow linear approximation of the exponential are used; see Figure 11.4. With more data points, any non-linear curve-fitting algorithm can be used allowing $C_{aw}NO$ and $D_{aw}NO$ to be separated.

11.3.4 The Högman and Meriläinen algorithm (HMA)

The Högman and Meriläinen algorithm (HMA) was first described in 1999 and is a non-linear model[18,19] with a set of three exhalation flow rates of typical values of 10 mL/s (low), 100 mL/s (medium), and 300 mL/s (high), see Figure 11.5. The low limit is presently under investigation. The high flow rate previously was 500 mL/s, at

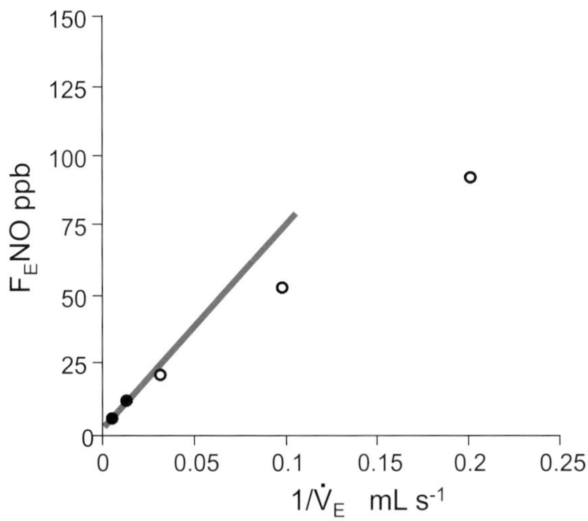

FIGURE 11.4
C_ANO can be estimated by the Pietropaoli-technique if flow rates above 200 mL/s are used. The intercept on the *Y*-axis is the C_ANO. $J_{aw}NO$ is the slope of the line.

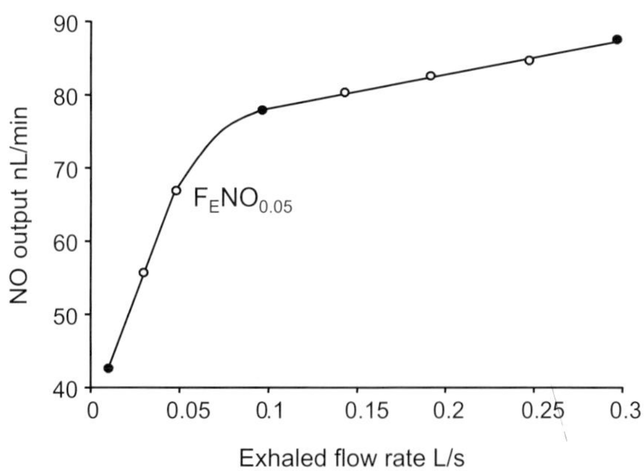

FIGURE 11.5
The Högman-Meriläinen Algorithm (HMA). To create the non-linear flow-volume curve at least tree flow rates are needed and these flow rates should be at both ends and one in the middle of the curve. $F_ENO_{0.05}$ is not one of these flow rates.

which it was difficult to find a plateau; this can be lowered to 300 mL/s but no lower.[27] Therefore, it can be stated that the so-called flow independent NO parameters are actually flow dependent. These high flow rates might still be too high for children to perform. The $\dot{V}_{NO}$ versus flow rate is plotted and a line set through the medium and

Table 11.2 NO parameters obtained from the different NO models generated by multiple flow rates. Note that the Tsoukias and George model needs an additional flow rate to measure $F_ENO_{0.05}$ while this value can be calculated with the HMA model.

Model	Range of Flow Rates (mL/s)	C_ANO	$J'_{aw}NO$	$J_{aw}NO$	$C_{aw}NO$	$D_{aw}NO$
Tsoukias and George 2CM[15]	Two VC at 100–500	×	×	–	–	–
Silkoff technique[21]	Nine breaths at 4–1550	×	×	×	×	×
Pietropaoli technique[20]	Six breaths at 6–1355	×	×	×	×	×
Högman-Meriläinen algorithm[19]	Three breaths at 10, 100, 300	×	×	×	×	×

high flow rates gives estimates of C_ANO and $J_{aw}NO$. $D_{aw}NO$ and $C_{aw}NO$, are then separated from $J_{aw}NO$ by using all three flow rates employing an iterative algorithm with a third-order correction term added to the first-order linear approximation of the exponential in Eqs. (11.2) and (11.3).

A specially added feature of this approach is to use an algorithm to test if the measured set of data points is mathematically consistent with the model. This is to avoid the potential problem of standard curve fitting algorithms when applied to three data points only, when a good fit can be found to Eq. (11.2), even if some of the data points are corrupted by measurement errors of technical origin. The condition required for data allowing reliable separation of $D_{aw}NO$ and $C_{aw}NO$, can be written in a general form as in Eq. (11.6) (see Table 11.2):

$$\frac{F_ENO_{low} - F_ENO_{medium}}{F_ENO_{medium} - F_ENO_{high}} < \frac{\dot{V}_{E,high}}{\dot{V}_{E,low}} \cdot \frac{\dot{V}_{E,medium} - \dot{V}_{E,low}}{\dot{V}_{E,high} - \dot{V}_{E,medium}} \tag{11.6}$$

If this is not the case, the data set suffers from measurement errors either in the flow or NO concentration, or both. Additionally, it should naturally be required that C_ANO be positive, since a concentration of a gas cannot be negative!

11.4 CORRECTIONS FOR AXIAL BACK DIFFUSION

During the past few years the role of axial back diffusion of NO at low flows from airway generation close to the alveolar compartment has been under investigation.[28–31]

A trumpet-shaped geometry for the airways is utilized instead of a constant diameter tube, as used in earlier models. Even a multi-compartment model has been suggested.[32] It still remains to be confirmed if the single or double trumpet model is enough to assess the magnitude of axial diffusion, because the relationships between the local airway wall area and the airway cross section vary a lot in the real fractal airway tree from a single tube even if its cross section is matched with the sum of all the branches.

Axial back diffusion models presently in use are those by Condorelli et al.[28] and by Kerckx et al.[29] The first one is a model combining 2CM with a consideration of axial diffusion and the trumpet shape of the airway tree. Four different flow rates between 100 and 250 mL/s are used to estimate $J_{aw}NO$ and C_ANO. The model predicts that a plot of $\dot{V}_{NO}$ versus $\dot{V}_E$ produces a linear relationship in which the slope is equal to $C_ANO + J'_{aw}NO \cdot 0.00078$ and the intercept is equal to $J'_{aw}NO/1.7$, according to Eqs. (11.7) and (11.8):

$$J'_{aw}NO_{corrected} = 1.7 \cdot J_{aw}NO \quad (11.7)$$

$$C_ANO_{corrected} = C_ANO - \left(J_{aw}NO/740\right) \quad (11.8)$$

Kerckx et al. also use the 2CM but generate the regression line with flow rates of 50, 175, and 300 mL/s. They conclude that the C_ANO should be corrected according to Eq. (11.9):

$$C_ANO_{corrected} = (C_ANO - 0.08 \cdot F_ENO_{0.05})/0.92 \quad (11.9)$$

Endogenous NO from the upper and lower airways is transported into the alveolar region. When the alveolus extends with inhaled gas the process of diffusion of O_2 and CO_2 takes place. This endogenous inhalation of NO in the ppb range is known to increase oxygenation.[33] Exogenous inhalation of NO in the ppm range has also been shown to improve oxygenation,[34,35] but most effectively during pulsed delivery in the beginning of the breath.[36] Hence, NO is transported together with the highest O_2 concentration to reach the pulmonary vessels, dilate them, and thereby facilitate the gas exchange. NO is metabolized and an on-off dilatation is achieved. Continuous NO inhalation will cause endothelin-1 (constriction agent) to increase,[37] while pulsed NO delivery does not affect its level.[38] Seen from a physiological viewpoint, the movement of NO by axial diffusion is therefore not an optimal solution. During exhalation when efficient gas exchange has taken place, dilatation of vessels by NO axial diffusion will only cause inhomogeneity of the lung.[36] Since NO values at multiple flow rates are obtained without breath hold for the extended NO analysis, axial diffusion is not taken into considerations. Another reason for not using the corrections is that they frequently give negative C_ANO, especially with increased $F_ENO_{0.05}$, as can be seen in Table 11.3. This can be caused by the high flow rates at which C_ANO was calculated and where axial diffusion might not be significant. Another reason is that back diffusion must (physiologically) be different from individual to individual and especially altered during disease, and so a fixed correction might be inappropriate.

Table 11.3 Data obtained with three flow rates 10, 100, and 350 mL/s with the HMA method in three subjects with low, normal, and high $F_ENO_{0.05}$. Corrections for axial back diffusion made by the formulae described by Condorelli et al. correction 1[28] and by Kerckx et al. correction 2.[29] With increasing $F_ENO_{0.05}$ and $J'_{aw}NO$ the corrections become negative.

Subjects	$F_ENO_{0.05}$ (ppb)	$J_{aw}NO$ (nL/s)	C_ANO (ppb)	C_ANO (ppb) Correction 1	C_ANO (ppb) Correction 2
A	8	0.4	1.0	0.5	0.4
B	19	1.3	1.4	0.0	−0.1
C	47	2.3	1.5	−1.6	−2.4

11.5 LIMITATIONS

Understanding the limitations of the different theoretical models is important, but it is also very important to know the limitations of the equipment. Some analyzers are built for environmental research (ppm not ppb) and are too slow to be used for modeling, while other analyzers are built for clinical use for high NO concentrations but with low accuracy at low NO concentrations. Extended NO analyses demand an analyzer with an accuracy of 0.1 ppb, which adequately measures low NO concentration at high flow rates. An accuracy of 0.1 mL/s for low flow rates and 1 mL/s for high flow rates is necessary for the flow measurements. The instruments should be easy to calibrate and to verify both the NO signal and flow rate and to have a view of the performance. When using linear regression, R^2 close to 1 will assure the investigator that the data are sound. It is not acceptable to use the wrong flow rates for the chosen model. Most often the model is used with flow rates lower than are recommended; including 50 mL/s, a value that gives a steeper slope and thereby a falsely high C_ANO and a lower intercept with a falsely low $J_{aw}NO$.[22]

Vital capacity (VC) maneuvers should be avoided since high pressures in the lung can cause a release of NO in animals.[39] It has been shown that performing a VC maneuver in humans results in a $F_ENO_{0.1}$ of 12 ppb with the time to reach a plateau of 16 s. The corresponding values after a deep breath were 12 ppb and 7 s.[18] There was no statistical difference in F_ENO between the maneuvers, but the time to reach a plateau was significantly longer, due to a down sloping curve before the plateau was reached. This has to be considered when automated programs are designed that often limit the time for exhalation. Avoidance of VC maneuvers is in keeping with the recommendations that NO measurements should be performed before other lung function measurements.[13]

As with all NO measurements the lips must be closed around the mouthpiece in order to avoid leaks. The system has to be tight, including tubing and fittings. For the measurements at low flow a proper exhaled pressure resistance is required in order to avoid contamination with nasal NO.

11.6 VALUES FROM NON-SMOKING HEALTHY SUBJECTS

The extended NO analysis (HMA) has been applied to a random sample of the Swedish population.[40] Based on these results, it was suggested that the values for healthy subjects should be considered to fall between the following ranges: $F_ENO_{0.05}$ 10–30 ppb; $C_{aw}NO$ 50–250 ppb; $D_{aw}NO$ 5–15 mL/s; $J_{aw}NO$ 0.8–1.6 nL/s; and C_ANO 0–4 ppb. Interestingly, this study showed a correlation between age and C_ANO in women ($r = 0.51$, $p = 0.001$) but not in men. Explanations for the higher C_ANO could possibly then be that the uptake of the inhaled NO is lower, i.e. that there is a ventilation/perfusion mismatch followed by inhomogeneity of the lung with age. Further capillary blood volume in women older than 40 years of age decreases in comparison to younger women.[41] The same research group has also shown that the diffusing capacity declines in a linear fashion with age.[42] In the study by Gelb et al. using the Tsoukias and George NO model and the Condorelli correction, C_ANO was significantly increased in a group of subject above 60 years of age while $J_{aw}NO$ was similar for all ages.[43] The values were in line with the study by Högman et al.[40] except for the $J_{aw}NO$ values which were about 70 % increased in the study by Gelb et al. This is explained by the corrected values suggested by Condorelli et al.[28]

11.7 THE USEFULNESS OF EXTENDED NO ANALYSIS

While extended NO analysis is unlikely to become a diagnostic test for respiratory diseases of itself the analysis will definitely add more information in the diagnostic process and in evaluating treatment, especially corticosteroid treatment in asthma. The recent Cochrane report stated that defining the dose of inhaled corticosteroids (ICS) based on the $F_ENO_{0.05}$ value cannot be routinely supported.[8] It has also been said that reference values will not have a role, since clinically stable patients have generally higher values.[44] That $F_ENO_{0.05}$ is high in eosinophilic asthma is due to an increased $J_{aw}NO$, which is the product of $C_{aw}NO$ and $D_{aw}NO$. The treatment with ICS will decrease $C_{aw}NO$ without affecting $D_{aw}NO$,[21] resulting in a still elevated $J'_{aw}NO$ and a $F_ENO_{0.05}$ higher than reference values. A key feature in eosinophilic rhinitis and asthma is an elevated $D_{aw}NO$.[45] Hence, if the $D_{aw}NO$ remains high despite treatment with ICS this will hinder the $F_ENO_{0.05}$ value to reach reference values. Instead of using $F_ENO_{0.05}$ as a guide for ISC treatment, the $C_{aw}NO$ value could possibly be beneficial.

Due to the simplicity and the instant results from the extended analysis, NO measurements are attractive from a patient's viewpoint. For example the C_ANO values seem to add to the understanding of the severity of asthma[9,10] and diseases such as COPD,[19,46] alveolitis[47], and scleroderma.[48] In Obstructive Sleep Apnea Syndrome the C_ANO but not the $J_{aw}NO$ is reduced, and this low C_ANO is associated with an increased risk of hypertension.[49] An increase in C_ANO could be interpreted as systemic engagement with circulating mediators causing iNOS expression in the alveolar region. In support of this theory is the effect of oral prednisone on the decrease of C_ANO.[50] Another explanation, most likely in COPD but also in severe asthma, is a reduced uptake of inhaled NO due to ventilation/perfusion mismatch. C_ANO has been found to be increased in systemic sclerosis and non-invasive NO analysis could possibly be used to assess the extent of interstitial lung disease[48] and be a tool for early detection of this disease.[51]

The modeling of NO production both in asthma and COPD has shown many similarities. With the results from the extended NO analysis two groups of COPD patients could clearly be seen,[19] one with increased $D_{aw}NO$, the characteristic feature of atopic asthma,[45] and one group with decreased $D_{aw}NO$. Notably, non-atopic rhinitis, non-atopic asthma, and healthy individuals have similar values of the NO parameters from the extended NO analysis.[45] Modeling of NO production in the lung is one way to see such differences. In the light of the discussion in the European Respiratory Society about focusing on pheno- and endo-types in respiratory diseases, instead of the diagnosis of asthma and COPD, extended NO analysis could be such a tool, either the linear model or the non-linear model. That monitoring of NO is here to stay is clear, but we do need to know how and when to use this tool. NO can be important in indicating the risk of new-onset asthma,[52] as well as lung function decline in asthma.[53] Atopic rhinitis patients have increased exhaled NO[54] due to an increased $D_{aw}NO$ and not an up-regulation of NO production in the airway wall.[19] Perhaps the development of atopic asthma could be prevented if there was a way to normalize $D_{aw}NO$ in rhinitis.

11.8 CONCLUSIONS

In clinical practice we need simple models/methods that will give us repeatable results. The extended NO analysis will give additional information about the production of NO in the lung, compared to just one NO concentration at one flow rate. NO measurement in exhaled breath is a non-invasive method that gives instant results, gives added value in the diagnostic procedure and is a valuable tool for following the effects of anti-inflammatory treatments.

To date, there has not been a task force on the modeling of NO, like that for the recommendations on the measurement of exhaled NO.[13] The abbreviations of the NO parameters have been agreed on[16] and until we get recommendations the researchers and clinicians must follow the flow rates used for the different models and which are also applicable for the models used for correction of different NO parameters. These NO parameters are indeed flow dependent.

References

1. Gaston B, Drazen JM, Loscalzo J, Stamler JS. The biology of nitrogen oxides in the airways. *Am J Respir Crit Care Med* 1994;**149**:538–51.
2. Ricciardolo FL. Multiple roles of nitric oxide in the airways. *Thorax* 2003;**58**:175–82.
3. Ricciardolo FL, Sterk PJ, Gaston B. Folkerts G. Nitric oxide in health and disease of the respiratory system. *Physiol Rev* 2004;**84**:731–65.

4. Ludviksdottir D, Janson C, Högman M, Hedenström H, Björnsson E, Boman G. Exhaled nitric oxide and its relationship to airway responsiveness and atopy in asthma. BHR-Study Group. *Respir Med* 1999;**93**:552–6.
5. Guo FH, Comhair SA, Zheng S, Dweik RA, Eissa NT, Thomassen MJ, et al. Molecular mechanisms of increased nitric oxide (NO) in asthma: evidence for transcriptional and post-translational regulation of NO synthesis. *J Immunol* 2000;**164**:5970–80.
6. Kobzik L, Bredt DS, Lowenstein CJ, Drazen J, Gaston B, Sugarbaker D, et al. Nitric oxide synthase in human and rat lung: immunocytochemical and histochemical localization. *Am J Respir Cell Mol Biol* 1993;**9**:371–7.
7. Pucci S, Incorvaia C. Allergy as an organ and a systemic disease. *Clin Exp Immunol* 2008;**153**(Suppl. 1):1–2.
8. Petsky HL, Cates CJ, Li A, Kynaston JA, Turner C, Chang AB. Tailored interventions based on exhaled nitric oxide versus clinical symptoms for asthma in children and adults. *Cochrane Database Syst Rev* 2009; CD006340.
9. Lehtimäki L, Kankaanranta H, Saarelainen S, Turjanmaa V, Moilanen E. Increased alveolar nitric oxide concentration in asthmatic patients with nocturnal symptoms. *Eur Respir J* 2002;**20**:841–5.
10. Lehtimäki L, Kankaanranta H, Saarelainen S, Turjanmaa V, Moilanen E. Peripheral inflammation in patients with asthmatic symptoms but normal lung function. *J Asthma* 2005;**42**:605–9.
11. Högman M, Strömberg S, Schedin U, Frostell C, Hedenstierna G, Gustafsson LE. Nitric oxide from the human respiratory tract efficiently quantified by standardized single breath measurements. *Acta Physiol Scand* 1997;**159**:345–6.
12. Silkoff PE, McClean PA, Slutsky AS, Furlott HG, Hoffstein E, Wakita S, et al. Marked flow-dependence of exhaled nitric oxide using a new technique to exclude nasal nitric oxide. *Am J Respir Crit Care Med* 1997;**155**:260–7.
13. ATS/ERS recommendations for standardized procedures for the online and offline measurement of exhaled lower respiratory nitric oxide and nasal nitric oxide, 2005. *Am J Respir Crit Care Med* 2005;**171**:912–30.
14. Fick A. Braunschweig: druck und Verlag von Friedrich Vieweg und Sohn Die Medizinische. *Physik* 1856.
15. Tsoukias NM, George SC. A two-compartment model of pulmonary nitric oxide exchange dynamics. *J Appl Physiol* 1998;**85**:653–66.
16. George SC, Högman M, Permutt S, Silkoff PE. Modeling pulmonary nitric oxide exchange. *J Appl Physiol* 2004;**96**:831–9.
17. Verbanck S, Kerckx Y, Schuermans D, de Bisschop C, Guenard H, Naeije R, et al. The effect of posture-induced changes in peripheral nitric oxide uptake on exhaled nitric oxide. *J Appl Physiol* 2009;**106**:1494–8.
18. Högman M, Drca N, Ehrstedt C, Meriläinen P. Exhaled nitric oxide partitioned into alveolar, lower airways and nasal contributions. *Respir Med* 2000;**94**:985–91.
19. Högman M, Holmkvist T, Wegener T, Emtner M, Andersson M, Hedenström H, et al. Extended NO analysis applied to patients with COPD, allergic asthma and allergic rhinitis. *Respir Med* 2002;**96**:24–30.
20. Pietropaoli AP, Perillo IB, Torres A, Perkins PT, Frasier LM, Utell MJ, et al. Simultaneous measurement of nitric oxide production by conducting and alveolar airways of humans. *J Appl Physiol* 1999;**87**:1532–42.

21. Silkoff PE, Sylvester JT, Zamel N, Permutt S. Airway nitric oxide diffusion in asthma: Role in pulmonary function and bronchial responsiveness. *Am J Respir Crit Care Med* 2000;**161**:1218–28.
22. Högman M, Meriläinen P. Extended NO analysis in asthma. *J Breath Res* 2007;**1**:024001.
23. Roy K, Borrill ZL, Starkey C, Hazel AL, Morris J, Vestbo J, et al. Use of different exhaled nitric oxide multiple flow rate models in COPD. *Eur Respir J* 2007;**29**:651–9.
24. Mahut B, Louis B, Zerah-Lancner F, Delclaux C. Validity criteria and comparison of analytical methods of flow-independent exhaled NO parameters. *Respir Physiol Neurobiol* 2006;**153**:148–56.
25. Tsoukias NM, Tannous Z, Wilson AF, George SC. Single-exhalation profiles of NO and CO2 in humans: effect of dynamically changing flow rate. *J Appl Physiol* 1998;**85**:642–52.
26. Hyde RW, Geigel EJ, Olszowka AJ, Krasney JA, Forster Jr RE, Utell MJ, et al. Determination of production of nitric oxide by lower airways of humans-theory. *J Appl Physiol* 1997;**82**:1290–6.
27. Holmkvist T, Meriläinen P, Högman M. The maximum flow required for the extended NO analysis. *Am J Respir Crit Care Med* 2003;**167**:A53.
28. Condorelli P, Shin HW, Aledia AS, Silkoff PE, George SC. A simple technique to characterize proximal and peripheral nitric oxide exchange using constant flow exhalations and an axial diffusion model. *J Appl Physiol* 2007;**102**:417–25.
29. Kerckx Y, Michils A, Van Muylem A. Airway contribution to alveolar nitric oxide in healthy subjects and stable asthma patients. *J Appl Physiol* 2008;**104**:918–24.
30. Kerckx Y, Van Muylem A. Axial distribution heterogeneity of nitric oxide airway production in healthy adults. *J Appl Physiol* 2009;**106**:1832–9.
31. Shin HW, George SC. Impact of axial diffusion on nitric oxide exchange in the lungs. *J Appl Physiol* 2002;**93**:2070–80.
32. Shelley DA, Puckett JL, George SC. Quantifying proximal and distal sources of NO in asthma using a multicompartment model. *J Appl Physiol* 2010;**108**:821–9.
33. Lundberg JO, Settergren G, Gelinder S, Lundberg JM, Alving K, Weitzberg E. Inhalation of nasally derived nitric oxide modulates pulmonary function in humans. *Acta Physiol Scand* 1996;**158**:343–7.
34. Pepke-Zaba J, Higenbottam TW, Dinh-Xuan AT, Stone D, Wallwork J. Inhaled nitric oxide as a cause of selective pulmonary vasodilatation in pulmonary hypertension. *Lancet* 1991;**338**:1173–4.
35. Rossaint R, Falke KJ, Lopez F, Slama K, Pison U, Zapol WM. Inhaled nitric oxide for the adult respiratory distress syndrome. *N Engl J Med* 1993;**328**:399–405.
36. Heinonen E, Nyman G, Meriläinen P, Högman M. Effect of different pulses of nitric oxide on venous admixture in the anaesthetized horse. *Br J Anaesth* 2002;**88**:394–8.
37. Pearl JM, Nelson DP, Raake JL, Manning PB, Schwartz SM, Koons L, et al. Inhaled nitric oxide increases endothelin-1 levels: a potential cause of rebound pulmonary hypertension. *Crit Care Med* 2002;**30**:89–93.
38. Grubb TL, Hogman M, Edner A, Frendin JH, Heinonen E, Malavasi LM, et al. Physiologic responses and plasma endothelin-1 concentrations associated with abrupt cessation of nitric oxide inhalation in isoflurane-anesthetized horses. *Am J Vet Res* 2008;**69**:423–30.
39. Strömberg S, Lönnqvist PA, Persson MG, Gustafsson LE. Lung distension and carbon dioxide affect pulmonary nitric oxide formation in the anaesthetized rabbit. *Acta Physiol Scand* 1997;**159**:59–67.

40. Högman M, Lafih J, Meriläinen P, Bröms K, Malinovschi A, Janson C. Extended NO analysis in a healthy subgroup of a random sample from a Swedish population. *Clin Physiol Funct Imaging* 2009;**29**:18–23.
41. Stam H, Versprille A, Bogaard JM. The components of the carbon monoxide diffusing capacity in man dependent on alveolar volume. *Bull Eur Physiopathol Respir* 1983;**19**:17–22.
42. Stam H, Hrachovina V, Stijnen T, Versprille A. Diffusing capacity dependent on lung volume and age in normal subjects. *J Appl Physiol* 1994;**76**:2356–63.
43. Gelb AF, George SC, Camacho F, Fraser C, Flynn Taylor C, Shakkottai S. Increased nitric oxide concentrations in the small airway of older normal subjects. *Chest* 2011;**139**:368–75.
44. Reddel HK, Taylor DR, Bateman ED, Boulet LP, Boushey HA, Busse WW, et al. An official American Thoracic Society/European Respiratory Society statement: asthma control and exacerbations: standardizing endpoints for clinical asthma trials and clinical practice. *Am J Respir Crit Care Med* 2009;**180**:59–99.
45. Högman M, Malinovschi A, Norbäck D, Janson C. Added value with extended NO analysis in atopy and asthma. *Clin Physiol Funct Imaging* 2011;**31**:294–9.
46. Brindicci C, Ito K, Resta O, Pride NB, Barnes PJ, Kharitonov SA. Exhaled nitric oxide from lung periphery is increased in COPD. *Eur Respir J* 2005;**26**:52–9.
47. Lehtimäki L, Kankaanranta H, Saarelainen S, Hahtola P, Järvenpää R, Koivula T, et al. Extended exhaled NO measurement differentiates between alveolar and bronchial inflammation. *Am J Respir Crit Care Med* 2001;**163**:1557–61.
48. Tiev KP, Cabane J, Aubourg F, Kettaneh A, Ziani M, Mouthon L, et al. Severity of scleroderma lung disease is related to alveolar concentration of nitric oxide. *Eur Respir J* 2007;**30**:26–30.
49. Foresi A, Leone C, Olivieri D, Cremona G. Alveolar-derived exhaled nitric oxide is reduced in obstructive sleep apnea syndrome. *Chest* 2007;**132**:860–7.
50. Gelb AF, Taylor CF, Nussbaum E, Gutierrez C, Schein A, Shinar CM, et al. Alveolar and airway sites of nitric oxide inflammation in treated asthma. *Am J Respir Crit Care Med* 2004;**170**:737–41.
51. Wuttge DM, Bozovic G, Hesselstrand R, Aronsson D, Bjermer L, Scheja A, et al. Increased alveolar nitric oxide in early systemic sclerosis. *Clin Exp Rheumatol* 2010;**28**:S5-9 .
52. Bastain TM, Islam T, Berhane KT, McConnell RS, Rappaport EB, Salam MT, et al. Exhaled nitric oxide, susceptibility and new-onset asthma in the Children's Health Study. *Eur Respir J* 2011;**37**:523–31.
53. van Veen IH, Ten Brinke A, Sterk PJ, Sont JK, Gauw SA, Rabe KF, et al. Exhaled nitric oxide predicts lung function decline in difficult-to-treat asthma. *Eur Respir J* 2008;**32**:344–9.
54. Henriksen AH, Sue-Chu M, Holmen TL, Langhammer A, Bjermer L. Exhaled and nasal NO levels in allergic rhinitis: relation to sensitization, pollen season and bronchial hyperresponsiveness. *Eur Respir J* 1999;**13**:301–6.

CHAPTER

12 Carbon Monoxide as an Exhaled Biomarker of Pulmonary Diseases

Stefan W. Ryter

Pulmonary and Critical Care Medicine, Brigham and Women's Hospital, Harvard Medical School, 75 Francis Street, Boston, MA 02115, USA

12.1 INTRODUCTION

The measurement of selected organic and inorganic components present in human exhaled breath may find practical clinical applications in diagnostics and therapeutic monitoring. The primary advantage of exhaled breath analysis over other clinical testing regimens is the non-invasive nature of the measurements. Exhaled analyte profiles could theoretically provide information on the presence or severity of lung or systemic diseases or the effectiveness of anti-inflammatory drug treatments.[1–6] The analytical approaches to exhaled breath testing include the detection of permanent gases or volatile organic compounds,[6] and/or the testing of biochemical or biophysical properties of exhaled breath condensate.[7]

Among the many possibilities for exhaled breath monitoring, a number of studies have focused on the low molecular weight diatomic gases present in breath, namely nitric oxide (NO) and carbon monoxide (CO). Exhaled nitric oxide (eNO) has been intensely studied in the context of human respiratory diseases, including asthma and chronic obstructive pulmonary disease (COPD) and has been suggested as a marker of inflammation.[8–12] Similar to eNO, exhaled carbon monoxide (eCO) levels have also been monitored in asthma/airways diseases and COPD. Furthermore, eCO has been employed to monitor cigarette smoking.[11–15] Additional focus areas for exhaled gas research involving eCO/eNO measurements include cystic fibrosis, diabetes, surgery, and critical illness, among others.[11–16] This chapter will focus primarily on the progress and potential applications of eCO measurement in lung disease. The sources of environmental and endogenous CO will be discussed from the perspective of breath monitoring. Emphasis will be placed on model studies in asthma and COPD.

12.2 CHEMICAL AND BIOCHEMICAL PROPERTIES OF CO

The physical properties of CO, including its lack of odor, color, and taste, increase the dangerous potential of this gas in accidental exposure situations.[17] CO is often compared to its cognate breath gas NO, due to apparent similarities with respect to

Volatile Biomarkers. http://dx.doi.org/10.1016/B978-0-44-462613-4.00012-X

structure, size, and molecular weight (CO, $M_R = 28.01$ g/mol, NO, $M_R = 30.006$ g/mol). Both CO and NO exist in gaseous form at ambient temperature and pressure (boiling points, CO: −191.5 °C, NO: −151.8 °C),[17,18] and are soluble in water (CO: 0.0015 mol/l, NO: 0.0033 mol/l, at 1.013 bar and 0 °C).[19] Despite these similarities, some important differences between these gases are evident.[20] CO is an inherently stable non-radical with little physiological reactivity with respect to non-iron compounds. CO is used as a reducing agent in industrial applications, though the biological significance of such reactions is limited.[21] On the other hand, NO contains one unpaired electron, which permits it to react in metal-catalyzed redox reactions under physiological conditions.[22]

CO and NO are similar in their capacity to act as heme ligands, and both gases can bind with certain heme-containing proteins at the heme iron. CO will bind only to reduced (ferrous) iron centers, whereas NO may bind to both ferrous and ferric hemes.[20] The reaction of NO with hemoglobin has a number of possible products. NO reacts with oxygenated hemoglobin to form methemoglobin and nitrate, and this reaction is regarded as a major mechanism for NO detoxification.[23] The formation and biological significance of nitrosyl hemoglobin and S-nitroso-hemoglobins have been described elsewhere.[24] CO binds readily to hemoglobin in the bloodstream under physiological conditions, to form a stable but reversible adduct carboxyhemoglobin (HbCO). This interaction occurs with ~250 times the affinity of molecular oxygen.[17] Binding of CO to hemoglobin impairs the release of the remaining bound O_2, further impairing O_2 delivery to tissues. The resulting hypoxemia can lead to death in cases of prolonged or excessive exposure.[25] CO can also bind the heme iron of mitochondrial cytochrome *c*: oxidase, resulting in reduced enzymatic activity and inhibition of cellular respiration. This metabolic toxicity of CO is not likely to be relevant to CO toxicity *in vivo*, which primarily involves hypoxemia.[25]

12.3 ENVIRONMENTAL SOURCES OF CO

Environmental sources can account for a significant fraction of the CO found in human exhaled breath. Automobile exhaust, burning organic material, and the inefficient combustion of fossil fuels represent the major sources of CO in the ambient air. Indoor levels of CO range from 0.5–5 parts-per-million (ppm) but may reach higher values (up to 30 ppm) at point sources of combustion. CO is a major urban air pollutant and health hazard, such that CO levels near high traffic areas can reach 10–50 ppm.[26] The World Health Organization guidelines recommend CO gas exposure <10 ppm for 8 h and 90 ppm for periods not exceeding 15 min.[26] The National Institute for Occupational Safety and Health (NIOSH, USA) sets exposure limits to a time weighted average of 35 ppm, and a ceiling at 200 ppm. The Occupational Safety and Health Administration (OSHA, USA) sets a permissible exposure limit at 50 ppm.[27] Cigarette smoke is another major source of ambient CO, which may yield up to 20 mg of CO per cigarette, depending on the type of cigarette.[28]

12.4 ENDOGENOUS SOURCES OF CO: THE HEME OXYGENASE ENZYME SYSTEM

It has long been recognized that a considerable amount of CO is produced endogenously in the body as the byproduct of hemoglobin turnover.[29,30] The rate of CO production in humans has been estimated at ~0.4 ml (18 μmol) CO per hour.[31] Patients with hemolytic anemia displayed increased rates of CO production up to ~3.6 ml/h (143 μmol/h).[32] Endogenously produced CO arises primarily from active heme metabolism catalyzed by the heme oxygenase (HO, E.C. 1:14:99:3) enzymes.[33,34] While the systemic turnover of hemoglobin is believed to represent the major source of CO production, in principle, heme derived from other hemoprotein sources can contribute at a cellular level.[35] A smaller fraction of systemic CO production (~ 15%) can originate from non-heme sources such as lipid peroxidation, or cytochrome p450-dependent drug metabolism.[36]

HO activity provides the rate-limiting step in the heme degradative pathway. HO oxidizes heme at the α-methene bridge in a reaction requiring molecular oxygen (O_2) and electrons from NADPH cytochrome p450 reductase. The reaction produces equimolar amounts of CO, iron, and bilirubin-IXα, the latter which is reduced to biliverdin-IXα by NADPH: biliverdin reductase.[33,34] HO exists as two major isoforms (HO-1, HO-2).[35] The inducible form of heme oxygenase (HO-1), a major cellular stress protein, responds to transcriptional regulation by multiple forms of chemical and physical stress. These include, but are not limited to, oxidants, heavy metals, ultraviolet radiation, thiol-reactive substances, and extreme fluctuations in ambient oxygen tension.[37] HO-1 represents a general marker of oxidative or pro-inflammatory stress, and consistently its elevated expression has been observed in various lung injury and sepsis models.[38–40] HO-2 is a constitutive isoform expressed highly in testis, liver, brain, and vascular endothelium.[41]

On the basis of the properties of the HO enzyme system, it can be predicted that the endogenous production of CO has a basal as well as stress-inducible component, both which depend on the bioavailability of heme. Coincidentally, heme acts as an inducer of HO-1 expression, as well as the major enzymatic substrate. The eCO generation attributable to the HO system is predicted to increase in response to systemic stress. A significant fraction of eCO can arise de novo in the airways and nasal passages.[42,43] Thus, eCO is believed to originate as the product of inducible HO-1 activity in the airway and nasal epithelium, as well as in alveolar macrophages, endothelial cells, and other lung cell types, as the consequence of local inflammation or oxidative stress[42,43] (Figure 12.1). CO in the airways may also arise by diffusion from the carboxyhemoglobin (HbCO) pool in the pulmonary circulation.[44] Elevations in HbCO have been correlated to increase in environmental CO,[45] and are also responsive to cigarette smoke inhalation.[46] In the absence of environmental exposure, increased HbCO may reflect increased HO activity in peripheral tissues, and increased systemic hemoglobin turnover.[45]

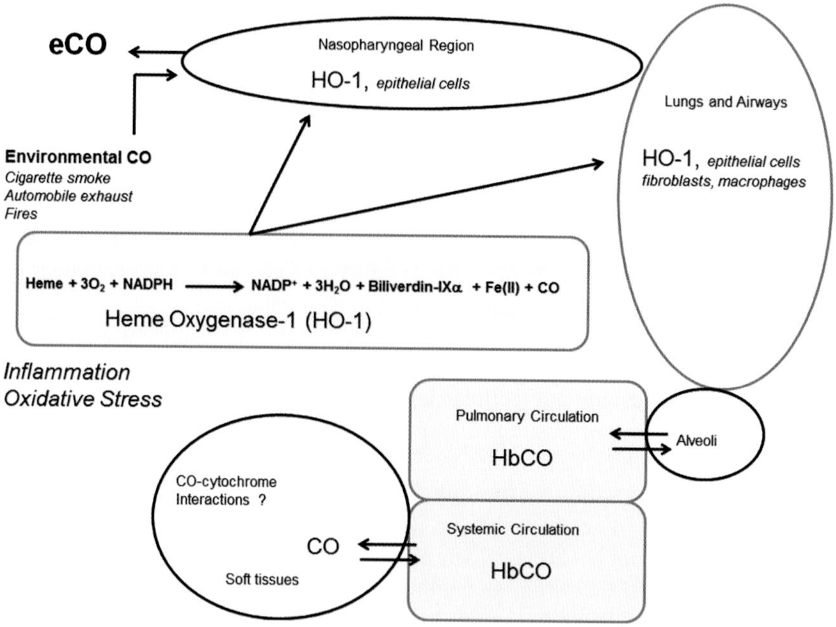

FIGURE 12.1

Schematic for the biodistribution of CO. Exhaled CO (eCO) is a component of the exhaled breath of humans. CO can be inspired as the consequence of environmental or occupational exposure to smoke or vehicle emissions. CO also evolves naturally in the body as the product of heme metabolism catalyzed by heme oxygenase (HO; E.C. 1:14:99:3) enzymes. The inducible form of this enzyme, HO-1, responds to regulation by pro-inflammatory stress. Thus, increased HO-1 activity in the airways as the consequence of airway inflammation may account in part for changes in eCO. CO may also evolve from systemic hemoglobin turnover. In the systemic circulation, CO exists mainly as carboxyhemoglobin (HbCO) and diffuses passively from the lung through the alveoli. CO in the systemic tissues may impact cellular signaling and homeostatic processes through interactions with cellular hemoproteins.

12.5 SIGNALING PROPERTIES OF CO

Recent studies suggest that the interactions of CO with intracellular targets can result in activation of cellular signaling pathways. Of these, the activation of guanylate cyclase (sGC) represents a major mechanism. CO can bind to the heme iron of sGC, inducing a conformational change leading to increased synthesis of guanosine 3′,-5′-monophosphate (cGMP). While both CO and NO bind to the heme iron of sGC, CO assumes a hexa-coordinate ligand structure, whereas the binding of NO is pentacoordinate, and involves the displacement of the axial histidine ligand of sGC.[47] Experimental evidence indicates that NO activates sGC *in vitro* and corresponding vasodilatory action *in vivo* with far greater potency.[48] Nevertheless, cGMP has been implicated in the vascular and neural effects of CO.[49–51] Another candidate mechanism for the vasoregulatory effects of CO involves activation of calcium-dependent potassium channels K_{ca} in vascular smooth muscle cells.[52] These interactions may be relevant to airway epithelia during inhalation and exhalation of CO. In various model studies, CO exposure has been reported to modulate the activation of mitogen activated protein kinases (MAPK), which are critical for the transduction or pro-inflammatory and/or stress signaling. [53–57] In macrophages, CO was shown to augment p38 MAPK activation under pro-inflammatory conditions.[53]

12.6 CYTO- AND TISSUE-PROTECTIVE EFFECTS OF CO

CO has been shown to exert cyto- and tissue-protective properties when applied at low concentration, and to inhibit cell death caused by apoptosis-inducing ligands in endothelial cells.[55] Furthermore, a potent anti-inflammatory effect of this gas was demonstrated *in vitro* and in animal models of endotoxin exposure.[53] In *in vitro* experiments, exposure to CO gas inhibited the secretion of pro-inflammatory cytokines in macrophages stimulated with bacterial endotoxins,[53] as well as inhibited Toll-like receptor (TLR) trafficking and activation.[58] This gas has also been shown to exert anti-proliferative effects *in vitro*, with respect to the proliferation of vascular smooth muscle.[59] Collectively, these anti-inflammatory, antiapoptotic, and antiproliferative effects provide the basis for a proposed therapeutic use of this gas. In rodent models, low concentration CO has demonstrated protective effects in ischemia/reperfusion injury,[56,57] vascular injury,[60] graft rejection,[60–62] and acute lung injury induced by high oxygen stress (hyperoxia).[54] In rodent models of ventilator-induced lung injury, inhaled CO reduced neutrophil granulocyte infiltration to the lung parenchyma and subsequent alveolar edema formation.[63–65] The tissue protective properties and proposed therapeutic applications of CO have been reviewed extensively elsewhere.[39,40]

12.7 METHODS FOR BREATH CO DETECTION

Early studies relying on measurements of eCO were compromised by relatively insensitive methods for detection of this gas. The eCO levels in exhaled breath are most commonly measured with electrochemical (chemiluminescence) technology.[63–68] The values thus obtained correlate with parallel gas chromatographic analysis and these sensors are sensitive in the 1–500 ppm range. Current analytical devices are portable which makes them ideal for clinical use.[69] More recent CO detection systems suitable for clinical measurements include a gas sensor adapted from a controlled potential electrolysis method, which is reportedly sensitive to 0.1 ppm.[69] CO may also be detected using experimental infrared laser spectroscopic methods (e.g. cavity leak out spectroscopy, CALOS; and integrated cavity output spectroscopy, ICOS).[70–73] A recent study has applied the CALOS technique to the detection of ^{13}CO, as a means for eCO measurements. Using this technique, the authors describe acute exercise-dependent, as well as long term natural variations in eCO output in healthy humans.[74] Recent studies also describe an infrared laser spectroscopic system for multi-component breath testing. Using the latter system, the authors have reported 1-s detection precisions of 0.5–0.8 ppb for NO, CO, and N_2O with an instrument response time of less than 1 s. These technologies, however, are currently available only in experimental settings.[75]

12.8 EXHALED CO IN HUMAN DISEASES

A number of studies have reported associations between eCO levels and various disease states. Modulations in eCO levels have been observed in critically ill or post-surgical patients and those with pulmonary diseases associated with inflammation, including COPD, asthma, cystic fibrosis, and bacterial infections.[13–16] The potential application of eCO as a predictive tool for assessing pulmonary inflammation in these disease contexts is discussed below.

12.8.1 Asthma

Asthma is defined as a chronic inflammatory disease of the airways resulting in airflow limitation. The airflow limitation is partially reversible either spontaneously or by therapy. The persistent inflammation associated with this disease causes airway hypersensitivity to allergens and irritants such as smoke, pollens, or dust. The pathological airway changes observed in asthma include eosinophil granulocyte predominant inflammation, edema, bronchoconstriction, and increased thick mucus secretion.[76] eCO values were found to be significantly increased in non-steroid treated asthmatic patients compared with healthy subjects, whereas these values returned to baseline in patients receiving corticosteroid therapy.[42] Acute exacerbations of asthma caused increases in measured eCO associated with a reduction in peak expiratory flow rate (PEFR). These changes were found to be reversible by oral glucocorticoid treatment.[77] Furthermore, eCO levels in asthma patients on β-agonists were decreased in response to a 4 week treatment with inhaled corticosteroids.[78] In this study, decreased eCO values correlated with decreases in sputum eosinophil numbers.[78] In a study of patients with atopic asthma, eCO values were found not to correlate with lung functional measurements in a cross-sectional analysis. However a longitudinal trend toward correlation of eCO values with bronchial reactivity was noted.[79] In allergic asthmatics, eCO increased in response to allergen challenge that preceded the peak decline in lung function as assessed by forced expiratory volume in 1 s (FEV_1).[80] The eCO values in these patients were not sensitive to stimuli that reduced FEV_1, including histamine, or to subsequent recovery by β-agonists, indicating that eCO values were independent of airway function.[80]

In patients with bronchial asthma, eCO levels were found to correlate with arterial HbCO levels.[81,82] The HbCO levels were found to increase at the time of exacerbation, and to decline to control levels in response to oral steroid treatment.[81] The changes in HbCO concentrations were significantly correlated with FEV_1.

Furthermore, arteriovenous HbCO differences were found to be higher in patients with bronchial asthma than those in control subjects.[82]

HO-1 expression has also been examined in the alveolar or airway macrophages of human asthmatics. Increased HO-1 expression was reported in alveolar macrophages of patients with recent asthma exacerbations relative to healthy controls or patients undergoing steroid treatment.[42] This observation was associated with increased bilirubin levels in the induced sputum, and correlated with reported increases in eCO values in non-steroid treated asthmatics.[42] However, another study has also reported no change in HO-1 expression in airway macrophages and epithelium of mild asthmatics relative to healthy subjects, and no further change with extended inhaled steroid treatment.[83]

Increased eCO and eNO have been recorded in children with persistent asthma, relative to healthy controls. In this study, eNO, but not eCO, was significantly higher in patients with infrequent episodic asthma.[84] In another study, eCO was found to increase during acute exacerbations in children, which was reversible by combination therapies.[85] In childhood asthma patients, eCO values were determined to be independent of expiratory flow rate.[86] Exhaled CO levels were also found to correlate with allergic seasonal rhinitis, which returned to control values at the end of the allergen season.[87,88] In a mouse model of allergic airway inflammation, eCO values were reported to increase during the 24–72 h interval after the final allergen challenge.[89]

Recent studies have examined differential effects of methacholine and allergen challenges on the levels of eCO and eNO in a cohort of adult subjects with atopic asthma. The authors found that bronchospasm negatively modulates eCO and eNO values, whereas the inflammatory stimulus of allergen exposure increases eNO.[90]

In spite of the aforementioned positive reports supporting eCO measurement as a biomarker of asthma exacerbation and response to steroid therapy, several studies have reported a lack of eCO changes in asthmatics.[79,83,91] In one such study, no change in eCO was recorded in asthmatics following 30 days of inhalation corticosteroid therapy, despite decreases in airway eosinophil content and bronchial responsiveness to metacholine.[83] A recent meta-analysis of the literature concluded that eCO levels generally increase in asthmatics relative to normal patients, but only partially reflect disease severity. Furthermore, the authors concluded that eCO levels do not distinguish between degrees of disease control or between steroid-treated and steroid-free patients in cross-sectional studies. However, eCO levels were concluded to be generally responsive to corticosteroid treatment. eCO may continue to represent a potentially useful biomarker of airway inflammation and oxidative stress in asthma in non-smokers.[92]

12.8.2 Chronic obstructive pulmonary disease, COPD

COPD is defined by consensus as "a preventable and treatable disease state characterized by airflow limitation that is not fully reversible, usually progressive, and associated with an abnormal inflammatory response of the lungs to noxious particles or gases, primarily caused by cigarette smoking".[93] The Global Initiative for Chronic

Obstructive Pulmonary Disease (GOLD) has established a four point scale for classifying disease severity, with stage 0 indicating smokers with no loss of lung function, while stage 3–4 indicating severe airflow limitation.[93] In the developed nations, cigarette smoking remains a primary cause of COPD, which develops in ~15–20% of smokers. COPD is a complex disease with unclear pathogenesis, though a number of mechanisms have been proposed. Environmental and occupational exposures to particulate air pollution, chronic microbiological infections, and asthma, may all contribute to COPD progression.[94] Aberrant and persistent airway inflammation is a major component of COPD, characterized by the influx of macrophages and neutrophils.[95] Alveolar macrophages and epithelial cells represent primary targets of CS exposure which react to the oxidant burden by producing pro-inflammatory cytokines, chemokines, and growth factors.[96] Increased oxidative stress associated with smoking may promote disequilibrium between tissue protease and anti-protease activities.[97] Deregulated protease activity in turn contributes to tissue degeneration characteristic of emphysema.[98] Programmed cell death mechanisms (i.e. apoptosis) may also play contributory roles in the disease.[98–100]

Cigarette smoke (CS) is a complex mixture consisting of gas and tar phases that contain ~4,500 chemical constituents, including free radicals, oxidants, aldehydes, aromatic hydrocarbons, as well as the diatomic gases NO and CO, which are generated during combustion.[101] A number of studies have examined the potential for CO as a breath marker of smoking status and/or COPD. In smokers without lung function impairment, measured eCO levels have been found to increase dramatically, relative to non-smokers. Chronic smokers may display eCO values up to 20–25 ppm.[102] This value correlates with the rate of cigarette smoking, and also varies with smoking habit or type of cigarette (i.e. filtered on non-filtered).[102,103] eCO is therefore a useful diagnostic tool to indicate smoking or to monitor smoking cessation.[42,104,105] Two studies of outpatient smokers suggested optimal cutoff values ranging from 6 to 6.5 ppm; values above this would indicate recurrent smoking with 90–96% sensitivity and 80–93% specificity.[104,106] However, in a survey of non-smokers living in a densely populated area, 13% of non-smoking subjects recorded eCO values of 7 ppm or higher.[107] In asthma and COPD, airway inflammation contributes to eCO values, which makes it difficult to define optimal cutoff values. Therefore, CO monitoring in smoking remains controversial, with some investigators suggesting that cutoff values should be adjusted among subpopulations, in particular for those subjects that also concurrently have asthma or COPD.[105–108]

In early studies on the measurement of eCO in COPD, patients with COPD were reported to have generally higher eCO values than non-smokers without COPD. Among those patients with COPD, those who were currently smoking at the time of study had higher eCO values than those that were former smokers.[12] eCO values in this study did not correlate with lung function measurements assessed by FEV_1, nor with eNO values. It should be noted that the measured eNO values were also higher in COPD patients than controls; however, eNO levels were reported higher among COPD non-smokers than COPD smokers. eNO values apparently negatively correlated with lung function as assessed by FEV_1.[12]

A recent study of a Brazilian population of smokers, all of which exhibited urinary cotinine levels >50 ng/mL, reported a higher median eCO value among those smokers with COPD vs. smokers without COPD. However, these differences were not significant after normalization for age, gender, and cotinine levels.[109] In a population survey of smokers from the Copenhagen Heart Study, eCO values were found to have no correlation to lung function or lung function decline in smokers.[103] Arterial HbCO levels were measured in COPD patients in comparison to patients without airways disease. Ex-smoking COPD patients with stable disease displayed higher arterial HbCO levels than healthy ex-smoking patients. The increases in HbCO correlated with COPD disease severity with greater eCO levels recorded in severe COPD, relative to moderately severe disease. Furthermore, HbCO levels further increased in patients with COPD exacerbations relative to stable disease.[110] Increased HbCO correlated with increased eCO at moderate stages of COPD severity (GOLD 2–3) but not at advanced stage (GOLD 4). The reason for this discrepancy is unclear, but may reflect declined lung/airways production of CO or decreased alveolar function and CO diffusion capacity at GOLD 4, in combination with increased systemic CO production.[110]

Further studies have examined eCO in alpha-1 antitrypsin (α1-AT) deficiency. α1-AT is a genetic cause of emphysema associated with increased neutrophil elastase activity and lung tissue destruction which arises independently of smoking status. In α1-AT deficiency patients, eCO levels were found to be decreased compared to those of healthy subjects or those with smoking-associated COPD.[111] The reasons for this observation, proposed by the authors to include more pronounced alveolar destruction or impaired CO diffusion capacity in α1-AT deficiency, remain unclear. Since α1-AT deficiency is also associated with pulmonary inflammation, the results of this study do not support a direct correlation between eCO and lung or airways inflammation.[111]

HO-1, the enzyme primarily responsible for endogenous CO production, has been implicated in the pathogenesis of COPD. HO-1 is associated with cellular defense mechanisms against oxidative stress, and in this regard potentially plays a role in cellular protection against CS-induced cell death.[112,113] HO-1 is potently induced by cigarette smoke extract (CSE) in human and rodent epithelial cells.[112,113] Recent studies have shown that HO-1 gene transfer can protect against emphysema development in the elastase model.[114] The role of HO-1 in clinical COPD is however, less well understood. Elevated levels of HO-1 were reported in the alveolar spaces of chronic smokers with and without COPD relative to non-smokers.[115] Increased HO-1 positive inflammatory cells were found during exacerbations of severe COPD relative to stable disease.[116] In contrast, recent studies report decreased HO-1 expression in alveolar macrophages of severe COPD patients relative to those of smokers without lung function impairment.[117] In a study which controlled for smoking status, ex-smoking COPD patients were reported to have reduced HO-1 expression in alveolar macrophages relative to healthy ex-smokers.[118] These observations would suggest that elevations of eCO observed in COPD are not necessarily consistent with HO-1 activation, as reported for asthmatics.[42] Recent mechanistic studies indicate that the Nrf2 system, the major transcriptional regulator of HO-1, may be inhibited during

COPD.[119,120] These data provide a possible mechanistic explanation for apparent down regulation of HO-1 in severe COPD. Further studies are needed to define the relationship between eCO and HO-1 in COPD.

12.8.3 Cystic fibrosis, CF

CF, a life-threatening lung disease, is a genetic disorder caused by mutation in the gene encoding the cystic fibrosis transmembrane conductance regulator (CFTR). CF is characterized by abnormal mucous buildup in the airways, which can cause pulmonary damage associated with secondary infections.[121] Measured eCO levels were found to be higher in untreated vs. oral steroid-treated CF patients.[122] Furthermore, eCO increased in patients during exacerbations of CF, correlating to deterioration of FEV_1, with normalization of the eCO levels after treatment.[123] CO levels correlate with exhaled ethane, a product of lipid peroxidation that serves as an indirect marker of oxidative stress. Both eCO and exhaled ethane were elevated in non steroid-treated CF patients, when compared to healthy controls, and these values were decreased by steroid treatment.[124] Increased eCO was detected in children with CF, relative to control patients. In addition to the inflammatory and oxidative stress responses to infection in these patients, eCO may respond to hypoxia. Exercise testing further increased eCO in CF children and correlated with the degree of oxyhemoglobin desaturation, suggestive of an increased HO-1 expression in CF patients during hypoxic states induced by exercise.[125] It should be noted that several studies have reported no significant fluctuation of eCO in CF patients relative to healthy controls.[86,91]

12.8.4 Infectious disease

In patients with pneumonia, higher HbCO levels can be measured at the onset of illness, with decreasing values to control levels after antibiotic treatment.[81] Elevated eCO levels were reported in lower respiratory tract infections, and these were decreased by antibiotic therapy.[126] Median eCO levels were higher among patients with bronchioectasis, a condition related to increased neutrophil inflammation, regardless of steroid treatment. However, eCO did not correlate with HbCO in this condition.[127] Furthermore, eCO levels in upper respiratory tract infections were higher when compared to healthy controls.[86,88,128]

12.8.5 Lung transplantation

Bronchiolitis obliterans syndrome (BOS) is the leading cause of death after lung transplantation. BOS affects 50–60% of all cases within the first 5 years following surgery with a ~30% mortality rate. The pathology of BOS involves mononuclear cell-predominant inflammation and scarring of the small airways resulting in altered lung function. Preventive therapeutic interventions are now available, but airflow limitation is not detectable early on during the course of the disease. The helium or nitrogen single breath washout technique is currently used to predict early BOS but the low sensitivity of the test limits is clinical value.[129] Increased eCO and eNO levels were detected in 65 patients with BOS.[130] When eCO and eNO was measured in combination with helium single breath washout it was highly discriminative of early (stage 0–1) BOS. The authors concluded that the combined testing has a high negative predictive value and it can be used as a non-invasive screening tool to exclude BOS following transplantation. More recently, increased eCO levels were shown to be correlated with elevated neutrophil granulocyte count and pro-inflammatory cytokine level in the bronchoalveolar lavage fluid of lung transplant patients.[131]

12.8.6 Critical care medicine

The potential application of eCO for the non-invasive monitoring of critically ill patients has also been explored. Elevated eCO values have been measured in mechanically ventilated critical care subjects. In a clinical study of 29 patients on mechanical ventilation with eight healthy non-smoking controls, average eCO values were higher in mechanically ventilated patients relative to healthy subjects, and correlated with arterial HbCO levels, as well as with serum total and indirect bilirubin levels.[69] HbCO also correlated with total bilirubin measurements. The patient group had variable respiratory function and disease severity. The authors concluded that heme metabolic activity was elevated among ventilated critical care patients versus healthy controls. eCO did not generally correlate with or predict inflammation or disease severity but a trend of higher eCO values was associated with survival.[69] Another study of 30 ventilated patients also recorded an increase in eCO values when compared to a control group, whereas no correlations were found between eCO and arterial or venous HbCO.[132] Furthermore, eCO in mechanically ventilated patients was reported to be transiently elevated by increases in oxygen intake.[133] Elevated eCO values were also recorded in mechanically ventilated critically ill patients with sepsis, relative to ventilated but non-septic controls. Elevated eCO values on the first day of treatment correlated with the probability of patient survival.[134] In a study involving surgical patients, all non-smokers, who had undergone general or spinal anesthesia, elevated mean eCO levels, as well as elevated arterial HbCO levels, were observed in post-surgical patients on the day after surgery relative to preoperative values, or later during recovery.[135] It has been debated whether this phenomenon indicates metabolism of volatile anesthetics but no differences in mean eCO levels were recorded between post-surgical patients that had received inhalation vs. spinal anesthesia.[136]

12.9 CONCLUSIONS

Carbon monoxide (CO) appears as one of the many molecules that can be measured in the exhaled breath of humans. The monitoring of CO, NO, and other volatiles in the breath has the advantage over other biomarker-based clinical tests in that these methods are non-invasive and practical for bedside measurements. CO is a common contaminant of outdoor and indoor air and therefore measurements of CO must be carefully controlled and account for ambient levels. While CO is typically present in breath at ppm levels, a range that can be detected with electrochemical apparatus, instrumentations with greater sensitivity based on laser spectroscopy are currently under development.

eCO has been proposed as a general marker of inflammation, and may also selectively reflect increased oxidative stress in the lung. Measured eCO levels have been reported to change in several inflammatory lung diseases, and respond generally to anti-inflammatory therapies. However, eCO is elevated in the exhaled breath of smokers; therefore, eCO is not likely to be a reliable marker of inflammation or of any disease state in patients who smoke. Despite improvements in the standardization and sensitivity of methods to detect eCO, the predictive value of eCO measurements as a diagnostic tool in human diseases remains uncertain. Further clinical studies may find select applications for this marker. Further basic research studies in this area may provide additional insight into human respiratory physiology and metabolism.

References

1. Amann A, Spanel P, Smith D. Breath analysis: the approach towards clinical applications. *Mini Rev Med Chem* 2007;**7**:115–29.
2. Risby TH, Solga SF. Current status of clinical breath analysis. *Appl Phys B-Lasers O* 2006;**85**:421–6.
3. Kharitonov SA, Barnes PJ. Biomarkers of some pulmonary diseases in exhaled breath. *Biomarkers* 2002;**7**:1–32.
4. Kharitonov SA, Barnes PJ. Exhaled markers of inflammation. *Curr Opin Allergy Clin Immunol* 2001;**1**:217–24.
5. Dweik RA, Amann A. Exhaled breath analysis: the new frontier in medical testing. *J Breath Res* 2008;**2**:030301.
6. Buszewski B, Kesy M, Ligor T, Amann A. Human exhaled air analytics: biomarkers of diseases. *Biomed Chromatogr* 2007;**21**:553–66.
7. Kazani S, Israel E. Exhaled breath condensates in asthma: diagnostic and therapeutic implications. *J Breath Res* 2010;**4**:047001.
8. Barnes PJ, Dweik RA, Gelb AF, Gibson PG, George SC, Grasemann H, et al. Exhaled nitric oxide in pulmonary diseases: a comprehensive review. *Chest* 2010;**138**:682–92.
9. Taylor DR, Pijnenburg MW, Smith AD, De Jongste JC. Exhaled nitric oxide measurements: clinical application and interpretation. *Thorax* 2006;**61**:817–27.
10. Choi J, Hoffman LA, Rodway GW, Sethi JM. Markers of lung disease in exhaled breath: nitric oxide. *Biol Res Nurs* 2006;**7**:241–55.
11. Antus B, Horvath I. Exhaled nitric oxide and carbon monoxide in respiratory diseases. *J Breath Res* 2007;**1**:024002.
12. Montuschi P, Kharitonov SA, Barnes PJ. Exhaled carbon monoxide and nitric oxide in COPD. *Chest* 2001;**120**:496–501.
13. Ryter SW, Sethi JM. Exhaled carbon monoxide as a biomarker of inflammatory lung disease. *J Breath Res* 2007;**1**:026004.

14. Horvath I, MacNee W, Kelly FJ, Dekhuijzen PN, Phillips M, Doring G, et al. Haemoxygenase-1 induction and exhaled markers of oxidative stress in lung diseases, summary of the ERS research seminar in Budapest, Hungary, September, 1999. *Eur Respir J* 2001;**18**:420–30.
15. Gajdocsy R, Horvath I. Exhaled carbon monoxide in airway diseases: from research findings to clinical relevance. *J Breath Res* 2010;**4**:047102.
16. Morimatsu H, Takahashi T, Matsusaki T, Hayashi M, Matsumi J, Shimizu H, et al. An increase in exhaled CO concentration in systemic inflammation/sepsis. *J Breath Res* 2010;**4**:047103.
17. Von Burg R. Carbon monoxide. *J Appl Toxicol* 1999;**19**:379–86.
18. Budavari S. *The Merck index: an encyclopedia of chemicals, drugs, and biologicals*. 11th ed. Rahway, NJ, USA: Merck; 1989.
19. *Gas Encylopedia – Air Liquide*. <http://encyclopedia.airliquide.com/ Encyclopedia.asp?GasID=41>; 2011 [accessed 28.02.12].
20. Hartsfield CL. Cross talk between carbon monoxide and nitric oxide. *Antioxid Redox Signal* 2002;**4**:301–7.
21. Elschenbroich C, *Organometallics*. 3rd, completely rev. and extended ed. 2006, Weinheim: Wiley-VCH.
22. Wink DA, Mitchell JB. Chemical biology of nitric oxide: Insights into regulatory, cytotoxic, and cytoprotective mechanisms of nitric oxide. *Free Radic Biol Med* 1998;**25**: 434–56.
23. Kim-Shapiro DB, Schechter AN, Gladwin MT. Unraveling the reactions of nitric oxide, nitrite, and hemoglobin in physiology and therapeutics. *Arterioscler Thromb Vasc Biol* 2006;**26**:697–705.
24. Stamler JS, Singel DJ, Loscalzo J. Biochemistry of nitric oxide and its redox-activated forms. *Science (New York, NY)* 1992;**258**:1898–1902.
25. Gorman D, Drewry A, Huang YL, Sames C. The clinical toxicology of carbon monoxide. *Toxicology* 2003;**187**:25–38.
26. Raub J. Environmental Health Criteria 213. Carbon Monoxide Second edition. 1999. Geneva, Switzerland: World Health Organization. p. 70–100.
27. Dolinay T, Choi AMK, Ryter SW. Can inhalation carbon monoxide be utilized as a therapeutic modality in human diseases? In: Amann A, Smith D, editors. *Breath analysis: for clinical diagnosis and therapeutic monitoring*. Singapore: World-Scientific; 2005. p. 203–36.
28. Calafat AM, Polzin GM, Saylor J, Richter P, Ashley DL, Watson CH. Determination of tar, nicotine, and carbon monoxide yields in the mainstream smoke of selected international cigarettes. *Tob Control* 2004;**13**:45–51.
29. Sjostrand T. The formation of carbon monoxide by *in vitro* decomposition of haemoglobin in bile pigments. *Acta Physiol Scand* 1952;**26**:328–33.
30. Sjostrand T. Formation of carbon monoxide in connexion with haemoglobin catabolism. *Nature* 1951;**168**:1118–9.
31. Coburn RF, Blakemore WS, Forster RE. Endogenous carbon monoxide production in man. *J Clin Invest* 1963;**42**:1172–8.
32. Coburn RF, Williams WJ, Kahn SB. Endogenous carbon monoxide production in patients with hemolytic anemia. *J Clin Invest* 1966;**45**:460–8.
33. Tenhunen R, Marver HS, Schmid R. The enzymatic conversion of heme to bilirubin by microsomal heme oxygenase. *Proc Natl Acad Sci USA* 1968;**61**: 748–55.

34. Tenhunen R, Marver HS, Schmid R. Microsomal heme oxygenase. Characterization of the enzyme. *J Biol Chem* 1969;**244**:6388–94.
35. Maines MD. The heme oxygenase system: a regulator of second messenger gases. *Annu Rev Pharmacol Toxicol* 1997;**37**:517–54.
36. Vreman HJ, Wong RJ, Stevenson DK. Carbon monoxide in breath, blood, and other tissues. In: Penney DG, editor. *Carbon monoxide toxicity*. Boca Raton, FL, USA: CRC Press; 2000. p. 19–60.
37. Keyse SM, Tyrrell RM. Heme oxygenase is the major 32-kDa stress protein induced in human skin fibroblasts by UVA radiation, hydrogen peroxide, and sodium arsenite. *Proc Natl Acad Sci USA* 1989;**86**:99–103.
38. Ryter SW, Tyrrell RM. The heme synthesis and degradation pathways: role in oxidant sensitivity.Heme oxygenase has both pro- and antioxidant properties. *Free Radic Biol Med* 2000;**28**:289–309.
39. Ryter SW, Alam J, Choi AM. Heme oxygenase-1/carbon monoxide: from basic science to therapeutic applications *Physiol Rev* 2006;**86**:583–650.
40. Ryter SW, Choi AM. Heme oxygenase-1/carbon monoxide: novel therapeutic strategies in critical care medicine. *Curr Drug Targets* 2010;**11**:1485–94.
41. Maines MD, Trakshel GM, Kutty RK. Characterization of two constitutive forms of rat liver microsomal heme oxygenase. Only one molecular species of the enzyme is inducible. *J Biol Chem* 1986;**261**:411–9.
42. Horvath I, Donnelly LE, Kiss A, Paredi P, Kharitonov SA, Barnes PJ. Raised levels of exhaled carbon monoxide are associated with an increased expression of heme oxygenase-1 in airway macrophages in asthma: a new marker of oxidative stress. *Thorax* 1998;**53**:668–72.
43. Andersson JA, Uddman R, Cardell LO. Carbon monoxide is endogenously produced in the human nose and paranasal sinuses. *J Allergy Clin Immunol* 2000;**105**:269–73.
44. Coburn RF. The carbon monoxide body stores. *Ann NY Acad Sci* 1970; **174**: 11–22.
45. Rudra CB, Williams MA, Sheppard L, Koenig JQ, Schiff MA, Frederick I.O, et al. Relation of whole blood carboxyhemoglobin concentration to ambient carbon monoxide exposure estimated using regression. *Am J Epidemiol* 2010;**171**:942–51.
46. Russell MA. Blood carboxyhaemoglobin changes during tobacco smoking. *Postgrad Med J* 1973;**49**:684–7.
47. Stone JR, Marletta MA. Soluble guanylate cyclase from bovine lung: activation with nitric oxide and carbon monoxide and spectral characterization of the ferrous and ferric states. *Biochemistry* 1994;**33**:5636–40.
48. Furchgott RF, Jothianandan D. Endothelium-dependent and -independent vasodilation involving cyclic GMP: relaxation induced by nitric oxide, carbon monoxide and light. Blood Vessels 1991;**28**:52–61.
49. Durante W, Johnson FK, Johnson RA. Role of carbon monoxide in cardiovascular function. *J Cell Mol Med* 2006;**10**:672–86.
50. Verma A, Hirsch DJ, Glatt CE, Ronnett GV, Snyder SH. Carbon monoxide: a putative neural messenger. *Science (New York, NY)* 1993;**259**:381–4.
51. Zakhary R, Poss KD, Jaffrey SR, Ferris CD, Tonegawa S, Snyder SH. Targeted gene deletion of heme oxygenase 2 reveals neural role for carbon monoxide. *Proc Natl Acad Sci USA* 1997;**94**:14848–53.

52. Wang R. Resurgence of carbon monoxide: an endogenous gaseous vasorelaxing factor. *Can J Physiol Pharmacol* 1998;**76**:1–15.
53. Otterbein LE, Bach FH, Alam J, Soares M, Tao Lu H, Wysk M, et al. Carbon monoxide has anti-inflammatory effects involving the mitogen-activated protein kinase pathway. *Nat Med* 2000;**6**:422–8.
54. Otterbein LE, Otterbein SL, Ifedigbo E, Liu F, Morse DE, Fearns C, et al. MKK3 mitogen-activated protein kinase pathway mediates carbon monoxide-induced protection against oxidant-induced lung injury. *Am J Pathol* 2003;**163**:2555–63.
55. Brouard S, Otterbein LE, Anrather J, Tobiasch E, Bach FH, Choi AM, et al. Carbon monoxide generated by heme oxygenase 1 suppresses endothelial cell apoptosis. *J Exp Med* 2000;**192**:1015–26.
56. Zhang X, Shan P, Alam J, Davis RJ, Flavell RA, Lee PJ. Carbon monoxide modulates Fas/Fas ligand, caspases, and Bcl-2 family proteins via the p38alpha mitogen-activated protein kinase pathway during ischemia-reperfusion lung injury. *J Biol Chem* 2003;**278**:22061–70.
57. Zhang X, Shan P, Otterbein LE, Alam J, Flavell RA, Davis RJ, et al. Carbon monoxide inhibition of apoptosis during ischemia-reperfusion lung injury is dependent on the p38 mitogen-activated protein kinase pathway and involves caspase 3. *J Biol Chem* 2003;**278**:1248–58.
58. Nakahira K, Kim HP, Geng XH, Nakao A, Wang X, Murase N, et al. Carbon monoxide differentially inhibits TLR signaling pathways by regulating ROS-induced trafficking of TLRs to lipid rafts. *J Exp Med* 2006;**203**:2377–89.
59. Morita T, Mitsialis SA, Koike H, Liu Y, Kourembanas S. Carbon monoxide controls the proliferation of hypoxic vascular smooth muscle cells. *J Biol Chem* 1997;**272**:32804–9.
60. Otterbein LE, Zuckerbraun BS, Haga M, Liu F, Song R, Usheva A, et al. Carbon monoxide suppresses arteriosclerotic lesions associated with chronic graft rejection and with balloon injury. *Nat Med* 2003;**9**:183–90.
61. Sato K, Balla J, Otterbein L, Smith RN, Brouard S, Lin Y, et al. Carbon monoxide generated by heme oxygenase-1 suppresses the rejection of mouse-to-rat cardiac transplants. *J Immunol* 2001;**166**:4185–94.
62. Song R, Kubo M, Morse D, Zhou Z, Zhang X, Dauber JH, et al. Carbon monoxide induces cytoprotection in rat orthotopic lung transplantation via anti-inflammatory and anti-apoptotic effects. *Am J Pathol* 2003;**163**:231–42.
63. Dolinay T, Szilasi M, Liu M, Choi AM. Inhaled carbon monoxide confers antiinflammatory effects against ventilator-induced lung injury. *Am J Respir Crit Care Med* 2004;**170**:613–20.
64. Hoetzel A, Dolinay T, Vallbracht S, Zhang Y, Kim HP, Ifedigbo E, et al. Carbon monoxide protects against ventilator-induced lung injury via PPAR-gamma and inhibition of Egr-1. *Am J Respir Crit Care Med* 2008;**177**:1223–32.
65. Hoetzel A, Schmidt R, Vallbracht S, Goebel U, Dolinay T, Kim HP, et al. Carbon monoxide prevents ventilator-induced lung injury via caveolin-1. *Crit Care Med* 2009;**37**:1708–15.
66. Vreman HJ, Stevenson DK, Oh W, Fanaroff AA, Wright LL, Lemons JA, et al. Semi-portable electrochemical instrument for determining carbon monoxide in breath. *Clin Chem* 1994;**40**:1927–33.
67. Vreman HJ, Baxter LM, Stone RT, Stevenson DK. Evaluation of a fully automated end-tidal carbon monoxide instrument for breath analysis. *Clin Chem* 1996;**42**:50–6.

68. Kharitonov SA, Barnes PJ. Exhaled markers of pulmonary disease. *Am J Respir Crit Care Med* 2001;**163**:1693–722.
69. Morimatsu H, Takahashi T, Maeshima K, Inoue K, Kawakami T, Shimizu H, et al. Increased heme catabolism in critically ill patients: correlation among exhaled carbon monoxide, arterial carboxyhemoglobin, and serum bilirubin IXalpha concentrations. *Am J Physiol Lung Cell Mol Physiol* 2006;**290**: L114–9.
70. Engel GS, Drisdell WS, Keutsch FN, Moyer EJ, Anderson JG. Ultrasensitive near-infrared integrated cavity output spectroscopy technique for detection of CO at 1.57 microm: new sensitivity limits for absorption measurements in passive optical cavities. *Appl Opt* 2006;**45**:9221–9.
71. Popp A, Muller F, Kuhnemann F, Schiller S, von Basum G, Dahnke H, et al. Ultra-sensitive mid-infrared cavity leak-out spectroscopy using a cw optical parametric oscillator. *Appl Phys B-Lasers O* 2002;**75**:751–4.
72. Murtz M, Halmer D, Horstjann M, Thelen S, Hering P. Ultra sensitive trace gas detection for biomedical applications. *Spectrochim Acta A Mol Biomol Spectrosc* 2006;**63**:963–9.
73. Fritsch T, Hering P, Murtz M. Infrared laser spectroscopy for online recording of exhaled carbon monoxide-a progress report. *J Breath Res* 2007;**1**:014002.
74. Sowa M, Murtz M, Hering P. Mid-infrared laser spectroscopy for online analysis of exhaled CO. *J Breath Res* 2010;**4**:047101.
75. Shorter JH, Nelson DD, Barry McManus J, Zahniser MS, Milton DK. Multicomponent breath analysis with infrared absorption using room-temperature quantum cascade lasers. *IEEE Sens J* 2009;**10**:76–84.
76. Wenzel SE. Asthma: defining of the persistent adult phenotypes. *Lancet* 2006;**368**:804–13.
77. Yamaya M, Sekizawa K, Ishizuka S, Monma M, Sasaki H. Exhaled carbon monoxide levels during treatment of acute asthma. *Eur Respir J* 1999;**13**: 757–60.
78. Zayasu K, Sekizawa K, Okinaga S, Yamaya M, Ohrui T, Sasaki H. Increased carbon monoxide in exhaled air of asthmatic patients. *Am J Respir Crit Care Med* 1997;**156**:1140–3.
79. Pearson P, Lewis S, Britton J, Fogarty A. Exhaled carbon monoxide levels in atopic asthma: a longitudinal study. *Respir Med* 2005;**99**:1292–6.
80. Paredi P, Leckie MJ, Horvath I, Allegra L, Kharitonov SA, Barnes PJ. Changes in exhaled carbon monoxide and nitric oxide levels following allergen challenge in patients with asthma. *Eur Respir J* 1999;**13**:48–52.
81. Yasuda H, Yamaya M, Yanai M, Ohrui T, Sasaki H. Increased blood carboxyhaemoglobin concentrations in inflammatory pulmonary diseases. *Thorax* 2002;**57**:779–83.
82. Yasuda H, Sasaki T, Yamaya M, Ebihara S, Maruyama M, Kanda A, et al. Increased arteriovenous carboxyhemoglobin differences in patients with inflammatory pulmonary diseases. *Chest* 2004;**125**:2160–8.
83. Lim S, Groneberg D, Fischer A, Oates T, Caramori G, Mattos W, et al. Expression of heme oxygenase isoenzymes 1 and 2 in normal and asthmatic airways: effect of inhaled corticosteroids. *Am J Respir Crit Care Med* 2000;**162**:1912–8.
84. Uasuf CG, Jatakanon A, James A, Kharitonov SA, Wilson NM, Barnes PJ. Exhaled carbon monoxide in childhood asthma. *J Pediatr* 1999;**135**:569–74.
85. Ohara Y, Ohrui T, Morikawa T, He M, Yasuda H, Yamaya M, et al. Exhaled carbon monoxide levels in school-age children with episodic asthma. *Pediatr Pulmonol* 2006;**41**:470–4.
86. Beck-Ripp J, Latzin P, Griese M. Exhaled carbon monoxide is not flow dependent in children with cystic fibrosis and asthma. *Eur J Med Res* 2004;**9**:518–22.

87. Monma M, Yamaya M, Sekizawa K, Ikeda K, Suzuki N, Kikuchi T, et al. Increased carbon monoxide in exhaled air of patients with seasonal allergic rhinitis. *Clin Exp Allergy* 1999;**29**:1537–41.
88. Andersson JA, Uddman R, Cardell LO. Increased carbon monoxide levels in the nasal airways of subjects with a history of seasonal allergic rhinitis and in patients with upper respiratory tract infection. *Clin Exp Allergy* 2002;**32**: 224–7.
89. Sethi JM, Choi AM, Calhoun WJ, Ameredes BT. Non-invasive measurements of exhaled NO and CO associated with methacholine responses in mice. *Respir Res* 2008;**9**:45.
90. Sethi JM, White AM, Patel SA, Dinella JV, Calhoun WJ, Choi AM. Bronchoprovocation testing in asthma: effect on exhaled monoxides. *J Breath Res* 2010;**4**:047104.
91. Zetterquist W, Marteus H, Johannesson M, Nordval SL, Ihre E, Lundberg JO, et al. Exhaled carbon monoxide is not elevated in patients with asthma or cystic fibrosis. *Eur Respir J* 2002;**20**:92–9.
92. Zhang J, Yao X, Yu R, Bai J, Sun Y, Huang M, et al. Exhaled carbon monoxide in asthmatics: a meta-analysis. *Respir Res* 2010;**11**:50.
93. Rabe KF, Hurd S, Anzueto A, Barnes PJ, Buist SA, Calverley P, et al. Global strategy for the diagnosis, management, and prevention of chronic obstructive pulmonary disease: GOLD executive summary. *Am J Respir Crit Care Med* 2007;**176**:532–55.
94. Barnes PJ, Shapiro SD, Pauwels RA. Chronic obstructive pulmonary disease: molecular and cellular mechanisms. *Eur Respir J* 2003;**22**:672–88.
95. Yao H, Rahman I. Current concepts on the role of inflammation in COPD and lung cancer. *Curr Opin Pharmacol* 2009;**9**:375–83.
96. Rutgers SR, Timens W, Kauffman HF, Postma DS. Markers of active airway inflammation and remodelling in chronic obstructive pulmonary disease. *Clin Exp Allergy* 2001;**31**:193–205.
97. Bowler RP, Barnes PJ, Crapo JD. The role of oxidative stress in chronic obstructive pulmonary disease. *COPD* 2004;**1**:255–77.
98. Tuder RM, Yoshida T, Arap W, Pasqualini R, Petrache I. State of the art. Cellular and molecular mechanisms of alveolar destruction in emphysema: an evolutionary perspective. *Proc Am Thorac Soc* 2006;**3**:503–10.
99. Park JW, Ryter SW, Choi AM. Functional significance of apoptosis in chronic obstructive pulmonary disease. *COPD* 2007;**4**:347–53.
100. Henson PM, Vandivier RW, Douglas IS. Cell death, remodeling, and repair in chronic obstructive pulmonary disease? *Proc Am Thorac Soc* 2006;**3**:713–7.
101. Church DF, Pryor WA. Free-radical chemistry of cigarette smoke and its toxicological implications. *Environ Health Perspect* 1985;**64**:111–26.
102. Hung J, Lin CH, Wang JD, Chan CC. Exhaled carbon monoxide level as an indicator of cigarette consumption in a workplace cessation program in Taiwan. *J Formos Med Assoc* 2006;**105**:210–3.
103. Fabricius P, Scharling H, Lokke A, Vestbo J, Lange P. Exhaled CO, a predictor of lung function? *Respir Med* 2007;**101**:581–6.
104. Middleton ET, Morice AH. Breath carbon monoxide as an indication of smoking habit. *Chest* 2000;**117**:758–63.
105. Chatkin J, Fritscher L, de Abreu C, Cavalet-Blanco D, Chatkin G, Wagner M, et al. Exhaled carbon monoxide as a marker for evaluating smoking abstinence in a Brazilian population sample. *Prim Care Respir J* 2007;**16**:36–40.

106. Deveci SE, Deveci F, Acik Y, Ozan AT. The measurement of exhaled carbon monoxide in healthy smokers and non-smokers. *Respir Med* 2004;**98**:551–6.
107. Jones AY, Lam PK. End-expiratory carbon monoxide levels in healthy subjects living in a densely populated urban environment. *Sci Total Environ* 2006;**354**:150–6.
108. Sato S, Nishimura K, Koyama H, Tsukino M, Oga T, Hajiro T, et al. Optimal cutoff level of breath carbon monoxide for assessing smoking status in patients with asthma and COPD. *Chest* 2003;**124**:1749–54.
109. Chatkin G, Chatkin JM, Aued G, Petersen GO, Jeremias ET, Thiesen FV. Evaluation of the exhaled carbon monoxide levels in smokers with COPD. *J Bras Pneumol* 2010;**36**:332–8.
110. Yasuda H, Yamaya M, Nakayama K, Ebihara S, Sasaki T, Okinaga S, et al. Increased arterial carboxyhemoglobin concentrations in chronic obstructive pulmonary disease. *Am J Respir Crit Care Med* 2005;**171**:1246–51.
111. Machado RF, Stoller JK, Laskowski D, Zheng S, Lupica JA, Dweik RA, et al. Low levels of nitric oxide and carbon monoxide in alpha 1-antitrypsin deficiency. *J Appl Physiol* 2002;**93**:2038–43.
112. Slebos DJ, Ryter SW, van der Toorn M, Liu F, Guo F, Baty CJ, et al. Mitochondrial localization and function of heme oxygenase-1 in cigarette smoke-induced cell death. *Am J Respir Cell Mol Biol* 2007;**36**:409–17.
113. Kim HP, Wang X, Chen ZH, Lee SJ, Huang MH, Wang Y, et al. Autophagic proteins regulate cigarette smoke-induced apoptosis: protective role of heme oxygenase-1. *Autophagy* 2008;**4**:887–95.
114. Shinohara T, Kaneko T, Nagashima Y, Ueda A, Tagawa A, Ishigatsubo Y. Adenovirus-mediated transfer and overexpression of heme oxygenase 1 cDNA in lungs attenuates elastase-induced pulmonary emphysema in mice. *Hum Gene Ther* 2005;**16**:318–27.
115. Maestrelli P, El Messlemani AH, De Fina O, Nowicki Y, Saetta M, Mapp C, et al. Increased expression of heme oxygenase (HO)-1 in alveolar spaces and HO-2 in alveolar walls of smokers. *Am J Respir Crit Care Med* 2001;**164**:1508–13.
116. Tsoumakidou M, Tzanakis N, Chrysofakis G, Siafakas NM. Nitrosative stress, heme oxygenase-1 expression and airway inflammation during severe exacerbations of COPD. *Chest* 2005;**127**:1911–8.
117. Maestrelli P, Paska C, Saetta M, Turato G, Nowicki Y, Monti S, et al. Decreased haem oxygenase-1 and increased inducible nitric oxide synthase in the lung of severe COPD patients. *Eur Respir J* 2003;**21**:971–6.
118. Slebos DJ, Kerstjens HA, Rutgers SR, Kauffman HF, Choi AM, Postma DS. Haem oxygenase-1 expression is diminished in alveolar macrophages of patients with COPD. *Eur Respir J* 2004;**23**:652–3; author reply 653.
119. Goven D, Boutten A, Lecon-Malas V, Marchal-Somme J, Amara N, Crestani B, et al. Altered Nrf2/Keap1-Bach1 equilibrium in pulmonary emphysema. *Thorax* 2008;**63**:916–24.
120. Mercado N, Thimmulappa R, Thomas CM, Fenwick PS, Chana KK, Donnelly LE, et al. Decreased histone deacetylase 2 impairs Nrf2 activation by oxidative stress. *Biochem Biophys Res Commun* 2011;**406**:292–8.
121. Ratjen F, Doring G. Cystic fibrosis. *Lancet* 2003;**361**:669–81.
122. Paredi P, Shah PL, Montuschi P, Sullivan P, Hodson ME, Kharitonov SA, et al. Increased carbon monoxide in exhaled air of patients with cystic fibrosis. *Thorax* 1999;**54**:917–20.
123. Antuni JD, Kharitonov SA, Hughes D, Hodson ME, Barnes PJ. Increase in exhaled carbon monoxide during exacerbations of cystic fibrosis. *Thorax* 2000;**55**:138–42.

124. Paredi P, Kharitonov SA, Leak D, Shah PL, Cramer D, Hodson ME, et al. Exhaled ethane is elevated in cystic fibrosis and correlates with carbon monoxide levels and airway obstruction. *Am J Respir Crit Care Med* 2000;**161**:1247–51.
125. Horvath I, Borka P, Apor P, Kollai M. Exhaled carbon monoxide concentration increases after exercise in children with cystic fibrosis. *Acta Physiol Hung* 1999;**86**:237–44.
126. Biernacki WA, Kharitonov SA, Barnes PJ. Exhaled carbon monoxide in patients with lower respiratory tract infection. *Respir Med* 2001;**95**:1003–5.
127. Horvath I, Loukides S, Wodehouse T, Kharitonov SA, Cole PJ, Barnes PJ. Increased levels of exhaled carbon monoxide in bronchiectasis: a new marker of oxidative stress. *Thorax* 1998;**53**:867–70.
128. Yamaya M, Sekizawa K, Ishizuka S, Monma M, Mizuta K, Sasaki H. Increased carbon monoxide in exhaled air of subjects with upper respiratory tract infections. *Am J Respir Crit Care Med* 1998;**158**:311–4.
129. Van Muylem A, Verbanck S, Estenne M. Monitoring the lung periphery of transplanted lungs. *Respir Physiol Neurobiol* 2005;**148**:141–51.
130. Van Muylem A, Knoop C, Estenne M. Early detection of chronic pulmonary allograft dysfunction by exhaled biomarkers. *Am J Respir Crit Care Med* 2007;**175**:731–6.
131. Vos R, Cordemans C, Vanaudenaerde BM, De Vleeschauwer SI, Schoonis A, Van Raemdonck DE, et al. Exhaled carbon monoxide as a noninvasive marker of airway neutrophilia after lung transplantation. *Transplantation* 2009;**87**:1579–83.
132. Scharte M, Bone HG, Van Aken H, Meyer J. Increased carbon monoxide in exhaled air of critically ill patients. *Biochem Biophys Res Commun* 2000;**267**:423–6.
133. Zegdi R, Caid R, Van De Louw A, Perrin D, Burdin M, Boiteau R, et al. Exhaled carbon monoxide in mechanically ventilated critically ill patients: influence of inspired oxygen fraction. *Intensive Care Med* 2000;**26**:1228–31.
134. Zegdi R, Perrin D, Burdin M, Boiteau R, Tenaillon A. Increased endogenous carbon monoxide production in severe sepsis. *Intensive Care Med* 2002;**28**:793–6.
135. Hayashi M, Takahashi T, Morimatsu H, Fujii H, Taga N, Mizobuchi S, et al. Increased carbon monoxide concentration in exhaled air after surgery and anesthesia. *Anesth Analg* 2004;**99**:444–8, table of contents.
136. Adachi T, Hirota K, Hara T, Sasaki Y, Hara Y. Exhaled carbon monoxide levels change in relation to inspired oxygen fraction during general anesthesia. *Anesth Analg* 2007;**105**:696–9.

CHAPTER

13

Exhaled Nitric Oxide in Clinical Practice: Recent Advances and New Challenges

Ildikó Horváth* **and Peter J. Barnes**†

**Department of Pulmonology, Semmelweis University, Budapest, Hungary*

†Section of Airway Disease, National Heart and Lung Institute, Imperial College London, London, UK

13.1 INTRODUCTION

Nitric oxide (NO) is one of the tiniest and most exiting mediators of the human body. Its presence and multiple functions all over the body has made this molecule one of the most studied both under physiological and pathophysiological conditions.[1–4] Despite detailed knowledge on NO biosynthesis and its regulation, together with the diverse actions mediated by NO, there are still several new discoveries and uncovered areas in relation to its role.[5–7] Research focusing on the role of NO in the respiratory system includes a wide range of studies, including functions such as cell homeostasis, growth and regulation, adaptation to different environmental conditions, participation in stress response and host defence.[8–10] The molecule is recognized to have a key role in almost all respiratory functions, including airway and vascular function, immune regulation, and neurotransmission. Recent discoveries in this field facilitate the reframing of the concepts of development and treatment of different pulmonary diseases, including chronic obstructive pulmonary disease (COPD) and lung cancer.[11,12] With the finding that NO can be detected in exhaled breath, a new field of research emerged from functional and drug development studies and paved the way toward the discovery of other exhaled biomarkers.

13.2 EXHALED NITRIC OXIDE

NO was first detected in exhaled breath in 1991.[13] After this ground-braking discovery it was shown that the fractional NO concentration in exhaled breath (F_ENO) is elevated in asthmatic patients compared to healthy subjects and this observation led to an exponentially growing field of research.[13–15] Over the past two decades the results of more than 2500 studies have been published on "exhaled nitric oxide" that can be found by a Pubmed search. This area of research covered wide range of technical aspects that enabled the field to bloom.

Volatile Biomarkers. http://dx.doi.org/10.1016/B978-0-44-462613-4.00013-1

13.3 TECHNICAL ASPECTS OF F_ENO MEASUREMENT

F_ENO measurement has been standardized and the American Thoracic Society (ATS) together with the European Respiratory Society (ERS) established guidelines for the measurement both for adults and children.[16,17] The recommendations have detailed information on the technical requirements for NO analyzers and careful instructions for on- and offline F_ENO determination, for multiple-flow measurements, and also for the use of this technique under special conditions (i.e. infants, patients in intensive care units). Furthermore, they also contain details concerning on measurement of nasal NO. The use of these recommendations has facilitated research and scientific discussion in the field as data from different laboratories became comparable.

Briefly, for single F_ENO measurement the subject should be seated, with the mouthpiece at proper height and position. The subject without wearing nose clip inserts a mouthpiece and inhales over 2–3 s through the mouth to total lung capacity (TLC), or near TLC, if TLC is not possible, and then exhales immediately against a mouthpiece pressure (5 cm H_2O or higher, but less than 20 cm H_2O). This pressure ensures velum closure to exclude contamination from nasal NO. As the value of F_ENO is flow-dependent, it was important to agree on a given flow rate that could be recommended for standard measurement. The recommended expiratory flow rate is 0.05 L/s. This flow was chosen on the basis of studies showing the best discriminative value for asthma, but it is noted in the recommendation that other flow rates might serve better for other purposes and emphasis was placed on the requirement to provide an exact flow rate to facilitate potential comparisons between different studies. The length of an exhalation is approximately 10 s to reach a stable NO plateau, but at least a 6 s long exhalation is required for acceptable measurement. This value is expressed in parts-per-billion (ppb; equivalent to nL/L). Use of NO free gas for inhalation is recommended, because the concentration of environmental NO is variable and may reach concentrations that can influence the detected F_ENO. At the time of the writing of the recommendations the most widely used instruments for F_ENO measurement were those that used the ozone/NO_2-based chemiluminescence principle. Therefore, the recommendations were given for this type of equipment. It was acknowledged that if new equipment became available, a new set of recommendations might be required.

As chemiluminescence is very sensitive, but rather robust and expensive for clinical purposes, there has been a lot of effort to create cheaper and smaller, portable devices. It should be acknowledged that these newer, simplified equipments have been developed on the basis of the current recommendations and until now the recommendations have not been counterchecked against them.

The first handheld analyzer, the NIOX MINO®, reached the market in 2006 and made F_ENO measurement widely available to clinicians.[18,19] This device uses an electrochemical approach and its results have been compared with those of the "gold-standard" chemiluminescence equipments. The F_ENO values obtained by these devices correlated well with values obtained by stationary chemiluminescence devices.[18–22] This result is of note, because the gold-standard determination of F_ENO is based on two recordings and reading of the plateau part of the signal, while for the handheld equipments a single automatic evaluation is used. However, despite significant correlations between the readings the studies also found the agreement between measurements to be only moderate and pointed out that the devices are not interchangeable.[20–24] While short-term repeatability of this measurement was found to be acceptable and inside clinically acceptable limits, it was demonstrated that its accuracy changed over time for some sensors, thus emphasizing the need for regular quality control procedures when using these instruments.[25] Other portable devices have appeared on the market in the last few years, some of which have shown comparable results to the NIOX MINO.[24,26] Handheld devices have also been used for children and it was shown that they can be successfully used for children from the age of 4 years, but their applicability is limited by a lower rate of success and poorer accuracy for this age-group.[27]

Other techniques, such as NO determination by laser-based approaches, have also been studied and require systematic evaluation.[28,29] Altogether, it is time to revisit the technical recommendations and include these sensor-based and laser-based instruments as well as the chemiluminescence instruments.

Another technical issue concerns the comparison of multiple-flow online measurements with single flow determinations. By measuring F_ENO at multiple-flow rates (10, 30, 50, 100, and 200 mL/s) followed by an analysis using mathematical models to separate the different compartments of the respiratory system, thus F_ENO from the alveolar region and the airway wall, and the related diffusing capacity, and flux can be estimated.[30–36] There are different approaches to the calculations and there is a need for standardization of this type of measurements to ensure comparable and reproducible results.

When considering factors determining the variability in F_ENO, biological as well as methodological factors need to be considered. Genetic factors (polymorphisms in NO producing enzymes), age, height, gender, atopy, and race are well-established factors influencing F_ENO even in asymptomatic subjects, while the influence of factors such as BMI is still under debate.[37–42] Environmental factors, including air pollution, cause alterations in F_ENO.[43] Various life-style factors, including smoking, smoke exposure, and diet, are associated with changes in F_ENO. New lines of evidence clearly demonstrate that not only environmental exposure or life-style changes can have significant effect on F_ENO, but epigenetic regulation also may occur on a daily basis. Therefore, is not surprising that the normal range of F_ENO is wide. The variability arising from these factors needs to be taken into account when F_ENO measurements are used for clinical decision making. It is clearly evident from the published studies that F_ENO can be as high as 100 ppb even in healthy subjects and the distribution of F_ENO is skewed in the healthy population.[44] This makes the interpretation of single measurements of F_ENO rather difficult. One of the potential choices would be to use reference values and from them calculate and express individual F_ENO as a percentage of a reference value, an approach widely used for lung function testing. However, this approach has not been widely used and does not solve the issue of high F_ENO values in healthy individuals. The other option is to define "cut-off" values to discriminate between healthy and disease groups. In the F_ENO studies the latter approach has been chosen most of the time. In the clinical setting, this is the recommended approach, but the evidence supporting the choice of a given F_ENO is not solid.[44]

13.4 F_ENO IN CLINICAL DECISION MAKING

There is now convincing evidence that exhaled NO measurements may contribute to clinical decision making in the care of asthmatic patients. An official clinical practice guideline of the American Thoracic Society for interpretation of exhaled nitric oxide levels for clinical applications was published in 2011.[39] This guideline provides a practical evidence-based approach for the clinical use of F_ENO. The guideline addresses the following questions: Why should a F_ENO test be obtained? Can F_ENO be used to discriminate between patients and healthy subjects? Can F_ENO be used to monitor airway inflammation? How should a F_ENO measurement be interpreted and reported? The answers given to these questions are evidence-based and decisions on the recommendations were made by using the GRADE criteria[45,46] for quality assessment of the available evidence. This step was followed by consensus-based decision making on the recommendations and their strength. This approach helps the user to understand the strength and the limitations of each recommendation. The publication of this document has great importance. One important point is the support for clinical use of F_ENO measurements. The second important point is that by summarizing current evidence the document highlights the areas where further research is required. Since this guideline was published only a few months before the writing of this chapter, it is too early to evaluate the clinical uptake and usefulness of these recommendations. In the following sections the application of F_ENO in different airway diseases will be summarized.

13.4.1 Asthma diagnosis and monitoring

Asthma is a clinical diagnosis based on symptoms and lung function tests. Chronic eosinophilic airway inflammation is defined as a key element of the disease in most cases and steroids are well-established treatment choices to control the disease. Increase in F_ENO is associated with eosinophilic airway inflammation based on studies showing a correlation between differential eosinophil cell count in induced sputum and F_ENO.[47–49] Several studies confirmed that treatment with oral or inhaled corticosteroids dose-dependently reduces F_ENO.[37,38] Similarly, selective inhibition of inducible nitric oxide synthase also inhibits F_ENO, confirming the effect of steroids.[50]

The current guideline recommends the measurement of F_ENO in the diagnosis of eosinophil airway inflammation and also in the prediction of steroid responsiveness in patients with respiratory symptoms suggestive of airway inflammation. These recommendations clearly indicate that the measurement is not appropriate for diagnosing asthma, a disease with different phenotypes with only one being eosinophilic airway inflammation. The guidelines do appropriately emphasize that F_ENO is also influenced by factors other than sputum eosinophilia. Dissociation of F_ENO and sputum inflammatory phenotype was demonstrated recently in a study that described a rapidly changing sputum inflammatory phenotype in children with asthma.[51] Studies concentrating on the diagnostic utility of F_ENO in asthma revealed the need for reevaluation of traditional lung function tests in the diagnostic algorithms, as they did not produce as good predictive value as F_ENO in steroid-naive, non-smoking asthmatic patients. Data in subgroups of asthmatic patients, such as patients with non-eosinophilic airway inflammation or without airway inflammation, however, do not show similar convincing results. In smokers, smoke exposure influences F_ENO both in healthy and asthmatic subjects and therefore caution needs to be taken when interpreting F_ENO in this patient group.[52–54] The guidelines do not provide details for this situation. Altogether, the guideline-based recommendation for F_ENO to support the diagnosis of asthma is weak. Based on moderate quality of evidence, there are increasing number of studies assessing the cost-effectiveness of F_ENO in diagnostic work-up.[55,56]

Nevertheless, it is important to highlight the great clinical potential of a non-invasive, easy to measure, online biomarker to provide reliable prediction of steroid responsiveness. Several lines of evidence indicate that exhaled NO can be used to monitor the effect of anti-inflammatory medication in asthma. Measurement of exhaled NO may also be used to monitor compliance with inhaled corticosteroid therapy, as high levels in patients who are prescribed inhaled steroids might indicate a failure to take the medication. Although there has been great expectation of the clinical utility of exhaled NO measurement, longitudinal studies have found conflicting results.[57–61] Asthma treatment strategies based on the measurement of exhaled nitric oxide did not result in significant reduction in asthma exacerbations compared with guideline-based treatment choices. The meta-analysis of the studies therefore concluded that there is insufficient evidence to recommend F_ENO for routine clinical practice.[62] This statement is in contrast with the recommendation of the ATS, based on low quality evidence, which strongly recommends the use of F_ENO in monitoring airway inflammation in patients with asthma.[44] A detailed analysis of the differences in study designs help to design further studies that are more comparable with each other and may help to further explore this field.[63] Furthermore, studies on asthmatic subgroups, such as pregnant asthmatics, are opening new avenues for the test.[23,64,65] The potential place of F_ENO as an asthma outcome has to be determined. At present, it is listed as a supplementary measure together with sputum eosinophils, urinary leukotrienes, and IgE.[66]

13.4.2 Chronic obstructive pulmonary disease (COPD)

Measurement of exhaled NO in COPD is less well documented. An increase in exhaled NO in COPD patients is correlated with increased numbers of eosinophils, an increased bronchodilators response, and steroid responsiveness.[67–70] This suggests that exhaled NO could potentially be used in COPD phenotyping and predicting treatment response.[71] F_ENO may also aid in differentiating asthma from COPD in elderly patients. Using the multiple-flow technique, it has been found that peripheral (alveolar) NO is increased in COPD and therefore might be useful in assessing the effects of anti-inflammatory treatments on peripheral lung inflammation.

13.4.3 Primary ciliary dyskinesia

Primary ciliary dyskinesia (PCD) is a rare genetic disease characterized by defective motility of cilia resulting in impaired mucociliary clearance and infertility in most cases. In the lungs, PCD is associated with chronic airway inflammation, recurrent infections, and bronchiectasis. In PCD patients both the exhaled and the nasal NO levels are very low, and combination of such low values is not seen in any other conditions.[72] Therefore, the simultaneous measurement of exhaled and nasal NO may serve as a simple screening tool for PCD in clinical practice.

13.4.4 Cystic fibrosis

Cystic fibrosis is also a genetic disease associated with intense neutrophilic inflammation in the airways leading to a fibro-cystic degeneration of the lung parenchyma. Decreased exhaled NO concentrations are seen in this disease, but the measurement of exhaled NO has a limited role in the monitoring of CF patients.[73]

13.4.5 Lung transplantation

Lung transplantation is a complex procedure and may result in several complications. Among the pulmonary problems, infection and rejection (acute or chronic) are probably the most troublesome. Pulmonary infections may increase exhaled NO,[74] while the relationship between acute rejection and exhaled NO is less clear. With regard to chronic rejection, also known as bronchiolitis obliterans syndrome, an increase in exhaled NO, even in the early, subclinical state, has been demonstrated.[75] Hence, elevation of exhaled NO levels in lung transplant patients may be a warning sign of pulmonary complication, and one should carefully consider the use of other invasive techniques to detect the underlying cause.

13.4.6 Chronic cough

Measurement of F_ENO may be useful in differentiating the causes of chronic cough, since an elevated value indicates that the underlying cause is asthma and the need for a trial of oral or inhaled steroids.

13.5 CONCLUSIONS AND DIRECTIONS FOR FUTURE AREAS OF RESEARCH

Measurement of exhaled NO is non-invasive, simple, provides immediate results, and is reproducible. Its potential use in the diagnosis and monitoring of asthma has been carefully studied and evaluated, while in other respiratory disorders less is known about the potential clinical usefulness of this measurement.[37] Recent developments, including multi-flow analysis, help to better understand the location of NO formation in the lungs. At the same time, this measurement does not provide a full insight into the complex production-reaction pathways related to NO production. NO rapidly reacts at the site of its release and a whole range of different NO-derived molecules can be detected. Furthermore, by exhaled breath condensate (EBC) sampling it has been shown that pH may influence exhaled NO concentration and, therefore, pH changes occurring during acute asthma exacerbation may be responsible—at least to some extent—for the observed increase in exhaled NO.[76] These lines of evidence, together with the influential effect of environmental NO, NO produced out of the lungs[77] suggest that proper caution needs to be taken when interpreting exhaled NO concentration.

References

1. Moncada S, Palmer RM, Higgs EA. Nitric oxide: physiology, pathophysiology, and pharmacology. *Pharmacol Rev* 1991;**43**:109–42.
2. Moncada S, Bolanos JP. Nitric oxide, cell bioenergetics and neurodegeneration. *J Neurochem* 2006;**97**:1676–89.
3. Horváth I, Sándor NT, Ruttner Z, McLaughlin AC. Role of nitric oxide in regulating cerebrocortical oxygen consumption and blood flow during hypercapnia. *J Cereb Blood Flow Metab* 1994;**14**:503–9.
4. Chattopadhyay M, Kodela R, Olson KR, Kashfi K. NOSH-aspirin (NBS-1120), a novel nitric oxide- and hydrogen sulfide-releasing hybrid is a potent inhibitor of colon cancer cell growth *in vitro* and in a xenograft mouse model. *Biochem Biophys Res Commun* 2012;**419**:523–8.
5. Bird IM. Endothelial nitric oxide synthase activation and nitric oxide function: new light through old windows. *J Endocrinol* 2011;**210**:239–41.
6. Rafikov R, Fonseca FV, Kumar S, Pardo D, Darragh C, Elms S, et al. eNOS activation and NO function: structural motifs responsible for the posttranslational control of endothelial nitric oxide synthase activity. *J Endocrinol* 2011;**210**:271–84.
7. Aranda E, López-Pedrera C, De La Haba-Rodriguez JR, Rodriguez-Ariza A. Nitric oxide and cancer: the emerging role of S-nitrosylation. *Curr Mol Med* 2012;**12**:50–67.
8. Moreno L, Perez-Vizcaino F, Harrington L, Faro R, Sturton G, Barnes PJ, et al. Pharmacology of airways and vessels in lung slices in situ: role of endogenous dilator hormones. *Respir Res* 2006;**7**:111.
9. Zhang X, Moilanen E, Lahti A, Hamalainen M, Giembycz MA, Barnes PJ, et al. Regulation of eosinophil apoptosis by nitric oxide: role of c-Jun-N-terminal kinase and signal transducer and activator of transcription 5. *J Allergy Clin Immunol* 2003;**112**:93–101.
10. Marcelin G, Aldridge JR, Duan S, Ghoneim HE, Rehg J, Marjuki H, et al. Fatal outcome of pandemic H1N1 2009 influenza virus infection is associated with immunopathology and impaired lung repair, not enhanced viral burden, in pregnant mice. *J Virol* 2011;**85**: 11208–19.

11. Seimetz M, Parajuli N, Pichl A, Veit F, Kwapiszewska G, Weisel FC, et al. Inducible NOS inhibition reverses tobacco-smoke-induced emphysema and pulmonary hypertension in mice. *Cell* 2011;**147**:293–305.
12. Fujita S, Masago K, Hatachi Y, Fukuhara A, Hata A, Kaji R, et al. Genetic polymorphisms in the endothelial nitric oxide synthase gene correlate with overall survival in advanced non-small-cell lung cancer patients treated with platinum-based doublet chemotherapy. *BMC Med Genet* 2010;**11**:167.
13. Gustafsson LE, Leone AM, Persson MG, Wiklund NP, Moncada S. Endogenous nitric oxide is present in the exhaled air of rabbits, guinea pigs and humans. *Biochem Biophys Res Commun* 1991;**181**:852–7.
14. Alving K, Weitzberg E, Lundberg JM. Increased amount of nitric oxide in exhaled air of asthmatics. *Eur Respir J* 1993;**6**:1368–70.
15. Kharitonov SA, Yates D, Robbins RA, Logan-Sinclair R, Shinebourne EA, Barnes PJ. Increased nitric oxide in exhaled air of asthmatic patients. *Lancet* 1994;**343**:133–5.
16. ATS/ERS recommendations for standardized procedures for the online and offline measurement of exhaled lower respiratory nitric oxide and nasal nitric oxide. *Am J Respir Crit Care Med* 2005;**171**:912–30.
17. Baraldi E, de Jongste JC. Measurement of exhaled nitric oxide in children, 2001. *Eur Respir J* 2002;**20**:223–37.
18. Alving K, Janson C, Nordvall L. Performance of a new hand-held device for exhaled nitric oxide measurement in adults and children. *Respir Res* 2006;**7**:67.
19. Menzies D, Nair A, Lipworth BJ. Portable exhaled nitric oxide measurement: comparison with the "gold standard" technique. *Chest* 2007;**131**:410–4.
20. Selby A, Clayton B, Grundy J, Pike K, Drew K, Raza A, et al. Are exhaled nitric oxide measurements using the portable NIOX MINO repeatable? *Respir Res* 2010;**11**:43.
21. Chládková J, Havlínová Z, Chyba T, Krcmová I, Chládek J. Analysis of single-breath profiles of exhaled nitric oxide in children with allergy and asthma: guideline-derived plateau concentrations compared to results of automatic evaluation by two analyzers. *J Asthma* 2008;**45**:820–6.
22. Kim SH, Moon JY, Kwak HJ, Kim SI, Park DW, Kim JW, et al. Comparison of two exhaled nitric oxide analyzers: the NIOX MINO hand-held electrochemical analyzer and the NOA280i stationary chemiluminescence analyzer. *Respirology* 2012.
23. Tamási L, Bohács A, Bikov A, Andorka C, Rigó Jr J, Losonczy G, et al. Exhaled nitric oxide in pregnant healthy and asthmatic women. *J Asthma* 2009;**46**:786–91.
24. Antus B, Horváth I, Barta I. Assessment of exhaled nitric oxide by a new hand-held device. *Respir Med* 2010;**104**:1377–80.
25. Taylor DR, Palmay R, Cowan JO, Herbison GP. Long term performance characteristics of an electrochemical nitric oxide analyser. *Respir Med* 2011;**105**;211–7.
26. Kapande KM, McConaghy LA, Douglas I, McKenna S, Hughes JL, McCance DR, et al. Comparative repeatability of two handheld fractional exhaled nitric oxide monitors. *Pediatr Pulmonol* 2011.
27. Kalliola S, Malmberg P, Rito T, Pelkonen AS, Mäkelä MJ. Can we use portable nitric oxide analyzer in young children? *Pediatr Pulmonol* 2011;**46**:627–31.

28. Shorter JH, Nelson DD, McManus JB, Zahniser MS, Sama SR, Milton DK. Clinical study of multiple breath biomarkers of asthma and COPD (NO, CO_2, CO and N_2O) by infrared laser spectroscopy. *J Breath Res* 2011;**5**:037108.
29. Mandon J, Högman M, Merkus PJ, van Amsterdam J, Harren FJ, Cristescu SM. Exhaled nitric oxide monitoring by quantum cascade laser: comparison with chemiluminescent and electrochemical sensors. *J Biomed Opt* 2012;**17**:017003.
30. Högman M, Drca N, Ehrstedt C, Meriläinen P. Exhaled nitric oxide partitioned into alveolar, lower airways and nasal contributions. *Respir Med* 2000;**94**:985–91.
31. Tsoukias NM, Shin HW, Wilson AF, George SC. A single-breath technique with variable flow rate to characterize nitric oxide exchange dynamics in the lungs. *J Appl Physiol* 2001;**91**:477–87.
32. Jorres RA. Modelling the production of nitric oxide within the human airways. *Eur Respir J* 2000;**16**:555–60.
33. Högman M, Meriläinen P. Extended NO analysis in asthma. *J Breath Res* 2007;**1**:024001.
34. Brindicci C, Ito K, Barnes PJ, Kharitonov SA. Differential flow analysis of exhaled nitric oxide in patients with asthma of differing severity. *Chest* 2007;**131**:1353–62.
35. McCurdy MR, Sharafkhaneh A, Abdel-Monem H, Rojo J, Tittel FK. Exhaled nitric oxide parameters and functional capacity in chronic obstructive pulmonary disease. *J Breath Res* 2011;**5**:016003.
36. Högman M, Lafih J, Meriläinen P, Bröms K, Malinovschi A, Janson C. Extended NO analysis in a healthy subgroup of a random sample from a Swedish population. *Clin Physiol Funct Imaging* 2009;**29**:18–23.
37. Barnes PJ, Dweik RA, Gelb AF, Gibson PG, George SC, Grasemann H, et al. Exhaled nitric oxide in pulmonary diseases: a comprehensive review. *Chest* 2010;**138**:682–92.
38. Antus B, Horváth I. Exhaled nitric oxide and carbon monoxide in respiratory diseases. *J Breath Res* 2007;**1**:024002.
39. Yamamoto M, Tochino Y, Chibana K, Trudeau JB, Holguin F, Wenzel SE. Nitric oxide and related enzymes in asthma: relation to severity, enzyme function and inflammation. *Clin Exp Allergy* 2012;**42**:760–8.
40. Dahgam S, Nyberg F, Modig L, Naluai AT, Olin AC. Single nucleotide polymorphisms in the NOS2 and NOS3 genes are associated with exhaled nitric oxide. *J Med Genet* 2012;**49**:200–5.
41. Torrone D, Kuriakose J, Moors K, Jiang H, Niedzwiecki M, Perera F, et al. Reproducibility and intraindividual variation over days in buccal cell DNA methylation of two asthma genes, interferon gamma (IFNgamma) and inducible nitric oxide synthase (iNOS). *Clin Epigenetics* 2012;**4**:3.
42. Spanier AJ, Kahn RS, Hornung RW, Wang N, Sun G, Lierl MB, et al. Environmental exposures, nitric oxide synthase genes, and exhaled nitric oxide in asthmatic children. *Pediatr Pulmonol* 2009;**44**:812–9.
43. Cornell AG, Chillrud SN, Mellins RB, Acosta LM, Miller RL, Quinn JW, et al. Domestic airborne black carbon and exhaled nitric oxide in children in NYC. *J Expo Sci Environ Epidemiol* 2012;**22**:258–66.
44. Dweik RA, Boggs PB, Erzurum SC, Irvin CG, Leigh MW, Lundberg JO, et al. An official ATS clinical practice guideline: interpretation of exhaled nitric oxide levels (F_ENO) for clinical applications. *Am J Respir Crit Care Med* 2011;**184**:602–15.

45. Guyatt G, Vist G, Falck-Ytter Y, Kunz R, Magrini N, Schunemann H. An emerging consensus on grading recommendations? *Evid Based Med* 2006;**11**:2–4.
46. Schunemann HJ, Jaeschke R, Cook DJ, Bria WF, El-Solh AA, Ernst A, et al. An official ATS statement: grading the quality of evidence and strength of recommendations in ATS guidelines and recommendations. *Am J Respir Crit Care Med* 2006;**174**:605–14.
47. Jatakanon A, Lim S, Kharitonov SA, Chung KF, Barnes PJ. Correlation between exhaled nitric oxide, sputum eosinophils, and methacholine responsiveness in patients with mild asthma. *Thorax* 1998;**53**:91–5.
48. Horváth I, Donnelly LE, Kiss A, Kharitonov SA, Lim S, Chung KF, et al. Combined use of exhaled hydrogen peroxide and nitric oxide in monitoring asthma. *Am J Respir Crit Care Med* 1998;**158**1042–6.
49. Berry MA, Shaw DE, Green RH, Brightling CE, Wardlaw AJ, Pavord ID. The use of exhaled nitric oxide concentration to identify eosinophilic airway inflammation: an observational study in adults with asthma. *Clin Exp Allergy* 2005;**35**:1175–9.
50. Hansel TT, Kharitonov SA, Donnelly LE, Erin EM, Currie MG, Moore WM, et al. A selective inhibitor of inducible nitric oxide synthase inhibits exhaled breath nitric oxide in healthy volunteers and asthmatics. *Faseb J* 2003;**17**:1298–300.
51. Fleming L, Tsartsali L, Wilson N, Regamey N, Bush A. Sputum inflammatory phenotypes are not stable in children with asthma. *Thorax* 2012.
52. Horváth I, Donnelly LE, Kiss A, Balint B, Kharitonov SA, Barnes PJ. Exhaled nitric oxide and hydrogen peroxide concentrations in asthmatic smokers. *Respiration* 2004;**71**:463–8.
53. McSharry CP, McKay IC, Chaudhuri R, Livingston E, Fraser I, Thomson NC. Short and long-term effects of cigarette smoking independently influence exhaled nitric oxide concentration in asthma. *J Allergy Clin Immunol* 2005;**116**:88–93.
54. Malinovschi A, Backer V, Harving H, Porsbjerg C. The value of exhaled nitric oxide to identify asthma in smoking patients with asthma-like symptoms. *Respir Med* 2012;**106**:794–801.
55. Berg J, Lindgren P. Economic evaluation of $F_{E}NO$ measurement in diagnosis and 1-year management of asthma in Germany. *Respir Med* 2008;**102**:219–31.
56. Price D, Berg J, Lindgren P. An economic evaluation of NIOX MINO airway inflammation monitor in the United Kingdom. *Allergy* 2009;**64**:431–8.
57. Pijnenburg MW, Bakker EM, Hop WC, De Jongste JC. Titrating steroids on exhaled nitric oxide in children with asthma: a randomized controlled trial. *Am J Respir Crit Care Med* 2005;**172**:831–6.
58. Smith AD, Cowan JO, Brassett KP, Herbison GP, Taylor DR. Use of exhaled nitric oxide measurements to guide treatment in chronic asthma. *N Engl J Med* 2005;**352**:2163–73.
59. Shaw DE, Berry MA, Thomas M, Green RH, Brightling CE, Wardlaw AJ, et al. The use of exhaled nitric oxide to guide asthma management: a randomized controlled trial. *Am J Respir Crit Care Med* 2007;**176**:231–7.
60. Szefler SJ, Mitchell H, Sorkness CA, Gergen PJ, O'Connor GT, Morgan WJ, et al. Management of asthma based on exhaled nitric oxide in addition to guideline-based treatment for inner-city adolescents and young adults: a randomised controlled trial. *Lancet* 2008;**372**:1065–72.
61. de Jongste JC, Carraro S, Hop WC, Baraldi E. Daily telemonitoring of exhaled nitric oxide and symptoms in the treatment of childhood asthma. *Am J Respir Crit Care Med* 2009;**179**:93–7.

62. Petsky HL, Cates CJ, Lasserson TJ, Li AM, Turner C, Kynaston JA, et al. A systematic review and meta-analysis: tailoring asthma treatment on eosinophilic markers (exhaled nitric oxide or sputum eosinophils). *Thorax* 2012;**67**:199–208.
63. Gibson PG. Using fractional exhaled nitric oxide to guide asthma therapy: design and methodological issues for ASthma TReatment ALgorithm studies. *Clin Exp Allergy* 2009;**39**:478–90.
64. Tamási L, Horváth I, Bohács A, Müller V, Losonczy G, Schatz M. Asthma in pregnancy—immunological changes and clinical management. *Respir Med* 2011;**105**:159–64.
65. Powell H, Murphy VE, Taylor DR, Hensley MJ, McCaffery K, Giles W, et al. Management of asthma in pregnancy guided by measurement of fraction of exhaled nitric oxide: a double-blind, randomised controlled trial. *Lancet* 2011;**378**:983–90.
66. Szefler SJ, Wenzel S, Brown R, Erzurum SC, Fahy JV, Hamilton RG, et al. Asthma outcomes: biomarkers. *J Allergy Clin Immun* 2012;**129**:S9–23.
67. Maziak W, Loukides S, Culpitt S, Sullivan P, Kharitonov SA, Barnes PJ. Exhaled nitric oxide in chronic obstructive pulmonary disease. *Am J Respir Crit Care Med* 1998;**157**:998–1002.
68. Papi A, Romagnoli M, Baraldo S, Braccioni F, Guzzinati I, Saetta M, et al. Partial reversibility of airflow limitation and increased exhaled NO and sputum eosinophilia in chronic obstructive pulmonary disease. *Am J Respir Crit Care Med* 2000;**162**:1773–7.
69. Dummer JF, Epton MJ, Cowan JO, Cook JM, Condliffe R, Landhuis CE, et al. Predicting corticosteroid response in chronic obstructive pulmonary disease using exhaled nitric oxide. *Am J Respir Crit Care Med* 2009;**180**:846–52.
70. Antus B, Barta I, Horváth I, Csiszer E. Relationship between exhaled nitric oxide and treatment response in COPD patients with exacerbations. *Respirology* 2010;**15**:472–7.
71. Brindicci C, Ito K, Resta O, Pride NB, Barnes PJ, Kharitonov SA. Exhaled nitric oxide from lung periphery is increased in COPD. *Eur Respir J* 2005;**26**:52–9.
72. Horváth I, Loukides S, Wodehouse T, Csiszer E, Cole PJ, Kharitonov SA, et al. Comparison of exhaled and nasal nitric oxide and exhaled carbon monoxide levels in bronchiectatic patients with and without primary ciliary dyskinesia. *Thorax* 2003;**58**:68–72.
73. Hubert D, Aubourg F, Fauroux B, Trinquart L, Sermet I, Lenoir G, et al. Exhaled nitric oxide in cystic fibrosis: relationships with airway and lung vascular impairments. *Eur Respir J* 2009;**34**:117–24.
74. Antus B, Csiszer E, Czebe K, Horváth I. Pulmonary infections increase exhaled nitric oxide in lung transplant recipients: a longitudinal study. *Clin Transplant* 2005;**19**:377–82.
75. Gabbay E, Walters EH, Orsida B, Whitford H, Ward C, Kotsimbos TC, et al. Post-lung transplant bronchiolitis obliterans syndrome (BOS) is characterized by increased exhaled nitric oxide levels and epithelial inducible nitric oxide synthase. *Am J Respir Crit Care Med* 2000;**162**:2182–7.
76. Antus B, Barta I, Kullmann T, Lazar Z, Valyon M, Horváth I, et al. Assessment of exhaled breath condensate pH in exacerbations of asthma and chronic obstructive pulmonary disease: a longitudinal study. *Am J Respir Crit Care Med* 2010;**182**:1492–7.
77. Salonen I, Huttunen K, Hirvonen MR, Dufva J, Groundstroem K, Dufva H, et al. Exhaled nitric oxide and atherosclerosis. *Eur J Clin Invest* 2012.

PART

E

Clinical Breath Tests

CHAPTER

14

An Update on ^{13}C-Breath Tests: The Transition to Acceptability into Clinical Practice

Anil S. Modak

Cambridge Isotope Laboratories Inc., 50 Frontage Road, Andover, MA 01810, USA

14.1 INTRODUCTION

Detection of endogenous volatile organic compounds resulting from various disease states has been one of the primary diagnostic tools used by physicians since the days of Hippocrates. With the advent of blood testing, biopsies, X-rays, and CT scans the use of breath in clinical practice to detect malignancies receded dramatically. The modern era of breath testing commenced in 1971, when Nobel Prize winner Linus Pauling[1] demonstrated that human breath is complex, containing well over 200 different volatile organic compounds (VOCs) in picomolar concentrations (around one part in a trillion). Some of these breath VOCs have more recently been researched as markers of disease. More than 3000 VOCs, both endogenous and exogenous, have now been identified in samples of exhaled human breath[2] and several have been specifically associated with disease processes.

Breath tests have been researched since the 1970s using both endogenous components in breath as well as exogenous gases generated by enzymatic metabolism of suitably labeled probes in areas such as personalized medicine, infectious diseases, monitoring disease progression and cancer risk prediction. Despite extensive research over the last four decades only a few breath tests have been approved by the FDA while a number are in the research and development phase. These can be sub classified into two groups, one that utilizes constitutive VOCs generated due to prevalent disease states and the other uses ^{13}C substrate as an *in vivo* probe for detection of gastrointestinal bacteria.

1. Detection of VOCs in human breath to identify oxidative stress caused by disease
   - NO breath test[3] to detect and monitor asthma (Aerocrine).
   - *Heartsbreath* test[4] for heart transplant rejection (Menssana Research, Inc.).

Volatile Biomarkers. http://dx.doi.org/10.1016/B978-0-44-462613-4.00014-3

2. $^{13}CO_2$ as an *in vivo* biomarker in human breath to diagnose enzyme deficiencies or infection after ingestion of ^{13}C substrates to help physicians personalize medication

  - urea-^{13}C breath test[5] for diagnosing *Helicobacter pylori* infection (Meretek Inc.).

14.2 HISTORY OF ^{13}C BREATH TESTS

Antoine Lavoisier, the father of modern chemistry, and his colleague Pierre Simon LaPlace were the first to analyze carbon dioxide in guinea pig breath and later found carbon dioxide in human breath, a product of combustion of oxygen and food. Carbon dioxide can easily be measured in exhaled breath, since it is present at approximately 5% by volume.

Stable isotope ^{13}C-labeled compounds have been widely used as diagnostic probes in research laboratories for almost four decades. In 1973 Lacroix et al.[6] were the first to use ^{13}C-labeled glucose as a substrate for metabolic study in humans. This was the beginning of modern day ^{13}C-breath testing utilizing a ^{13}C-labeled substrate to generate $^{13}CO_2$ in breath. Prior to this report, similar investigations were conducted using the radioactive isotope ^{14}C instead of the stable isotope ^{13}C. Subsequent to the wide availability of low-cost ^{13}C-substrates and the development of non-dispersive isotope selective IR spectrometry (NDIRS), research groups from around the world have increasingly used these isotopic tracers in nutritional, medical, and veterinary research. The scope of these diagnostic ^{13}C-stable isotope probes is now being expanded to provide precise evaluations of the presence or absence of etiologically significant changes in metabolism due to a specific disease or the lack of a specific enzyme or drug–drug interaction. This concept exploits the use of the ^{13}C-label incorporated at the appropriate site into a selected substrate specifically designed for the targeted enzyme (foreign or inherent) in a discrete metabolic pathway. The enzyme-substrate interaction results in the release of $^{13}CO_2$ in the expired breath. The subsequent quantification of $^{13}CO_2$ is an indirect *in vivo* indicator of the pharmacokinetics and enzyme activity. A number of ^{13}C-substrates have been used for monitoring a certain physiological function or evaluating enzyme activity.

The viability of administering an oral dose of a ^{13}C-substrate and procuring metabolic or diagnostic information from its metabolism and conversion to $^{13}CO_2$ is attractive due to its non-invasive nature. The salient features of the ^{13}C-breath test are that they are non-invasive, non-radioactive, safe, simple, and effective. Breath tests have the potential to be useful screening methods that can be applied at point of care settings (e.g. hospitals and physicians' offices) and do not require patients to wait for hours or days for results. The non-invasive method is especially attractive for infants, children, pregnant lactating women, and seniors, all averse to the use of needles. Such breath tests have the potential to be companion diagnostic tests for identifying responders and non responders for drugs metabolized by various polymorphic CYP450 enzymes. Breath tests using stable isotope substrates have now been researched for four decades with a number of them promising significant improvement in clinical utility as non-invasive substitutes for existing invasive tests being currently used in clinical practice. Despite the obvious benefits of non-invasive, functional *in vivo* phenotype breath tests only one, the urea-^{13}C-breath test (UBT), has been approved by the FDA.

14.3 STANDARDIZATION OF INSTRUMENTATION AND BREATH COLLECTION BAGS

The widespread research of ^{13}C-breath tests in the 1980s and early 1990s was limited due to the expense and non-portability of the mass spectrometers (IRMS) required for the measurement of $^{13}CO_2/^{12}CO_2$ ratios in breath samples. The mass spectrometers were available only at major pharmaceutical companies and universities but not at point-of-care facilities such as hospitals and clinics. In the last decade, non dispersive IR spectrometers have been developed and approved by the FDA for evaluating "delta over baseline" (DOB) values in breath samples. The POCone $^{13}CO_2$ breath analyzer (see Figure 14.1) is a miniaturized bench top (small and lightweight) point-of-care spectrometer that can be used in the clinical setting for patient convenience and improved efficiency. The price is around \$12,000. The DOB measurements correlate exceedingly well to the gold-standard GC-MS values (typically to $\pm 0.3‰$). All isotope abundances are expressed as ratios in delta notation as per mil (parts per thousand) differences relative to the Vienna Pee Dee Belemnite standard (V-PDB). The measurement time per sample has been reduced to 2 min. Robust breath bags (aluminum lined of volumes 1.3 L and 300 mL) used for collection of breath samples have been designed with a two-way valve to fit the spectrometer. Recently, an AS-10 autosampler was approved by the FDA for use with the POCone spectrometer, which is capable of analyzing multiple breath samples (up to 10 in one cycle). Handheld laser spectrometers currently being developed could be the eventual point-of-care instrument for measurement of DOB values in breath samples.

14.4 BREATH TESTS DURING THE PERIOD 2005–2011

The ^{13}C-breath tests prior to 2005 were covered extensively in two book chapters[7,8] and several review articles.[9–15] The present chapter comprehensively reviews the significant advances in the field since 2005.

14.4.1 Urea breath test (UBT)

Helicobacter pylori is a global pathogen infecting approximately 50% of the world population.[16] Infection by this bacterium could potentially result in various gastroduodenal diseases such as peptic ulcer, functional dyspepsia, MALT lymphoma, and

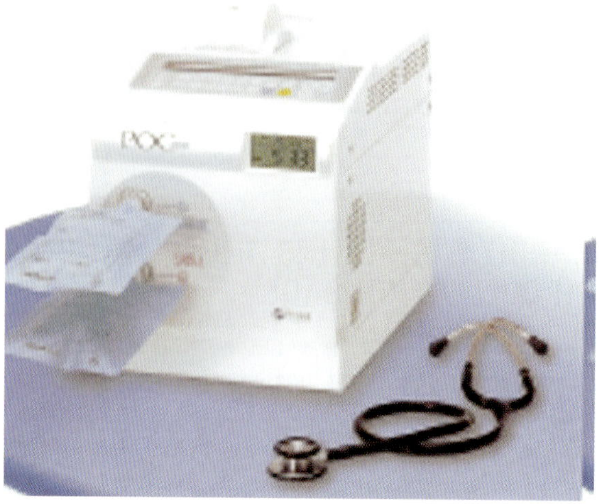

FIGURE 14.1
$^{13}CO_2$ Breath analyser POCone® Infra red spectrometer.

distal gastric cancer. The detection of *H. pylori* infection can be established by a non-invasive (^{13}C) urea breath test (UBT), the only ^{13}C-breath test approved by the FDA in 1996, along with other undesirable non-invasive methods such as the stool antigen test and serology. *H. pylori* infection can also be diagnosed by invasive techniques such as histology, rapid urease test, and culture tests. *H. pylori* eradication therapy is mandatory when peptic ulcers and gastric MALT lymphoma are diagnosed. The UBT has received a lot of attention in a number of review papers,[17–22] which detail the clinical utility of the diagnostic test for reliably diagnosing *H. pylori* infection rapidly (20 min) and non-invasively. A few papers[23–25] compare the accuracy of the UBT to several other invasive diagnostic tests such as gastric biopsies performed during endoscopy (rapid urease test, histology, culture) and a large number of papers[26–45] have focused on the efficacy of eradication therapy and the selection of alternate therapies. First-line therapy is generally a combination of a proton pump inhibitor with clarithromycin and amoxicillin or metronidazole, while second-line therapy uses levofloxacin and rifabutin as good rescue antibiotics. A number of recent papers on UBT focus on antibiotic resistance and optimizing therapies for eradication. The UBT is clinically useful in detection as well as monitoring the eradication of infection following antibiotic therapy. Ineffective, incompetent marketing of the UBT, inadequate reimbursement compared to endoscopy and a lack of patient and physician knowledge of the clinical utility of the UBT has resulted in poor acceptance in routine clinical practice in the US.

14.4.2 Gastric emptying breath tests

The conventional gold standard for measurement of gastric emptying of solids worldwide is Scintigraphy. It is by far the most commonly used method with consensus protocols for the measurement of gastric emptying recently published.[46] However, scintigraphy is not generally available at all major healthcare centers, and involves radiation exposure, precluding its use in pregnant patients and children. Breath tests using ^{13}C-substrates like ^{13}C-octanoic acid (^{13}C-OABT) or the edible ^{13}C-enriched blue-green alga, Spirulina platensis for solid gastric emptying or ^{13}C-sodium acetate breath test for liquid gastric emptying have been extensively researched.[47] These substrates are rapidly absorbed in the proximal small intestine, metabolized in the liver with the generation of $^{13}CO_2$ which is exhaled rapidly with low interindividual variability. Thus, $^{13}CO_2$ exhalation reflects gastric emptying of food intake.[48]

The most significant reason why non-invasive breath tests have not yet replaced the radioactive Scintigraphy test for gastric emptying is the relatively long time required (240 min) for the breath test with multiple breath sample collection. Researchers have now reduced the number of sample collections to just five time points (45, 90, 120, 150, and 180 min) post ingestion of ^{13}C-substrate making the test more user friendly. Various mathematical calculations have been evaluated to get a tighter correlation of the Scintigraphy results with gastric emptying tests; however, larger multicenter studies should be carried out preferably including Scintigraphic data.[49–54] The Szarka linear regression[55] and Viramontes[56] generalized linear regression methods require relatively few breath samples (5) and the analysis provides the closest agreement with $t_{1/2}$ results based on scintigraphy. These methods provide the best analyses of breath $^{13}CO_2$ excretion in stable isotope gastric emptying breath tests.

The effects of body positions,[57] gelatinization,[58] various diet ingredients such as aperitif,[59] peppermint oil,[60] coffee,[61] sucralose[62] and in various disease states such as functional dyspepsia,[63,64] celiac disease,[65] cholecystolithiasis,[66] diabetes,[67] obesity,[68] Hirschsprung disease[69] critically ill[70] and GERD[64,71] on the gastric emptying rates have also been researched. Delayed liquid gastric emptying in both children[72] and adults[73] has been studied using ^{13}C-sodium acetate as a substrate.

14.4.3 Liver function breath tests

The ^{13}C breath tests for evaluating cytosolic, mitochondrial, and microsomal hepatic reserves (liver function) for a variety of liver diseases have been researched most widely by hepatologists over the last three decades.[9,74,75] The complexity of the liver, its diverse functions and the severity of liver diseases makes it extremely tricky to use one specific ^{13}C-labeled substrate for determining liver metabolic function. Blood tests (ALT, AST, bilirubin levels) and scoring indexes have been routinely utilized by physicians to identify the severity of liver damage and to monitor patients with liver disease. Specific functional phenotype breath tests have been recently investigated for cytosolic[76–79] (phenylalanine-^{13}C), microsomal[80–91] (aminopyrine-^{13}C, CYP450 enzymes dextromethorphan-^{13}C for CYP2D6, pantoprazole-^{13}C for CYP2C19, methacetin-^{13}C for chronic liver disease and CYP1A2, and caffeine-^{13}C for CYP1A2), and mitochondrial function[92–100] (methionine-^{13}C, ketocaproic acid-^{13}C, sodium octanoate-^{13}C, pyruvate-^{13}C).

14.4.4 Breath tests for pancreatic insufficiency

Diseases involving the pancreas like cystic fibrosis and pancreatic cancer lead to abnormal pancreatic function, tests for which have had a central role in the study of pancreatic disease for more than 70 years. The gold-standard test is the invasive secretin stimulation test, which measures the ability of the pancreas to respond to the hormone secretin. The small intestines produce secretin in the presence of partially digested food. The other tests for pancreatic function are non-invasive fecal elastase test which measures the enzyme elastase, and invasive computed tomography (CT) scan with contrast dye, abdominal ultrasound, endoscopic retrograde cholangiopancreatography, endoscopic ultrasound, and magnetic resonance cholangiopancreatography. The ^{13}C-mixed triglyceride (1,3-distearoyl, 2-[^{13}C]-octanoyl glycerol) test (^{13}C-MTG-BT) is a non invasive alternative for evaluation of the exocrine pancreatic function test by comparing the $^{13}CO_2$ breath excretion with the lipase output in the duodenum.[101–105] ^{13}C-MTG-BT has more recently been researched to assess and monitor the efficacy of oral enzyme substitution therapy in patients with chronic pancreatitis.[106–112] ^{13}C-dipeptide has also been used as a substrate to evaluate pancreatic exocrine secretion.[113,114]

14.4.5 Breath tests for CYP450 enzymes

The metabolism of ingested xenobiotics is clinically significant in that it minimizes risk and optimizes therapeutic benefits. A majority of the drugs approved by the FDA are metabolized by phase I enzymes. Stable isotope-labeled xenobiotics can be used to provide rapid *in vivo* phenotype assessment of phase I enzymes. Three rapid, non-invasive, single time point phenotype breath tests have been researched using [^{13}C]-pantoprazole,[88] [^{13}C]-dextromethorphan,[89] and [^{13}C]-methacetin[115,116] for assessing polymorphic CYP2C19, CYP2D6, and CYP1A2 enzyme activity. *O*-Dealkylation of all these substrates in the liver leads to $^{13}CO_2$ and reflects enzyme activity of these polymorphic enzymes. These tests can easily be carried out at the point-of-care settings (clinics or hospitals) and would enable physicians to prescribe personalized therapy for each individual by selecting the ideal medication at the right dose for optimal efficacy of xenobiotics metabolized by these enzymes.

The consequence of low or no CYP2D6 activity on the formation of the highly estrogenic tamoxifien metabolite, endoxifen, and thus the treatment response in breast cancer patients has heightened attention on the potential importance of determining the CYP2D6 phenotype prior to initiation of treatment. The [^{13}C]-dextromethorphan-breath test (DM-BT) is currently being studied for personalizing endocrine therapy for breast cancer patients. The CYP2D6 enzyme also affects the metabolism of antidepressants, neuroleptics, β-blockers, and analgesics like the pro-drugs tramadol and codeine. This enzyme also appears to be one of the most important polymorphic drug metabolizing enzymes in causing adverse drug reactions.

Emerging evidence suggests that inter-individual and inter-ethnic differences in CYP2C19 activity influence the therapeutic response of drugs, such as proton pump inhibitors (PPIs), clopidogrel, cyclophosphamide, and thalidomide. The [^{13}C]-pantoprazole breath test (Ptz-BT) offers advantages over existing genotype and phenotype approaches in predicting or assessing CYP2C19 activity *in vivo*, particularly in effectively distinguishing poor metabolizers (PMs) from intermediate metabolizers (IMs) and extensive metabolizers (EMs) of CYP2C19, because it can be performed rapidly, noninvasively and at a single time point breath collection post ingestion of substrate.[88] The Ptz-BT has already been used to illustrate its clinical utility for personalizing PPI[117] and clopidogrel[118] therapy.

Cytochrome P450 1A2 (CYP1A2), an hepatic enzyme, plays an important role in metabolizing drugs and xenobiotics such as paracetamol, theophylline, caffeine, and clozapine. Methacetin is rapidly metabolized primarily by CYP1A2 in the liver and in small doses lacks toxicity making it an ideal substrate for evaluating cytochrome P450 enzyme activity. In healthy liver cells, ^{13}C-methacetin is rapidly absorbed and *O*-dealkylated in the first pass metabolic pathway by CYP1A2 into acetaminophen and $^{13}CO_2$. A single time point breath collection 15 min post ingestion of ^{13}C-methacetin is capable of identifying patients with acute or chronic liver disease or lowered CYP1A2 enzyme activity.[119–126]

14.4.6 ^{13}C-Glucose breath tests

Insulin resistance contributes to the pathophysiology of type-2 diabetes, obesity, metabolic syndrome, and cardiovascular diseases. Therefore, quantifying insulin sensitivity/resistance in humans is of great significance for epidemiological studies and clinical and basic science investigations for eventual use in clinical practice. Fasting blood glucose tolerance test (GTT), insulin tolerance test (K ITT), insulin suppression test, and the hyperinsulinemic-euglycemic clamp technique are the typical tests carried out in clinical practice. The ^{13}C-glucose breath test has been researched as a rapid, non-invasive alternative to assess insulin resistance in both adults[127–129] and children.[130]

14.4.7 Miscellaneous breath tests

Uracil-2-[^{13}C] breath test (Ura-BT). Dihydropyrimidine Dehydrogenase (DPD) is the rate limiting enzyme involved in the catabolism of pyrimidines like thymidine and uracil. DPD is also the main enzyme involved in the degradation of structurally related compounds like 5-Fluorouracil (5-FU), a widely used anticancer drug. A decrease in DPD enzyme activity results in toxicity to 5-FU in cancer patients. The Ura-BT[131–134] has been developed using a single time-point determination at 50 min, to rapidly identify DPD-deficient individuals prior to 5-FU therapy. The Ura-BT has also been useful to predict the effect of the antitumor drug S1 by enabling physicians to optimize the amount of DPD inhibitor.[135] A number of patients present with 5-fluorouracil-associated toxicity despite normal DPD enzyme activity, suggesting possible deficiencies in dihydropyrimidinase (DHP), encoded by the DPYS gene, and/or beta-ureidopropionase (BUP-1), encoded by the UPB1 gene. The dihydrouracil-^{13}C breath test[136] confirmed that the 1635delC mutation does in fact reduce *in vivo* DHP activity.

Hypercapnia is a condition in COPD patients where the body retains CO_2 in the blood leading to elevated paCO_2 levels. The ^{13}C-sodium bicarbonate breath test enables pulmonogists to noninvasively detect patients with hypercapnia and to monitor them following therapy.[137]

Small bowel bacterial overgrowth (SBBO) is found in a wide variety of adult diseases associated with intestinal failure and dysfunction, including short bowel syndrome and other conditions following massive bowel resection, dysmotility disorders, and inflammatory bowel disease. SBBO is an important cause of malabsorption and thus may contribute to diarrhea and malnutrition. Cultures of intestinal aspirates is the gold-standard test to diagnose SSBO. The lactose-[(^{13})C]-ureide breath test (LUBT)[138] has been used to diagnose SBBO and to compare it with the glucose hydrogen breath test (GHBT). The absence or deficiency of the mucosal sucrase-isomaltase enzyme leads to congenital sucrase-isomaltase deficiency (CSID). Specific diagnosis requires upper gastrointestinal biopsy with evidence of low to absent sucrase enzyme activity and normal histology. The hydrogen breath test (BT) is useful, but is not specific for confirmation of CSID. The ^{13}C-sucrose labeled BT is an accurate and specific non-invasive confirmatory test for CSID and for monitoring patients on enzyme replacement therapy like sucraid.[139] The ^{13}C-sucrose-BT has also been used as a non-invasive biomarker of environmental gut health.[140] A Scandinavian group of researchers have investigated the use of ^{13}C-sorbitol and ^{13}C-xylose as non-invasive substrates for breath tests to diagnose celiac disease.[141,142]

14.4.8 Animal studies

A number of ^{13}C-breath test studies have been researched in animals or animal models since 2005. These include breath tests with different ^{13}C-substrates for liver function,[143–146] gastric emptying,[147,148] cancer,[149] and pancreatic exocrine insufficiency.[150,151]

14.5 THE FUTURE OF ^{13}C-BREATH TESTS

$^{13}CO_2$ in breath can be a useful biomarker for phenotype tests to evaluate activity of various polymorphic enzymes which metabolize most xenobiotics. The paucity of ^{13}C-breath tests in routine clinical practice needs to be urgently addressed by obtaining FDA approval through pre-market approval (PMA) applications for the promising diagnostic tests, currently in the research and development stages, to rapidly evaluate enzyme activity.

There are a number of scientific, developmental, regulatory and marketing barriers to acceptance and integration of breath tests into routine clinical practice.[152,153] Financial success of the diagnostic breath tests will depend entirely on how well the test is marketed to physicians, healthcare organizations, payers (reimbursement) and insurance companies, and most importantly to patients, the eventual beneficiaries. Emphasis needs to be placed on making the breath tests user friendly with a maximum of two breath sample collections post ingestion of ^{13}C-labeled substrate, similar to the UBT, to make inroads to acceptability into clinical practice.

References

1. Pauling L, Robinson AB, Teranishi R, Cary P. Quantitative analysis of urine vapor and breath by gas-liquid partition chromatography. *Proc Natl Acad Sci USA* 1971;**68**:2374–6.
2. Phillips M, Herrera J, Krishnan S, Zain M, Greenberg J, Cataneo RN. Variation in volatile organic compounds in the breath of normal humans. *J Chromatogr B Biomed Sci Appl* 1999;**729**:75–88.
3. Silkoff PE, Carlson M, Bourke T, Katial R, Ogren E, Szefler SJ. The Aerocrine exhaled nitric oxide monitoring system NIOX is cleared by the US Food and Drug Administration for monitoring therapy in asthma. *J Allergy Clin Immunol* 2004;**114**:1241–56.
4. Phillips M, Boehmer JP, Cataneo RN, Cheema T, Eisen HJ, Fallon JT, et al. Heart allograft rejection: detection with breath alkanes in low levels (the HARDBALL study). *J Heart Lung Transplant* 2004;**23**:701–8.
5. Graham DY, Klein PD, Evans DJ, Evans Jr DG, Alpert LC, Opekun AR, et al. *Campylobacter pylori* detected noninvasively by the ^{13}C-urea breath test. *Lancet* 1987;**1**:1174–7.
6. Lacroix M, Mosora F, Pontus M, Lefebvre P, Luyckz A, Lopez-Habib G. Glucose naturally labeled with carbon-13: use for metabolic studies in man. *Science* 1973;**181**: 445–6.
7. Modak A. ^{13}C breath tests: transition from research to clinical practice. In: Amann A, Smith D, editors. *Breath analysis for clinical diagnosis and therapeutic monitoring*. Singapore: World Scientific; 2005. p. 457–78.
8. Veereman-Wauters G. Breath testing in pediatrics. In: Mulberg AE, Silber SA, Van den Anker JN, editors. *Pediatric drug development: concepts and applications*. Hoboken, NJ: John Wiley and Sons; 2009. p. 481–500.
9. Modak AS. Stable isotope breath tests in clinical medicine: a review. *J Breath Res* 2007;**1**:014003.
10. Braden B. Methods and functions: breath tests. *Best Pract Res Clin Gastroenterol* 2009;**23**:337–52.
11. Modak AS. ^{13}C breath tests in personalized medicine: fiction or reality? *Expert Rev Mol Diagn* 2009;**9**:805–15.
12. Braden B, Lembcke B, Kuker W, Caspary WF. ^{13}C-breath tests: current state of the art and future directions. *Dig Liver Dis* 2007;**39**:795–805.
13. Parra MD, Martinez JA. Nutritional aspects of breath testing based on stable isotopes. *Nutr Rev* 2006;**64**:338–47.
14. Modak AS. Single time point diagnostic breath tests: a review. *J Breath Res* 2009;**4**: 017002.
15. Gisbert JP, Gonzalez-Lama Y. Breath tests in the diagnosis of gastrointestinal diseases. *Gastroenterol Hepatol* 2005;**28**:407–16.
16. Peek Jr RM, Blaser MJ. *Helicobacter pylori* and gastrointestinal tract adenocarcinomas. *Nat Rev Cancer* 2002;**2**:28–37.

17. Ohara S. Diagnostic methods for *H. pylori* infection. *Nihon Rinsho* 2009;**67**:2279–83.
18. Fischbach W. *Helicobacter pylori. Internist (Berl)* 2009;**50**:979–86;quiz 987.
19. Raymond J, Kalach N. *Helicobacter pylori* infection in children. *Rev Prat* 2006;**56**:5–12.
20. Gisbert JP, Abraira V. Accuracy of *Helicobacter pylori* diagnostic tests in patients with bleeding peptic ulcer: a systematic review and meta-analysis. *Am J Gastroenterol* 2006;**101**:848–63.
21. Koletzko S. Noninvasive diagnostic tests for *Helicobacter pylori* infection in children. *Can J Gastroenterol* 2005;**19**:433–9.
22. Bourke B, Ceponis P, Chiba N, Czinn S, Ferraro R, Fischbach L, et al. Canadian Helicobacter study group consensus conference: update on the approach to *Helicobacter pylori* infection in children and adolescents—an evidence-based evaluation. *Can J Gastroenterol* 2005;**19**:399–408.
23. Sfarti C, Stanciu C, Cojocariu C, Trifan A. ^{13}C-urea breath test for the diagnosis of *Helicobacter pylori* infection in bleeding duodenal ulcer. *Rev Med Chir Soc Med Nat Iasi* 2009;**113**:704–9.
24. Calvet X, Sanchez-Delgado J, Montserrat A, Lario S, Ramirez-Lazaro MJ, Quesada M, et al. Accuracy of diagnostic tests for *Helicobacter pylori*: a reappraisal. *Clin Infect Dis* 2009;**48**:1385–91.
25. Elitsur Y, Tolia V, Gilger MA, Reeves-Garcia J, Schmidt-Sommerfeld E, Opekun AR, et al. Urea breath test in children: the United States prospective, multicenter study. *Helicobacter* 2009;**14**:134–40.
26. Ghadir MR, Shafaghi A, Iranikhah A, Pakdin A, Joukar F, Mansour-Ghanaei F. Furazolidone, amoxicillin and omeprazole with or without bismuth for eradication of *Helicobacter pylori* in peptic ulcer disease. *Turk J Gastroenterol* 2011;**22**:1–5.
27. Furuta T, Kato M, Sugimoto M, Sasaki M, Kamoshida T, Furukawa K, et al. Triple therapy with ecabet sodium, amoxicillin and lansoprazole for 2 weeks as the rescue regimen for *H. pylori* infection. *Intern Med* 2011;**50**:369–74.
28. Furuta T, Sugimoto M, Kodaira C, Nishino M, Yamade M, Uotani T, et al. The dual therapy with 4 times daily dosing of rabeprazole and amoxicillin as the 3rd rescue regimen for eradication of *H. pylori. Hepatogastroenterology* 2010;**57**:1314–9.
29. Tas A, Akbal E, Kocak E, Koklu S. Moxifloxacin-tetracycline-lansoprazole triple therapy for first-line treatment of *Helicobacter pylori* infection: a prospective study. *Helicobacter* 2011;**16**:52–4.
30. Lee VW, Chau TS, Chan AK, Lee KK, Waye MM, Ling TK, et al. Pharmacogenetics of esomeprazole or rabeprazole-based triple therapy in *Helicobacter pylori* eradication in Hong Kong non-ulcer dyspepsia Chinese subjects. *J Clin Pharm Ther* 2010;**35**:343–50.
31. Koizumi W, Tanabe S, Nakatani K, Ishido K, Nishimura K, Azuma M, et al. Quadruple therapy with ecabet sodium, omeprazole, amoxicillin and metronidazole is effective for eradication of *Helicobacter pylori* after failure of first-line therapy (KDOG0201 Study). *J Clin Pharm Ther* 2010;**35**:303–7.
32. Gao XZ, Qiao XL, Song WC, Wang XF, Liu F. Standard triple, bismuth pectin quadruple and sequential therapies for *Helicobacter pylori* eradication. *World J Gastroenterol* 2010;**16**:4357–62.
33. Mamori S, Higashida A, Kawara F, Ohnishi K, Takeda A, Senda E, et al. Age-dependent eradication of *Helicobacter pylori* in Japanese patients. *World J Gastroenterol* 2010;**16**:4176–9.

34. Ozdil B, Akkiz H, Bayram S, Bekar A, Akgollu E, Sandikci M. Influence of CYP2C19 functional polymorphism on *Helicobacter pylori* eradication. *Turk J Gastroenterol* 2010;**21**:23–8.
35. Molina-Infante J, Perez-Gallardo B, Fernandez-Bermejo M, Hernandez-Alonso M, Vinagre G, Duenas C, et al. Clinical trial: clarithromycin vs. levofloxacin in first-line triple and sequential regimens for *Helicobacter pylori* eradication. *Aliment Pharmacol Ther* 2010;**31**:1077–84.
36. Alazmi WM, Buhaimed W, Al-Mekhaizeem K, Siddique I. Efficacy of standard triple therapy in the treatment of *Helicobacter pylori* infection: experience from Kuwait. *Dig Dis Sci* 2010;**55**:3120–3.
37. De Francesco V, Zullo A, Perna F, Giorgio F, Hassan C, Vannella L, et al. *Helicobacter pylori* antibiotic resistance and ^{13}C-urea breath test values. *J Med Microbiol* 2010;**59**: 588–91.
38. Pellicano R, Oliaro E, Fagoonee S, Astegiano M, Berrutti M, Saracco G, et al. Clinical and biochemical parameters related to cardiovascular disease after *Helicobacter pylori* eradication. *Int Angiol* 2009;**28**:469–73.
39. Ma J, Liu W, Zhang L, Pan K, Zhao H, Zhou T, et al. A placebo-controlled trial of 10-day bismuth-based quadruple therapy to eradicate *Helicobacter pylori* infection; a pilot study for the large Linqu County trial. *Eur J Gastroenterol Hepatol* 2010;**22**:597–601.
40. Sanches B, Coelho L, Moretzsohn L, Vieira Jr G. Failure of *Helicobacter pylori* treatment after regimes containing clarithromycin: new practical therapeutic options. *Helicobacter* 2008;**13**:572–6.
41. Nocon M, Kuhlmann A, Leodolter A, Roll S, Vauth C, Willich SN, et al. Efficacy and cost-effectiveness of the ^{13}C-urea breath test as the primary diagnostic investigation for the detection of *Helicobacter pylori* infection compared to invasive and non-invasive diagnostic tests. *GMS Health Technol Assess* 2009;**5**:Doc14.
42. Scaccianoce G, Zullo A, Hassan C, Gentili F, Cristofari F, Cardinale V, et al. Triple therapies plus different probiotics for *Helicobacter pylori* eradication. *Eur Rev Med Pharmacol Sci* 2008;**12**:251–6.
43. Buzas GM, Szeles I. Interpretation of the ^{13}C-urea breath test in the choice of second- and third-line eradication of *Helicobacter pylori* infection. *J Gastroenterol* 2008;**43**:108–14.
44. Niv Y, Hazazi R. *Helicobacter pylori* recurrence in developed and developing countries: meta-analysis of ^{13}C-urea breath test follow-up after eradication. *Helicobacter* 2008;**13**:56–61.
45. Zullo A, Perna F, Ricci C, Hassan C, Tampieri A, De Francesco V, et al. ^{13}C-urea breath test values and *Helicobacter pylori* eradication. *Dig Dis Sci* 2008;**53**:370–4.
46. Abell TL, Camilleri M, Donohoe K, Hasler WL, Lin HC, Maurer AH, et al. Consensus recommendations for gastric emptying scintigraphy: a joint report of the American Neurogastroenterology and Motility Society and the Society of Nuclear Medicine. *Am J Gastroenterol* 2008;**103**:753–63.
47. Sanaka M, Yamamoto T, Kuyama Y. Retention, fixation, and loss of the [^{13}C] label: a review for the understanding of gastric emptying breath tests. *Dig Dis Sci* 2008;**53**: 1747–56.
48. Ghoos YF, Maes BD, Geypens BJ, Mys G, Hiele MI, Rutgeerts PJ, et al. Measurement of gastric emptying rate of solids by means of a carbon-labeled octanoic acid breath test. *Gastroenterology* 1993;**104**:1640–7.

49. Odunsi ST, Camilleri M, Szarka LA, Zinsmeister AR. Optimizing analysis of stable isotope breath tests to estimate gastric emptying of solids. *Neurogastroenterol Motil* 2009;**21**:706–e38.
50. Verbeke K. Will the ^{13}C-octanoic acid breath test ever replace scintigraphy as the gold standard to assess gastric emptying. *Neurogastroenterol Motil* 2009;**21**:1013–6.
51. Keller J, Andresen V, Wolter J, Layer P, Camilleri M. Influence of clinical parameters on the results of ^{13}C-octanoic acid breath tests: examination of different mathematical models in a large patient cohort. *Neurogastroenterol Motil* 2009;**21**:1039–83.
52. Clegg ME, Shafat A. Procedures in the ^{13}C octanoic acid breath test for measurement of gastric emptying: analysis using Bland-Altman methods. *Scand J Gastroenterol* 2010;**45**:852–61.
53. Glerup H. Bluhme H, Villadsen GE, Rasmussen K, Ejskjaer N, Dahlerup JF. Gastric emptying: a comparison of three methods. *Scand J Gastroenterol* 2007;**42**:1182–6.
54. Sanaka M, Nakada K, Nosaka C, Kuyama Y. The Wagner-Nelson method makes the ^{13}C-breath test comparable to radioscintigraphy in measuring gastric emptying of a solid/liquid mixed meal in humans. *Clin Exp Pharmacol Physiol* 2007;**34**:641–4.
55. Szarka LA, Camilleri M, Vella A, Burton D, Baxter K, Simonson J, et al. A stable isotope breath test with a standard meal for abnormal gastric emptying of solids in the clinic and in research. *Clin Gastroenterol Hepatol* 2008;**6**:635–43.
56. Viramontes BE, Kim DY, Camilleri M, Lee JS, Stephens D, Burton DD, et al. Validation of a stable isotope gastric emptying test for normal, accelerated or delayed gastric emptying. *Neurogastroenterol Motil* 2001;**13**:567–74.
57. Ikeda T, Inamori M, Fujisawa N, Iwasaki T, Akiyama T, Akimoto K, et al. Effects of body positions on gastric emptying with enteral nutrition: a crossover study using a continuous real time ^{13}C breath test (BreathID system). *Hepatogastroenterology* 2008;**55**:1905–7.
58. Kawasaki N, Suzuki Y, Urashima M, Nakayoshi T, Tsuboi K, Tanishima Y, et al. Effect of gelatinization on gastric emptying and absorption. *Hepatogastroenterology* 2008;**55**: 1843–45.
59. Inamori M, Iida H, Endo H, Hosono K, Akiyama T, Yoneda K, et al. Aperitif effects on gastric emptying: a crossover study using continuous real-time ^{13}C breath test (BreathID System). *Dig Dis Sci* 2009;**54**:816–8.
60. Inamori M, Akiyama T, Akimoto K, Fujita K, Takahashi H, Yoneda M, et al. Early effects of peppermint oil on gastric emptying: a crossover study using a continuous real-time ^{13}C breath test (BreathID system). *J Gastroenterol* 2007;**42**:539–42.
61. Akimoto K, Inamori M, Iida H, Endo H, Akiyama T, Ikeda T, et al. Does postprandial coffee intake enhance gastric emptying?: a crossover study using continuous real time ^{13}C breath test (BreathID system). *Hepatogastroenterology* 2009;**56**:918–20.
62. Ma J, Bellon M, Wishart JM, Young R, Blackshaw LA, Jones KL, et al. Effect of the artificial sweetener, sucralose, on gastric emptying and incretin hormone release in healthy subjects. *Am J Physiol Gastrointest Liver Physiol* 2009;**296**:G735–9.
63. Fruehauf H, Steingoetter A, Fox MR, Kwiatek MA, Boesiger P, Schwizer W, et al. Characterization of gastric volume responses and liquid emptying in functional dyspepsia and health by MRI or barostat and simultaneous ^{13}C-acetate breath test. *Neurogastroenterol Motil* 2009;**21**:697–e37.
64. Shindo T, Futagami S, Hiratsuka T, Horie A, Hamamoto T, Ueki N, et al. Comparison of gastric emptying and plasma ghrelin levels in patients with functional dyspepsia and non-erosive reflux disease. *Digestion* 2009;**79**:65–72.

65. Rocco A, Sarnelli G, Compare D, de Colibus P, Micheli P, Somma P, et al. Tissue ghrelin level and gastric emptying rate in adult patients with celiac disease. *Neurogastroenterol Motil* 2008;**20**:884–90.
66. Vignolo MC, Savassi-Rocha PR, Coelho LG, Soares MP, Cardoso-Junior A, Barbosa TF, et al. Gastric emptying before and after cholecystectomy in patients with cholecystolithiasis. *Hepatogastroenterology* 2008;**55**:850–4.
67. Matsumoto M, Yoshimura R, Akiho H, Higuchi N, Kobayashi K, Matsui N, et al. Gastric emptying in diabetic patients by the (13)C-octanoic acid breath test: role of insulin in gastric motility. *J Gastroenterol* 2007;**42**:469–74.
68. Cardoso-Junior A, Coelho LG, Savassi-Rocha PR, Vignolo MC, Abrantes MM, de Almeida AM, et al. Gastric emptying of solids and semi-solids in morbidly obese and non-obese subjects: an assessment using the ^{13}C-octanoic acid and ^{13}C-acetic acid breath tests. *Obes Surg* 2007;**17**:236–41.
69. Medhus AW, Bjornland K, Emblem R, Husebye E. Liquid and solid gastric emptying in adults treated for Hirschsprungs disease during early childhood. *Scand J Gastroenterol* 2007;**42**:34–40.
70. Nguyen NQ, Chapman M, Fraser RJ, Ritz M, Bryant LK, Butler R, et al. Long-standing type II diabetes mellitus is not a risk factor for slow gastric emptying in critically ill patients. *Intensive Care Med* 2006;**32**:1365–70.
71. Hoffman I, Tertychnyy A, Ectors N, De Greef T, Haesendonck, N. Tack J. Duodenogastroesophageal reflux in children with refractory gastro-esophageal reflux disease. *J Pediatr* 2007;**151**:307–11.
72. Jones BL, Pearce RE, Abdel-Rahman SM, Friesen CA, James LP, Kearns GL. Characterization of delayed liquid gastric emptying in children by the ^{13}C-acetate breath test. *J Breath Res* 2009;**3**:047004.
73. Goetze O, Fox M, Kwiatek MA, Treier R, Schwizer W, Thumshirn M, et al. Effects of postgastric ^{13}C-acetate processing on measurement of gastric emptying: a systematic investigation in health. *Neurogastroenterol Motil* 2009;**21**:1047–e85 .
74. Qu XH, Huang XL, Cai MC, Chen ZW, Lu LG, Yan WL, et al., ^{13}C breath tests: a useful tool to quantify hepatic reserve function]. *Zhonghua Gan Zang Bing Za Zhi* 2009;**17**: 318–20.
75. Ilan Y. Review article: the assessment of liver function using breath tests. *Aliment Pharmacol Ther* 2007;**26**:1293–302.
76. Zhang GS, Bao ZJ, Zou J, Yin SM, Huang YQ, Huang H, et al. Clinical research on liver reserve function by ^{13}C-phenylalanine breath test in aged patients with chronic liver diseases. *BMC Geriatr* 2010;**10**:23.
77. Moran S, Gallardo-Wong I, Rodriguez-Leal G, McCollough P, Mendez J, Castaneda B, et al. L-[1-^{13}C]phenylalanine breath test in patients with chronic liver disease of different etiologies. *Isotopes Environ Health Stud* 2009;**45**:192–7.
78. Wutzke KD, Wigger M. Effect of alcohol consumption on the liver detoxication capacity as measured by [^{13}C2]aminopyrine and L-[1-^{13}C]phenylalanine breath tests. *Isotopes Environ Health Stud* 2009;**45**:185–91.
79. Wada M, Wada Y, Uchiyama M, Kajiwara M, Takatori K. ^{13}C-phenylalanine breath test correlates with liver fibrosis in postoperative biliary atresia. *Pediatr Int* 2007;**49**:836–41.
80. Freeman RB, Dixon M, Horth B, Melanson AM, Palladino MB, Kinzel J, et al. L-[1-^{13}C] phenylalanine breath test for monitoring hepatic function after living donor liver transplant surgery. *J Breath Res* 2007;**1**:026002.

81. Forestier J, Dumortier J, Guillaud O, Ecochard M, Roman S, Boillot O, et al. Noninvasive diagnosis and prognosis of liver cirrhosis: a comparison of biological scores, elastometry, and metabolic liver function tests. *Eur J Gastroenterol Hepatol* 2010;**22**:532–40.
82. Lalazar G, Ilan Y. Assessment of liver function in acute or chronic liver disease by the methacetin breath test: a tool for decision making in clinical hepatology. *J Breath Res* 2009;**3**:047001.
83. Li HX, Wang JP, Yang Y, Zhang J, Feng QL. The value of studying liver function reserve in hepatic carcinoma by ^{13}C-methacetin breath test. *Zhonghua Nei Ke Za Zhi* 2009;**48**: 383–7.
84. Lock JF, Taheri P, Bauer S, Holzhutter HG, Malinowski M, Neuhaus P, et al. Interpretation of non-invasive breath tests using ^{13}C-labeled substrates–a preliminary report with superset ^{13}C-methacetin. *Eur J Med Res* 2009;**14**:547–50.
85. Wutzke KD, Forberger A, Wigger M. Effect of alcohol consumption on the liver detoxication capacity as measured by [^{13}C]methacetin- and [methyl-^{13}C]methionine-breath tests. *Isotopes Environ Health Stud* 2008;**44**:219–26.
86. Lalazar G, Pappo O, Hershcovici T, Hadjaj T, Shubi M, Ohana H, et al. A continuous ^{13}C methacetin breath test for noninvasive assessment of intrahepatic inflammation and fibrosis in patients with chronic HCV infection and normal ALT. *J Viral Hepat* 2008;**15**:716–28.
87. Dinesen L, Caspary WF, Chapman RW, Dietrich CF, Sarrazin C, Braden B. ^{13}C-methacetin-breath test compared to also noninvasive biochemical blood tests in predicting hepatic fibrosis and cirrhosis in chronic hepatitis C. *Dig Liver Dis* 2008;**40**:743–8.
88. Desta Z, Modak A, Nguyen PD, Lemler SM, Kurogi Y, Li L, et al. Rapid identification of the hepatic cytochrome P450 2C19 activity using a novel and noninvasive ^{13}C-pantoprazole breath test. *J Pharmacol Exp Ther* 2009;**329**:297–305.
89. Leeder JS, Pearce RE, Gaedigk A, Modak A, Rosen DI. Evaluation of a ^{13}C-dextromethorphan breath test to assess CYP2D6 phenotype. *J Clin Pharmacol* 2008;**48**: 1041–51.
90. Schmilovitz-Weiss H, Niv Y, Pappo O, Halpern M, Sulkes J, Braun M, et al. The ^{13}C-caffeine breath test detects significant fibrosis in patients with nonalcoholic steatohepatitis. *J Clin Gastroenterol* 2008;**42**:408–12.
91. Goetze O, Selzner N, Fruehauf H, Fried M, Gerlach T, Mullhaupt B. ^{13}C-methacetin breath test as a quantitative liver function test in patients with chronic hepatitis C infection: continuous automatic molecular correlation spectroscopy compared to isotopic ratio mass spectrometry. *Aliment Pharmacol Ther* 2007;**26**:305–11.
92. Grattagliano I, Lauterburg BH, Palasciano G, Portincasa P. ^{13}C-breath tests for clinical investigation of liver mitochondrial function. *Eur J Clin Invest* 2010;**40**:843–50.
93. Wu IC, Ohsawa I, Fuku N, Tanaka M. Metabolic analysis of ^{13}C-labeled pyruvate for noninvasive assessment of mitochondrial function. *Ann NY Acad Sci* 2010;**1201**:111–20.
94. Duro D, Fitzgibbons S, Valim C, Yang CF, Zurakowski D, Dolan M, et al. ^{13}C-Methionine breath test to assess intestinal failure-associated liver disease. *Pediatr Res* 2010;**68**: 349–54.
95. Mawatari H, Inamori M, Fujita K, Yoneda M, Iida H, Endo H, et al. The continuous real-time ^{13}C-octanoate breath test for patients with nonalcoholic steatohepatitis using the BreathID system. *Hepatogastroenterology* 2009;**56**:1436–8.
96. Duro D, Duggan C, Valim C, Bechard L, Fitzgibbons S, Jaksic T, et al. Novel intravenous ^{13}C-methionine breath test as a measure of liver function in children with short bowel syndrome. *J Pediatr Surg* 2009;**44**:236–40;discussion 240.

97. Banasch M, Emminghaus R, Ellrichmann M, Schmidt WE, Goetze O. Longitudinal effects of hepatitis C virus treatment on hepatic mitochondrial dysfunction assessed by ^{13}C-methionine breath test. *Aliment Pharmacol Ther* 2008;**28**:443–49.
98. Candelli M, Miele L, Armuzzi A, Nista EC, Pignataro G, Fini L, et al. ^{13}C-Methionine breath tests for mitochondrial liver function assessment. *Eur Rev Med Pharmacol Sci* 2008;**12**:245–9.
99. Candelli M, Armuzzi A, Miele L, Nista EC, Pignataro G, Zileri Dal Verme L, et al. Gender affects ^{13}C-ketoisocaproic acid breath test. *Eur Rev Med Pharmacol Sci* 2007;**11**:401–6.
100. Crujeiras AB, Parra D, Goyenechea E, Abete I, Gonzalez-Muniesa P, Martinez JA. Energy restriction in obese subjects impact differently two mitochondrial function markers. *J Physiol Biochem* 2008;**64**:211–9.
101. Braden B. ^{13}C breath tests for the assessment of exocrine pancreatic function. *Pancreas* 2010;**39**:955–9.
102. Jonderko K, Dus Z, Szymszal M, Kasicka-Jonderko A, Blonska-Fajfrowska B. Normative values for the ^{13}C-mixed triglyceride breath test in two age groups. *Med Sci Monit* 2009;**15**:CR255–9.
103. Moran-Villota S, Arteaga ME, Rodriguez-Leal GA, Medina-Zavala M, Gallardo-Wong I, Milke-Garcia P, et al. Evaluation of lipid digestion using the ^{13}C-mixed triglyceride breath test in patients with chronic pancreatitis. *Rev Gastroenterol Mex* 2007;**72**:202–6.
104. Schneider AR, Hammerstingl R, Heller M, Povse N, Murzynski L, Vogl TJ, et al. Does secretin-stimulated MRCP predict exocrine pancreatic insufficiency?: A comparison with noninvasive exocrine pancreatic function tests. *J Clin Gastroenterol* 2006;**40**:851–5.
105. Slater C, Preston T, Weaver LT. Acetate correction for postabsorption metabolism does not improve the [^{13}C]mixed triacylglycerol breath test. *J Pediatr Gastroenterol Nutr* 2006;**43**:666–72.
106. Nakamura H, Morifuji M, Murakami Y, Uemura K, Ohge H, Hayashidani Y, et al. Usefulness of a ^{13}C-labeled mixed triglyceride breath test for assessing pancreatic exocrine function after pancreatic surgery. *Surgery* 2009;**145**:168–75.
107. Nakamura H, Murakami Y, Uemura K, Hayashidani Y, Sudo T, Ohge H, et al. Reduced pancreatic parenchymal thickness indicates exocrine pancreatic insufficiency after pancreatoduodenectomy. *J Surg Res* 2011;**171**:473–8.
108. Nakamura H, Murakami Y, Morifuji M, Uemura K, Hayashidani Y, Sudo T, et al. Analysis of fat digestive and absorptive function after subtotal gastrectomy by a ^{13}C-labeled mixed triglyceride breath test. *Digestion* 2009;**80**:98–103.
109. Suzuki M, Tanaka K, Ohtani K, Kitamura K, Kudo T, Shoji H, et al. Estimation of postoperative fat absorption using the ^{13}C mixed-triglyceride breath test in children with choledochal cyst. *Eur J Pediatr* 2009;**168**:35–8.
110. Herzog DC, Delvin EE, Albert C, Marcotte JE, Pelletier VA, Seidman EG. ^{13}C-labeled mixed triglyceride breath test (^{13}C MTG-BT) in healthy children and children with cystic fibrosis (CF) under pancreatic enzyme replacement therapy (PERT): a pilot study. *Clin Biochem* 2008**41**:1489–92.
111. Rapoport SI, Maev IV, Kucheriavyi Iu A, Shubina NA, Semenova NV, Stukova N, et al. Experience with the use of ^{13}C-breath test for the evaluation of the efficacy of enzyme substitution therapy with microencapsulated pancreatinin in patients with severe primary pancreatic insufficiency. *Klin Med (Mosk)* 2009;**87**:53–7.
112. Dominguez-Munoz JE, Iglesias-Garcia J, Vilarino-Insua M, Iglesias-Rey M. ^{13}C-mixed triglyceride breath test to assess oral enzyme substitution therapy in patients with chronic pancreatitis. *Clin Gastroenterol Hepatol* 2007;**5**:484–8.

113. Ishii Y, Kohno T, Ito A, Suzuki S, Takayama T, Asai S. Evaluation of pancreatic exocrine secretion using ^{13}C-dipeptide (benzoyl-L-tyrosyl-[1-^{13}C]alanine) breath test: focusing on pancreatoduodenectomy cases. *Pancreas* 2007;**35**:313–9.
114. Ishii Y, Kohno T, Ito A, Suzuki S, Takayama T, Asai S. Measurement of extra-pancreatic secretory function by ^{13}C-dipeptide breath test. *Transl Res* 2007;**149**:298–303.
115. Braun M, Pappo O, Zuckerman E, Sulkes J, Kitai Y, Cohen O, et al. The innovative continuous BreathID (R) test system predicts the extent of hepatic injury in patients with nonalcoholic fatty liver disease. *Gastroenterology* 2006;**130**:A821.
116. Lalazar G, Adar T, Ilan Y. Point-of-care continuous ^{13}C-methacetin breath test improves decision making in acute liver disease: results of a pilot clinical trial. *World J Gastroenterol* 2009;**15**:966–72.
117. Furuta T, Kodaira C, Nishino M, Yamade M, Sugimoto M, Ikuma M, et al. ^{13}C-pantoprazole breath test to predict CYP2C19 phenotype and efficacy of a proton pump inhibitor, lansoprazole. *Aliment Pharmacol Ther* 2009;**30**:294–300.
118. Furuta T, Iwaki T, Umemura K. ^{13}C-pantoprazole breath test as a predictor of the antiplatelet function of clopidogrel. *Eur J Clin Pharmacol* 2010;**66**:457–63.
119. Candelli M, Pompili M, Suppressa P, Lenato GM, Bosco G, Rapaccini GL, et al. Liver involvement in hereditary hemorrhagic telangiectasia: can breath test unmask impaired hepatic first-pass effect? *Intern Emerg Med* 2011.
120. Kasicka-Jonderko A, Loska D, Jonderko K, Kaminska M, Blonska-Fajfrowska B. Interference of acute cigarette smoking with ^{13}C-methacetin breath test. *Isotopes Environ Health Stud* 2011;**47**:34–41.
121. Ilan Y. A fourth dimension in decision making in hepatology. *Hepatol Res* 2010;**40**: 1143–54.
122. Schneider A, Caspary WF, Saich R, Dietrich CF, Sarrazin C, Kuker W, et al. ^{13}C-methacetin breath test shortened: 2-point-measurements after 15 minutes reliably indicate the presence of liver cirrhosis. *J Clin Gastroenterol* 2007;**41**:33–7.
123. Palmieri VO, Grattagliano I, Minerva F, Pollice S, Palasciano G, Portincasa P. Liver function as assessed by breath tests in patients with hepatocellular carcinoma. *J Surg Res* 2009;**157**:199–207.
124. Hendrichova M, Malek F, Koprivova H, Vranova J, Ostadal P, Kratka K, et al. Correlation of NT-proBNP with metabolic liver function as assessed with ^{13}C-methacetin breath test in patients with acute decompensated heart failure. *Int J Cardiol* 2010;**144**:321–2.
125. Holtmeier J, Leuschner M, Schneider A, Leuschner U, Caspary WF, Braden B. ^{13}C-methacetin and ^{13}C-galactose breath tests can assess restricted liver function even in early stages of primary biliary cirrhosis. *Scand J Gastroenterol* 2006;**41**:1336–41.
126. Liu YX, Huang LY, Wu CR, Cui J. Measurement of liver function for patients with cirrhosis by ^{13}C-methacetin breath test compared with Child-Pugh score and routine liver function tests. *Chin Med J (Engl)* 2006;**119**:1563–6.
127. Mizrahi M, Lalazar G, Adar T, Raz I, Ilan Y. Assessment of insulin resistance by a ^{13}C glucose breath test: a new tool for early diagnosis and follow-up of high-risk patients. *Nutr J* 2010;**9**:25.
128. Singal P, Janghorbani M, Schuette SA, Chisholm R, Mather KJ. Intra-individual variability of CO breath isotope enrichment compared to blood glucose in the oral glucose tolerance test. *Diabetes Technol Ther* 2010;**12**:947–53.
129. Banerjee D, Vikram N, Mishra P, Bhatt R, Prakash S, Misra A. Correlation of a ^{13}C-glucose breath test with surrogate markers of insulin resistance in urban and rural Asian Indians. *Metab Syndr Relat Disord* 2009;**7**:215–9.

130. Jetha MM, Nzekwu U, Lewanczuk RZ, Ball GD. A novel, non-invasive ^{13}C-glucose breath test to estimate insulin resistance in obese prepubertal children. *J Pediatr Endocrinol Metab* 2009;**22**:1051–9.
131. Ezzeldin HH, Acosta EP, Mattison LK, Fourie J, Modak A, Diasio RB. ^{13}C-5-FU breath test current status and future directions: a comprehensive review. *J Breath Res* 2009;**3**:047002.
132. Mattison LK, Fourie J, Hirao Y, Koga T, Desmond RA, King JR, et al. The uracil breath test in the assessment of dihydropyrimidine dehydrogenase activity: pharmacokinetic relationship between expired $^{13}CO_2$ and plasma 2-^{13}C-dihydrouracil. *Clin Cancer Res* 2006;**12**:549–55.
133. Saif MW, Mattison L, Carollo T, Ezzeldin H, Diasio RB. Dihydropyrimidine dehydrogenase deficiency in an Indian population. *Cancer Chemother Pharmacol* 2006;**58**:396–401.
134. Mattison LK, Fourie J, Desmond RA, Modak A, Saif MW, Diasio RB. Increased prevalence of dihydropyrimidine dehydrogenase deficiency in African-Americans compared with Caucasians. *Clin Cancer Res* 2006;**12**:5491–5.
135. Ishii Y, Suzuki S, Takahashi Y, Takayama T, Asai S. Can the 2-^{13}C-uracil breath test be used to predict the effect of the antitumor drug S-1? *Cancer Chemother Pharmacol* 2010;**66**:333–43.
136. Thomas HR, Ezzeldin HH, Guarcello V, Mattison LK, Fridley BL, Diasio RB. Genetic regulation of dihydropyrimidinase and its possible implication in altered uracil catabolism. *Pharmacogenet Genomics* 2007;**17**:973–87.
137. Modak AS, Reid JM, Williams LA, Hobbs GR. A simple non-invasive method to detect and monitor hypercapnia: the sodium ^{13}C-bicarbonate breath test Isotopes. *Environ Health Stud* 2007;**43**:23–9.
138. Berthold HK, Schober P, Scheurlen C, Marklein G, Horre R, Gouni-Berthold I, et al. Use of the lactose-^{13}C-ureide breath test for diagnosis of small bowel bacterial overgrowth: comparison to the glucose hydrogen breath test. *J Gastroenterol* 2009;**44**:944–51.
139. Robayo-Torres CC, Opekun AR, Quezada-Calvillo R, Villa X, Smith EO, Navarrete M, et al. ^{13}C-Breath tests for sucrose digestion in congenital sucrase isomaltase-deficient and sacrosidase-supplemented patients. *J Pediatr Gastroenterol Nutr* 2009;**48**:412–8.
140. Ritchie BK, Brewster DR, Davidson GP, Tran CD, McNeil Y, Hawkes JS, et al. ^{13}C-Sucrose breath test: novel use of a noninvasive biomarker of environmental gut health. *Pediatrics* 2009;**124**:620–6.
141. Tveito K, Hetta AK, Askedal M, Brunborg C, Sandvik L, Loberg EM, et al. A novel one-hour ^{13}C-sorbitol breath test versus the H2-sorbitol breath test for assessment of coeliac disease. *Scand J Gastroenterol* 2009;**44**:813–9.
142. Tveito K, Brunborg C, Sandvik L, Loberg EM, Skar V. ^{13}C-xylose and ^{14}C-xylose breath tests for the diagnosis of coeliac disease. *Scand J Gastroenterol* 2008;**43**:166–73.
143. Shalev T, Aeed H, Sorin V, Shahmurov M, Didkovsky E, Ilan Y, et al. Evaluation of the ^{13}C-octanoate breath test as a surrogate marker of liver damage in animal models. *Dig Dis Sci* 2010;**55**:1589–98.
144. Silva S, Wyse CA, Goodfellow MR, Yam PS, Preston T, Papasouliotis K, et al. Assessment of liver function in dogs using the ^{13}C-galactose breath test. *Vet J* 2010;**185**:152–6.
145. Shirin H, Aeed H, Shalev T, Sorin V, Stavinski S, Shahmurov M, et al. Utility of a ^{13}C-methacetin breath test in evaluating hepatic injury in rats. *J Gastroenterol Hepatol* 2008;**23**:1762–8.

146. Danicke S, Beineke A, Goyarts T, Valenta H, Beyer M, Humpf HU. Effects of a Fusarium toxin-contaminated triticale, either untreated or treated with sodium metabisulphite ($Na_2S_2O_5$, SBS), on weaned piglets with a special focus on liver function as determined by the ^{13}C–methacetin breath test. *Arch Anim Nutr* 2008;**62**:263–86.
147. Sugiyama E, Inada M, Kunizaki J, Tobita K, Yoshida T, Kashimoto M, et al. Desirable pharmacokinetic properties of ^{13}C-uracil as a breath test probe of gastric emptying in comparison with ^{13}C-acetate and ^{13}C-octanoate in rats. *Scand J Gastroenterol* 2009;**44**:1067–75.
148. Matsumoto K, Kimura H, Tashima K, Uchida M, Horie S. Validation of ^{13}C-acetic acid breath test by measuring effects of loperamide, morphine, mosapride, and itopride on gastric emptying in mice. *Biol Pharm Bull* 2008;**31**:1917–22.
149. Tooley KL, Howarth GS, Lymn KA, Butler RN. Optimization of the non-invasive ^{13}C-sucrose breath test in a rat model of methotrexate-induced mucositis. *Cancer Chemother Pharmacol* 2010:**65**;913–21.
150. Kohno T, Ito A, Hosoi I, Hirayama J, Shibata K. Synthetic ^{13}C-dipeptide breath test for the rapid assessment of pancreatic exocrine insufficiency in rats. *Scand J Gastroenterol* 2007;**42**:992–9.
151. Uchida M, Mogami O. Usefulness of breath test for evaluating pancreatic exocrine function using N-benzoyl-L-tyrosyl-1-^{13}C-L-alanine sodium in non-invasive and conscious rats. *Biol Pharm Bull* 2008;**31**:785–8.
152. Modak AS. Barriers to overcome for transition of breath tests from research to routine clinical practice. *J Breath Res* 2011;**5**:030202.
153. Modak AS. Breath biomarkers for personalized medicine. *Per Med* 2010;**7**:643–53.

PART

F

Development and Use of Sensors

CHAPTER

15 Sensors for Exhaled Gas Analysis: An Analytical Review

Gennadii Kamarchuk*, Alexander Pospelov†, and Ievgeniia Kushch‡

*B. Verkin Institute for Low Temperature Physics and Engineering, National Academy of Sciences of Ukraine, 47 Lenin Avenue, Kharkov 61103, Ukraine

†National Technical University "Kharkov Polytechnic Institute", 21 Frunze Str., Kharkov 61002, Ukraine

‡Institute for Children and Adolescents Health Care, National Academy of Medical Sciences of Ukraine, 52-A 50 let VLKSM Avenue, Kharkov 61153, Ukraine

15.1 INTRODUCTION. SENSORS AS A PROSPECTIVE TOOL FOR IMPLEMENTATION OF FUNDAMENTAL FINDINGS IN THE AREA OF EXHALED GAS ANALYSIS IN THE CLINICAL SETTING

The problem of breath gas analysis has attracted much attention recently due to the development of efficient methods for non-invasive diagnostics of the human physiological state.[1–3] Exhaled air is a complex mixture of exogenous and endogenous gases; some of them are being produced locally in the respiratory airways, gastro-esophageal tract, and mouth cavity while its composition is also affected by environmental and behavioral factors like smoking[4] and air pollution.[5] As of today, up to 600 of different volatile organic compounds (VOCs) have been identified in human breath, the concentrations of some being dependent on the metabolic state.[1,6–11] Detailed analysis of these compounds is difficult made even more challenging due to mutual interaction of the exhaled volatiles. A wide range of techniques for breath gas analysis is currently being exploited. Mass spectrometry, gas chromatography-mass spectrometry (GC-MS), and infrared (IR) spectroscopy[12–17] are at the forefront. Although these conventional techniques yield adequate results, they are hampered by a number of restrictive disadvantages, being costly, sophisticated, and bulky. The necessity of breath sample collection and pre-processing of samples for some of the techniques, as well as the need of highly skilled personnel to provide technical maintenance, hinder their widespread use in clinical practice.

Therefore simple and inexpensive technologies enabling online, real-time breath analysis avoiding collection and pre-processing of breath samples are desirable for breath gas analysis. Specifically designed gas sensors have become available in the last decade. Simple non-invasive diagnostic tests based on novel sensory technologies have proven to be inexpensive, easy-to-use, and reliable alternatives for breath-gas analyses.[18–30]

Volatile Biomarkers. http://dx.doi.org/10.1016/B978-0-44-462613-4.00015-5

The primary focus of sensor development has been toward the most widely studied volatiles, such as nitric oxide, hydrogen, ammonia, acetone, carbon oxides, etc. which have proven to be reliable disease markers.[7,27,31–37] Ethanol vapor is of a special importance, as its detection is not only of medical, but also of social importance. Currently, a large variety of portable sensors for gaseous ethanol is being exploited (see, for instance, Refs. 3,38–43). Resistivity of the material constituting the transforming element of the sensors changes when exposed to alcohol vapor. They are designed for application in the chemical, pharmacological, and food-processing industries, in commerce, and in medicine. Another example of sensory breath applications concerns the detection and quantification of hydrocarbons, carbon dioxide and oxygen, which are monitored by portable handheld gas analyzers for individual use that sound an alarm when the level of these gases exceeds safe concentrations limits.[3,20,22,44]

Metal-oxide materials,[45] conducting polymers,[46] carbon nanotubes,[47] organic dielectrics,[48,49] organic conductors (or synthetic metals),[50,51] etc. among prospective materials[52] have been used in new high-sensitive sensor devices for breath analysis. Additional functionalization is able to enhance the sensor properties of the above materials.[22,53–55]

Nanotechnology opens new unlimited possibilities for the creation of unique sensors for breath gas analysis. Therefore, efforts are being made to improve the crucial parameters of these sensor devices by a serious know-how approach combining solid state physics and material science. Qualitatively, new sensor characteristics can be discovered by using advanced fundamental benchmarks that have not yet been applied. Taking this into account, for instance, we proposed a recent novel approach[56] in this field based on the application of point-contact spectroscopy[57] to sensor techniques. Utilization of new fundamental properties of point contacts in combination with advanced functional materials provides devices with excellent and unprecedented characteristics (see, for instance, Table 1 in Ref. 58).

Introduction of multisensory technologies of the “electronic nose” type is one of promising trends of sensors application for breath gas analysis.[19,24,59,60] The electronic nose has turned into a powerful instrument for a wide variety of biomedical applications[61] since the time it was conceived.[62] Unlike the conventional approach to sensors development, which mostly involved improving selectivity to specific compounds, problem solving in the area of electronic nose technologies deals with application of multiple arrays of relatively nonselective sensors.[63,64] Development of this technique has become possible due to enhancement of computing power that enables real-time processing of the multiparametric information. It is possible to obtain data of reasonable accuracy both on the composition of complex gaseous media and the concentrations of its separate components. Selectivity of single sensors toward the detected substances is not of primary importance. Rather, it is important that sensors in electronic noses should have essentially similar sensitivities to the different compounds in the mixture.

Devices based on chemical sensors should be compact, light weight, self-contained (can be battery-powered) units that are easy to use and read, accurate, reliable, reproducible, and totally non-invasive. One of the great things about nanotech sensors is that they typically consume very little power—micro- or even nano-watts. This makes such technologies perfect for integration into battery-powered wireless networks for multiple/simultaneous/quick screening and diagnosis of a patient's status. Sensor techniques hold great promise in breath analysis diagnosis.[65]

15.2 SENSORY METROLOGICAL PERFORMANCE ESSENTIAL FOR EXHALED BREATH MEASUREMENTS

A broad spectrum of gaseous breath compounds offers the opportunities for development of analytical methods for diagnosing human health. Conventional approaches to breath analysis involve the analysis of each volatile compound in the exhaled breath. When sensors-based techniques are used for breath analysis, it is the sensitivity, selectivity, stability, time of response, and recovery time that are the main characteristics considered.

It is important that sensors have high sensitivity due to the low concentrations of the most diagnostically valuable volatile compounds. Aside from sensitivity, the sensor's limit of detection (LOD) is another important parameter. LOD is the lowest detectable concentration of the analyte that can be reliably measured by a sensor; it indicates the ability of the device to work reliably at a given minimal concentration. LOD depends on the variability of the background sensor signal. The variability/variance, in turn, is conditioned both by the detection principle and the technological finesse of the device. Let us assume, for instance, that the sensor's output signal is the resistance of a gas sensitive layer. Then the background signal is a certain resistance $(R_{\mathrm{a}})_{\mathrm{i}}$ registered by the transducer (where R stands for resistance, the subscript "a" stands for ambient air, and "i" represents the number of measurement) under conditions when the sensor is placed into an environment not containing the analyte. After multiple measurements it is possible to estimate the variance σ of the acquired results. Then, according to the theory of errors, it is possible to affirm with 99.7% probability that a measurement result within the range of $(R_{\mathrm{a}})_{\mathrm{m}} \pm 3\sigma$, where $(R_{\mathrm{a}})_{\mathrm{m}}$ is an arithmetic mean value of the background signal, will be obtained in the absence of analyte. Then measured values beyond these limits are indicative of the presence of the volatile compound to be analyzed. Concentration of the compound causing a sensor response at the level of $R = (R_{\mathrm{a}})_{\mathrm{m}} \pm 3\sigma \pm \varepsilon$, where ε is an infinitesimal quantity, can be considered to be its detection limit, namely, LOD.

The LOD of the majority of modern solid state chemical sensors, manufactured as a rule on the basis of conventional bulk structures, vary from hundreds of ppb (parts-per-billion) to dozens of ppm (parts-per-million). Their nanostructure analogs have already reached that level of sensitivity and demonstrate a stability and fast enhancement of their performance.[33] Therefore, both types of sensory elements can be efficiently applied to the detection of breath gas compounds.

Here is a simple example. It is known that determination of breath ammonia level can be helpful in fast non-invasive diagnosis and monitoring of kidney and liver disorders, chronic dyspepsia, and gastric ulcers.[1] The breath gas of humans suffering from gastric ulcers contains the products of the vital metabolism of the anaerobe bacteria *Helicobacter pylori*,[15,66] which is now recognized to be a basic etiologic factor of these ulcers. Ammonia produced by *H. pylori* metabolites can be a biomarker of this infection. One should note that the usual ammonia concentration in the breath gas of ulcerous patients is about 2–10 ppm.[67] The concentration of ammonia in the breath of patients with renal failure can reach 1.5–2 ppm.[68] The breath of patients with hepatic encephalopathy has ammonia at a level of 0.5–1 ppm.[69] At the same time breath gas of healthy people may contain hundreds of ppb of this compound.[70] It is obvious that modern sensory devices can detect diagnostically valuable amounts of ammonium in human exhaled air. For instance, the LOD for ammonium of the device described in Ref. 19 is 0.6 ppm. The chemiresistive sensor utilized in Ref. 71 detects sub-ppm concentrations of ammonia. Point heterocontact single walled carbon nanotubes-based sensors have a LOD for ammonia below 120 ppb.[58] It is easily seen that sensors with such specifications can be used for the development of portable devices for diagnostic conditions, where ammonia may be a biomarker.

It is essential to distinguish the sensitivity of the sensor per se in a basic scheme without additional amplifiers and the sensitivity of the setup, which can be potentially achieved by combining the sensors with different ancillary equipment. In the first instance sensitive elements are mostly suitable for the development of portable hand-held devices. In the latter case sensory devices are more appropriate for use in the research environment due to larger dimensions and higher cost. For instance, point-contact single walled carbon nanotube-based sensors[58] in a basic measurement scheme, where a simple multimeter is used for the measurement demonstrate sensitivity to ammonium at levels below 120 ppb. Such point contact sensors may be efficiently used for the development of tools for the diagnosis of conditions that result in elevated of exhaled ammonia. Such a detection device will virtually meet requirements of portability and will be available for mass clinical applications. TCNQ derivatives-based point contact nanosensors for diagnostics of peptic ulcer disease and the detection of volatile products of *H. pylori* activity can serve as an example of this approach.[26,72] The combination of point contact sensor with a point contact spectrometer provides another option of sensors utilization largely suitable for research purposes. Point contact spectrometers have unique sensitivity allowing significantly increasing sensitivity of a sensor device. Combining point-contact sensor with a point-contact spectrometer one can achieve ultra-high sensitivity of the setup, because point-contact spectrometers can register signals as low as 1 nV and the signal-to-noise ratio is excellent. The sensitivity of point-contact sensors[58] can be increased by more than six orders of magnitude when they are combined with a point contact spectrometer, recognizing the fact that the value of voltage bias used for resistance measurement is 1–10 mV.[58] A similar algorithm is used when solid sensitive elements are combined with different modifications of optical spectroscopy; in this case the sensitivity of the setup greatly increases. For instance, quantum cascade laser-based breath sensor platforms employing a quartz-enhanced photoacoustic spectroscopy technique reported in Ref. 73 has a detection sensitivity for exhaled ammonia of

<10 ppbv (part-per-billion by volume) level with 1 s time resolution. Unfortunately, optical equipment that makes use of lasers is often expensive and large.

Sensor selectivity allows the detection of specific certain compounds of a gaseous mixture. This selectivity is of importance in breath analysis due to the very complex nature of the mixtures. The presence of many compounds complicates quantitative and qualitative detection of compounds or products in a homologous series. Sensor selectivity plays a key role when the conventional approach is used to detect single breath volatiles. High selectivity of solid state sensors can be achieved by the special combination of various conductive materials tuned to a desired group of analyte molecules. For instance, in order to provide high selectivity for semiconductor elements it is possible to apply carbon nanotubes blended with different types of sensory polymers.[74]

The reproducibility of sensor signal to identical analyte concentration within its working capacity period indicates the sensor stability. As a rule, high stability results from well-developed technology and accepts that there is a strictly limited period of reliable exploitation of the sensor device until the next calibration. Stability of sensors used for medical diagnostics is essential, as it allows medical staff to focus mainly on the interpretation of the records and optimization of diagnostic and treatment procedures rather than on equipment checking and maintenance.

The definition of response time is the time period for a device to undergo signal (resistance) change from 10% to 90% of the value in the equilibrium state on exposure to an analyte. Response time depends on the approach used for the signal registration, peculiarities of the sensor construction, properties and composition of the medium to be analyzed, temperature, difference of the concentrations of the components in the medium, and the direction of the concentration change (increase or decrease). This definition is acceptable for conventional sensors with a bell-shaped response curve and for so-called indicators (signaling devices) working on the "yes-no" principle. It is considered that the faster signal growth under gas action, the more efficient is the sensor. Response times to various breath gas components may vary from several seconds to several tens of minutes depending on sensor type.[33,71,75,76]

The spectral response type is typical for TCNQ derivative-based point-contact sensors.[26] Therefore, a different efficacy criterion is applied in this case, namely, the behavior of the response curve to the action of a complex gaseous medium like breath gas, resembles a point-contact spectrum. For instance, as it is well known from point-contact spectroscopy[77] that a change of an initial segment of the point-contact spectrum is according to the function $f(x) \sim x^2$ and follows Debye's law,[78] which reflects the density distribution of quantum states of atomic vibrations (quasiparticles) in the crystal lattice. Further, the course of the point-contact spectrum becomes nonmonotonic and depends on the type and amount of quasiparticles (phonons, magnons, etc.) participating in the interaction. Similar behavior is typical of the response curve of TCNQ derivatives-based point-contact sensors.[26] The signal response time dependence has a nonmonotonic view, which may reflect differing interactions of the various breath gas components with the point-contact gas sensitive material (see Section 16.4).

In this case, other time parameters characterize the efficacy of sensor, namely, the time necessary for the point-contact multistructure to start responding to a breath gas and the interval during which the gas sensitive substance could entirely interact with the components of gaseous medium. Such response curve can reflect quantitative and qualitative composition (profile) of exhaled gas and provide much information for its analysis. It is important for the sensor response and time of interaction with the analyzed medium to be appropriate for diagnostic applications. TCNQ derivatives-based point-contact sensors demonstrate fast response to the complex action of breath gas and are able to follow breath profiles in a regimen of one-minute-long exposure.[26]

A similar approach can be applied to a recovery curve. As is known, the sensor recovery time is that which a physical system comprised of a large number of particles requires to achieve thermodynamic equilibrium. When applied to exhaled gas analysis both the sensory element and properties of the gaseous medium to be analyzed must be considered. Therefore, recovery time is determined by the properties of these subsystems and the specificities of their interaction. Recovery time of conventional sensors is only a constrained period between two consequent measurements with a recovery curve of a monotonic character. The speed of the signal change and signal level predetermines the recovery time. The shorter the recovery time and the more monotonic the recovery, the more advanced a sensor is considered to be. Modern sensors applied to breath gas analysis have, for example, recovery times of 10–20 s,[45] 1–5 min,[79,80] and of 2–10 min.[30] For point-contact sensors, the nonmonotonic nature of interaction of the sensor's gas-sensitive matter with the components of the gaseous medium and their subsequent desorption is a more important property of the recovery curve. Such behavior is typical of a spectroscopic type, when the response time depends on the exhaled gas profile, related to metabolic changes in health and disease, which opens new wide possibilities for the development of new diagnostics tools in breath analysis. The duration of the recovery period is also important for performing subsequent measurements. Point-contact sensor recovery times last for 1–3 min depending on the application, which enables them to be used for diagnostic purposes including medical applications and real-time breath analysis.

15.3 TYPES OF SENSORS USED FOR EXHALED GAS ANALYSIS

All the numerous sensor devices may be classified into three groups, namely, physical, chemical, and biological sensors.[81–83] Physical sensors are designed to characterize quantitatively parameters like temperature, pressure, magnetic field, and forces beyond the sphere of chemical studies. Chemical sensors selectively respond to the pre-determined chemical object. Biologically sensitive devices based on immobilized biological materials combine biological components with physical-chemical detectors. They are able to transform the biological activity of those systems into a measurable signal when exposed to certain physical actions or chemical compounds. In this review we first discuss chemical sensors.

In many fields, including the development of modern diagnosis by breath analysis, a variety of chemical sensors find important application, with the transducer (sensitive element) being the principal element of such a device. The design of the sensitive element basically determines the operational characteristics of the whole device. Chemical sensors may be divided into optical, mass-sensitive, electrochemical, and semiconductor types. The operation of optical sensors is based at the registration of optical effects, which develop during interaction with the substance to be analyzed.[84–93] For instance, glass fibers coated by a chemical active substance are used in optical fiber sensors. Fluorescent dyers immobilized in a polymer matrix are often used as an active substance. The changes in the fluorescence spectrum of the dye develop during interaction of the coating with the gas components. The radiation interacts with the chemical coating in a specific way while propagating along the optic fiber. This allows registering changes of spectral lines and detecting substances to be analyzed. For example, the response to CO_2 can be measured by monitoring the absorbance spectrum.[86,88]

There are two types of mass-sensitive sensors used in breath analysis, the quartz crystal microbalance (QCM) and the surface acoustic wave (SAW) device.[46,94–98] QCM is characterized by its own resonant frequency during excitation by an alternating current. The mass of the crystal is one of the parameters defining its resonant frequency. Gas molecules are adsorbed on the crystal's surface during sensor exposure to a gaseous medium, changing its mass and resonant frequency. It is possible to achieve a selective adsorption of the gas mixture components using certain polymers, metal oxides, and other coatings.[99–102] SAW sensors work in a similar way to QCM on the principle that a change in the mass of the piezoelectric sensor coating due to gas absorption results in a change in the resonant frequency on exposure to a vapor. Similarly, the QCM sensor produces a wave that travels through the bulk of the sensor while the SAW device produces a surface wave that travels along the surface of the sensor.

Electrochemical sensors are the most developed and widely used group of chemical sensors that register changes of the electronic properties of the sensitive element depending on interaction between the electrode system and the analyzed compound in breath.[41,75,103–119] A classical version of the electrode system is represented by two electrodes divided by an electrolyte. Different types of electrochemical sensors are distinguished depending on the nature of the output signal, namely potentiometric (indicates the electrode potential), amperometric (the current in the sensory circuit at a certain potential), conductormetric (the electrolyte conductivity), and impedansometric (complex resistance) types. Electrolytes in the electrode system can be liquid, solid, and gelatinized. Selectivity is achieved both by means of constructive approaches (specially selected electrode materials, membranes) and by using different detection regimens (range and type of the applied potential changes).

A wide variety of sensor devices are based on semiconductor characteristics changing due to absorption or adsorption of analyte molecules onto the surface of a transducer (semiconductor electrochemical sensitive element). This type of sensor is easy to produce because modern screen-printing or chemical vapor deposition techniques can be used. Semiconductor sensors have found wide application in breath analysis[25,37,120–127] because of their high sensitivity, reversibility of the processes involved, durability of operation, low cost, small dimensions, and simplicity of use.

Metal-oxide-semiconductor field-effect transistor (FET)-based sensitive elements, simultaneously registering the process of interaction with analyte and enhancing the response signal are another type of semiconductor sensor.[47,128,129] Chemical reactions of VOCs in the active zone of the sensor accelerated by catalytically active metals are the basis of FET sensor operation. Mobile products of reaction, diffusing through the FET sensor gate, change the electrical parameters of the transistor, which results in the generation of an enhanced analytical signal.

The mechanism of signal formation in conventional semiconductor sensors also works reliably at the level of single atoms and molecules; therefore it is exploited in the majority of the nanostructure sensor elements.

Taking this into account it seems reasonable to focus on the fundamental principles of the semiconductor sensors, which is crucial to an understanding of their application to breath analysis. To illustrate the working principle of these devices we focus on the example of a gas sensor in contact with air and whose operation is based on the variability of conductivity (Figure 15.1).

Oxygen molecules of the room air interact with the conducting sample surface and are adsorbed on it. This process causes a modification of a superficial layer in the sample and changes its conductive properties. The subsequent action of an analyte gas can lead to a reaction resulting in a full or partial replacement of absorbed oxygen by the analyte gas. Therefore, there is a redistribution of the density of electron states in the superficial layer and the local electrical conductivity varies proportionally to the amount of analyte gas.

We can evaluate the variation in conductivity of the sensor as it interacts with the gaseous agent. It is known that the electric resistance of an homogeneous conductor

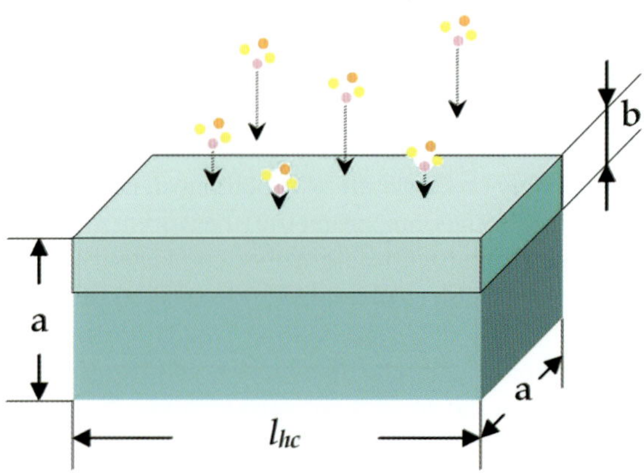

FIGURE 15.1
Principle of operation of conductive gas sensor. l_{hc}—length of homogeneous conductor (hc); $S = a^2$—cross-section area; b—the depth of gas action.

(hc) can be described by

$$R_0 = \rho l_{hc}/S, \tag{15.1}$$

where S is the cross-sectional area of the conductor, l_{hc} its length, and ρ the resistivity of the material. Assume that the conductor is a parallelepiped of square cross-section S with a side equal to a. Interaction of the conductor surface with an external agent can result in a change of the cross-section area S through which electric current flows. If b is the depth through which the gas interacts, then $S = a(a - b)$ and the conductor resistance will be equal to

$$R_{int} = \rho l_{hc}/S = \rho l_{hc}/a(a - b) \tag{15.2}$$

The resistance ratio before and after interaction is

$$R_0/R_{int} = 1 - b/a \tag{15.3}$$

We can recognize some limiting cases from the above Eq. (15.3). Firstly, if $b \to a$, the variation of the conductor resistance greatly changes, i.e. $R_{int} \gg R_0$. In this case the response of the sensitive element to the action of the external agent should be observable experimentally. Secondly, if $a \gg b$ we have $R_0/R_{int} \approx 1$ and it will be difficult to observe the response of the conductor to the external action. Thus, the conductive sample will respond to the action of an external agent if the thickness of the superficial layer, whose conductivity varies after the interaction, is close to or equal to the cross-sectional area of the sample, i.e. correlation $b \to a$ is realized.

In the case of gas media analysis the highest sensitivity can be realized for thin samples because the gas action is not quite so deep. Actually, nanostructures sensors have a greater prospect as supersensitive devices for analysis of breath constituents present at extremely low concentration. For instance, atom-sized point-contact nanosensors, operating on a sensitivity quantum edge, demonstrate superb performance.[56] The interaction of an atom-sized contact with only one molecule of gas results in a resistivity change at least as high as $12.9/n$ kΩ, n being the number of atoms in the contact. Such a value is easily registered by the simplest measuring instruments.

Development of technologies that exploit recent advances in the areas of physics, chemistry, biology, and medicine makes growing demands on the analytical techniques used for environmental and processing monitoring. Therefore, greater attention is given to the creation of gas sensors with good metrological characteristics. Traditional chemical sensors provide good results and to some extent comply with performance specifications required for monitoring. Methods for their manufacture are well established and the devices are debugged and thoroughly tested, enabling different companies to produce large batches of reliable devices. It should be noted that less and less reserve remains in the development of conventional chemical sensors, which now verge on their technical frontiers. Therefore, the main aspirations and expectations of developers of sensor devices are directed toward investigation of nano-objects and their potential applications.

A wide spectrum of nanosensors is known. Nanotubes, nanorods, nanobelts, nanowires, point contacts-based sensors are among the most actively investigated.[58,74,130]

The operational principles of the majority of nanosensors are based on the same patterns as conventional semiconductor sensors. The advantages of nanosensors are largely due to high surface-to-volume ratio, porosity, small diameter, and the reactions between the oxygen species O^{2-}, O^{-}, and O_2^{-} on the surface and samples. Carbon, oxides of metals, conductive organic and inorganic polymers, different semiconductors, and synthetic metals are traditionally used for nanosensor production. Those materials are able to efficiently adsorb gaseous analytes resulting in changing electronic properties under action of the gaseous media, which is of obvious importance for breath analysis.[45,125,131–135] Substances used for these purposes can demonstrate specific catalytic activity accelerating chemical reactions between VOCs in the active sensor's zone.

15.4 DETECTION PRINCIPLES AND CONCEPTS INVOLVED IN BREATH ANALYSIS USING SENSORS

Any intervention changes a system under study, breaking stable bonds and leading to an imbalance in a concerted subsystem. A so-called active experiment, when a system under study itself becomes an arena of an investigation process generating results aggravated by an unavoidable inaccuracy. Such inaccuracies are a result of the fact that analogs of a target object but not a target itself is investigated. Precision of interpretation of such experiment's results depends on the depth of investigative interventions. The more complex a system is, the stricter such a relation is. In this regard, data about the state of a living organism, obtained non-invasively, are most reliable and failsafe. That is why it is not surprising that medical diagnostics based on breath analysis are making rapid progress during the last decades. Breath is a valuable biological material because its constituents effectively reflect the metabolic profile of the human organism. Any deviation of this profile in multidimensional concentration space can provide evidence of the pathology of the organism and for the treatment of adverse clinical states.

During the development of breath diagnostic methods two aspects are very important, namely, technical and medical. According to the technical aspect it is necessary to obtain a reliable picture of the metabolic profile. The medical aspect is aimed at an adequate analysis of this profile. Considering the technical aspect it is necessary to point out two alternative paradigms which are used to obtain the metabolic profile. The first paradigm is based on an additive synthesis of the metabolic profile from breath gas components. In the framework of this approach the concept of disease markers exists. According to the second paradigm a metabolic profile is a quite complicated function both of the nature of breath gas constituents and the correlation of their concentrations. Researchers who support this paradigm consider that the most correct analysis is to obtain a sensor image of the metabolic profile as a multicharacteristic response curve which results from the direct action of breath on a special integrant sensor. In this case reliable diagnostic conclusions can be made by revealing correlations between characteristic parameters of the sensor image and results of conventional medical investigations. It should be noted that the concept of the integral sensor image removes the problem of the highly selective analysis of breath gas. Instead of this, the problem of nature and a structural organization of the gas-sensitive layer appears. The latter should provide a representation of a projection of the multidimensional concentration figure on the coordinate plate "electric signal-time."

"Electronic nose" analyzers are multisensor systems which consist of a set of relatively nonselective sensory elements.[61,63,136] The problem is that response to the action of the every gas is strictly individual. Substance recognition is performed after device training. The training procedure consists of recording the sensor array responses to various specific gaseous media. The library of responses obtained is stored in the device memory. The device indicates the presence of the corresponding gaseous component or a group of components if a similar response or response combination is found in the library. Such an analysis supposes an additive formation of the metabolic profile and therefore it works in the framework of the concept of markers. At the same time a multisensor system "electronic nose" allows an integral sensor image of the metabolic profile to be obtained, i.e. it can be an instrument for the realization of the second paradigm as referred to in the previous paragraph.

The main problem when working in the framework of both the first and the second approach (paradigm) is that immediately after exhalation there are multiple interactions among the constituents of a complicated medium like exhaled breath. Also, the interaction times for different components of breath with the sensor can essentially vary. It is quite natural that the breath mixture can undergo transformation on its way from its source to its exhalation into atmosphere. Additionally, it should be taken into consideration that the interaction rate can be much higher because part of reactions can happen by autocatalysis, when the output products are able to serve as catalyst for further processes. Such effects may have an impact on data of measurements, especially in the example when determining particular components of the complex gas mixture. In this case a direct record of the integral characteristics of a complicated object under study makes it possible to obtain more reproducible results.

Taking into account above statements one could consider an original approach to the investigation and analysis of complex gas mixtures like breath. This approach is based on the utilization of the point-contact multistructure.[72] This suggests an effective approach to breath investigations. We applied this approach to breath analysis using TCNQ derivatives-based point-contact sensors.[26,137] Using a combination of the modified technique for sensor analysis of composite gas mixture and the unique properties of the gas-sensitive point-contact matrix, a complex dynamic of the interaction between sensor material and volatile compounds of exhaled air has been observed. A gas sensitive continuum of a mesoscopic point-contact multistructure of TCNQ salts is a set of a large number of resistant elements, namely point contacts,[57] formed by the contacts of the lateral surfaces of the needle-shaped crystals of the TCNQ salt. The point contacts of this structure correspond to the contacts obtained by the displacement technique[138] known to be one of the most efficient and reliable methods in point contact spectroscopy.[57] The resistance of each of these elements changes in the presence of gases because of the point contact gas-sensitive effect discovered by our group.[56,58] An assembly of point-contact sensitive elements connected to each other and located between the electric current feeding plates of a sensor forms a mesoscopic point-contact multistructure, which provides significant enhancement of an output signal and an increased selectivity to the gases under analysis.

Usually the response curve of a chemical sensor in contact with gaseous media is bell-shaped, and consists of two monotonic segments, an initial segment corresponding to the period of exposition characterized by a monotonic growth of the registered signal up to its maximum value, while after the end of gas exposure there is a monotonic decrease of response signal corresponding to the relaxation period.[18,21,139] The TCNQ-based point-contact sensors demonstrate a more complicated structure of the response curve on exposure to the breath gas.[26,72] The typical dependence of voltage drop (or resistance change) on such a sensor with time during the exposure period for healthy volunteers (Figure 15.2) is characterized by (i) the presence of a maximum on the response curve and (ii) saturation of the response signal. The response curve area may indicate that the sensor has completely interacted with the exhaled air components and that the sensitive substance transits to a quasi-equilibrium state. Additionally, the

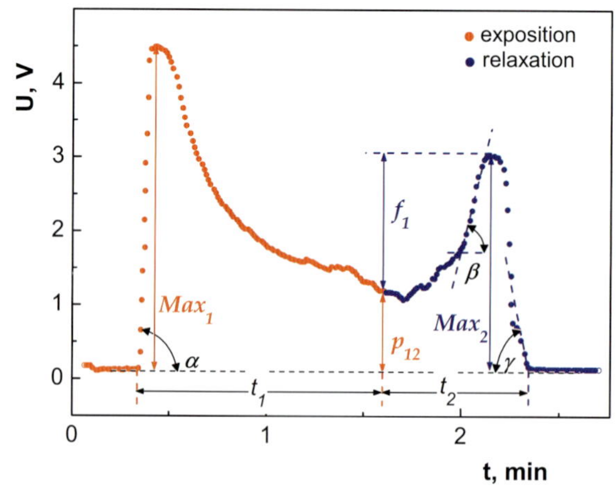

FIGURE 15.2

Characteristic view of a TCNQ compound-based point-contact multistructure sensor response curve. *U*—sensor voltage, *t*—time. Parameters of the sensor response curve: absolute value of the exposition maximum (Max_1); absolute value of the relaxation maximum (Max_2); ordinate of the final segment of the exposition phase (p_{12}); ratio between the height of relaxation maximum above the exposition signal's final level and its absolute value [$(Max_1 - p_{12})/Max_1$]; relation of the height of the relaxation maximum above the level of the relaxation to the absolute value of this maximum (f_1/Max_2); slope of the initial segment of the exposition phase ($x = \tan\alpha$); slope of the initial segment of the relaxation phase ($y = \tan\beta$); slope of the end segment of the relaxation phase ($z = \tan\gamma$); difference between the maxima of the speed of velocities change at the beginning and at the end of exposition ($g = x - z$); time of exposition (t_1); time of relaxation (t_2). Data taken from Ref. 26.

appearance of a maximum in the relaxation curve has been registered after the end of exposure to exhaled air.

Prior to this work, such a behavioral pattern of the response curves had been unknown in sensor engineering. Due to the complex non-monotonic structure the response curves of the new type point-contact sensors are much more informative than those of traditional and nanostructure chemical sensors. Registration of breath profile by point-contact multistructures allows simple diagnostics without the need to determine particular components of breath gas. The proposed method should be very effective because it bases on an advanced approach to solutions of analytical problems by means of point-contact spectroscopy.[57] This is basically the opposite of the conventional approach of determining separate biomarkers, which is widely being exploited now. In this way, registration of point-contact sensor responses to the action of human breath together with further statistical analysis of the obtained profile can provide a fine and easy-to-use method for non-invasive medical diagnosis, which is similar to the procedure of electrocardiogram interpretation.

Implementation of the original characteristics of TCNQ-based point-contact sensors has already resulted in interesting output during medical research on human breath analysis, which is useful for the development of non-invasive diagnosis of gastric diseases.[26] It was ascertained that the sensor response depends on the variations in the exhaled breath of different people and differs for patients having peptic ulcers (Figure 15.3). Thus, breath-gas sensors based on point contacts of TCNQ salts were able to detect typical resistive behavior for patients having this pathology.

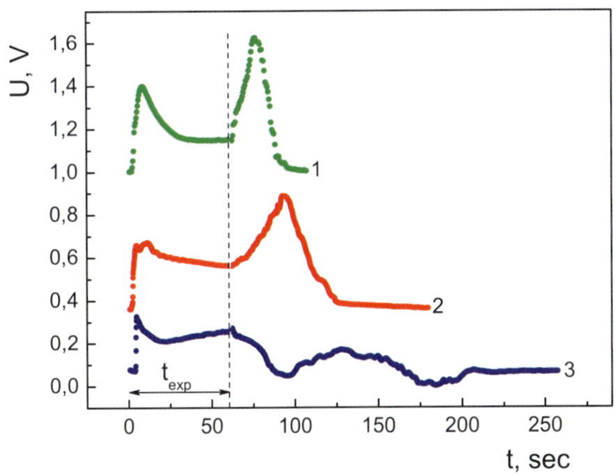

FIGURE 15.3

Response curves of TCNQ-based point-contact sensors due to the action of breath gas of different study participants. Curve 1—healthy volunteer without *Helicobacter pylori* infection; curve 2—*Helicobacter pylori*-positive patient with erosive changes of gastric mucosa; curve 3—*Helicobacter pylori*-positive patient with a bleeding gastric ulcer. *U*—sensor voltage, *t*—time, t_{exp}—time of exposure.

As an example of a potential clinical application of the proposed sensor approach to breath analysis we demonstrate some of the results of a phase I clinical trial, conducted in Ukraine. We estimated the values of relaxation time t_2 and the ratio Max_1/Max_2 in adolescents with different endoscopy results. We found that the mean relaxation time was significantly longer in subjects with a severe gastric and duodenal inflammation characterized by presence of ulcerative and erosive damage of mucous membrane compared to a healthy control group ($p<0.05$). The exhaled air of patients with milder forms of the disease and of the control group caused faster sensor relaxation after their interaction. It is known that chronic active inflammation in the gastric mucosa involves several interleukins (IL-8, IL-10, and IFN-gamma), known as immunological markers of the blood serum.[140] Hence, it is possible that extensive gastric inflammation would lead to an increased release of some of the volatile inflammatory biomarkers in breath causing differences in the sensor response.

15.5 MEDICAL APPLICATIONS OF SENSORY BREATH ANALYSIS

15.5.1 Compounds-specific sensors

The most widely used sensors are for breath alcohol detection from which blood alcohol levels can be obtained. A large variety of sensor devices have been developed to detect intoxicated drivers.[38,39,42,141–143] Measuring ethanol by means of electrochemical-sensor-based portable analyzer is recommended during endo-urological surgical interventions both during spontaneous breathing and in mechanically ventilated patients undergoing inhalation anaesthesia, in order to detect and quantify the irrigation fluid absorbed.[144,145]

The role of the fractional exhaled nitric oxide (F_ENO) as a diagnostic and management tool for asthma and other inflammatory airways diseases is increasingly being recognized by national and international professional bodies. Guidelines already include recommendations for the measurement of exhaled NO for monitoring airway inflammation by GINA (Global Initiative for Asthma) and the National Asthma Expert Panel, part of the National Heart Lung and Blood Institute (NHLBI, US) (also British, Dutch, Finnish, Spanish, Swedish). The guidelines issued by the American Thoracic Society and the European Respiratory Society (ATS/ERS 2005) provide guidance as to how exhaled NO measurements should be carried out and include such parameters as the exhalation flow rate (50 mL/s), the correct exhalation time, and the method for analysis of the NO content.[146] Stationary chemiluminescence analyzers are the "gold standard" technique for F_ENO measurement in the market today. Although excellent for specialist and research applications they have several drawbacks, especially investment cost, user friendliness, transportability, requirements for frequent calibration, and technical maintenance making them unsuitable in the primary health care setting and for individual use.[28] However, sensor devices have a large potential for the clinical routine management of inflammatory airway diseases.

F_ENO sensors are among the most numerous representatives of the compounds-specific sensors for exhaled gas analysis. Elevated levels of F_ENO are described by sensor breath analysis in patients with variety of conditions followed by airway inflammation, including asthma,[147–149] untreated bronchiectasis, seasonal and allergic rhinitis,[105,147,150,151] viral respiratory illnesses, ciliary diskinesia, cystic fibrosis, and chronic obstructive pulmonary disease (COPD).[28] Levels of exhaled NO are reduced in smokers, thus F_ENO sensitive sensors can be used to monitor smoking behavior[152,153] and the efficacy of corticosteroid treatment.

In the clinical environment F_ENO measurements are performed by various handheld electrochemical sensors miniaturized but with preserved performance, which simplifies the measurement and yet adheres to ATS/ERS standards.[28,113,116,118,150,154–156] Specially developed sampling and analysis technology allows compensation for the relatively slow response properties of these electrochemical sensors and achieve clinically acceptable agreement between the stationary and portable systems measurement.[147–160] Some of the handheld NO analyzers are suitable for nasal F_ENO measurements[105,151,161] and may be useful in differentiating hyperplasic eosinophil rhinosinusitis from chronic unspecific rhinosinusitis. Moreover, nasal F_ENO may be used to monitor the clinical course of chronic rhinosinusitis with polyps.[151] Electronic noses can also be used to discriminate asthma patients from healthy control groups, although the accuracy of classification of the severity of the disease is less reliable.[61,162]

However, despite the undeniable benefits of NO measurements in the clinical setting, NO sensor devices are not so widely used as might be expected. This is mostly due to some limitations, such as the following:[28]

- the high cost;[163]
- the dimension of the devices;
- technical maintenance, including calibration;
- reproducibility. It has been reported that the absolute reproducibility of exhaled nitric oxide measurements is poor, but acceptable when the values are categorized as low, normal, intermediate, or high. One measurement is therefore sufficient when using categorizing exhaled nitric oxide values; but a mean of at least two measurements is required for absolute values.[113,164]

Hydrogen breath tests are widely used to explore the pathophysiology of functional gastrointestinal disorders, this being one of the oldest sensory breath tests approved by the US Food and Drug Administration (FDA). The peak of research activity in this subject was at the end of the last century (2000). At that time, breath hydrogen (H_2) analysis, as used in gastroenterologic function tests, required a stationary analysis system equipped with a gas chromatograph or an electrochemical sensor cell. Therefore, much effort has been directed to the development and validation of portable devices.[110,111,165] As a result, considerably cheaper pocket-sized industrial breath H_2-analyzers have been introduced into the clinical environment[31,110,111,165–169] for the detection of lactose intolerance, oro-caecal, and duodeno-caecal transit times of different carbohydrate test meals, carbohydrate malabsorbtion detection, carbohydrate breakdown deficiency detection, bacterial overgrowth diagnosis, evaluation of colonic flora, small intestinal bacterial overgrowth, and irritable bowel syndrome.[170,171] Separate studies recommend performing of breath hydrogen analysis in premature infants as a diagnostic tool for necrotizing enterocolitis.[172] Modern tendencies in this area are focused at development of sensory tools with higher sensitivity which are able to detect H_2 in human breath samples which coexist with various gases such as methane (CH_4) and carbon monoxide (CO)[19] being also at a different operating principle. More recently thermoelectric hydrogen sensors have been proposed for breath hydrogen analysis.[31] Portable CH_4 analyzers are of diagnostic importance in patients with malabsorption and bacterial overgrowth.[167]

Detecting exhaled carbon dioxide (CO_2) in exhaled breath is an essential aspect of clinical medicine and surgery. This has stimulated the development and clinical implementation of several capnography sensors.[3,173] Monitoring of respiratory carbon dioxide indicates how effectively this respiratory gas is eliminated from the pulmonary system and how effectively it is transported through the vascular system. Knowing the levels of exhaled CO_2 can help in the assessment of disease severity and the response to treatment. CO_2 monitoring is of importance in anaesthetized patients whose respiratory activity may be repressed by anaesthetics and other medications resulting in an elevated CO_2 concentration in blood (hypercapnia). Nowadays, capnography is a routine procedure in emergency medicine and intensive care units.[174] Mass spectrometry and infrared (IR) optical technique are considered to be the gold standards for respiratory gas analysis, but as mentioned before, they have some disadvantages such as high cost and size, etc. The new breath sensor solutions overcome these limitations.[24,175–179]

Possible influences of the exhaled gas humidity and time of sensor response are amongst the drawbacks of sensor CO_2 monitoring. Humidity effects may be avoided by using CO_2 selective sensor,[88,180] by complete removing humidity from the exhaled breath before it reaches the sensor, for example, using a molecular sieve, or by using a reference sensor that does not respond to CO_2 but responds to humidity. Fast CO_2 changes during breathing may be registered by sensors with short response time (<1 s), or by means of mathematical models which can predict the fast CO_2 input based on the sensor's slow output.[181] A prospective application of carbon dioxide sensors is breath sampling control for medical applications to provide breath samples of well-defined and reproducible composition.[182,183]

Acetone is a ketone body produced by the liver during fatty-acid metabolism. Together with other ketones it is used as an energy source when glucose is not readily available. The breath acetone concentration slowly increases in diabetes or the fasting state.[9] Exhaled acetone concentration varies in healthy people and on an average does not exceed 1 ppm.[184,185] The level of acetone in the breath gas of insulin-dependent patients with diabetes mellitus may significantly elevate compared to healthy controls and exceed the level of 1.8 ppm.[186] For this reason, breath acetone may be a useful candidate as a breath marker in diabetes. In fact, it has been studied extensively and shown to be correlated with blood glucose.[187–189] However, a large number of physiological and pathophysiological conditions, influencing acetone production, including biological variability and diet,[190] that disqualify acetone alone as a specific marker of blood glucose and the assessment of the degree of ketoacidosis. Thus diabetes monitoring using breath analysis would be more reliable, if the profile of other volatile biomarkers in exhaled breath is taken into account.[191] Nevertheless, some effort is currently being focused at development of portable sensors for acetone detection.[35,36,45,134,192–194]

Ammonia is a major constituent of exhaled breath and a product of protein catabolism.[185] Ammonia is released during decomposition of amino acids mostly in the gut and is then converted to urea by the liver. Exhaled ammonia has attracted attention of breath analysis specialists due to the possibility of its application to haemodialysis monitoring,[7,68] diagnosis of hepatic encephalopathy,[7,69] detection of *H. pylori* infection,[7,67,195] and the investigation of halitosis.[7,196] Significant elevation of exhaled breath ammonia after protein-rich meals has been seen.[197] Ammonia in the oral cavity may be useful for assessing halitosis and oral hygiene; as one of the well-established products of protein decay and protein degradation, methyl mercaptan has been shown to be correlated with ammonia.[196] The normal range for human breath ammonia varies from 50 to 2000 ppb;[7] therefore efficient analytical techniques for breath ammonia detection must have a limit of detection of several tens of ppb.[198]

The QCM[23] and the liquid-film conductivity sensors[199] are among chemical sensors that are proposed as potential breath ammonia monitors. The QCM sensor depicted in Ref. 23 is capable of measuring breath ammonia concentrations as low as 0.1 ppm up to 10 ppm with an accuracy of ± 0.1 ppm and the liquid-film conductivity sensor has a LOD as low as 18 ppb.[199] Electrochemical and semiconductor portable ammonium detectors are being recommended and utilized in the clinical setting for diagnosing *H. pylori* infection.[71,75,166,200,201] However, recent publications report a low specificity and sensitivity for these methods,[202] probably due to the fact that large amounts of ammonia are being produced locally in the mouth cavity, in particular, in halitosis patients. Therefore, the gold standard ^{13}C breath test is currently recognized as more reliable, but also being a more costly tool for diagnosing *H. pylori* infection.

The detection of carbon oxide (CO) in exhaled breath by sensors is of interest for monitoring neonatal jaundice, the rate of heamolysis, identification of aplastic crises, indirectly assessing the rate of vaso-occlusion and sequestration in sickle cell disease, and CO intoxication.[109,203]

15.5.2 Sensors for breath profiling

The efforts of many research groups are focused on revealing exhaled biomarkers of various cancers in order to develop non-invasive methods for their early detection.[204–212] This is complicated for many reasons; in particular, since the metabolism of cancer cells can greatly alter during their lifetime, and many other diseases can alter the breath composition. Taking into account the role of oxidative stress and other factors, the joint efforts of the researchers have been directed toward the identification of volatile organic compounds biomarkers. Despite the joint efforts of scientists in this field, specific biomarkers of certain tumors has not yet been exploited in the clinical setting and at the moment breath-based diagnosis of cancer can only be performed only on the basis of the profile of the exhaled breath. Therefore, breath sensor development for this purpose is being developed separately from the fundamental approaches and the concepts of electronic noses and sensor arrays are been offered.

Most of the electronic nose studies are focused on differentiating between breath biomarker profiles of cancer patients, patients with other pulmonary nonmalignant diseases, smokers and healthy people (proof-of-principle studies). Di Natale et al. were among the first researchers to demonstrate these possibilities from a study of post-operative patients with lung cancer using sensor array technology.[213] Similar results were obtained for patients with bronchogenic carcinoma.[214] In the latter study, participants were classified into patients who had undergone lung cancer resection and a non-cancer group, the breath profile data being obtained by another type of electronic nose comprising 32 separate polymer composite sensors. The authors concluded that the exhaled breath of patients with lung cancer has specific peculiarities that allow the identification of malignant disease using the electronic nose. Several other proof-of-principle studies using sensor arrays have demonstrated the possibility of distinguishing between different types of lung cancer at different stages.[215–218] It is, however, important to note that some of the subjects in the above studies were diagnosed with advanced disease. Hence, it is imperative that future studies should be targeted at diagnosing patients who are in the early stages of the disease where they are still amenable to curative surgery.

Since the development of lung cancer is much more frequent in COPD patients and smokers than in relatively healthy individuals, attention should be directed toward the differences in exhaled biomarker profiles between these patient/volunteer groups. Therefore, another focus of electronic nose research is the differentiation of pulmonary malignancies from other nonmalignant conditions such as COPD, bronchial asthma, pneumonia, and smoking that are commonly followed by chronic inflammation and oxidative stress and which can change breath profiles in a way that can mislead breath diagnosis of cancer.[218–220] Actually, the neoplastic process itself also induces a pro-inflammatory state and hence contributes to the oxidative stress burden.[221] Dragonieri et al. showed that breath VOC-profiles of non-small cell lung cancer (NSCLC) patients obtained using an electronic nose clustered distinctly from those of COPD subjects and from healthy controls. Similar promising results were obtained by D'Amico et al.[220] where patients with NSCLC, other lung diseases, and healthy controls were correctly differentiated by analyzing breath profiles by a novel artificial olfaction system. Another study[218] using an electronic nose device based on an FET and resistive sensors compared breath profiles of patients with bronchial asthma, COPD, pneumonia, lung cancer, and healthy controls. The highest sensitivity and specificity of the method was obtained for pneumonia—90% and 98%, but for lung cancer—74% and 95%, respectively. The authors concluded that the electronic nose device is able to discriminate among different lung diseases with sufficiently good accuracy. The possibility of differential diagnosis of obstructive airway diseases such as COPD and asthma and risk assessment in asymptomatic smokers using electronic nose molecular profiling of exhaled air was shown in the study of Fens.[222]

Dragonieri and coauthors found evidence to distinguish breath profiles of patients with severe and mild asthma from those of nonsmoking control subjects.[223]

A series of electronic nose studies has been conducted by the group of Haick and co-workers.[47,60,79,80,224,225] Most of the diagnostically valuable results were obtained using arrays of nanosensors based on assemblies of 5-nm gold nanoparticles coated with different organic compounds that are sensitive to various organic vapors including those of alcohols, benzenes, and alkenes.[226] Using a combination of solid phase microextraction-gas chromatography/mass spectrometry, VOCs that appear in the breath of NSCLC patients and in the control media were identified. Such data were used for the creation and training of highly sensitive, cross-reactive arrays of sensors based on different chemiresistors.[79,80] Using prepared samples in combination with a pattern recognition method they were able to recognize VOCs patterns emitted from different types of NSCLC cells and to distinguish the breath of lung cancer patients from the breath of healthy individuals. The nanosensor array was tested to see if it was possible to differentiate various cancers—lung, breast, colorectal, prostate; between primary cancers vs. metastases, and if it was possible to use this approach to check the efficacy of therapy and the response to treatment, and the detection of earlier recurrence in an annual routine survey.[225] Patients with benign breast conditions that are potential breast cancer precursors were differentiated from women with negative mammography and women with malignant lesions using electronic nose breath profiling.[60]

Proof-of-principle studies have been carried out to evaluate the possible application of an analyzing system based on sensor array to diabetes diagnosis,[227] renal diseases,[228] and a wider circle of pathologies[229] and pathogens.[230,231] Guo and coauthors tested a new specific breath analysis system and showed that it was able to distinguish breath samples from subjects suffering from diabetes, renal diseases, and airways inflammation from those provided by healthy controls.[229] Promising results were obtained in monitoring the progress of related therapies, especially in evaluating the efficacy of haemodialysis.

Hanson and coauthors performed an electronic nose study with the participation of mechanically ventilated patients in a surgical intensive care unit to diagnose ventilator-associated pneumonia.[232–234] They demonstrated that the electronic nose was able to discriminate between two groups of patients with different severity of inflammation. The data obtained by the electronic nose correlated well with a clinical pneumonia severity score and therefore could serve as a diagnostic tool for the prediction and diagnosis of ventilator-associated pneumonia and its severity. Similar electronic nose applications were investigated by Kai and coauthors who had already developed and tested a portable electronic nose system for home healthcare applications.[235] This system may distinguish between pneumonia and pneumoconiosis and could assist in quick monitoring of the course of respiratory diseases at home by family members. Another electronic nose was tested on patients in intensive-care units with different bacterial infections of their respiratory airways. Six different pathogens were differentiated, namely, *Pseudomonas aeruginosa*, *Acinetobacter baumannii*, *Klebsiella pneumoniae*, *Staphylococcus aureus*, and *Acinetobacter lwoffii*.[230]

A large number of electronic nose studies are devoted to finding sets of biomarkers typical of COPD, which is a common condition observed in chronic smokers and a concomitant background disease of lung cancer patients. Smoking is one of the essential COPD causative agents, and thus smokers with chronic lung inflammation often belong to a high cancer risk group. The early stages of malignant diseases in this case may be hidden by the previously existing background of chronic inflammatory lung disease. Despite that, there are some similarities in the clinical manifestations of COPD and other nonmalignant lung diseases, such as asthma, bronchiectasis, interstitial lung disease, and pneumonia, which present difficulties in differentiating these conditions in routine medical care. Therefore, describing a pattern of exhaled gas characteristic for COPD is a step toward the development of a lung cancer screening tool and other electronic nose applications.

Understanding the pathophysiology of the disease can assist in suggesting what might be present in COPD breath profiles. As the disease is a typical nonspecific inflammatory condition of respiratory airways, the exhaled gas of such patients would not contain specific volatiles, like, for instance, the by-products of bacterial activity in infectious diseases, the metabolites of specifically activated immune system in asthma and allergy, or the specific atypical cells metabolites in lung cancer. There is more than one morphological phenotype in COPD, making the description of a characteristic VOC pattern even more complicated. A COPD breath print is rather a cluster of the biomarkers, overlapping with those same compounds, but at different levels, seen for smokers, cancer, and other lung disease patients. That is clearly exemplified by F_ENO measurements: elevated F_ENO is observed in all of the aforementioned conditions (except for smoking for which F_ENO is decreased), but the F_ENO level change varies depending on the disease type and severity.[236] As mentioned above, Dragonieri and coauthors demonstrated in a cross-sectional study that an electronic nose can discriminate between patterns of the exhaled breath of patients with lung cancer from COPD patients and healthy controls.[219] These investigations were extended by Fens and coauthors.[222] They have shown that the electronic nose can differentiate breath profiles of patients with COPD and patients with asthma and discriminate them from asymptomatic smoking and nonsmoking controls. Capuano and coauthors contributed to the characterization of the exhaled breath profiles of patients with the different COPD phenotypes.[30] Quartz microbalance gas sensor arrays have been used to discriminate between COPD patients and to observe the efficacy of corticosteroid treatment.[30] These promising preliminary results could be prerequisites to investigate the performance of the electronic nose technology in clinical settings.[237]

Thus, the use of electronic noses for breath analysis offers several potential advantages but they have disadvantages as well. Their strengths include high sensitivity, ease of use, and portability. Their limitations include loss of sensitivity in the presence of water vapor or high concentrations of a single component, sensor drift and the difficulties in providing absolute calibration, the relatively short life of some sensors, necessity to do considerable method development work for each specific application, and the inability to obtain quantitative data. Nevertheless, electronic nose technology does appear to have provided very encouraging results from the various studies performed. New developments in statistical approaches and data sharing between electronic noses can facilitate this research area.[212]

Other sensors providing complex information about exhaled breath and the organism's state are the novel type point-contact nanosensors with a sensitive layer based on mesoscopic point-contact multistructure.[26,72] The origins of the different segments of the response curve of these sensors (see, for instance, Figures 15.2 and 15.3) are conditioned by the action of different components of the breath sample; therefore, the separate sectors of the output signal have different physical-chemical and, correspondingly, medical-physiological origins. This new theoretical concept had been confirmed by the results of a clinical sensor study, conducted in Ukraine, involving the participation of 116 adolescent patients suffering from gastric and duodenal inflammation of different severity and 30 healthy controls. Patients with functional dyspepsia, chronic gastritis, gastroduodenitis, gastric and duodenal ulcer were enrolled. Profound inflammation with destruction of the gut mucosa was endoscopically diagnosed and histologically confirmed in 20 subjects; the rest of the patient cohort had moderate and mild superficial (catarrhal) alterations of the stomach and duodenum. All patients had their blood analyzed for the presence of *H. pylori* infection and underwent pH monitoring. Healthy controls were tested for the presence of *H. pylori* only; gastroduodenal disease was excluded basing on the results of interview and physical examination.

The response of the new point-contact multistructure sensors was registered during one-minute long exposure to the exhaled breath of the participants. After exposure had ceased, the sensors recovered to their initial state in ambient air and a short while later the signal registration was discontinued. The ratio of exposure and recovery maxima, the duration of exposure time and the relaxation time of the sensor during the recovery period were estimated. It was seen that the mean recovery time was significantly longer against the background of severe gastric and duodenal inflammation. Namely, patients with more severe gastritis and duodenitis, characterized by presence of mucosal destruction in view of erosive and ulcerative changes, had longer recovery time of sensors compared to patients with milder disease forms, presented by superficial gastritis and functional dyspepsia (1.7 ± 0.3 and 1.1 ± 0.04, correspondingly, $p < 0.05$). Conversely, the sensors recovered almost twice as fast under the action of breath gas of patients with mild forms of the disease and of controls. The observed effect could be due to the action of the products of mucosal destruction on the sensor's sensitive layer, forming more stable bounds and the consequential slower desorption from the surface.

Another parameter of the sensor output, namely the ratio of the exposure to recovery maxima (Max_1/Max_2), correlated well with the level of fasting gastric pH, estimated in ranks, and scored from 1 to 20 from the total severe hypochlorhydria to subtotal moderate hypochlorhydria ($r = -0.38$; $p < 0.05$) according to the following equation:[238]

$$\text{Fasting gastric pH (ranked)} = 14.75 - 1.33(\text{Max}_1/\text{Max}_2)$$

The Max_1/Max_2 ratio was not influenced by different *H. pylori* status of the patients.

The nonsingular clinical correlations of the TCNQ derivatives-based point-contact sensor output signal illustrates the complicated nature of the sensor's sensitive layer interaction with breath gas, leading to appearance of the information-rich spectra, which may provide more clinically relevant information than conventional gas sensors and provide data similar to breath profiling. The phenomenon needs further investigation in different groups of patients and laboratory testing with different VOCs. The new point-contact sensors could serve as a useful tool for non-invasive diagnosis and monitoring of destructive changes of gastric and duodenal mucosa and the level of fasting gastric pH in primary care.

15.6 CONCLUDING REMARKS

Based on this literature review, sensor analysis of exhaled gas is now in its developmental stage characterized by intensive studies and the creation of experimental prototype devices. An essential and indisputable advantage of the sensor approach at the current stage of development and testing is the provision of large amounts of statistical data which can provide development of relevant criteria for a reliable evaluation of the obtained results. The currently available devices still require comprehensive testing and further updating. Practical experience shows that even generally recognized FDA procedures for registering devices does not guarantee unambiguous breath analysis in a large group of patients with the same pathology. This is clearly exemplified by one of the most advanced and widely used sensors for nitric oxide detection.[28,164] The complexity of breath gas composition, differences between experimental techniques and modes, and possible deviations from standard regimen for sensor exploitation may lead to discrepancies between results obtained by different research groups. This creates a need for wide agreement for the use of new sensors under various conditions and thorough fine-tuning of the methodology of sensor analysis of exhaled gas. Composite devices such as sensor arrays of the electronic nose type are among the promising tools for non-invasive diagnostics. However, the current technologies and their manufacture are not yet readily available, and each device is in a way unique with non-reproducible batch-to-batch properties. Further research and development is required before such devices can confidently be introduced into the clinical setting. Another problem is that breath-print patterns (profiles) essentially depend on the methods of breath collection and sampling. Therefore, unification of breath sampling techniques is an essential step toward future wide-scale application of electronic nose technology.

Nonetheless, we have enough data that indicate the likelihood of the ultimate successful development of breath sensors that will have eliminated their existing pitfalls. This retrospective review of the investigations devoted to development of methodology and methods for non-invasive medical diagnosis using exhaled air analysis by sensors are, in general, encouraging and some general remarks can be made in this review:

1. The analysis of exhaled gas of animals and humans using sensors is a unique and highly informative biological medium that can provide valuable information

for the evaluation of the state of the organisms. Stable, high information flow in the field breath gas analysis offers wide possibilities for non-invasive medical diagnostics.

2. Chemical sensors allow information to be obtained on the composition and concentrations of exhaled gas compounds, being the potent instruments for monitoring the organism's state. Constructional properties and performance of these devices provide good prospects for their medical applications.
3. Considering all the sensor types that have been applied to breath analysis, the devices working on electrical and electrochemical principles currently prevail in the field.
4. Polysensory systems of the electronic nose-type have been most extensively developed in the field of breath gas sensors.
5. Nanostructure devices providing the possibility of a marked increase in workability and performance of sensors are acquiring an increasing importance in exhaled gas analysis. The combination of nanostructure gas sensitive layers with modern detection/informatics principles and treatment of results is a most promising approach, solving a number of technical problems involved in the creation of sensor-based diagnostic tools.
6. Analysis of an individual's breath metabolic profile obtained by the new generation of point-contact nanostructure sensor matrices can provide diagnostically valuable information on the state of the human organism.

Acknowledgments

The authors acknowledge Dr. E. Faulques, Dr. L. Kamarchuk, and Dr. A. Yeremenko for fruitful discussions.

References

1. Amann A, Smith D. *Breath analysis for clinical diagnosis and therapeutic monitoring*. Singapore: World Scientific; 2005.
2. Dweik RA, Amann A. Exhaled breath analysis: the new frontier in medical testing. *J Breath Res* 2008;**2**:3030301.
3. Paschke KM, Mashir A, Dweik RA. Clinical applications of breath testing. *F1000 Med Rep* 2010;**2**:56.
4. Kushch I, Schwarz K, Schwentner L, Baumann B, Dzien A, Schmid A, et al. Compounds enhanced in a mass spectrometric profile of smokers' exhaled breath versus non-smokers as determined in a pilot study using PTR-MS. *J Breath Res* 2008;**2**:026002.
5. van Amsterdam JG, Hollander A, Snelder JD, Fischer PH, van Loveren H, Vos JG, et al. The effect of air pollution on exhaled nitric oxide of atopic and nonatopic subjects. *Nitric Oxide* 1999;**3**:492–5.
6. Kushch I, Arendacka B, Stolc S, Mochalski P, Filipiak W, Schwarz K, et al. Breath isoprene—aspects of normal physiology related to age, gender and cholesterol profile as determined in a proton transfer reaction mass spectrometry study. *Clin Chem Lab Med* 2008;**46**:1011–8.

7. Hibbard T, Killard AJ. Breath ammonia analysis: clinical application and measurement. *Crit Rev Anal Chem* 2011;**41**:21–35.
8. Grob NM, Aytekin M, Dweik RA. Biomarkers in exhaled breath condensate: a review of collection, processing and analysis. *J Breath Res* 2008;**2**:037004.
9. Miekisch W, Schubert JK, Noeldge-Schomburg GF. Diagnostic potential of breath analysis—focus on volatile organic compounds. *Clin Chim Acta* 2004;**347**:25–39.
10. Mashir A, Dweik RA. Exhaled breath analysis: the new interface between medicine and engineering. *Adv Powder Technol* 2009;**20**:420–5.
11. Phillips M, Basa-Dalay V, Bothamley G, Cataneo RN, Lam PK, Natividad MP, et al. Breath biomarkers of active pulmonary tuberculosis. *Tuberculosis (Edinb)* 2010;**90**: 145–51.
12. Smith D, Španěl P. Selected ion flow tube mass spectrometry (SIFT-MS) for on-line trace gas analysis. *Mass Spectrom Rev* 2005;**24**:661–700.
13. Phillips M, Herrera J, Krishnan S, Zain M, Greenberg J, Cataneo RN. Variation in volatile organic compounds in the breath of normal humans. *J Chromatogr B* 1999;**729**:75–88.
14. Ishimaru M, Yamada M, Nakagawa I, Sugano S. Analysis of volatile metabolites from cultured bacteria by gas chromatography/atmospheric pressure chemical ionization-mass spectrometry. *J Breath Res* 2008;**2**:037021.
15. Graham DY, Klein PD, Evans Jr DJ, Evans DG, Alpert LC, Opekun AR, et al. *Campylobacter pylori* detected noninvasively by the 13C-urea breath test. *Lancet* 1987;**1**: 1174–7.
16. Modak A. ^{13}C breath tests: transition from research to clinical practice. In: Amann A, Smith D, editors. *Breath analysis for clinical diagnosis and therapeutic monitoring*. Singapore: World Scientific; 2005; p. 457–78.
17. von Basum G, Dahnke H, Halmer D, Hering P, Murtz M. Online recording of ethane traces in human breath via infrared laser spectroscopy. *J Appl Physiol* 2003;**95**:2583–90.
18. Kuzmych O, Allen BL, Star A. Carbon nanotube sensors for exhaled breath components. *Nanotechnology* 2007;**18**:375505.
19. de Lacy Costello BPJ, Ewen RJ, Ratcliffe NM. A sensor system for monitoring the simple gases hydrogen, carbon monoxide, hydrogen sulfide, ammonia and ethanol in exhaled breath. *J Breath Res* 2008;**2**:037011.
20. Burke CS, Moore JP, Wencel D, Maccraith BD. Development of a compact optical sensor for real-time, breath-by-breath detection of oxygen. *J Breath Res* 2008;**2**:037012.
21. Gelperin A, Johnson AT. Nanotube-based sensor arrays for clinical breath analysis. *J Breath Res* 2008;**2**:037015.
22. Higgins C, Wencel D, Burke CS, MacCraith BD, McDonagh C. Novel hybrid optical sensor materials for in-breath O_2 analysis. *Analyst* 2008;**133**:241–7.
23. Ishida H, Satou T, Tsuji K, Kawashima N, Takemura H, Kosaki Y, et al. The breath ammonia measurement of the hemodialysis with a QCM-NH_3 sensor. *Biomed Mater Eng* 2008;**18**:99–106.
24. Hunter GW, Dweik RA. Applied breath analysis: an overview of the challenges and opportunities in developing and testing sensor technology for human health monitoring in aerospace and clinical applications. *J Breath Res* 2008;**2**:037020 [7 pp.].
25. Haick H, Hakim M, Patrascu M, Levenberg C, Shehada N, Nakhoul F, et al. Sniffing chronic renal failure in rat model by an array of random networks of single-walled carbon nanotubes. *ACS Nano* 2009;**3**:1258–66.
26. Kushch IG, Korenev NM, Kamarchuk LV, Pospelov AP, Alexandrov YL, Kamarchuk GV. Sensors for breath analysis: an advanced approach to express diagnostics and monitoring

of human diseases. In: Mikhalovsky S, Khajibaev A, editors. *Biodefence*. Berlin: Springer Science+Business Media B.V.; 2011. p. 63–75 [chapter 7].

27. Antus B, Horvath I, Barta I. Assessment of exhaled nitric oxide by a new hand-held device. *Respir Med* 2010;**104**:1377–80.
28. Maniscalco M, Lundberg JO. Hand-held nitric oxide sensor NIOX MINO® for the monitoring of respiratory disorders. *Expert Rev Respir Med* 2010;**4**:715–21.
29. Montuschi P, Santonico M, Mondino C, Pennazza G, Mantini G, Martinelli E, et al. Diagnostic performance of an electronic nose, fractional exhaled nitric oxide, and lung function testing in asthma. *Chest* 2010;**137**:790–6.
30. Capuano R, Santonico M, Martinelli E, Pennazza G, Paolesse R, Bergamini A. COPD diagnosis by a gas sensor array. *Procedia Engineering, Eurosensors XXIV Conference* 2010;**5**:484–7.
31. Nishibori M, Shin W, Izu N, Itoh T, Matsubara I. Sensing performance of thermoelectric hydrogen sensor for breath hydrogen analysis. *Sensor Actuat B—Chem* 2009;**137**:524–8.
32. Timmer B, Olthius W, van den Berg A. Ammonia sensors and their applications—a review. *Sensor Actuat B-Chem* 2005;**107**:666–77.
33. Huang J, Wan Q. Gas sensors based on semiconducting metal oxide one-dimensional nanostructures. *Sensors* 2009;**9**:9903–24.
34. Fine GF, Cavanagh LM, Afonja A, Binions R. Metal oxide semi-conductor gas sensors in environmental monitoring. *Sensors* 2010;**10**:5469–502.
35. Maksimovich N, Vorotyntsev V, Nikitina N, Kaskevich O, Karabun P, Martynenko F. Adsorption semiconductor sensor for diabetic ketoacidosis diagnosis. *Sensor Actuat B-Chem* 1996;**36**:419–21.
36. Ryabtsev SV, Shaposhnick AV, Lukin AN, Domashevskaya EP. Application of semiconductor gas sensors for medical diagnostics. *Sensor Actuat B—Chem* 1999;**B59**:26–9.
37. Gouma P, Kalyanasundaram K, Yun X, Stanacevic M, Wang LS. Nanosensor and breath analyzer for ammonia detection in exhaled human breath. *IEEE Sens J* 2010;**10**:49–53.
38. Emerson V, White P. Alcohol analysis. In: *Crime scene to court. The essentials of forensic science*, The Royal Society of Chemistry; 2004. p. 350–76.
39. Jones AW, Andersson L. Determination of ethanol in breath for legal purposes using a five-filter infrared analyzer: studies on response to volatile interfering substances. *J Breath Res* 2008;**2**:026006.
40. Bishop SC, Johnson G, Smith L, Fiorentino DD, Garcia T, Garcia R. Manual versus automatic sampling variations of a preliminary alcohol screening device. *J Anal Toxicol* 2009;**33**:521–4.
41. Buckley TJ, Pleil JD, Bowyer JR, Davis JM. Evaluation of methyl tert-butyl ether (MTBE) as an interference on commercial breath-alcohol analyzers. *Forensic Sci Int* 2001;**123**:111–8.
42. Fell J, Compton C, Voas R. A Note on the use of passive alcohol sensors during routine traffic stops. *Traffic Injury Prevent* 2008;**9**:534–8.
43. Marques PR, McKnight AS. Field and laboratory alcohol detection with 2 types of transdermal devices. *Alcohol Clin Exp Res* 2009;**33**:703–11.
44. Yu YF, Pawliszyn J. On-line monitoring of breath by membrane extraction with sorbent interface coupled with CO_2 sensor. *J Chromatogr A* 2004;**1056**:35–41.
45. Righettoni M, Tricoli A, Pratsinis SE. Si:WO_3 sensors for highly selective detection of acetone for easy diagnosis of diabetes by breath analysis. *Anal Chem* 2010;**82**:3581–7.
46. Groves WA, Achutan C. Laboratory and field evaluation of a SAW microsensor array for measuring perchloroethylene in breath. *J Occup Environ Hyg* 2004;**1**:779–88.

47. Peng G, Tisch U, Haick H. Detection of nonpolar molecules by means of carrier scattering in random networks of carbon nanotubes: toward diagnosis of diseases via breath samples. *Nano Lett* 2009;**9**:1362–8.
48. Mitsubayashi K, Minamide T, Otsuka K, Kudo H, Saito H. Optical bio-sniffer for methyl mercaptan in halitosis. *Anal Chim Acta* 2006;**573–574**:75–80.
49. Silva LI, Freitas AC, Rocha-Santos TA, Pereira ME, Duarte AC. Breath analysis by optical fiber sensor for the determination of exhaled organic compounds with a view to diagnostics. *Talanta* 2011;**83**:1586–94.
50. Pyshkin O, Kamarchuk G, Yeremenko A, Kravchenko A, Pospelov A, Alexandrov Y, et al. Evidence for sensory effects of a 1D organic conductor under gas exposure *J Breath Res* 2011;**5**:016005.
51. Liao JY, Ho KC. A study of partially irreversible characteristics in a TTF-TCNQ gas sensing system. *Sensor Actuat B—Chem* 2008;**130**:343–50.
52. Nolte AJ, Stafford CM, Li T, Yoo PJ, Harding J, Lin-Gibson S, et al, editors. Soft matter, biological materials and biomedical materials–synthesis, characterization and applications. Part IV. Materials and Sensors for Biomedical Applications. In: Materials research society symposium proceedings of 2010 MRS fall meeting, vol. 1301. Cambridge: Cambridge University Press; 2011. p. 308.
53. Kolmakov A, Chen X, Moskovits M. Functionalizing nanowires with catalytic nanoparticles for gas sensing application. *J Nanosci Nanotechnol* 2008;**8**:111–21.
54. Mubeen S, Zhang T, Chartuprayoon N, Rheem Y, Mulchandani A, Myung NV. Sensitive detection of H_2S using gold nanoparticle decorated single-walled carbon nanotubes. *Anal Chem* 2010;**82**:250–7.
55. Johnson ATC, Khamis SM, Preti G, Kwak J, Gelperin A. DNA-coated nanosensors for breath analysis. *IEEE Sens J* 2010;**10**:159–66.
56. Kamarchuk GV, Pospelov AP, Yeremenko AV, Faulques EC, Yanson IK. Point-contact sensors: new prospects for a nanoscale-sensitive technique. *Europhys Lett* 2006;**76**: 575–81.
57. Naidyuk YuG, Yanson IK. Point-contact spectroscopy. New York: Springer; 2005.
58. Kamarchuk GV, Kolobov IG, Khotkevich AV, Yanson IK, Pospelov AP, Levitsky IA. New chemical sensors based on point heterocontact between single wall carbon nanotubes and gold wires. *Sensor Actuat B-Chem* 2008;**134**:1022–6.
59. Cheng ZJ, Warwick G, Yates DH, Thomas PS. An electronic nose in the discrimination of breath from smokers and non-smokers: a model for toxin exposure. *J Breath Res* 2009;**3**: 036003.
60. Shuster G, Gallimidi Z, Reiss AH, Dovgolevsky E, Billan S, Abdah-Bortnyak R, et al. Classification of breast cancer precursors through exhaled breath breast cancer. *Res Treat* 2011;**126**:791–6.
61. Wilson AD, Baietto M. Advances in electronic-nose technologies developed for biomedical applications. *Sensors* 2011;**11**:1105–76.
62. Shurmer H, Fard A, Barker J, Bartlett P, Dodd G, Hayat U. Development of an electronic nose. *Phys Technol* 1987;**18**:170.
63. Arshak K, Moore E, Lyons G, Harris F, Clifford S. A review of gas sensors employed in electronic nose applications. *Sensor Rev* 2004;**24**:181–98.
64. Zubchuk VI, Korzhov V, Stashkevich VF, Yaremko VS. Problems of express-diagnosis based on selective gas analyzers. *Electron Commun (Kiev, Ukraine)* 2008;**3–4**:115–7.
65. Davis CE, Frank M, Mizaikoff B, Oser H. The future of sensors and instrumentation for human breath analysis. *IEEE Sens J* 2010;**10**:3–6.

66. Megraud F, Lehours P. *Helicobacter pylori* detection and antimicrobial susceptibility testing. *Clin Microbiol Rev* 2007;**20**:280–322.
67. Penault C, Španěl P, Smith D. Detection of *Helicobacter pylori* infection by breath ammonia following urea ingestion. In: Amann A, Smith D, editors. *Breath analysis for medical diagnosis and therapeutic monitoring*. Singapore: World Scientific; 2005. p. 393–99.
68. Narasimhan LR, Goodman W, Patel CK. Correlation of breath ammonia with blood urea nitrogen and creatinine during hemodialysis. *Proc Natl Acad Sci U S A* 2001;**98**:4617–21.
69. DuBois S, Eng S, Bhattacharya R, Rulyak S, Hubbard T, Putnam D, et al. Breath ammonia testing for diagnosis of hepatic encephalopathy. *Dig Dis Sci* 2005;**50**:1780–4.
70. Smith D, Španěl P. The novel selected-ion flow tube approach to trace gas analysis of air and breath. *Rapid Commun Mass Spectrom* 1996;**10**:1183–98.
71. Dun CDR, Blac M, Cowell DC, Penault C, Ratcliffe NM, Spence R, et al. Ammonia vapour in the mouth as a diagnostic marker for *Helicobacter pylori* infection: preliminary "proof of principle" pharmacological investigations. *Br J Biomed Sci* 2001;**58**:66–75.
72. Pospelov AP, Kushch EG, Aleksandrov YL, Pletnev AM, Kamarchuk GV. Sensors of active type for analysis of breath. *Sensory Electron Microsyst Technol* 2007;**4**:49–54.
73. Lewicki R, Kosterev AA, Bakhirkin YA, Thomazy DM, Doty J, Dong L, et al. Real time ammonia detection in exhaled human breath with a quantum cascade laser based sensor. In: *2009 Conference on Lasers and Electro-Optics and Quantum Electronics and Laser Science Conference (Cleo/Qels 2009)*, Vols. 1–5; 2009. p. 710–1.
74. Sinha N, Ma J, Yeow JT. Carbon nanotube-based sensors. *J Nanosci Nanotechnol* 2006; **6**:573–90.
75. Kozlov AV, Evstratova YS, Novikova VP, Tkachenko EI, Khochinskaia OY. Gas analyzer "HelicoSense" as a new device for diagnosis of the *Helicobacter* infection. *Med Tekh (Meditsinskaya Tekhnika, Moscow)* 2006;**40**(3):44–6.
76. Fleischer M, Simon E, Rumpel E, Ulmer H, Harbeck M, Wandel M. Detection of volatile compounds correlated to human diseases through breath analysis with chemical sensors. *Sensor Actuat B-Chem* 2002;**83**:245–9.
77. Khotkevich AV, Yanson IK. *Atlas of point contact spectra of electron-phonon interactions in metals*. Boston: Kluwer Academic; 1995.
78. Ashcroft NW, Mermin ND. *Solid state physics*. New York: Holt; 1976.
79. Barash O, Peled N, Hirsch FR, Haick H. Sniffing the unique "odor print" of non-small-cell lung cancer with gold nanoparticles. *Small* 2009;**5**:2618–24.
80. Peng G, Tisch U, Adams O, Hakim M, Shehada N, Broza YY. Diagnosing lung cancer in exhaled breath using gold nanoparticles. *Nat Nanotechnol* 2009;**4**:669–73.
81. Fraden J. *Handbook of modern sensors, Physics, designs, and applications*. 3rd ed. New York: Springer; 2004.
82. Janata J. *Principles of Chemical Sensors*, 2nd ed. New York: Springer Science+Business Media; 2009.
83. Zhang X, Ju H, Wang J, editors. *Electrochemical sensors, biosensors and their biomedical applications*. Elsevier Inc. San Diego, New York, London: Academic Press; 2007. p. 616.
84. Arregui FJ, Liu YJ, Matias IR, Claus RO. Optical fiber humidity sensor using a nano Fabry-Perot cavity formed by the ionic self-assembly method. *Sensor Actuat B-Chem* 1999;**B59**:54–9.
85. Baldini F, Falai A, De Gaudio AR, Landi D, Lueger A, Mencaglia A, et al. Continuous monitoring of gastric carbon dioxide with optical fibers. *Sensor Actuat B-Chem* 2003; **90**:132–8.

86. Fernández-Sánchez JF, Cannas R, Spichiger S, Steiger R, Spichiger-Keller UE. Optical CO_2-sensing layers for clinical application based on pH-sensitive indicators incorporated into nanoscopic metal-oxide supports. *Sensor Actuat B—Chem* 2007;**128**:145–53.
87. Kudo H, Sawai M, Suzuki Y, Wang X, Gessei T, Takahashi D. Fiber-optic bio-sniffer (biochemical gas sensor) for high-selective monitoring of ethanol vapor using 335 nm UV-LED. *Sensor Actuat B-Chem* 2010;**147**:676–80.
88. Mills A, Lepre A, Wild L. Breath-by-breath measurement of carbon dioxide using a plastic film optical sensor. *Sensor Actuat B-Chem* 1997;**39**:419–25.
89. Robinson JK, Bollinger MJ, Birks JW. Luminol/H_2O_2 chemiluminescence detector for the analysis of nitric oxide in exhaled breath. *Anal Chem* 1999;**71**:5131–6.
90. Wen F, Zhang SC, Na N, Wu YY, Zhang XR. Development of a sensitive gas sensor by trapping the analytes on nanomaterials and in situ cataluminescence detection. *Sensor Actuat B-Chem* 2009;**B141**:168–73.
91. Hök B, Pettersson H, Andersson AK, Haasl S, Ekerlund P. Breath analyzer for alcolocks and screening devices. *IEEE Sens J* 2010;**10**:10–15.
92. Shorter JH, Nelson DD, McManus JB, Zahniser MS, Milton DK. Multicomponent breath analysis with infrared absorption using room-temperature quantum cascade lasers. *IEEE Sens J* 2010;**10**:76–84.
93. Kim SS, Young C, Vidakovic B, Gabram-Mendola SGA, Bayer CW, Mizaikoff B. Potential and challenges for mid-infrared sensors in breath diagnostics. *IEEE Sens J* 2010;**10**:145–158.
94. Groves WA, Zellers ET. Analysis of solvent vapors in breath and ambient air with a surface acoustic wave sensor array. *Ann Occup Hyg* 2001;**45**:609–23.
95. Fernandes DLA, Rolo TA, Oliveira JABP, Gornes MTSR. A new analytical system, based on an acoustic wave sensor, for halitosis evaluation. *Sensor Actuat B-Chem* 2009;**136**:73–9.
96. Rimeika R, Ciplys D, Poderys V, Rotomskis R, Shur MS. Fast-response surface acoustic wave humidity sensor based on hematoporphyrin film. *Sensor Actuat B-Chem* 2009;**137**:592–6.
97. Sivaramakrishnan S, Rajamani R, Smith CS, McGee KA, Mann KR, Yamashita N. Carbon nanotube-coated surface acoustic wave sensor for carbon dioxide sensing. *Sensor Actuat B-Chem* 2008;**132**:296–304.
98. Saraoglu HM, Kocan M. Determination of blood glucose level-based breath analysis by a quartz crystal microbalance sensor array. *IEEE Sens J* 2010;**10**:104–9.
99. Cui DF, Cai XX, Han JG, He ZH, Zhu HZ, Feng GP. Lipid LB films for room-temperature ethanol gas assay. *Sensor Actuat B-Chem* 1997;**B45**:229–32.
100. Pennazza G, Marchetti E, Santonico M, Mantini G, Mummolo S, Marzo G, et al. Application of a quartz microbalance based gas sensor array for the study of halitosis. *J Breath Res* 2008;**2**:017009.
101. Huang HH, Zhou J, Chen SY, Zeng L, Huang YP. A highly sensitive QCM sensor coated with Ag+-ZSM-5 film for medical diagnosis. *Sensor Actuat B-Chem* 2004;**101**:316–21.
102. Ito J, Nakamoto T, Uematsu H. Discrimination of halitosis substance using QCM sensor array and a preconcentrator. *Sensor Actuat B-Chem* 2004;**99**:431–6.
103. Hahn CE. Electrochemical analysis of clinical blood-gases, gases and vapours. *Analyst* 1998;**123**:57R–86R.
104. Minamide T, Mitsubayashi K, Jaffrezic-Renault N, Hibi K, Endo H, Saito H. Bioelectronic detector with monoamine oxidase for halitosis monitoring. *Analyst* 2005;**130**:1490–4.
105. Maniscalco M, Laurentiis Gd, Weitzberg E, Lundberg JO, Sofia M. Validation study of nasal nitric oxide measurements using a hand-held electrochemical analyser. *Eur J Clin Invest* 2008;**38**:197–200.

106. Vreman HJ, Stevenson DK, Oh W, Fanaroff AA, Wright LL, Lemons JA, et al. Semi-portable electrochemical instrument for determining carbon monoxide in breath. *Clin Chem* 1994;**40**:1927–33.
107. Carpagnano GE, Kharitonov SA, Foschino-Barbaro MP, Resta O, Gramiccioni E, Barnes PJ. Increased inflammatory markers in the exhaled breath condensate of cigarette smokers. *Eur Respir J* 2003;**21**:589–93.
108. Fritsch T, van Herpen M, von Basum G, Hering P, Murtz M. Is exhaled carbon monoxide level associated with blood glucose level? A comparison of two breath analyzing methods. *J Biomed Opt* 2008;**13**:34012.
109. Sylvester KP, Patey RA, Rafferty GF, Rees D, Thein SL, Greenough A. Exhaled carbon monoxide levels in children with sickle cell disease. *Eur J Pediatr* 2005;**164**:162–5.
110. Braden B, Braden CP, Klutz M, Lembcke B. Analysis of breath hydrogen (H_2) in diagnosis of gastrointestinal function: validation of a pocket breath H_2 test analyzer. *Z Gastroenterol* 1993;**31**:242–5.
111. Duan LP, Braden B, Clement T, Caspary WF, Lembcke B. Clinical evaluation of a miniaturized desktop breath hydrogen analyzer. *Z Gastroenterol* 1994;**32**:575–8.
112. Rumessen JJ, Kokholm G, Gudmand-Hoyer E. Methodological aspects of breath hydrogen (H_2) analysis. Evaluation of a H_2 monitor and interpretation of the breath H_2 test. *Scand J Clin Lab Invest* 1987;**47**:555–60.
113. Alving K, Janson C, Nordvall L. Performance of a new hand-held device for exhaled nitric oxide measurement in adults and children. *Respir Res* 2006;**7**:67.
114. Boot JD, de Ridder L, de Kam ML, Calderon C, Mascelli MA, Diamant Z. Comparison of exhaled nitric oxide measurements between NIOX MINO electrochemical and Ecomedics chemiluminescence analyzer. *Respir Med* 2008;**102**:1667–71.
115. Cobos Barroso N, Perez-Yarza EG, Sardon Prado O, Reverte Bover C, Gartner S, Korta Murua J. Exhaled nitric oxide in children: a noninvasive marker of airway inflammation. *Arch Bronconeumol* 2008;**44**:41–51.
116. Hemmingsson T, Linnarsson D, Gambert R. Novel hand-held device for exhaled nitric oxide-analysis in research and clinical applications. *J Clin Monit Comput* 2004;**18**: 379–87.
117. Sardon Prado O, Perez-Yarza EG, Aldasoro Ruiz A, Korta Murua J, Mintegui Aranburu J, Emparanza Knorr JI. Fractional exhaled nitric oxide: validation of a 6 second exhalation time with two different analysers. *Ann Pediatr (Barc)* 2008;**69**:221–6.
118. Schneider A, Tilemann L, Schermer T, Gindner L, Laux G, Szecsenyi J, et al. Diagnosing asthma in general practice with portable exhaled nitric oxide measurement—results of a prospective diagnostic study. *Respir Res*; 2009;**10**:15.
119. Farrell S, Barker ML, Gerlach RW. Overnight malodor effect with a 0.454% stabilized stannous fluoride sodium hexametaphosphate dentifrice. *Compend Contin Educ Dent* 2007;**28**:658–61 [quiz 662, 671].
120. Kuban P, Berg JM, Dasgupta PK. Durable microfabricated high-speed humidity sensors. *Anal Chem* 2004;**76**:2561–7.
121. Shimura M, Yasuno Y, Iwakura M, Shimada Y, Sakai S, Suzuki K, et al. A new monitor with a zinc-oxide thin film semiconductor sensor for the measurement of volatile sulfur compounds in mouth air. *J Periodontol* 1996;**67**:396–402.
122. Bengtsson J, Bengtson JP. The inspiratory to end-tidal oxygen difference during exercise. *Int J Clin Monit Comput* 1997;**14**:217–23.
123. Kirovskaya IA, Zemtsov AE. Breath gases analysis using sensors. In: *Proceeding of VII conference "Analytics of Siberia and Far East"*. Novosibirsk; 2004. p. 44.

124. Murata T, Rahardjo A, Fujiyama Y, Yamaga T, Hanada M, Yaegaki K, et al. Development of a compact and simple gas chromatography for oral malodor measurement. *J Periodontol* 2006;**77**:1142–7.
125. Neri G, Bonavita A, Micali G, Donato N. Design and development of a breath acetone MOS sensor for ketogenic diets control. *IEEE Sens J* 2010;**10**:131–6.
126. Lo Schiavo S, Livoti L, Calisto A, Bramanti A, Donato N, Latino M, et al. Novel sensing materials for breath analysis devices. In: *2010 Annual International Conference of the IEEE Engineering in Medicine and Biology Society (EMBC)*; 2010. pp. 670–3.
127. Krutovertsev S, Chuprin M, Ivanova O, Pislyakov A, Shevchenko A, Konovalov V, et al. Nitric oxide monitoring system based on ZnO sensors for exhaled air. *Procedia Eng.* 2010;**5**:244–7.
128. Sekiguchi T, Nakamura M, Kato M, Nishikawa K, Hokari K, Sugiyama T, et al. Immunological *Helicobacter pylori* urease analyzer based on ion-sensitive field effect transistor. *Sensor Actuat B-Chem* 2000;**67**:265–9.
129. Marinelli F, Dell'Aquila A, Torsi L, Tey J, Suranna GP, Mastrorilli P, et al. An organic field effect transistor as a selective NO(x) sensor operated at room temperature. *Sensor Actuat B-Chem* 2009;**140**:445–50.
130. Huang XJ, Choi YK. Chemical sensors based on nanostructured materials. *Sensor Actuat B-Chem* 2007;**122**:659–71.
131. Hongsith N, Choopun S. Enhancement of ethanol sensing properties by impregnating platinum on surface of ZnO tetrapods. *IEEE Sens J* 2010;**10**:34–8.
132. Hutter T, Ruschin S. Non-imaging optical method for multi-sensing of gases based on porous silicon. *IEEE Sens J* 2010;**10**:97–103.
133. Andre N, Druart S, Gerard P, Pampin R, Moreno-Hagelsieb L, Kezai T, et al. Miniaturized wireless sensing system for real-time breath activity recording. *IEEE Sens J* 2010;**10**: 178–84.
134. Wang L, Kalyanasundaram K, Stanacevic M, Gouma P. Nanosensor device for breath acetone detection. *Sensor Lett* 2010;**8**:709–12.
135. He Q, Zheng DY, Hu SS. Development and application of a nano-alumina based nitric oxide sensor. *Microchim Acta* 2009;**164**:459–64.
136. Wilson AD, Baietto M. Applications and advances in electronic-nose technologies. *Sensors* 2009;**9**:5099–148.
137. Kushch IG, Korenev NM, Kamarchuk LV, Pospelov AP, Alexandrov YL, Kamarchuk GV. Noninvasive recognision of gastroduodenal mucosal damage: new possibilities using sensor technique for breath gas analysis. *Hepatogastroenterology* 2008;**55**:OP115.
138. Chubov PN, Yanson IK, Akimenko AI. Electron-phonon interaction in aluminum point contacts. *Fizika Nizkikh Temp* 1982;**8**:64–80.
139. Kim YS, Ha SC, Kim K, Yang H, Choi SY, Kim YT, et al. Room-temperature semiconductor gas sensor based on nonstoichiometric tungsten oxide nanorod film. *Appl Phys Lett* 2005;**86**:213105–7.
140. Andersen LP, Holck S, Janulaityte-Gunther D, Kupcinskas L, Kiudelis G, Jonaitis L, et al. Gastric inflammatory markers and interleukins in patients with functional dyspepsia, with and without *Helicobacter pylori* infection. *FEMS Immunol Med Microbiol* 2005;**44**: 233–8.
141. Foss RD, Voas RB, Beirness DJ. Using a passive alcohol sensor to detect legally intoxicated drivers. *Am J Public Health* 1993;**83**:556–60.
142. Lestina DC, Lund AK. Laboratory evaluation of two passive alcohol sensors. *J Stud Alcohol* 1992;**53**:328–34.

143. Garzella C, Comini E, Bontempi E, Depero LE, Frigeri C, Sberveglieri G, et al. Sol-gel TiO_2 and W/TiO_2 nanostructured thin films for control of drunken driving. *Sensor Actuat B-Chem* 2002;**83**:230–7.
144. Gehring H, Nahm W, Baerwald J, Konecny E, Schmucker P. Breath alcohol analyzers with electrochemical sensory–accuracy of measurements in lung model ventilation. *Biomed Tech (Berl)* 1996;**41**:54–9.
145. Gehring H, Nahm W, Klotz KF, Knipper A, Zimmermann K, Baerwald J, et al. Measurement of expired alcohol concentrations with a new electrochemical sensor. A model investigation to determine interference with volatile anesthetics and clinical application. *Anaesthesist* 1996;**45**:154–62.
146. ATS/ERS recommendations for standardized procedures for the online and offline measurement of exhaled lower respiratory nitric oxide and nasal nitric oxide, 2005. *Am J Respir Crit Care Med* 2005;**171**:912–30.
147. Bodini A, Peroni D, Loiacono A, Costella S, Pigozzi R, Baraldi E, et al. Exhaled nitric oxide daily evaluation is effective in monitoring exposure to relevant allergens in asthmatic children. *Chest* 2007;**132**:1520–5.
148. Kostikas K, Papaioannou AI, Tanou K, Koutsokera A, Papala M, Gourgoulianis KI. Portable exhaled nitric oxide as a screening tool for asthma in young adults during pollen season. *Chest* 2008;**133**:906–13.
149. Tamasi L, Bohacs A, Bikov A, Andorka C, Rigo Jr J, Losonczy G, et al. Exhaled nitric oxide in pregnant healthy and asthmatic women. *J Asthma* 2009;**46**:786–91.
150. Vahlkvist S, Sinding M, Skamstrup K, Bisgaard H. Daily home measurements of exhaled nitric oxide in asthmatic children during natural birch pollen exposure. *J Allergy Clin Immunol* 2006;**117**:1272–6.
151. Weschta M, Deutschle T, Riechelmann H. Nasal fractional exhaled nitric oxide analysis with a novel hand-held device. *Rhinology* 2008;**46**:23–7.
152. Kharitonov SA, Robbins RA, Yates D, Keatings V, Barnes PJ. Acute and chronic effects of cigarette smoking on exhaled nitric oxide. *Am J Respir Crit Care Med* 1995;**152**: 609–12.
153. Pertseva TO, Gashynova KY, Efimova NO. The influence of tobacco smoking on the concentration of nitric oxide in exhaled air in patients with chronic obstructive pulmonary disease. *Ukrainian Pulm J (Kiev, Ukraine)* 2010;**3**:19–21.
154. Cano-Garcinuno A, Carvajal-Uruena I, Diaz-Vazquez CA, Dominguez-Aurrecoechea B, Garcia-Merino A, de Rodas PM, et al. Clinical correlates and determinants of airway inflammation in pediatric asthma. *J Investig Allergol Clin Immunol* 2010;**20**:303–10.
155. Matsunaga K, Hirano T, Kawayama T, Tsuburai T, Nagase H, Aizawa H, et al. Reference ranges for exhaled nitric oxide fraction in healthy Japanese adult population. *Allergol Int* 2010;**59**:363–7.
156. Matsunaga K, Hirano T, Akamatsu K, Koarai A, Sugiura H, Minakata Y, et al. Exhaled nitric oxide cutoff values for asthma diagnosis according to rhinitis and smoking status in Japanese subjects. *Allergol Int* 2011;**60**:331–7.
157. Menzies D, Nair A, Lipworth BJ. Portable exhaled nitric oxide measurement: Comparison with the "gold standard" technique. *Chest* 2007;**131**:410–4.
158. Khalili B, Boggs PB, Bahna SL. Reliability of a new hand-held device for the measurement of exhaled nitric oxide. *Allergy* 2007;**62**:1171–4.
159. Pisi R, Aiello M, Tzani P, Marangio E, Olivieri D, Chetta A. Measurement of fractional exhaled nitric oxide by a new portable device: comparison with the standard technique. *J Asthma* 2010;**47**:805–9.

160. Brooks CR, Brogan SB, van Dalen CJ, Lampshire PK, Crane J, Douwes J. Measurement of exhaled nitric oxide in a general population sample: a comparison of the Medisoft HypAir FE(NO) and Aerocrine NIOX analyzers. *J Asthma* 2011;**48**:324–8.
161. Hervas D, Milan JM, Garde J. Differences in exhaled nitric oxide in atopic children. *Allergol Immunopathol (Madr)* 2008;**36**:331–5.
162. Röck F, Barsan N, Weimar U. Electronic nose: current status and future trends. *Chem Rev* 2008;**108**:705–25.
163. Berg J, Lindgren P. Economic evaluation of FE(NO) measurement in diagnosis and 1-year management of asthma in Germany. *Respir Med* 2008;**102**:219–31.
164. Selby A, Clayton B, Grundy J, Pike K, Drew K, Raza A, et al. Are exhaled nitric oxide measurements using the portable NIOX MINO repeatable? *Respir Res* 2010;**11**:43.
165. Fleming SC. Evaluation of a hand-held hydrogen monitor in the diagnosis of intestinal lactase deficiency. *Ann Clin Biochem* 1990;**27**(Pt 5):499–500.
166. Nikolaev IN, Nozdrya DA. About the possibility of use of sensor-based gas analyzers for diagnostics of diseases using breath test method. *Phys Med (Russ Sci J)* 2006;**16**(2):15–20.
167. Urita Y, Watanabe T, Ishihara S, Maeda T, Sasaki Y, Kazuo H, et al. Breath hydrogen and methane levels in a patient with volvulus of the sigmoid colon. *J Breath Res* 2008;**2**:037025.
168. Chubarova AI, Mukhina YG, Korneva TI, Kushnirenko IA, Kurgasheva EK, D'yakonova GV, et al. Evaluation of lactose by determination of hydrogen concentration in exhaled breath. *The Practitioner (Moscow)* 2009;(08/09):4.
169. Chubarova AI, Mukhina YG, Korneva TI, Kushnirenko IA, Kurgasheva EK, D'yakonova GV, et al. Determination of hydrogen concentration in exhaled air of children with lactose intolerance. *Child Nutrition Issues (Moscow)* 2009;**4**:19.
170. Simrén M, Stotzer PO. Use and abuse of hydrogen breath tests. *Gut* 2006;**55**:297–303.
171. Saad RJ, Chey WD. Breath tests for gastrointestinal disease: the real deal or just a lot of hot air? *Gastroenterology* 2007;**133**:1763–6.
172. Rauter L, Startinig B, Mutz I. Determination of hydrogen in expiratory air of premature infants in suspected necrotizing enterocolitis. *Monatsschr Kinderheilkd* 1992;**140**:296–9.
173. Administration USFaD. *FDA Medical Devices*.
174. Nagler J, Krauss B. Capnography: a valuable tool for airway management. *Emerg Med Clin North Am* 2008;**26**:881–97 [vii].
175. Zislin BD, Chistyakov AV, Markov AV, Pochepko DV, Kontorovich MB, Bagin VA, et al. Monitoring of respiratory gases. *Intens Care J (Zhurnal Intensivnaya Terapiya, Ekaterinburg, Russsia)* 2005;**3**:8.
176. Folke M, Hok B. A new capnograph based on an electro acoustic sensor. *Med Biol Eng Comput* 2008;**46**:55–9.
177. Garey DM, Ward R, Rich W, Heldt G, Leone T, Finer NN. Tidal volume threshold for colorimetric carbon dioxide detectors available for use in neonates. *Pediatrics* 2008;**121**:e1524–7.
178. Hildebrandt T, Espelund M, Olsen KS. Evaluation of a transportable capnometer for monitoring end-tidal carbon dioxide. *Anaesthesia* 2010;**65**:1017–21.
179. Berggren M, Hosseini N, Nilsson K, Stenqvist O. Improved response time with a new miniaturised main-stream multigas monitor. *J Clin Monit Comput* 2009;**23**:355–61.
180. Ma W, Liu X, Pawliszyn J. Analysis of human breath with micro extraction techniques and continuous monitoring of carbon dioxide concentration. *Anal Bioanal Chem* 2006;**385**: 1398–408.
181. Sivaramakrishnan S, Rajamani R, Johnson BD. Dynamic model inversion techniques for breath-by-breath measurement of carbon dioxide from low bandwidth sensors. *Conf Proc IEEE Eng Med Biol Soc* 2009; **2009**:3039–42.

182. Miekisch W, Hengstenberg A, Kischkel S, Beckmann U, Mieth M, Schubert JK. Construction and evaluation of a versatile CO_2 controlled breath collection device. *IEEE Sens J* 2010;**10**:211–5.
183. Vautz W, Baumbach JI, Westhoff M, Züchner K, Carstens ETH, Perl T. Breath sampling control for medical application. *Int J Ion Mobility Spectrom* 2010;**13**:41–6.
184. Schwarz K, Pizzini A, Arendacka B, Zerlauth K, Filipiak W, Schmid A, et al. Breath acetone-aspects of normal physiology related to age and gender as determined in a PTR-MS study. *J Breath Res* 2009;**3**:027003.
185. Enderby B, Lenney W, Brady M, Emmett C, Španěl P, Smith D. Concentrations of some metabolites in the breath of healthy children aged 7–18 years measured using selected ion flow tube mass spectrometry (SIFT-MS). *J Breath Res* 2009;**3**:036001.
186. Deng C, Zhang J, Yu X, Zhang W, Zhang X. Determination of acetone in human breath by gas chromatography-mass spectrometry and solid-phase microextraction with on-fiber derivatization. *J Chromatogr B: Analyt Technol Biomed Life Sci* 2004;**810**:269–75.
187. Cao W, Duan Y. Breath analysis: potential for clinical diagnosis and exposure assessment. *Clin Chem* 2006;**52**:800–11.
188. Turner C, Parekh B, Walton C, Španěl P, Smith D, Evans M. An exploratory comparative study of volatile compounds in exhaled breath and emitted by skin using selected ion flow tube mass spectrometry. *Rapid Commun Mass Spectrom* 2008;**22**:526–32.
189. Wang CJ, Mbi A, Shepherd M. A study on breath acetone in diabetic patients using a cavity ringdown breath analyzer: exploring correlations of breath acetone with blood glucose and glycohemoglobin A1C. *IEEE Sens J* 2010;**10**:54–63.
190. Španěl P, Dryahina K, Rejskova A, Chippendale TW, Smith D. Breath acetone concentration; biological variability and the influence of diet. *Physiol Meas* 2011;**32**:N23–31.
191. Smith D, Španěl P, Fryer AA, Hanna F, Ferns GA. Can volatile compounds in exhaled breath be used to monitor control in diabetes mellitus? *J Breath Res* 2011;**5**:022001.
192. Lukash SI, Peretyat'ko VY. Computer system for acetone determination in exhaled breath. *Comput Facil Netw Syst (Kiev, Ukraine)* 2004;**3**:50–6.
193. Lazorenko YP, Konoplya MM, Sokolov VA, Misyura AG. Nanoscale structured films of hydrogel of cellulose derivatives for creation the sensors for measuring acetone content in products of exhaled air. *Bull Cherkasy Univ Chem Sci (Ukraine)* 2010;**175**: 37–43.
194. Meng G-F, Xiang Q, Pan Q-Y, Xu J-Q. The selective acetone detection based on Fe_3O_4 doped WO_3 nanorods. *Sensor Lett* 2011;**9**:128–31.
195. Kearney DJ, Hubbard T, Putnam D. Breath ammonia measurement in *Helicobacter pylori* infection. *Dig Dis Sci* 2002;**47**:2523–30.
196. van den Broek AM, Feenstra L, de Baat C. A review of the current literature on aetiology and measurement methods of halitosis. *J Dent* 2007;**35**:627–35.
197. Smith D, Španěl P, Davies S. Trace gases in breath of healthy volunteers when fasting and after a protein-calorie meal: a preliminary study. *J Appl Physiol* 1999;**87**:1584–8.
198. Aguilar AD, Forzani ES, Nagahara LA, Amlani I, Tsui R, Tao NJ. Breath ammonia sensor based on conducting polymer nanojunctions. *IEEE Sens J* 2008;**8**:269–73.
199. Toda K, Li J, Dasgupta PK. Measurement of ammonia in human breath with a liquid-film conductivity sensor. *Anal Chem* 2006;**78**: 7284–91.
200. Kornienko EA, Dmitrenko MA, Drozdova SN, Grigor'ev SV, Mamaev NA, Igumenova EL, et al. Comparative evaluation of methods for registration of ammonia concentration in throat cavity's air for *Helicobacter pylori* infection diagnosis. *Lab Diagn (St. Petersburg, Russia)* 2004;**2**:14–7.

201. Lazebnik LB, Mikhaylov AA, Morozov IA, Nikolaev IN, Nozdrya DA, Churikova AA. Estimation of efficiency of *Helicobacter pylori* diagnostics by means of an ammonia gas-analyzer on the basis of MIS-sensor in comparison with invasive diagnostics methods. *Clin Exp Gastroenterol* 2008;**8**:54–8.
202. Sukhanov AV, Pickersky IE, Ignatov AV, Serebrennikova EN. The informativity of the respiratory urea Helic test in the definition of seeding degree of *Helicobacter pylori* and pathological changes of stomach's mucous membrane: the comparison with gistological and cytological methods. *Siberian Med J (Irkutsk, Russia)* 2006;**7**:11–6.
203. Crocheton N, Machet E, Haouache H, Houdart E, Huat G, Claverot J, et al. Screening for carbon monoxide (CO) poisoning by general practitioners in home visits. *Presse Med* 2010;**39**:e29–34.
204. O'Neill HJ, Gordon SM, O'Neill MH, Gibbons RD, Szidon JP. A computerized classification technique for screening for the presence of breath biomarkers in lung cancer. *Clin Chem* 1988;**34**:1613–8.
205. Phillips M, Cataneo RN, Cummin AR, Gagliardi AJ, Gleeson K, Greenberg J, et al. Detection of lung cancer with volatile markers in the breath. *Chest* 2003;**123**:2115–23.
206. Phillips M, Cataneo RN, Ditkoff BA, Fisher P, Greenberg J, Gunawardena R, et al. Volatile markers of breast cancer in the breath. *Breast J* 2003;**9**:184–91.
207. Phillips M, Altorki N, Austin JH, Cameron RB, Cataneo RN, Greenberg J, et al. Prediction of lung cancer using volatile biomarkers in breath. *Cancer Biomark* 2007;**3**:95–109.
208. Poli D, Carbognani P, Corradi M, Goldoni M, Acampa O, Balbi B, et al. Exhaled volatile organic compounds in patients with non-small cell lung cancer: cross sectional and nested short-term follow-up study. *Respir Res* 2005;**6**:71.
209. Filipiak W, Sponring A, Mikoviny T, Ager C, Schubert J, Miekisch W, et al. Release of volatile organic compounds (VOCs) from the lung cancer cell line CALU-1 *in vitro*. *Cancer Cell Int* 2008;**8**:17.
210. Sponring A, Filipiak W, Mikoviny T, Ager C, Schubert J, Miekisch W, et al. Release of volatile organic compounds from the lung cancer cell line NCI-H2087 *in vitro*. *Anticancer Res* 2009;**29**:419–26.
211. Abela JE, Skeldon KD, Stuart RC, Padgett MJ. Exhaled ethane concentration in patients with cancer of the upper gastrointestinal tract—a proof of concept study. *Biosci Trends* 2009;**3**:110–4.
212. Horvath I, Lazar Z, Gyulai N, Kollai M, Losonczy G. Exhaled biomarkers in lung cancer. *Eur Respir J* 2009;**34**:261–75.
213. Di Natale C, Macagnano A, Martinelli E, Paolesse R, D'Arcangelo G, Roscioni C, et al. Lung cancer identification by the analysis of breath by means of an array of non-selective gas sensors. *Biosens Bioelectron* 2003;**18**:1209–18.
214. Machado RF, Laskowski D, Deffenderfer O, Burch T, Zheng S, Mazzone PJ, et al. Detection of lung cancer by sensor array analyses of exhaled breath. *Am J Respir Crit Care Med* 2005;**171**:1286–91.
215. Blatt R, Bonarini A, Calabro E, Della Torre M, Matteucci M, Pastorino U. Lung cancer identification by an electronic nose based on an array of MOS sensors. In: *2007 IEEE International Joint Conference on Neural Networks*, Vols. 1–6; 2007. p. 1423–8.
216. Mazzone PJ, Hammel J, Dweik R, Na J, Czich C, Laskowski D. Diagnosis of lung cancer by the analysis of exhaled breath with a colorimetric sensor array. *Thorax* 2007;**62**:565–8.

217. Tran VH, Chan HP, Thurston M, Jackson P, Lewis C, Yates D, et al. Breath analysis of lung cancer patients using an electronic nose detection system. *IEEE Sens J* 2010;**10**:1514–8.
218. Taivans I, Jurka N, Balode L, Bukovskis M, Kopeika U, Ogorodniks V, et al. Exhaled air analysis in patients with different lung diseases using artificial odour sensors. *Proc Latvian Acad Sci Sect B Natural Exact Appl Sci* 2009;**63**:210–3.
219. Dragonieri S, Annema JT, Schot R, van der Schee MP, Spanevello A, Carratu P, et al. An electronic nose in the discrimination of patients with non-small cell lung cancer and COPD. *Lung Cancer* 2009;**64**:166–70.
220. D'Amico A, Pennazza G, Santonico M, Martinelli E, Roscioni C, Galluccio G, et al. An investigation on electronic nose diagnosis of lung cancer. *Lung Cancer* 2010;**68**:170–6.
221. Chan HP, Lewis C, Thomas PS. Oxidative stress and exhaled breath analysis: a promising tool for detection of lung cancer. *Cancers* 2010;**2**:32–42.
222. Fens N, Zwinderman AH, van der Schee MP, de Nijs SB, Dijkers E, Roldaan AC, et al. Exhaled breath profiling enables discrimination of chronic obstructive pulmonary disease and asthma. *Am J Respir Crit Care Med* 2009;**180**:1076–82.
223. Dragonieri S, Schot R, Mertens BJ, Le Cessie S, Gauw SA, Spanevello A, et al. An electronic nose in the discrimination of patients with asthma and controls. *J Allergy Clin Immunol* 2007;**120**:856–62.
224. Peng G, Trock E, Haick H. Detecting simulated patterns of lung cancer biomarkers by random network of single-walled carbon nanotubes coated with nonpolymeric organic materials. *Nano Lett* 2008;**8**:3631–5.
225. Peng G, Hakim M, Broza YY, Billan S, Abdah-Bortnyak R, Kuten A, et al. Detection of lung, breast, colorectal, and prostate cancers from exhaled breath using a single array of nanosensors. *Br J Cancer* 2010;**103**:542–51.
226. Savage N. Early detection: spotting the first signs. *Nature* 2011;**471**:S14–5.
227. Yu J-B, Byun H-G, So M-S, Huh J-S. Analysis of diabetic patient's breath with conducting polymer sensor array. *Sensor Actuat B-Chem* 2005;**108**:305–8.
228. Lin YJ, Guo HR, Chang YH, Kao MT, Wang HH, Hong RI. Application of the electronic nose for uremia diagnosis. *Sensor Actuat B-Chem* 2001;**76**:177–80.
229. Guo D, Zhang D, Li N, Zhang L, Yang J. A novel breath analysis system based on electronic olfaction. *IEEE Trans Biomed Eng* 2010;**57**:2753–63.
230. Shih CH, Lin YJ, Lee KF, Chien PY, Drake P. Real-time electronic nose based pathogen detection for respiratory intensive care patients. *Sensor Actuat B-Chem* 2010;**148**: 153–7.
231. Chambers ST, Bhandari S, Scott-Thomas A, Syhre M. Novel diagnostics: progress toward a breath test for invasive *Aspergillus fumigatus*. *Med Mycol* 2011;**49**(Suppl. 1):S54–61.
232. Hockstein NG, Thaler ER, Torigian D, Miller Jr WT. Deffenderfer O, Hanson CW. Diagnosis of pneumonia with an electronic nose: correlation of vapor signature with chest computed tomography scan findings. *Laryngoscope* 2004;**114**:1701–5.
233. Hockstein NG, Thaler ER, Lin Y, Lee DD, Hanson CW. Correlation of pneumonia score with electronic nose signature: a prospective study. *Ann Otol Rhinol Laryngol* 2005;**114**:504–8.
234. Hanson CW, Thaler ER. Electronic nose prediction of a clinical pneumonia score: biosensors and microbes. *Anesthesiology* 2005;**102**:63–8.
235. Yu K, Wang YS, Yu J, Wang P. A portable electronic nose intended for home healthcare based on a mixed sensor array and multiple desorption methods. *Sensor Lett* 2011;**9**:876–83.

236. Louhelainen N, Myllarniemi M, Rahman I, Kinnula VL. Airway biomarkers of the oxidant burden in asthma and chronic obstructive pulmonary disease: current and future perspectives. *Int J Chron Obstruct Pulmon Dis* 2008;**3**:585–603.
237. Machado RF. Identifying chronic obstructive pulmonary disease and asthma by exhaled breath analysis: does the (e)Nose know? *Am J Respir Crit Care Med* 2009;**180**:1038–9.
238. Kushch IG. *Clinical evaluation of breath analysis using TCNQ-derivatives based sensors in adolescents with gastroduodenal pathology*. Kharkiv Medical Academy of Postgraduate Education (Ministry of Health of Ukraine): Kharkiv; 2008. p. 153.

CHAPTER

Arrays of Nanomaterial-Based Sensors for Breath Testing

16

Ulrike Tisch[*] and Hossam Haick[†]

[*] *The Department of Chemical Engineering, Technion-Israel Institute of Technology, Haifa 32000, Israel*

[†] *The Department of Chemical Engineering and Russell Berrie Nanotechnology Institute, Technion-Israel Institute of Technology, Haifa 32000, Israel*

16.1 INTRODUCTION

Modern breath testing requires fast, cost-effective, and high-throughput instrumentation that aids and augments the physicians' diagnosis.[1–5] Current technological advances could soon move diagnostic breath testing into standard medical practice, side-by-side with diagnostic imaging tests and conventional biomarker monitoring in blood, urine, and stool. Spectrometry and spectroscopy techniques are powerful tools for detecting volatile organic compounds (VOCs) that could serve as disease biomarkers.[2,6–8] However, to date, the use of these techniques has been impeded by the need for expensive equipment, the high levels of expertise required to operate such instruments, the lengthy time for analysis, and the need for preconcentration techniques. Significantly smaller, easier-to-use, and less expensive sensing techniques are needed before breath testing can become a clinical reality. Sensor arrays that are based on nanomaterials may answer these requirements, and are therefore very attractive for breath testing in clinical settings.[4,5,9–17]

In conventional sensing technology, a highly selective receptor is designed that specifically binds the analyte of interest.[18] This concept is appropriate for detecting a well-defined target analyte in the presence of a constant or controlled background. However, it is not particularly useful for the analysis and comparison of disease-specific patterns that stem from several relatively similar VOCs, because a separate, highly selective sensor would have to be synthesized for each of the constituent VOCs.[18] In practice, most sensors show some cross-reactivity to chemically or structurally similar compounds.

An alternative strategy utilizes arrays of broadly cross-reactive sensors that allow the identification of disease-related VOC patterns in exhaled breath.[19] Sensor arrays mimic the biological systems responsible for the mammalian sense of smell and are therefore often called *artificial olfactory systems* or *electronic noses*.[20] Every sensor

Volatile Biomarkers. http://dx.doi.org/10.1016/B978-0-44-462613-4.00016-7

in an array responds to all (or to a large subset) of the mixture compounds.[19,20] The constituent sensors must be sufficiently diverse to provide individually different responses to a given VOC mixture, but should not be strictly chemically selective. For breath testing, the sensor array elements should be sensitive to very low concentrations of the VOCs in exhaled breath in the presence of water vapor, because breath samples are fully humidified. Each sensor should respond rapidly to small changes in the concentrations of the breath VOCs, so that the sensor array output is specific to a given metabolic state. When removed from the sample, the sensors should relax rapidly to their baseline states. Alternatively, disposable sensor arrays could be used, if the device fabrication is reproducible and simple enough that large quantities of identical units could be manufactured at acceptable costs.

The combined responses of the sensor array elements are used to establish disease-specific response patterns or, in other words, the *breath print* of the disease, by applying suitable pattern recognition algorithms such as principal component analysis (PCA), supported vector machines (SVM), discriminant factor analysis (DFA), or neuronal network analysis.[19]

16.2 THE DESIGN OF THE SENSOR ARRAY

Identifying a breath print of a specific disease by applying statistical pattern recognition algorithms to the multivariant output of sensor arrays is essentially a *black box* approach to sensing. The particular pitfall of this method is the risk of false positive mixture identifications. This risk is especially high, if complex realistic samples such as exhaled breath are analyzed, using unsuitable sensors and/or very flexible pattern recognition methods.[19] Therefore, the successful implementation of nanomaterial-based sensor arrays for diagnostic breath testing calls for illuminating the black box surrounding the targeted breath sample and the designated sensor array (see schematics in Figure 16.1).[14,15] The first step in this endeavor is the identification of the VOCs that distinguish the disease in question from healthy states, as well as the identification of important background VOCs that stem from possible confounding factors, through complementary chemical analysis (e.g. spectrometry or spectroscopy techniques). This allows an intelligent design of the sensor array and, in addition, the preparation of *artificial* breath mixtures containing the key compounds that make up the disease-specific VOC patterns.

The sensor array should be designed in such a way that each constituent sensor detects all or part of the separate VOCs in the breath sample. When pre-selecting a set of sensors for a particular application, one should consider that the polarity of the breath VOCs often dominates the sensing mechanism in electrical transducers. Polar VOCs are typically detected through charge exchange with suitable polar sensing elements.[14,15,21] Non-polar VOCs are more challenging to sense.[13–15] Such compounds have to be detected indirectly via steric interaction with the sensing material, which, in turn, affects the electronic transport through the sensing element. Concrete examples of sensible choices of nanomaterials for sensing various disease-specific VOC patterns are presented in the following sections. The pre-selection of potentially

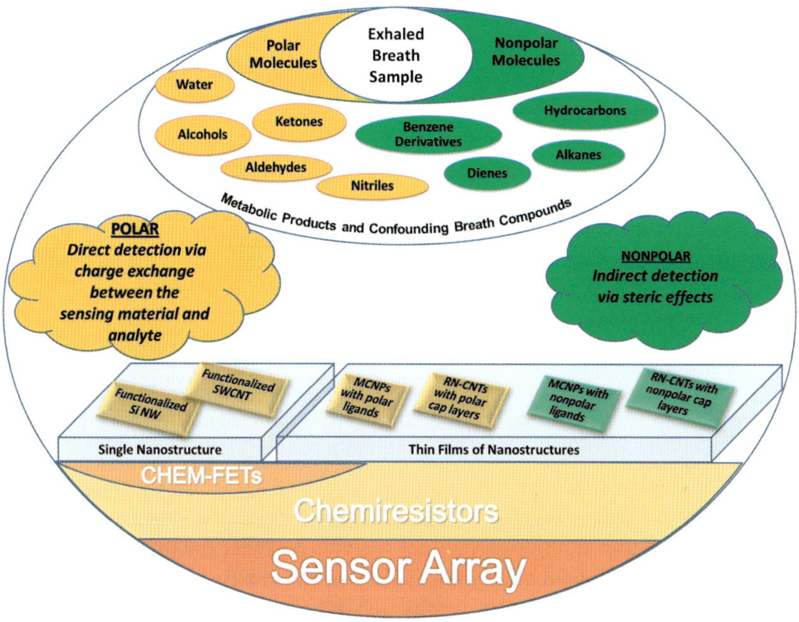

FIGURE 16.1
Schematics illustrating an intelligent approach for choosing the nanomaterial-based sensor array elements for a particular breath testing application, guided by the chemical analysis of the breath sample. Si NW, silicon nanowire; SWCNT, single-walled carbon nanotube; RN-CNTs, random networks of carbon nanotubes; MCNPs, molecularly capped metal nanoparticles; CHEM-FETs, chemically sensitive field-effect transistors.

suitable sensors is followed by the experimental determination of their detection limits, sensitivities, resolutions, and dynamic ranges for the most important VOC markers and also for water molecules. (NOTE: Exhaled breath is fully humidified at body temperature. The relative humidity of breath samples lies between 80% and 90%, depending on the sampling technique.) Then the most suitable sensors can be grouped into an array.

The discriminatory power of the sensor array can efficiently be tested and optimized by exposing it to the artificial breath mixtures that contain the key disease marker VOCs, before exposing it to actual breath samples that were collected from patients and controls. Using artificial breath mixtures during the initial stages of the sensor array design, evaluation, and optimization of the pattern recognition algorithm provides several advantages: (i) the signature of each individual VOC on the sensor array can be determined, (ii) the correlation between the sensor's sensitivity and specificity to an individual VOC and to the same compound within a VOC mixture can be studied, and (iii) the necessary iterative feedback during sensor optimization can be obtained without the intervention of (disruptive) parameters, for example, fluctuations in humidity, temperature, and contamination. If the statistical analysis of the sensor array's output yields good discrimination for the artificial breath mixtures, the next step is testing breath samples collected from patients. Otherwise, an iterative improvement of the sensor array should be attempted, until the sensor array successfully distinguishes between disease states and suitable controls.

16.3 NANOMATERIALS FOR SENSOR ARRAYS

Nanomaterials offer countless advantages for sensing applications. Most significantly, their small characteristic dimensions (1–100 nm) increase the active surface-to-volume ratio and generate novel interfaces, yielding excellent sensitivity as well as rapid response and recovery times. The tailorable chemical and physical properties of nanomaterials that can yield unusual target binding properties, and their structural robustness are other important advantages. The possibility to achieve synergetic properties by combining several organic and/or inorganic nanometric components is especially attractive for the sensing of breath VOCs.

This chapter focuses on the use of monolayer-capped metal nanoparticles (MCNPs), carbon nanotubes (CNTs), and semiconducting nanowires in sensor arrays that are based on electrical transduction. Excellent reviews on other nanomaterials that have been exploited for sensing applications, including those based on optical, piezoelectric, or electrochemical transducers, are available in the literature.[22–24]

16.4 CHEMIRESISTIVE MCNP FILMS FOR SENSOR ARRAYS

MCNP films are among the most suitable nanomaterial-based electrical transducers for sensing disease patterns that are dominated by non-polar VOCs.[14,15] MCNPs can easily be synthesized as colloids in almost any size, shape (e.g. spheres, rods, cubes, and structures with hexagonal, triangular, star-shaped, or branched outlines), and chemical compositions, using wet chemical methods such as metal salt reduction.[25] MNCP films behave as Ohmic conductors for sufficiently small voltages. In chemiresistors, the metal cores of the MCNPs (consisting either of a single metal or of an alloy of two or more metals) provide the electric conductivity. The MCNP ligands form a non-conductive organic matrix between the metal cores that provides selective binding sites for organic molecules. Adsorption of breath VOCs into the organic matrix can have two counteracting effects[14,15]:

- Three-dimensional swelling of the MNCP film, which, in turn, increases the interparticle tunneling distance for charge carriers and, hence, the film resistance.
- An increase of the permittivity of the organic matrix around the metal cores which, in turn, decreases the potential barriers between the metal cores and, consequently, the film resistance.

Currently, most MCNP-based chemiresistors are fabricated from *spherical* MCNPs. However, using films of spherical MCNPs for sensing non-polar breath VOCs can be a disadvantage because the voids between adjacent spherical MCNPs make up 30–60% of the total film volume. These voids can host VOC molecules during the exposure process, which does not contribute (or contributes only minimally) to the swelling-induced sensing signal. Moreover, the interface area between adjacent spherical MCNPs, at which the breath VOCs adsorb and induce changes in the resistivity, is relatively small compared with their total surface area. This could also suppress efficient electron transfer, thereby increasing the signal-to-noise ratio.[21,26,27] Films made of *cubic* MCNPs could help to overcome this limitation and could provide significantly higher sensitivity toward non-polar VOCs.[21,26,27]

The choice of the MCNP ligand is a critical step in the optimization of a sensor array for the detection of specific VOCs in the exhaled breath. Viability matching between specific capping ligands and the targeted breath VOCs must be taken into account to control the sensor-analyte interactions. The wide choice of available ligands (e.g. alkythiols,[26–30] alkylamines,[31] para-thiophenols,[32] carboxylates,[33] organo-dithiols,[34,35] cross-linked polyphenylene dendrimers,[36,37] and cross-linked tetraethyl orthosilicate and α-, ω-dithiolate[38]) allows tailoring the film's chemical selectivity to a specific application. Polar breath VOCs can be sensed using MCNP films with suitable polar ligands through charge exchange (see Figure 16.1). Non-polar VOCs, such as hydrocarbons (see Table 16.1), can be sensed indirectly via steric effects, through absorption into the organic matrix of MCNP films with hydrophobic functionalities (see Figure 16.1), which at the same time suppress the sensitivity to water vapor.[4] The organic ligands of the MCNPs also serve as well-defined organic spacers controlling the nearest neighbor separation. Uniform inter-particle distances in the MCNP films yield very good signal-to-noise ratios. Reproducible electrical properties and sensing characteristics of MCNP films can be achieved by controlling the film structure, using wet chemical deposition techniques (spin-coating, drop-casting, spray-coating, Langmuir-Blodgett techniques, etc.). These assemblies can also be interlinked by direct ligand-linker exchange reactions followed by precipitation or, in a more controlled way, by a layer-by-layer self-assembly procedure.[30,35,36,39]

Clinical studies on breath samples collected from 177 volunteers (see Figure 16.2a) have shown the ability of an array of 14 MCNP chemiresistors (see Figure 16.2(b–c)) to distinguish unambiguously between the four most widespread primary cancers in the western world (lung, colon, breast, and prostate cancers) in a single breath test.[5] The ligands of MCNPs were adjusted to the relatively wide range of marker VOCs of the four targeted cancers. For example, the sensors responded rapidly and fully reversibly to decane, trimethyl benzene, ethylbenzene, and heptanol (cf. Table 16.1) at concentrations between $Pa/Po = 0.001–0.005$.[4] Pa and Po are the partial pressure and the saturated vapor pressure of the analyte at room temperature, respectively. Most of the sensors showed a detection limit of 1–5 parts per billion (ppb) to these representative VOCs.[4] MCNPs functionalized with 4-methoxy-toluenethiol, 2-mercaptobenzoxazole, or 11-mercapto-1-undecanol have detection limits of 2–10 ppb upon exposure to acetaldehyde (a promising VOC for lung cancer[40]) and formaldehyde (a promising VOC for breast cancer[41]), much below the concentration levels of these VOCs in the exhaled breath of cancerous patients.[4] Similar rapid and fully reversible response characteristics were obtained when the sensors were exposed to the breath samples (see Figure 16.2d). The signals from all the sensors in

Table 16.1 Selected examples of breath VOC biomarkers of disease and confounding factors that have been reported in the literature.

	Compound Name	Type	Dipole Moment	Possible Origin
Polar breath VOCs	Acetonitrile	Nitrile	3.92D	Smoking marker[3,43]
	Acetone	Ketone	2.88D	Lung cancer, diabetes,[53] chronic kidney disease[46]
	Acetaldehyde	Aldehyde	2.7D	Breast cancer[40]
	Formaldehyde	Aldehyde	2.33D	Lung cancer[41]
	Hexanol	Alcohol	−1.54D	Lung cancer[4]
	Decanol	Alcohol	1.169D	—
	Octanol	Alcohol		Alzheimer's disease[16]
	Ethanol	Alcohol	1.691	Hepatic steatosis,[53] diabetes,[53] lung cancer[4]
Breath VOCs with very small dipole moments	Ethylbenzene	Benzene derivative	0.6D	Parkinson's disease,[17] diabetes[53]
	2-Methyl-1-butene	Unsaturated hydrocarbon (alkene)	0.51D	Smoking marker[43]
	Isoprene	Diene hydro-carbon	0.25D	Lung cancer,[53] chronic kidney disease[46]
	Styrene	Aromatic hydrocarbon	0.13D	Parkinson's disease,[17] lung cancer,[53] Alzheimer's disease,[16] chronic kidney disease[46]
	Benzene or methylated benzene	Aromatic hydrocarbon	~0.1D	Parkinson's disease,[17] Alzheimer's disease,[16] TB,[53] colorectal cancer[5]
Non-polar breath VOCs	Hexane and methylated hexane	Hydrocarbon	—	Head and neck cancer,[12] Alzheimer's disease,[16] lung cancer[5]
	Decane or methylated decane	Hydrocarbon	—	Lung Cancer,[53] head and neck cancer,[12] Alzheimer's disease,[16] breast cancer,[5] prostate cancer[5]
	Octane or methylated octane	Hydrocarbon	—	Parkinson's disease,[17] lung cancer,[53] Alzheimer's disease,[16] head and neck cancer,[12] chronic kidney disease[46]

the array were analyzed using standard PCA.[4,5,19,42] Figure 16.2e shows that well-separated clusters were obtained for the four different cancers and the healthy states. Remarkably, the results were not affected by confounding factors such as age, gender, intake of food additives, drug treatment, and smoking habits.

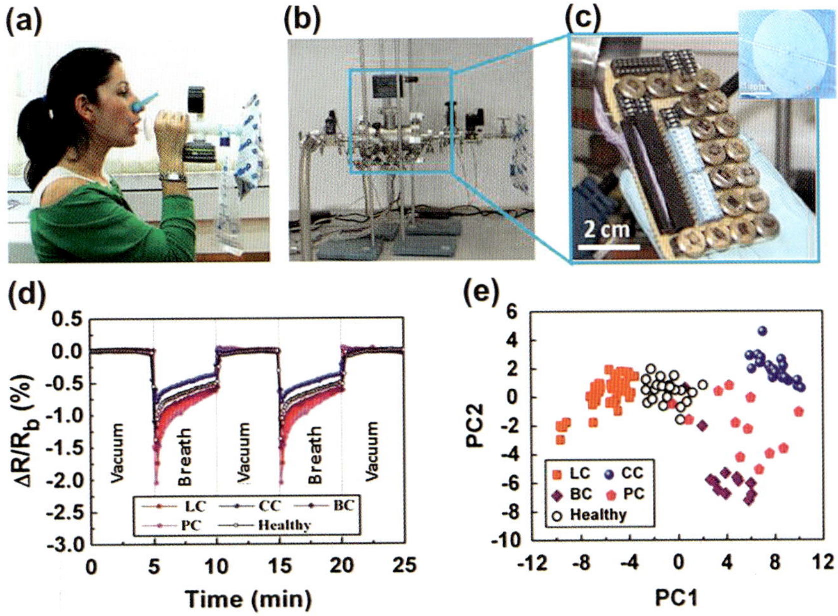

FIGURE 16.2
(a) Laboratory prototype of a handheld device that collects exhaled breath from cancer patients. (b) Exposure chamber with inlet for breath samples. (c) Array of cross-reactive molecularly capped metal nanoparticles chemiresistors for the diagnosis of breath samples—see inset for a scanning electron microscopy image of the sensor's template. (d) Typical responses, $\Delta\ R/R_b$, of one sensor functionalized with 2-ethylhexanethiol upon exposure to breath of patients with lung cancer (LC), colon cancer (CC), breast cancer (BC), and prostate cancer (PC), as well as a healthy controls, as representative examples. (e) PCA pattern recognition for LC, BC, CC, PC, and healthy controls. Reprinted with permission from Ref. 5. Copyright 2010 Nature.

A separate clinical study showed that a choice of five MCNPs with carefully adjusted ligands could identify and distinguish between two cancers that are closely related to tobacco smoking, head-and-neck-cancer (HNC) and lung cancer (LC).[12] In this endeavor, it was particularly important to minimize the sensitivity of the MCNP films to VOCs that stem from cigarette smoking, such as acetonitrile and 2-methyl-1-butene[3,43] (see Table 16.1). Figure 16.3 shows the PCA maps that were calculated from the sensing signals of the five sensors. Excellent separation was achieved between HNC patients and healthy controls (see Figure 16.3a), between LC patients and healthy controls (see Figure 16.3b), as well as between HNC and LC patients (see Figure 16.3c). Figure 16.3d shows that the same sensor configuration and statistical treatment showed no separation between smokers and non-smokers. The three test groups could also be distinguished in a one-step-data analysis (see Figure 16.3e).

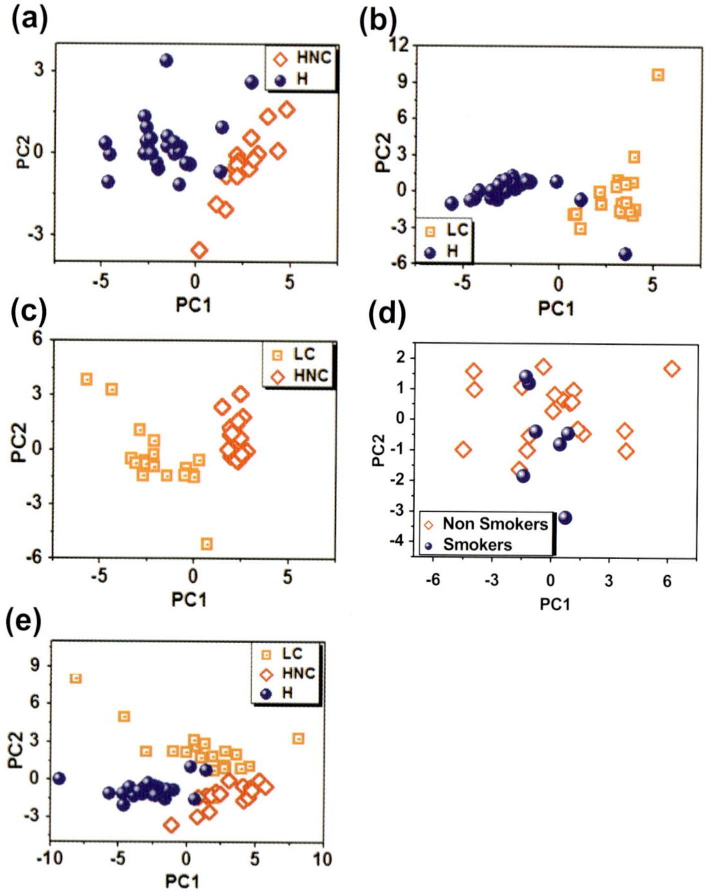

FIGURE 16.3

PCA maps that were calculated from the signals of five MCNP sensors to (a) head-and-neck cancer (HNC) and healthy sub-populations, (b) lung cancer (LC) and healthy (H) sub-populations, (c) HNC and LC, and (d) smokers and non-smokers and (e) all patients: HNC, LC, and healthy controls. Each patient is represented by one point in the plot. The different cancer states are clearly identified and separated, while the same statistical treatment showed no separation between smokers and non-smokers. Reprinted with permission from Ref. 12. Copyright 2011 Nature.

Further studies confirmed that MCNP sensor arrays can indeed be adjusted to the detection of the subtle VOC changes at the onset of cancer, which is vital for future screening applications. An array of seven MCNP sensors, containing six spherical Au MCNPs with different ligands and one cubic Pt MCNP, could detect benign breast conditions including micro-calcifications and benign tumors, that can be (potential) precursors of breast cancer.[44] The benign breast conditions could be distinguished both

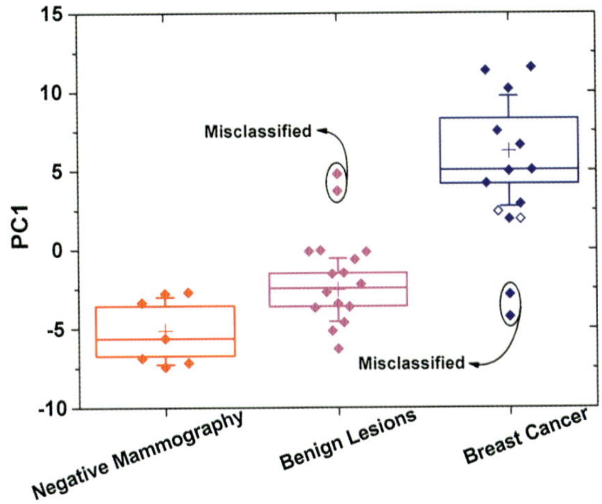

FIGURE 16.4
Graphical representation of the first principal component (PC1) that was obtained from a sensor array based on six spherical Au MCNPs and one cubic Pt MCNP, separating between women with negative mammography, women with benign breast conditions, and breast cancer patients. Each point represents one patient. The positions of the PC1 mean values are marked with +, the boxes correspond to their 95% confidence limits, and the error bars to the standard deviation of PC1. Reprinted with permission from Ref. 44. Copyright 2010 Springer.

from healthy states (negative mammography) and from the early stages of invasive breast cancer, using only the first principal component, PC1, and statistically significant separation ($p < 0.05$) could be achieved. Figure 16.4 represents the PC1 values that were calculated for the three study groups. Early detection of potential breast cancer precursors through breath testing could complement existing breast cancer screening programs and could make cost-effective breast cancer screening available for women who cannot be screened by mammography.

Other arrays of MCNP sensors were successfully applied to the detection and staging of chronic kidney disease (CKD) from exhaled breath.[45] Excellent distinction could be achieved, with an array of only two sensors, between early stage CKD and healthy states.[46] A breath test for the identification of early stage CKD would allow gross diagnosis of CKD by general practitioners, especially in remote areas where access to specialists is rare. Furthermore, an array of three sensors could distinguish between advanced and end-stage CKD with high accuracy (79% and 85%, respectively). Detecting progression from advanced to end-stage CKD is especially important, because it usually marks the onset of dialysis treatment.

16.5 SINGLE-WALLED CARBON NANOTUBES (SWCNTS) FOR SENSORS ARRAYS

Single-walled carbon nanotubes (SWCNTs) can be understood as single rolled-up graphene sheets that form cylinders with a diameter between 1 and 10 nm. SWCNTs possess an extremely high electrical conductivity similar to that of copper, but with the ability to carry much higher currents. They are either metallic or *p*-type semiconducting in air, making them attractive for chemiresistors and chemically sensitive field effect transistors (CHEM-FETs).[47] SWCNTs can also provide a convenient interface to standard microelectronic readout circuitry, since they are typically several microns long.[47] However, using pristine SWCNTs in sensor arrays has several drawbacks, due to their lack of chemical selectivity, their low sensitivity toward non-polar analytes, and their strong interaction with water molecules.[48]

Chemical selectivity can be achieved through functionalizing the SWCNTs with suitable organic molecules or by decorating them with metallic nanoparticles.[47] CHEM-FETs based on individual SWCNTs could serve as highly sensitive sensor array elements for polar breath compounds.[47] These devices, however, have notoriously low sensitivity toward non-polar VOCs. This could be attributed to absence of charge transfer between the adsorbed molecules and the SWCNTs, absence of pinning effect in the SWCNT energy levels, and/or failure to modify the Schottky barrier at the SWCNT/electrode contact.[13]

Using a hybrid structure consisting of a random network of carbon nanotubes (RN-CNT) that is capped with an organic layer may overcome this limitation and, simultaneously, yield devices that are simpler to fabricate.[13,48] In these devices, the organic cap-layers swell upon exposure to the non-polar analytes and, as a result, induce carrier-scattering at the SWCNT junctions, increasing the electrical resistance. Organically functionalized RN-CNTs are also suitable for sensing polar breath VOCs with increased sensitivity, if polar functionalities are used. Therefore, arrays of RN-CNTs functionalized with suitable organic cap-layers hold high potential for detecting mixtures of polar and non-polar breath VOCs.

The strong response of SWCNTs to background humidity tends to obscure small signals obtained from the breath VOCs of interest, especially from non-polar compounds.[48] A promising strategy to diminish the response to water vapor involves coating the RN-CNTs with organic cap layers that can serve as *on-chip* analyte-amplifiers for the VOCs of interest, and, simultaneously, can passivate the RN-CNT against the water molecules. Polycyclic aromatic hydrocarbons (PAHs) and their derivatives, among them hexa-*peri*-hexabenzocoronene (HBC-C_{12}), are particularly interesting for this application, because they: (i) organize into nano- or micro-columns with distinctive geometrical and morphological features; (ii) expose only their side groups to the VOCs; and /or (iii) hold promise for novel sensors based on single PAH nano- or micro-columns.[49,50] Additionally, PAH assemblies can be used as an adsorption layer having either semiconducting properties or relatively insulating properties.

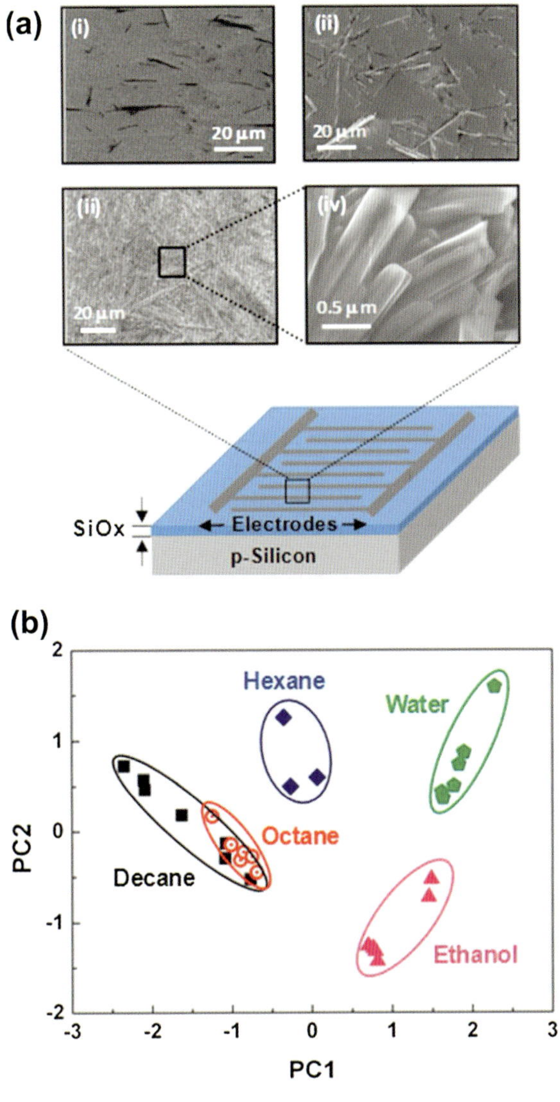

FIGURE 16.5

(a) Schematic representation of a chemiresistor based on random networks of carbon nanotubes (RN-CNTs) that are partially covered by hexa-*peri*-hexabenzocorone $-C_{12}$ layers (HBC-C_{12}), as well as scanning electron micrographs of RN-CNTs covered by different portions with HBC-C_{12} layers. This was achieved by drop-casting solutions with three different HBC-C_{12} concentrations, whereby a higher concentration in solution yielded a higher surface coverage: (i) 10^{-3} M HBC-C_{12} resulted in 10%, surface coverage (ii) 10^{-4} M in 30%, and (iii) 10^{-5} M in 90%. The surface coverage was determined through digital image processing from the different contrasts of covered and uncovered areas in the scanning electron micrographs; (iv) shows a higher magnification of the 90% film

FIGURE 16.5 (*Continued*)
coverage. Note that the RN-CNT is not visible at the relatively low magnification used to show the HBC-C_{12} films. (b) Principle component analysis of the responses of pristine RN-CNTs and RN-CNTs covered 10%, 30%, and 90% with an HBC-C_{12} cap layer, to several representative volatile organic compounds and to water vapor at $P_a/P_0 = 0.2$ (P_a and P_0 stand for partial pressure and saturated vapor pressure, respectively, of the analyte at room temperature). Every data point corresponds to one exposure to a specific analyte. Reprinted with permission from Ref. 51. Copyright 2010 Wiley.

Figure 16.5 shows a schematic representation of a hybrid RN-CNT/HBC-C_{12} sensor, as well as scanning electron micrographs of the HBC-C_{12} cap layers that cover different portions of the RN-CNT surface. The sensitivity and selectivity of the hybrid layers can be tuned by changing the HBC-C_{12} surface coverage. A choice of HBC-C_{12}/RN-CNT bilayers with different HBC-C_{12} surface coverage can form a sensor array with excellent discriminative abilities between polar and non-polar analytes, between different polar analytes, and between different non-polar analytes at very low concentrations in the presence of water vapor. As shown in Figure 16.5b, a small array of pristine RN-CNTs and RN-CNTs that are partially covered with HBC-C_{12} cap layers (10%, 30%, and 90% surface coverage) can robustly distinguish between several representative volatile organic compounds and water vapor.[51]

Accurate and reproducible sensing of representative polar and non-polar breath VOCs was demonstrated in the presence of a realistic humidity background (RH ~ 80%), using sensor arrays based on appropriate combinations of RN-CNTs that were functionalized with different synthetically designed PAH derivatives.[52] The discriminative power of the sensors array could be tailored by controlling the shape (e.g. semi-triangle or triangle shapes) and size (e.g. carbon number of 42–50) of the aromatic corona, and the functional substituents of the PAHs (e.g. alcohol, ester, carboxylic acid, etc.). Figure 16.6a shows the sensing response of several hybrid layers that incorporate different PAHs to simulated polar and non-polar breath VOCs at different concentrations, both in dry (RH ~ 5%) and humid (~ 80%) atmospheres (cf. Table 16.1). Typically, the PAH cap layers substantially improved the sensing signals, compared to pristine RN-CNTs, especially in humid atmospheres (see Figure 16.6a). A combination of three hybrid RN-CNT/PAH layers could distinguish between the ethanol, ethyl benzene, and octane in humid atmospheres with high accuracy, as shown in Figure 16.6b. Abnormal breath levels of these representative polar and non-polar VOCs can be indicative of a variety of diseases, such as diabetes, cancer, neurodegenerative diseases, and kidney disease (cf. Table 16.1).[4,12,16,17,46,53]

RN-CNTs can also be modified with metal nanoparticles[54] or cyclodextrin derivatives,[16,17] in order to provide chemical selectivity for sensor arrays. Cyclodextrins are particular interesting for sensing non-polar or weakly polar VOCs. They have a hydrophilic exterior and possess a considerably less hydrophilic cavity, so that they are able to form host-guest complexes with non-polar VOCs and VOCs with small dipole

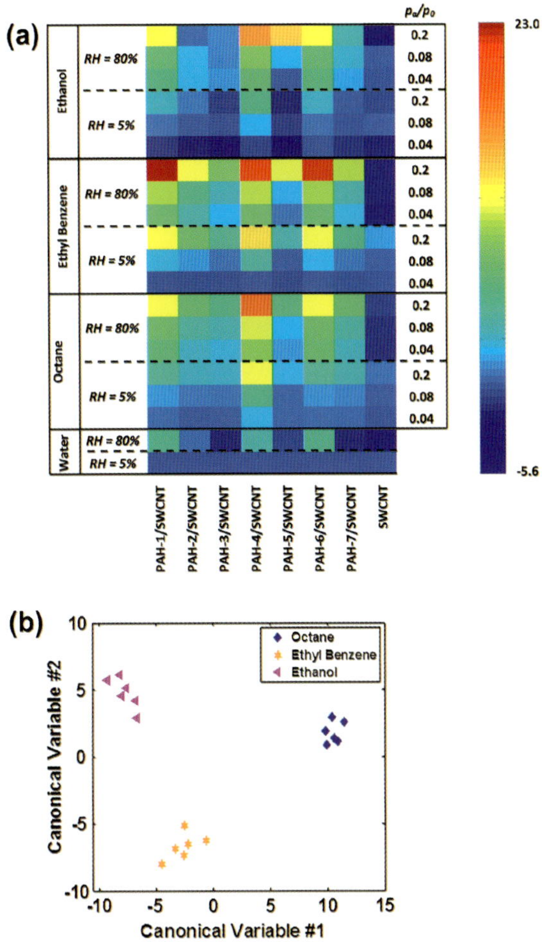

FIGURE 16.6

(a) Hot-plot representation of the sensing signals from different hybrid RN-CNT/PAH layers upon exposure to three representative VOCs exposures in dry (5%) and high humidity (80% RH) atmospheres. (b) DFA discrimination result between octane, ethyl benzene and ethanol measurements in high humidity atmospheres (80% RH), obtained with an array of three RN-CNT/PAH sensors. Reprinted with permission from Ref. 52. Copyright 2011 American Chemical Society.

moments. Integrating RN-CNTs functionalized with different organic functionalities and/or metal nanoparticles into a hybrid sensor array can create a versatile sensing tool that can be easily minimized in size. Lu et al. have recently demonstrated a chip

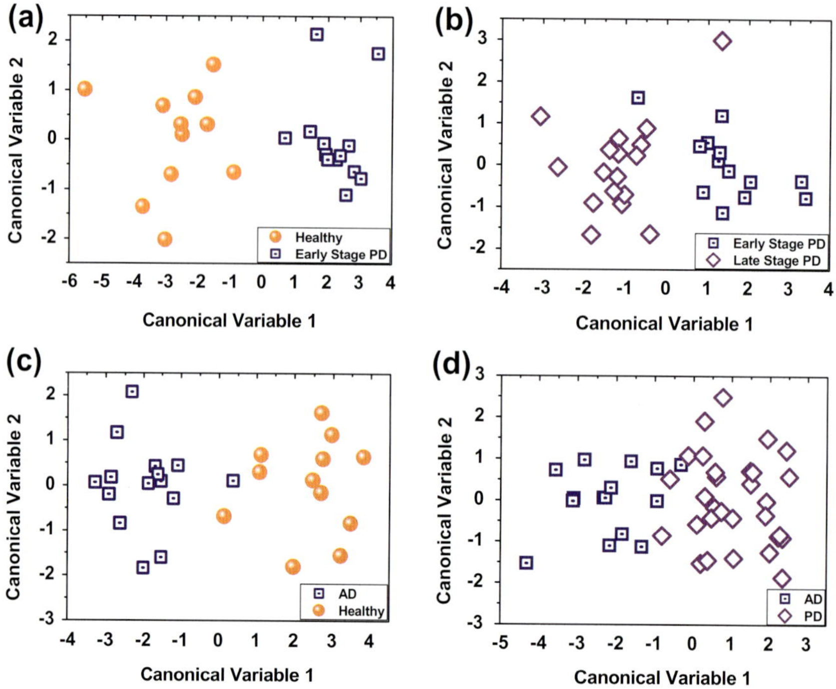

FIGURE 16.7

DFA plots of the first two canonical variables that were calculated from the responses of (a) three RN-CNT-based sensors to distinguish healthy and early-stage PD (Parkinson's disease) patient populations, (b) two RN-CNT sensors and one Pt MCNP sensor to distinguish early-stage and late-stage PD patient populations, (c) two RN-CNT sensors and one Au MCNP sensors to distinguish the AD (Alzheimer's disease) patient population and healthy controls, (d) the same three sensors to distinguish AD and PD patient populations. Each patient is represented by one point in plot. Reprinted with permission from Ref. 16. Copyright 2013 Future Medicine.

with 32 RN-CNT sensing elements that distinguishes acetone and benzene at parts per million (ppm) concentration levels, as well as several inorganic compounds.[54]

Combining RN-CNT sensors with MCNP sensors (cf. Section 16.4) in the same array opens additional possibilities for diagnostic breath testing. For example, a mixed MCNP and RN-CNT sensor array allowed identifying and distinguishing different neurogenetative diseases such as Parkinson's disease (PD) and Alzheimer's disease (AD).[16,17] Preliminary GC-MS studies have indicated that the breath VOC patterns of these diseases could stem from a mixture of polar and non-polar metabolomics products (see Table 16.1).[16,17] Considering the chemical nature of the detected VOCs, an array of sensors based either on spherical Au MCNPs, cubic Pt MCNPs or RN-CNT sensors functionalized with cyclodextrin derivatives was designed. The first two sensor types were designed to react mostly to the non-polar components of the disease-specific VOC patterns, whereas the third type targets polar molecules that are less polar than water. Figure 16.7 shows DFA plots of the first two canonical variables that were calculated from the responses of mixed RN-CNT- and MCNP sensors. A set of three RN-CNT-based sensors distinguished clearly between healthy and early-stage PD patient populations (see Figure 16.7a), and a set of two RN-CNT sensors and one Pt MCNP sensor distinguished early-stage and late-stage PD patient populations (see Figure 16.7b). Furthermore, the combination of two RN-CNT sensors and one Au MCNP sensor was able to distinguish the AD patient population from healthy controls (see Figure 16.7c), as well as the AD and PD patient populations from each other (see Figure 16.7d). The classification success rate was estimated through leave-one-out cross validation. For this purpose, DFA was computed using a training data set that excluded one test sample. After the DFA computation, the test sample was projected onto the two-dimensional DFA space that was calculated using the training set. Thereby the test sample was *blinded* against the DFA model, so that its class affiliation was unknown. All possibilities of leaving out one sample were tested, and the results were recorded in terms of correct and false classifications. From these, the accuracy[†] of the distinction between healthy and early-stage PD, between early-stage and late-stage PD, between healthy and AD and between AD and PD was determined as 96%, 80%, 85%, and 84%, respectively.[16]

16.6 SEMICONDUCTING NANOWIRES FOR CROSS-REACTIVE SENSORS

Si NWs offer unique properties for signal transduction of breath VOCs including: (i) easy fabrication of (p-type and n-type) crystalline Si NWs having nearly uniform diameter, length, and shape;[55–59] (ii) high surface-to-volume ratio, and thus a large surface that can be functionalized;[60,61] (iii) efficient conductivity and high carrier mobility,[58,60–62] yielding high sensitivity to analytes adsorbed on their surfaces[63]; and (vi) electrical properties of Si NWs that can be controlled by means of molecular engineering with organic molecules[42,61,63–67] (cf. also Refs 43,68).

Highly sensitive detection of inorganic polar gas analytes (e.g. N_2O, NO) has been reported,[69] but the detection of non-polar gas molecules still remains challenging. For example, while Si NW FETs exhibited detection limits down to 20 ppb NO, the same devices showed low signal-to-noise ratio at 1000 ppm of the non-polar VOC hexane[69] and almost no responses at 40 ppb hexane.[66] This is probably due to interference of the native oxide layer on the Si NW surface with the sensing process.

Recently, the sensitivity of oxide-coated Si NW FETs has been greatly improved, and detection of several tens of ppm of hexane, octane, and decane (potential disease marker VOCs in exhaled breath, cf. Table 16.1) was achieved by surface modification with non-crossed monolayers of silane molecules. The sensitivity towards non-polar

[†]The accuracy is defined as the ratio between the number of correctly classified samples and the total number of samples.

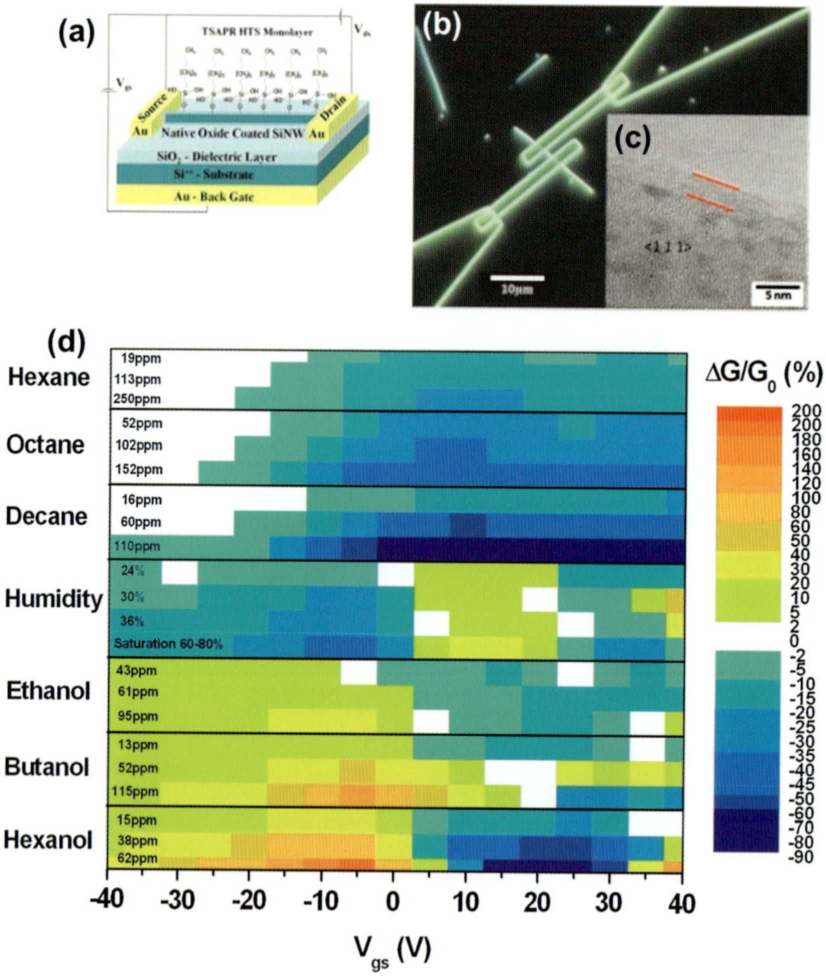

FIGURE 16.8

(a) Schematic illustration of a functionalized back gate Si NW FET sensor; (b) SEM image of a representative Si NW FET; and (c) TEM image of a representative Si NW having ~5 nm native oxide on its surface. (d) Counter map of the relative conductivity change at different gate voltages for several polar and non-polar VOCs. Reprinted with permission from Ref. 72. Copyright 2011 American Chemical Society.

VOCs was comparable to that towards polar VOCs (ethanol, butanol, hexanol which are potential disease marker VOCs in exhaled breath, cf. Table 16.1). Figure 16.8a and b show a schematic representation of the functionalized Si NW FET having 5 nm native surface oxide, and a scanning electron micrograph of the actual device, respectively. The counter map in Figure 16.8c shows the relative conductivity change calculated for the functionalized Si NW FET sensor upon exposure to polar and non-polar VOCs. The most pronounced result is the high sensitivity of the Si NW FET sensor to non-polar VOCs (below the threshold voltage). In the same experiment, a bare Si NW FET without a functional layer showed no response at all to the non-polar VOCs. The sensing of non-polar VOCs was ascribed to changes in the charged surface state density induced by conformational changes of the adsorptive monolayer due to adsorption of non-polar VOCs between the chains and/or the pinholes of the adsorptive monolayer. Increased sensitivity to non-polar molecules might also be achieved by adsorbing appropriate organic receptors directly to the surface of the Si NWs, after removing the oxide layer and attaching appropriate organic receptors that can cover most (>90%) of the Si surface sites.[66]

Another successful strategy for improving the sensitivity to non-polar VOCs relies on attaching appropriate organic receptors that cover most of the Si surface sites, without any intervening oxide.[66] Arrays of non-oxidized Si NW FETs were prepared, and each device was terminated with a different organic functionality. The different organic functionalities were designed to match the representative classes of lung cancer breath biomarkers. Among the attached functionalities were 2-(4-chlorobutyl)-1,3-dioxolane, 4-chlorobutyl chloroformate, 4-chlorobutyl benzoate, 1-chloro-4-phenylbutane, 4-chlorobutyl acetate, 4-chloro-1-butanol, and C_3-C_8 alkyl molecules. The array of sensors was exposed to mixtures that roughly simulate breath patterns of lung cancer and healthy states, in high humidity atmospheres and at room temperature. Based on the complementary GC-MS analysis of breath samples from lung cancer patients and healthy controls, the *cancerous* breath was simulated by a mixture of 40 ppb isoprene, 37 ppb hexanal, 19 ppb styrene, 15 ppb heptanal, 24 ppb 1,2,4-trimethyl benzene, and 22 ppb decane; the *healthy* breath was simulated by a mixture of 26 ppb isoprene, 20 ppb udecane, and 29 ppb decane.[66] PCA of the sensing signals showed clear discrimination between the simulated breath patterns of lung cancer and healthy states (see Figure 16.9). Complementary studies indicated that an improved distinction between healthy and lung cancer states is possible by increasing the diversity of the molecule-terminated, non-oxidized Si NW FETs array.

Apart from Si NWs, other semiconducting nanowires can also be used for sensor arrays. Chen et al. have combined different individual semiconductor nanowire FETs (In_2O_3, SnO_2, ZnO) and RN-CNT FETs into an array for sensing several inorganic gasses and ethanol at sub-ppm levels.[70] Integrated micromachined hotplates allowed precise control of the individual sensors' temperature and guaranteed optimal performance of all constituent sensors.[70,71] The selectivity of semiconducting nanowires could be improved by means of doping, as well as by functionalization with metal nanoparticles or organic cap layers.

16.7 SUMMARY AND CONCLUSIONS

MCNPs, SWCNTs, and semiconducting nanowires could form a solid foundation of diagnostic sensor arrays that could move breath testing in the near future forward into clinical practice. An intelligent choice of nanomaterials for a specific breath test should be based on the biomarkers' polarity that often dominates the biomarker-sensor interaction in electrical transducers. Polar disease-marker VOCs are usually detected directly through charge exchange with suitable polar ligands. Chemically sensitive field-effect transistors based on functionalized individual semiconducting nanowires or SWCNTs can serve as highly sensitive sensor array elements for polar breath compounds. Sensing non-polar breath VOCs, such as volatile hydrocarbons, is more difficult to achieve. Typically, nanomaterials embedded into an organic matrix with suitable functionalities are used to enhance the adsorption of the non-polar analyte molecules into the organic matrix, causing film swelling and affecting the electronic transport through the hybrid layer. Swellable MCNP films or RN-CNTs that are embedded in or beneath organic layers are a suitable choice for sensing non-polar biomarker patterns. These hybrid films can also be tailored for sensing polar breath VOCs, if polar functionalities are incorporated.

Sensors based on MCNPs and RN-CNT have already been applied to realistic breath testing, and pilot clinical studies have shown promising preliminary results for the identification of different cancers, including pre-cancerous states, for kidney disease and for neurodegenerative diseases. The feasibility test of devices based on semiconducting nanowires is still in the pre-clinical phase, and the results are encouraging. Very good detection limits for disease-specific VOC patterns can be achieved, because a few molecules are sufficient to change the electrical properties of the nano-sized sensing elements in electronic transducers. Easy, cost-effective, and high-throughput fabrication of chemisensitive nanomaterial-based electric

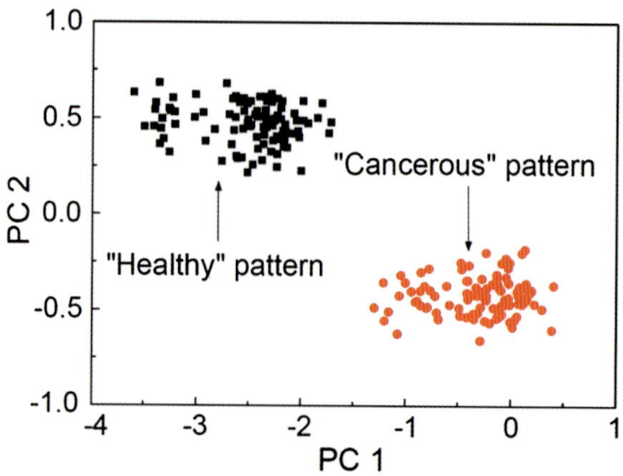

FIGURE 16.9

PCA of an array of molecule-terminated, non-oxidized Si NW FETs on exposure to a mixture that roughly simulates healthy and cancerous breath samples. *Cancerous* breath: 40 ppb isoprene, 37 ppb hexanal, 19 ppb styrene, 15 ppb heptanal, 24 ppb 1,2,4-trimethyl benzene, and 22 ppb decane. *Healthy* breath: 26 ppb isoprene, 20 ppb udecane, and 29 ppb decane. Reprinted with permission from Ref. 66. Copyright 2008 IEEE.

transducers can be achieved because the sensing signals of one-dimensional nanomaterials such as tubes or wires, as well as two-dimensional nanomaterial films, can be read by means of conventional microelectronic readout circuitry.

Nanomaterial-based sensor arrays for breath testing still face many challenges on the way toward clinical practice, among them the lack of precise knowledge about disease biomarkers that should be the basis for intelligent sensor design, the effect of the confounding high and variable content of water vapor in the breath samples on the sensing of biomarker VOCs at (sub) ppb concentrations, and the reproducibility of the device fabrication process.

Acknowledgments

Funding from the Marie Curie Excellence Grant of the FP-6, the European Research Council (ERC), and FP7-Health Programs are acknowledged.

References

1. Cao WQ, Duan YX. Current status of methods and techniques for breath analysis. *Crit Rev Anal Chem* 2007;**37**:3–13
2. Amann A, Spanel P, Smith D. Breath analysis: the approach towards clinical applications. *Mini Rev Med Chem* 2007;**7**:115–29.
3. Miekisch W, Schubert JK, Noeldge-Schomburg GF. Diagnostic potential of breath analysis—focus on volatile organic compounds. *Clin Chim Acta* 2004;**347**:25–39.
4. Peng G, Tisch U, Adams O, Hakim M, Shehada N, Broza YY, et al. Diagnosing lung cancer in exhaled breath using gold nanoparticles. *Nat Nanotechnol* 2009;**4**:669–73.
5. Peng G, Hakim M, Broza YY, Billan S, Abdah-Bortnyak R, Kuten A, et al. Detection of lung, breast, colorectal, and prostate cancers from exhaled breath using a single array of nanosensors . *Br J Cancer* 2010;**103**:542–51.
6. Pauling L, Robinson AB, Teranishi R, Cary P. Quantitative analysis of urine vapor and breath by gas-liquid partition chromatography. *Proc Natl Acad Sci U S A* 1971;**68**: 2374–6.
7. Amann A, Miekisch W, Pleil J, Risby T, Schubert W. Methodological issues of sample collection and analysis of exhaled breath. Horvath I, de Jongste JC European respiratory society monograph. *European Respiratory Society Lausanne* 2010;**49**:96–114.
8. Amann A, Poupart G, Telser S, Ledochowski M, Schmid A, Mechtcheriakov S. Applications of breath gas analysis in medicine. *Int J Mass Spectrom* 2004;**239**:227–33.
9. Barash O, Peled N, Hirsch FR, Haick H. Sniffing the unique odor print of non-small-cell lung cancer with gold nanoparticles. *Small* 2009; **5**:2618–24.
10. Haick H. Chemical sensors based on molecularly modified metallic nanoparticles. *J Phys D Appl Phys* 2007;**40**:7173–86.
11. Haick H, Hakim M, Patrascu M, Levenberg C, Shehada N, Nakhoul F, et al. Sniffing chronic renal failure in rat model by an array of random networks of single-walled carbon nanotubes. *Acs Nano* 2009;**3**:1258–66.
12. Hakim M, Billan S, Tisch U, Peng G, Dvrokind I, Marom O, et al. Diagnosis of head-and-neck cancer from exhaled breath. *Br J Cancer* 2011;**104**:1649–55.

13. Peng G, Tisch U, Haick H. Detection of nonpolar molecules by means of carrier scattering in random networks of carbon nanotubes: toward diagnosis of diseases via breath samples. *Nano Lett* 2009;**9**:1362–8.
14. Tisch U, Haick H. Sensors based on monolayer-capped metal nanoparticles. In: Korotcenkov G. editor. *Chemical sensors nanstructured materials*, Vol.2. 2010. p. 141–202.
15. Tisch U, Haick H. Nanomaterials for cross-reactive sensor arrays. *Mrs Bull* 2010;**35**: 797–803.
16. Tisch U, Schlesinger I, Ionescu R, Nassar M, Axelrod N, Robertman D, Tessler Y, Marmur A, Aharon-Peretz J, Haick H. Detection of Alzheimer's and Parkinson's Diseases from exhaled breath using nanomaterial-based sensors. *Nanomedicine* 2013; **8**:43–56.
17. Tisch U, Aluf Y, Ionescu R, Nakhleh M, Bassal R, Axelrod N, et al. Detection of asymptomatic nigrostriatal dopaminergic lesion in rats by exhaled air analysis using carbon nanotube sensors. *ACS Chem Neurosci* 2012;**3**:161-66.
18. Lavigne JJ, Anslyn EV. Sensing a paradigm shift in the field of molecular recognition: from selective to differential receptors. *Angew Chem Int Edit* 2001;**40**:3119–30.
19. Rock F, Barsan N, Weimar U. Electronic nose: current status and future trends. *Chem Rev* 2008;**108**:705–25.
20. Persaud K, Dodd G. Analysis of discrimination mechanisms in the mammalian olfactory system using a model nose. *Nature* 1982;**299**:352–5.
21. Dovgolevsky E, Konvalina G, Tisch U, Haick H. Mono layer-capped cubic platinum nanoparticles for sensing nonpolar analytes in highly humid atmospheres. *J Phys Chem C* 2010;**114**:14042–9.
22. Wang GQ, Wang YQ, Chen LX, Choo J. Nanomaterial-assisted aptamers for optical sensing. *Biosens Bioelectron* 2010;**25**:1859–68.
23. Lang HP, Hegner M, Gerber C. Cantilever array sensors. *Mater Today* 2005;**8**:30–6.
24. Sadik OA, Mwilu SK, Aluoch A. Smart electrochemical biosensors: from advanced materials to ultrasensitive devices. *Electrochim Acta* 2010;**55**:4287–95.
25. Burda C, Chen XB, Narayanan R, El-Sayed MA. Chemistry and properties of nanocrystals of different shapes. *Chem Rev* 2005;**105**:1025–102.
26. Dovgolevsky E, Tisch U, Haick H. Chemically sensitive resistors based on monolayer-capped cubic nanoparticles: towards configurable nanoporous sensors. *Small* 2009;**5**: 1158–61.
27. Dovgolevsky E, Haick H. Direct observation of the transition point between quasi-spherical and cubic nanoparticles in a two-step seed-mediated growth method. *Small* 2008;**4**: 2059–66.
28. Wohltjen H, Snow AW. Colloidal metal-insulator-metal ensemble chemiresistor sensor. *Anal Chem* 1998;**70**:2856–9.
29. Shiigi H, Yamamoto Y, Yakabe H, Tokonami S, Nagaoka T. Electrical property and water repellency of a networked monolayer film prepared from Au nanoparticles. *Chem Commun* 2003:1038–9.
30. Joseph Y, Guse B, Yasuda A, Vossmeyer T. Chemiresistor coatings from Pt- and Au-nanoparticle/nonanedithiol films: sensitivity to gases and solvent vapors. *Sensor Actuat B-Chem* 2004;**98**:188–95.
31. Joseph Y, Krasteva N, Besnard I, Guse B, Rosenberger M, Wild U, et al. Gold-nanoparticle/organic linker films: self-assembly, electronic and structural characterisation, composition and vapour sensitivity. *Faraday Discuss* 2004;**125**:77–97.

32. Evans SD, Johnson SR, Cheng YLL, Shen TH. Vapour sensing using hybrid organic-inorganic nanostructured materials. *J Mater Chem* 2000;**10**:183–8.
33. Leopold MC, Donkers RL, Georganopoulou D, Fisher M, Zamborini FP, Murray RW. Growth, conductivity, and vapor response properties of metal ion-carboxylate linked nanoparticle films. *Faraday Discuss* 2004;**125**:63–76.
34. Joseph Y, Peic A, Chen XD, Michl J, Vossmeyer T, Yasuda A. Vapor sensitivity of networked gold nanoparticle chemiresistors: importance of flexibility and resistivity of the interlinkage. *J Phys Chem C* 2007;**111**:12855–9.
35. Joseph Y, Guse B, Vossmeyer T, Yasuda A, Gold nanoparticle/organic networks as chemiresistor coatings: the effect of film morphology on vapor sensitivity. *J Phys Chem C* 2008;**112**: 12507–14.
36. Krasteva N, Besnard I, Guse B, Bauer RE, Mullen K, Yasuda A, et al. Self-assembled gold nanoparticle/dendrimer composite films for vapor sensing applications. *Nano Lett* 2002;**2**: 551–5.
37. Krasteva N, Fogel Y, Bauer RE, Mullen K, Joseph Y, Matsuzawa N, et al. Vapor sorption and electrical response of Au-nanoparticle-dendrimer composites. *Adv Funct Mater* 2007;**17**:881–8.
38. Dirk SM, Howell SW, Price BK, Fan HY, Washburn C, Wheeler DR, et al. Vapor sensing using conjugated molecule-linked Au nanoparticles in a silica matrix. *J Nanomater* 2009.
39. Joseph Y, Besnard I, Rosenberger M, Guse B, Nothofer HG, Wessels JM, et al. Self-assembled gold nanoparticle/alkanedithiol films: preparation, electron microscopy, XPS-analysis, charge transport, and vapor-sensing properties. *J Phys Chem B* 2003;**107**:7406–13.
40. Smith D, Wang TS, Sule-Suso J, Spanel P, El Haj A. Quantification of acetaldehyde released by lung cancer cells *in vitro* using selected ion flow tube mass spectrometry. *Rapid Commun Mass Sp* 2003;**17**:845–50.
41. Ebeler SE, Clifford AJ, Shibamoto T. Quantitative analysis by gas chromatography of volatile carbonyl compounds in expired air from mice and human. *J Chromatogr B* 1997; **702**:211–5.
42. Jie JS, Zhang WJ, Peng KQ, Yuan GD, Lee CS, Lee ST. Surface-dominated transport properties of silicon nanowires. *Adv Funct Mater* 2008;**18**:3251–7.
43. Bajtarevic A, Ager C, Pienz M, Klieber M, Schwarz K, Ligor M, et al. Noninvasive detection of lung cancer by analysis of exhaled breath. *BMC Cancer* 2009;**9**:348.
44. Shuster G, Gallimidi Z, Reiss AH, Dovgolevsky E, Billan S, Abdah-Bortnyak R, et al. Classification of breast cancer precursors through exhaled breath. *Breast Cancer Res Treat* 2011;**126**:791–6.
45. Devarajan P. Review: neutrophil gelatinase-associated lipocalin: a troponin-like biomarker for human acute kidney injury. *Nephrology (Carlton)* 2010;**15**:419–28.
46. Marom O, Nakhoul F, Tisch U, Shiban A, Abassi Z, Haick H. Gold nanoparticle sensors for detecting chronic kidney disease and disease progression. *Nanomedicine* 2012;**7**:639-50.
47. Kauffman DR, Star A. Carbon nanotube gas and vapor sensors. *Angew Chem Int Edit* 2008;**47**:6550–70.
48. Peng G, Trock E, Haick H. Detecting simulated patterns of lung cancer biomarkers by random network of single-walled carbon nanotubes coated with nonpolymeric organic materials. *Nano Lett* 2008;**8**:3631–5.
49. Kastler M, Pisula W, Wasserfallen D, Pakula T, Mullen K. Influence of alkyl substituents on the solution- and surface-organization of hexa-peri-hexabenzocoronenes. *J Am Chem Soc* 2005;4286–96.

50. Wu JS, Pisula W, Mullen K. Graphenes as potential material for electronics. *Chem Rev* 2007;**107**;718–47.
51. Zilberman Y, Tisch U, Shuster G, Pisula W, Feng XL, Mullen K, et al. Carbon nanotube/hexa-peri-hexabenzocoronene bilayers for discrimination between nonpolar volatile organic compounds of cancer and humid atmospheres. *Adv Mater* 2010;**22**: 4317–20.
52. Zilberman Y, Ionescu R, Feng XL, Mullen K, Haick H. Nanoarray of polycyclic aromatic hydrocarbons and carbon nanotubes for accurate and predictive detection in real-world environmental humidity. *Acs Nano* 2011;**5**:6743–53.
53. Modak AS. Breath biomarkers for personalized medicine. *Personalized Medicine* 2010; **7**:643–53.
54. Lu YJ, Partridge C, Meyyappan M, Li J. A carbon nanotube sensor array for sensitive gas discrimination using principal component analysis. *J Electroanal Chem* 2006;**593**: 105–10.
55. Cui Y, Duan XF, Hu JT, Lieber CM. Doping and electrical transport in silicon nanowires. *J Phys Chem B* 2000;**104**:5213–6.
56. Zhang YH, Tang YN, Wang N, Yu D.P, Lee CS, Bello I, et al. Silicon nanowires prepared by laser ablation at high temperature. *Appl Phys Lett* 1998;**72**:1835–7.
57. Hu JT, Odom TW, Lieber CM. Chemistry and physics in one dimension: synthesis and properties of nanowires and nanotubes. *Accounts Chem Res* 1999;**32**:435–45.
58. Schmidt V, Wittemann JV, Senz S, Gosele U. Silicon nanowires: a review on aspects of their growth and their electrical properties. *Adv Mater* 2009;**21**:2681–702.
59. Das Kanungo P, Kogler R, Nguyen-Duc K, Zakharov N, Werner P, Gosele U. Ex situ n and p doping of vertical epitaxial short silicon nanowires by ion implantation. *Nanotechnology* 2009;**20**:165706.
60. Thelander C, Agarwal P, Brongersma S, Eymery J, Feiner LF, Forchel A, et al. Nanowire-based one-dimensional electronics. *Materials Today* 2006;**9**:28–35.
61. Blase X, Fernandez-Serra MV. Preserved conductance in covalently functionalized silicon nanowires. *Phys Rev Lett* 2008;**100**:046802.
62. Cui Y, Zhong ZH, Wang DL, Wang WU, Lieber CM. High performance silicon nanowire field effect transistors. *Nano Lett* 2003;**3**:149–52.
63. Leao CR, Fazzio A, da Silva AJR. Si nanowires as sensors: choosing the right surface. *Nano Lett* 2007;**7**:1172–7.
64. Haick H, Hurley PT, Hochbaum AI, Yang PD, Lewis NS. Electrical characteristics and chemical stability of non-oxidized, methyl-terminated silicon nanowires. *J Am Chem Soc* 2006;**128**:8990–1.
65. Leu PW, Shan B, Cho KJ. Surface chemical control of the electronic structure of silicon nanowires: density functional calculations. *Phys Rev B* 2006;**73**:195320.
66. Assad O, Haick H, Chemically Sensitive Field Effect Transistors of Oxide-Free Silicon Nanowires—Towards Detection of Volatile Biomarkers of Cancer. In *2008 IEEE international symposium on industrial electronics*, vols. 1–5; 2008. p. 484–8.
67. Bashouti MY, Tung RT, Haick H. Tuning the electrical properties of si nanowire field-effect transistors by molecular engineering. *Small* 2009;**5**:2761–9.
68. Magid I, Burstein L, Seitz O, Segev L, Kronik L, Rosenwaks Y. Electronic characterization of Si(100)-bound alkyl monolayers using kelvin probe force microscopy. *J Phys Chem C* 2008;**112**:7145–50.

69. McAlpine MC, Ahmad H, Wang DW, Heath JR. Highly ordered nanowire arrays on plastic substrates for ultrasensitive flexible chemical sensors. *Nat Mater* 2007;**6**:379–84.
70. Chen PC, Ishikawa FN, Chang HK, Ryu K, Zhou C. A nanoelectronic nose: a hybrid nanowire/carbon nanotube sensor array with integrated micromachined hotplates for sensitive gas discrimination. *Nanotechnology* 2009;**20**:125503.
71. Santra S, Guha PK, Ali SZ, Hiralal P, Unalan HE, Covington JA. ZnO nanowires grown on SOI CMOS substrate for ethanol sensing. *Sensor Actuat B-Chem* 2010;**146**:559–65.
72. Paska Y, Stelzner T, Christiansen S, Haick H. Enhanced sensing of nonpolar volatile organic compounds by silicon nanowire field effect transistors. *Acs Nano* 2011;**5**:5620–6.

CHAPTER

Smart Sensor Systems for Human Health Breath Monitoring Applications

17

Gary W. Hunter[*], Jennifer C. Xu[*], A.M. Biaggi-Labiosa[*], Prabir K. Dutta[†], Suvra P. Mondal[†], Benjamin J. Ward[‡], D.B. Makel[‡], Chung-Chiun Liu[§], C.W. Chang[**], D. Laskowski[††], and Raed A. Dweik[††]

[*] *NASA Glenn Research Center, 21000 Brookpark Road, Cleveland, OH 44135, USA*

[†] *Ohio State University, Department of Chemistry, 100 West 18th Avenue, Columbus, OH 43210, USA*

[‡] *Makel Engineering, Inc., 15505 Neo Parkway, Cleveland, OH 44128, USA*

[§] *Department of Chemical Engineering, Case Western Reserve University, 10900 Euclid Ave., Cleveland, OH 44106, USA*

[**] *ASRC /NASA Glenn Research Center, 21000 Brookpark Road, Cleveland, OH 44135, USA*

[††] *Cleveland Clinic, Respiratory Institute & Pathobiology/Lerner Research Institute, Desk A90, 9500 Euclid Ave., Cleveland, OH 44195, USA*

17.1 INTRODUCTION

One approach to fully realize the potential of breath analysis as a diagnostic tool is the development of a portable, Smart Breath Health Diagnostic System (SBHDS) with comprehensive capabilities that can be used in the clinic or at home.[1] It is envisioned that, at any time or place as needed, a patient could breathe into this device and information regarding the patient's health would be obtained. While no single technique can provide complete diagnosis of a patient, a wide range of conditions may be diagnosed using the SBHDS. Table 17.1 shows a brief list of exhaled gaseous species and constituents that have been identified as potential disease markers and could be targeted by an SBHDS. This list is preliminary and a significant amount of work is necessary in order to correlate the chemical species biomarkers exhaled with underlying physiological state and potential diseases. Further, even given the correlation between a given exhaled chemical species or combination of species and a given disease state, there are notable technical challenges in measuring chemical species with both the selectivity and sensitivity needed in order to make the definitive assessment of the chemical composition of the breath. Nonetheless, the development of a SBHDS is viewed as a major component of the future of medicine.

The SBHDS needs to be compact, simple to use, energy efficient, able to store tests results, and capable of transmitting data. The overall idea is to provide a simple tool that can enhance diagnosis and treatment at an earlier stage and lower cost. This vision is being advanced today as improvements in measurement technology and miniaturized sensor systems are realized. Previously, we have discussed the technical

Volatile Biomarkers. http://dx.doi.org/10.1016/B978-0-44-462613-4.00017-9
2013 Published by Elsevier B.V.

Table 17.1 A sampling of the correlation between exhaled chemical species and constituents, physical ailment, and possible cause.[2–6]

Chemical Species Detected	Physical Ailment	Possible Cause
NO	Asthma, airway inflammation	Triggered condition due to irritants or exposures, lung inflammation
CO	Lung inflammation, hemolysis, smoke exposure	Asthma, hemolytic anemia, various exposures
Hydrocarbons	Lung cancer, other cancers, other lung or systemic diseases	Radiation exposure, previous smoking, metabolic changes/ abnormalities
H_2S	Airway inflammation	Lung inflammation
pH	Gastric acid reflux	GERD/peptic ulcer disease

approaches associated with the realization of a SBHDS.[1] These approaches include the application of silicon processing or Micro Electro Mechanical Systems (MEMS) to enable new sensor capabilities, and combined with the use of smart sensor technology, lead to an instrument that is a simple, cost-effective, and accurate diagnostic too.[7–12]

The first steps toward the SBHDS are ongoing. This chapter discusses this smart sensor technology approach to the SBHDS, and efforts to migrate this technology into breath monitoring applications as part of an effort funded by the State of Ohio Third Frontier Program. The emphasis is on monitoring inflammation in the airways of individuals with asthma via the measurement of nitric oxide (NO) in the low parts-per-billion, ppb, range, a known marker of inflammation in a subset of asthmatics.[13,14] This effort has evolved into a much smaller NO sensor system that will be much more cost effective than the present FDA approved monitors.[5] Future system capabilities for health care applications will be able to adapt and monitor many other compounds associated with disease as these compounds are identified. This chapter will: (1) Provide a brief review of MEMS techniques and Smart Sensor System technology; (2) Describe a demonstration of this technical approach for exercise monitoring; (3) Discuss a solid state sensor approach to monitoring NO in asthma, including a hardware demonstration; (4) Describe the effort to miniaturize the sensor system and the future impact to the biomedical field.

17.2 SMART SENSOR SYSTEMS

A *Smart Sensor System* is, at a minimum, the combination of a sensing element with processing capabilities provided by a microprocessor.[15] A more expansive view of a Smart Sensor System is a complete self-contained sensor system that includes the capabilities for data storage and processing, self-contained power, and an ability to transmit or display informative data to a user, operational system, or monitor station (e.g. a clinical data center). One technical direction that is the topic of this chapter is to combine this Smart Sensor System approach with microsensor platform technology.[7,9,16,17] This chemical microsensor technology is produced using MEMS technology. The advent of MEMS technology allows the production of sensors of decreased size, weight, and low power consumption. Batch processing is used to produce many of the same type of sensor in one fabrication run with precise control of sensor structure, while at the same time reducing costs of production. New sensor platforms provide very different types of information on the environment and these sensors are meant to have limited cross-sensitivity (i.e. be orthogonal in their response).[7] Integration of the information from these orthogonal sensors can provide increased whole-field information on the environment.[16,17]

An example of sensor system development that demonstrates the above technology trends is an integrated Smart Leak Detection System targeted for a range of launch vehicle propulsion systems.[8,9,18,19] The system shown in Figure 17.1 includes a microsensor array fabricated using microfabrication (MEMS) based technology, featuring two hydrogen sensors and a hydrocarbon sensor used to detect fuel leaks. The array has been incorporated with signal conditioning electronics, power, data storage, and telemetry. This postage stamp sized "Lick and Stick" type gas sensor technology can enable a matrix of leak detection sensors placed throughout a region with minimal size and weight. Other Smart Sensor Systems have been developed for fire detection, environmental monitoring, and emissions monitoring purposes. These sensor systems also include the capabilities for multiparameter detection, data processing, wireless communication, and if necessary, battery-powered operation.[16,17] The electronics can be programmed to provide the user with the specific information required on a regular basis, but can also provide further diagnostic information

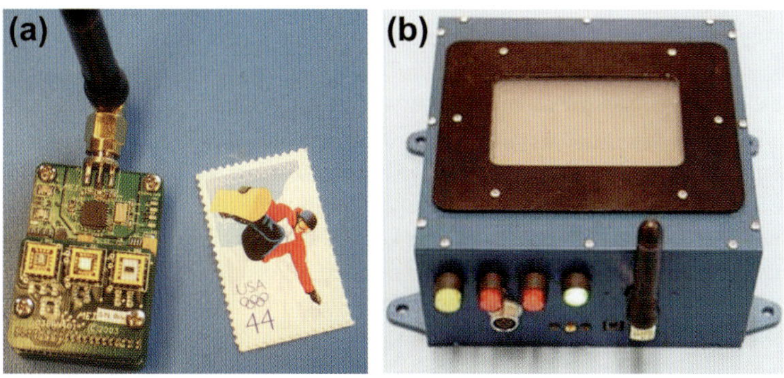

FIGURE 17.1
(a) A "Lick and Stick" Leak Detection System with two hydrogen sensors and a hydrocarbon sensor combined with supporting electronics and wireless communication. (b) A wireless hub that is an interface for multiple "Lick and Stick" wireless systems.

when needed. One approach is to place a number of these sensors in a region, and the resulting measurements are then fed into a central processing hub to allow an understanding of the region or environment (Figure 17.1b). This approach has been demonstrated previously for environmental monitoring applications.[17]

One of the fundamental features of this Smart Sensor System technology is that, if chosen properly in the development stage, base microsensor platform technology combined with a core Smart Sensor System hardware platform can address a range of applications. The overall approach is to work from this core technology base and adapt the Smart Sensor System to individual applications as necessary. This is one of the advantages of a Smart Sensor System: the ability to program new functions into processors and swap out module boards within the system allowing a wide range of adaptability.[16] In order to meet the needs of other applications using this technical foundation, further targeted development and adaptation of the system is needed. In particular, to address human health breath monitoring applications, further development of this technical foundation of base platform microsensor technology and Smart Sensor System technology has been ongoing.

17.3 BREATH MONITORING: SMART SENSOR SYSTEM DEVELOPMENT

The development described below first demonstrates the basic capability of a Smart Sensor System to measure oxygen (O_2) and carbon dioxide (CO_2), and the following sections then discuss the present development of a portable asthma monitoring system.

17.3.1 Exercise breath sensor system demonstration

A motivation for development of a breath monitoring system is related to the use of exercise equipment by astronauts. Astronauts are presently placed on a regular exercise program designed as a countermeasure to the physical degradation due to microgravity exposure. This exercise program is conducted in a period of time when the astronaut, whose work load is considerable, is already involved in health-related activities, and monitoring has limited impact on their schedule. Optimization of the astronaut's time, the effectiveness of the workout, and the concurrent information derived on the astronaut's health would be highly beneficial. The detection of O_2 and CO_2 are fundamental measurements for exercise monitoring. A non-intrusive system that can maximize the effect of the workout is an enabling technology for long-term space missions. This system would need to be easily integrated into astronaut exercise equipment, provide reliable data for long durations without recalibration, have little effect on vehicle mass or power consumption, and provide the data to either the astronaut or ground systems in an easy-to-use manner. A core feature of this approach is that it can also be applied to general patient exercise monitoring.

A proof-of-concept miniaturized metabolic gas monitor system for use in exercise feedback and astronaut health monitoring based on the integration of microsensor technology with Smart Sensor System hardware has been demonstrated as

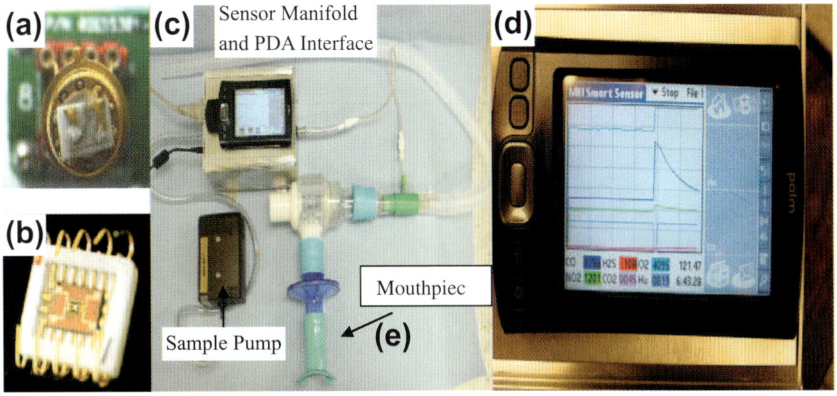

FIGURE 17.2
Exercise Breath Sensor System: (a) Packaged CO_2 Sensor; (b) Packaged O_2 sensor; (c) Picture of overall Exercise Breath Sensor System including a mouthpiece for breath collection, a sensor manifold with personal digital assistant (PDA) interface, and a mini sampling pump; (d) Sensor system user interface, which operates using PDA or a laptop.

described below. A prototype system was fabricated and tested using miniaturized solid-state O_2 and CO_2 sensors as the core sensing technique. Examples of the O_2 and CO_2 sensor technology is shown in Figure 17.2a and b, while the operational principles of the sensors are described elsewhere.[8,9,16,17] These core sensor technologies were then integrated into a system which provided an interface to the subject, as well as capabilities for processing and analyzing the data.

The overall system is shown in Figure 17.2c and includes a mouthpiece for breath collection, the sensor manifold with personal digital assistant (PDA) interface, and a mini sampling pump. The subject breathes into the mouthpiece and the breath is analyzed by sensors integrated into the manifold system. The sensor signals are processed and displayed on a handheld unit. The user interface for the system operates using a PDA as shown in Figure 17.2d, or could be modified for multiple possible user interfaces to allow interfacing with the computer or even wireless transmission to a central processing point (e.g. within a room, a hospital, or even a remote doctor's office).

The complete system was characterized, including patient testing, to evaluate the maturity and capability of microsensors to replace rack-sized lab equipment with a miniaturized metabolic gas monitor system. Figure 17.3 shows the experimental setup. The instrumentation includes the capability to monitor O_2 via an oxygen analyzer with a heated zirconia sensor, CO_2 via a near-IR detector, and a pneumotach. Additional physiological parameters such as blood pressure (BP), pulse rate, and others can also be tracked, but data from these devices were not used in this demonstration.

The subject's exhaled breath was fed into the microsensor based system shown in Figure 17.2c, in parallel to the benchtop instrumentation. Figure 17.4 compares representative data provided by the microsensors and the benchtop instrumentation,

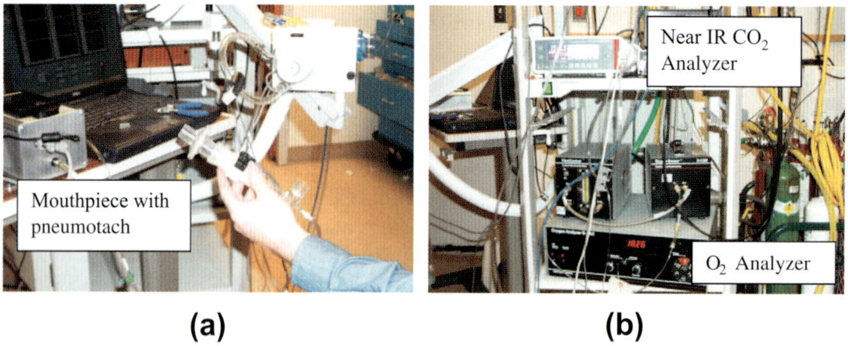

FIGURE 17.3

Test system used to compare the microsystem of Figure 17.2(c) to benchtop instrumentation. (a) Patient interface to benchtop instrumentation including mouthpiece with pneumotach. (b) Instrument rack-sized CO_2 and O_2 analyzer systems.

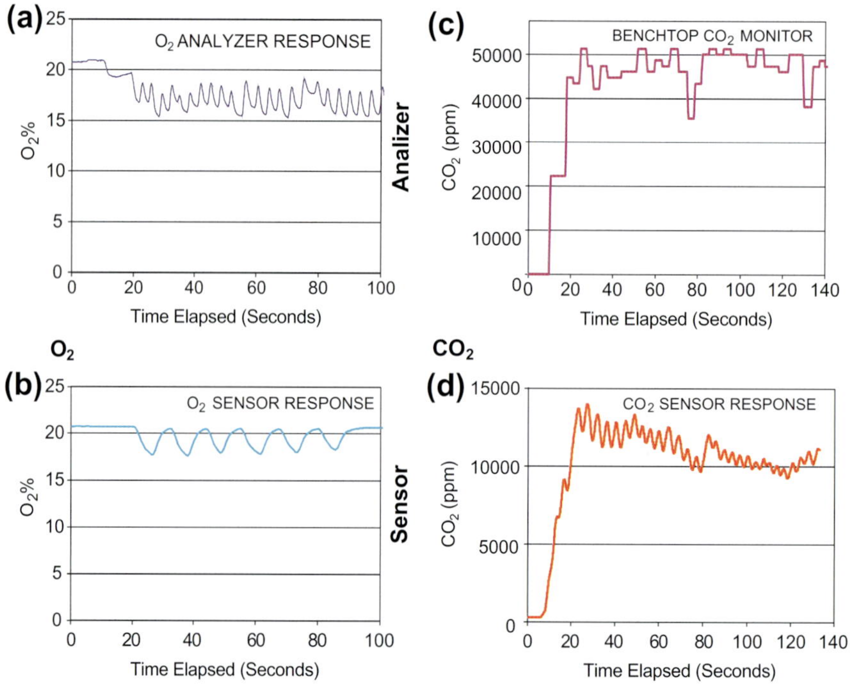

FIGURE 17.4

Testing results of prototype breath monitoring system at Cleveland Clinic and Makel Engineering, Inc. comparing clinical benchtop analyzer to microsensor system: (a) Benchtop O_2 Analyzer; (b) O_2 Sensor; (c) Benchtop CO_2 Analyzer; and d) CO_2 Sensor. The test results show that the CO_2 and O_2 sensors can provide breath to breath information at various respiratory rates to increase or decrease exhaled gas concentrations.

which is significantly larger and requires more power than microsensors. For both the microsensors and the benchtop instrumentation, CO_2 and O_2 concentrations were observed in the expected range of 0–5% and 15–20%, respectively. Further, changes with time in signal for these instruments are noted that correspond to breath-to-breath change. Overall, testing results demonstrated that the O_2 and CO_2 microsensors provide the functionality required for an exercise breath monitoring system, but at a significantly reduced size compared to that of traditional laboratory instrumentation. Thus, this approach using microsensor technology combined with smart hardware can, in principle, be used to replace rack-sized lab equipment with portable systems in human breath monitoring applications. It is this approach that is being used to address home health care asthma monitoring with a portable system.

17.4 HOME ASTHMA BREATH MONITORING TECHNOLOGY

One of the most promising potential uses of a portable miniaturized breath analyzer for improved health care is for monitoring asthmatics at home.[2,3,20–23] In particular, NO is involved in the pathogenesis of airway inflammation in asthma, and measuring exhaled NO has notable utility in monitoring the inflammatory component of asthma.[13,14,24,25] Exhaled NO is present in the breath at concentrations ranging from low ppb to well over 100 ppb. The concentration level of NO in exhaled breath provides information about airway inflammation that can otherwise only be obtained from more invasive tests. This is a major breakthrough in asthma care and has potential in the rapid assessment of airway inflammation for monitoring the response to anti-inflammatory therapy.[26] However, technological improvements are necessary to make equipment for measuring NO more portable, less expensive, and easier to maintain and calibrate for home health care applications.[2,24–28] Efforts toward such improvements are the subject of the following discussion. In particular, components of this development work include: (1) Identification of viable NO sensor technology with the required sensitivity and selectivity; (2) Integration of these sensors into a sensor systems to investigate their performance in breath monitoring applications; and (3) Miniaturization of the sensor and hardware technology in order to achieve a portable system.

17.4.1 Solid-state NO electrochemical cells

Standard electrochemical methods based on liquid electrolytes detect concentration levels in the parts-per-million (ppm) range, and thus are not sensitive enough for asthma monitoring applications. Since NO levels in exhaled breath are in the ppb range, sensitive methods utilizing chemiluminescence techniques have been used to detect these low levels given its high sensitivity. The chemiluminescence method depends on the reaction of NO with ozone to form NO_2 in an excited state, which then emits the extra energy as light that can be detected by a photomultiplier tube.[29] Several commercial chemiluminescence analyzers are available to measure NO levels. However, the chemiluminescence approach, while appropriate for some clinical settings, has multiple challenges. The technology and the devices are very expensive, their use is restricted to the physician's office, and they are not portable and therefore inappropriate for home testing. Thus, while mature approaches to NO detection exist, several technological improvements are still needed to make equipment for measuring NO more portable, less expensive, and easier to maintain and calibrate. In turn, a major reason that such a product does not presently exist for NO is a lack of adequate sensor technology. Thus, developing such miniaturized, portable sensors is needed to address this problem.[30]

Recent developments in solid-state NO_x sensor technology have enabled more sensitive measurements appropriate for breath monitoring applications. In particular, an approach developed by The Ohio State University (OSU), which is discussed in more detail elsewhere,[30–33] has been a major emphasis of our group's activities. This solid-state NO_x sensor design uses an oxygen-ion-conducting yttria-stabilized- zirconia (YSZ) suitable for mixed potential NO_x sensors at temperatures of 500–600 C. This chapter concentrates on the integration of this sensor into a hardware system and considerations necessary for its miniaturization for asthma monitoring applications.

The sensor design uses a planar potentiometric sensor comprised of yttria-stabilized zirconia electrolyte and metal oxide electrodes as shown in Figure 17.5. These are potentiometric sensors for total NO_x detection at high temperatures using a working electrode of tungsten trioxide (WO_3), and a Pt reference electrode. By using Pt-loaded zeolite Y (PtY) as a filter on the reference electrode, the signal magnitude can be increased. Pt wires are used to connect to the working and reference electrodes. The use of zirconia as an electrolyte implies higher temperature operation for the sensor.

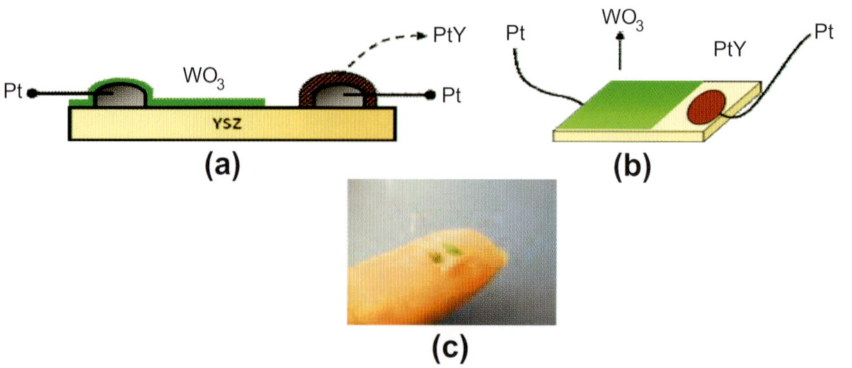

FIGURE 17.5

Basic design of a solid state electrochemical cell. (a) Side view showing the various components of the sensor: working electrode (WO_3), reference electrode (Pt covered with PtY), and connecting Pt wires. (b) Top view of resulting sensor. (c) Picture of the final sensor structure.

In summary, this approach features high sensitivity, no CO interference with use of an external catalytic filter and minimal O_2 interference, quick response/recovery time, good long-term stability, and total NO_x measuring capability making the sensor attractive for a range of applications.

A major feature of this OSU NO sensor design is that, like standard batteries, connecting multiple sensors in series results in a larger signal than that from a single sensor. This overall approach is illustrated in Figure 17.6. Figure 17.6a and b show an array of 2 and 10 sensors, respectively, connected in series. Although some size optimization can occur, connecting the sensors in series leads to a larger sensor structure overall, with a corresponding increase in operating power necessary. The response of 2 and 10 sensor arrays to a range of NO concentrations ranging from 7 to 125 ppb is shown in Figure 17.6c and d, respectively. The response to NO at a given concentration increases with the number of sensors connected in series. For example,

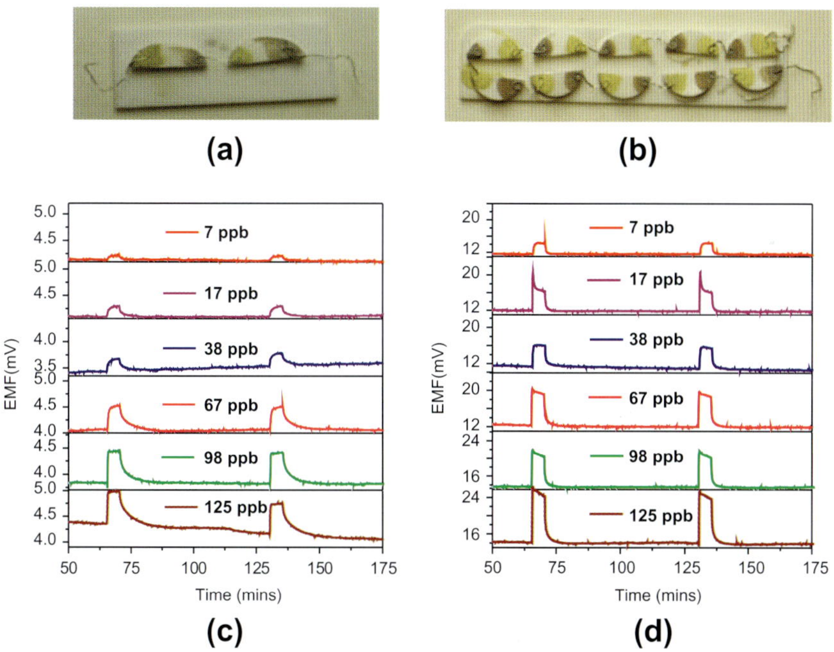

FIGURE 17.6
Photographs of (a) a 2 NO sensor array and (b) a 10 NO sensor array. Sensor response to 7–125 ppb NO concentrations for a (c) 2 sensor and a (d) 10 sensor array. Note the change in EMF (voltage) scale between Figure 17.6c and d. The sensor temperature is 600 C, gas flow rate of 500 sccm, and a background gas of 20% O_2. The data show that increasing the number of sensors increases the overall sensor array sensitivity.

while a 2 sensor array has a response of 0.3 mV with 38 ppb NO, as the number of sensors in the array increases to 10, the signal increases to 4.5 mV. Most importantly, Figure 17.6 shows the capability of an arrangement of multiple OSU NO sensors in series to detect NO down to a few ppb, i.e. at levels of interest for asthma monitoring.

17.4.2 Prototype hardware trials

These results suggest the potential use of the OSU NO sensor design as the foundation for a NO measuring device for home monitoring applications. In order to explore this potential, prototype hardware was fabricated which could accommodate sensor arrays such as that shown in Figure 17.6a and b. This hardware is shown in Figure 17.7, which include the core sensor electronics from the "Lick and Stick" technology shown above, a heater system to accommodate the sensor array, gas handling for circulation of breath, and an external display and interconnects to a laptop. The purpose of this hardware is simply to be able to verify the capability of the sensing approach to detect NO in breath at appropriate levels for asthma monitoring.

A series of tests were performed using this prototype system, including exposure to bagged samples containing calibrated NO mixtures and bottled gases. These tests include the use of an OSU 10 NO sensor array placed into the heat chamber of Figure 17.7 and heated to 600 C. Figure 17.8 shows the bottled gas results with a baseline in air, followed by a response to 100, 200, and 300 ppb of NO in air, followed by air. This test shows the basic capability of the components, including sensor and electronics systems, to measure NO at the basic levels required for use in asthma monitoring applications. However, such a system as shown in Figure 17.7 has limited portability and would not allow battery operation. To achieve a portable

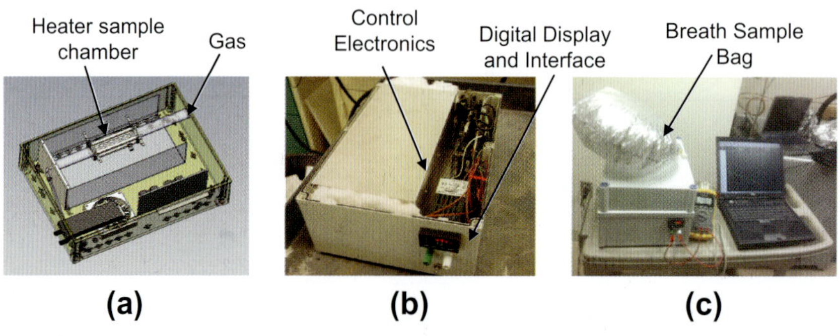

FIGURE 17.7
Prototype Hardware System. (a) Overall design including heater stage for an array of sensors such as that shown in Figure 17.6 as well as gas handling for circulation of breath, and an external display and interconnects to laptops. (b) Resulting hardware including core "Lick and Stick" electronics. (c) Prototype Hardware testing at the Cleveland Clinic including connection to breath gas sample bags.

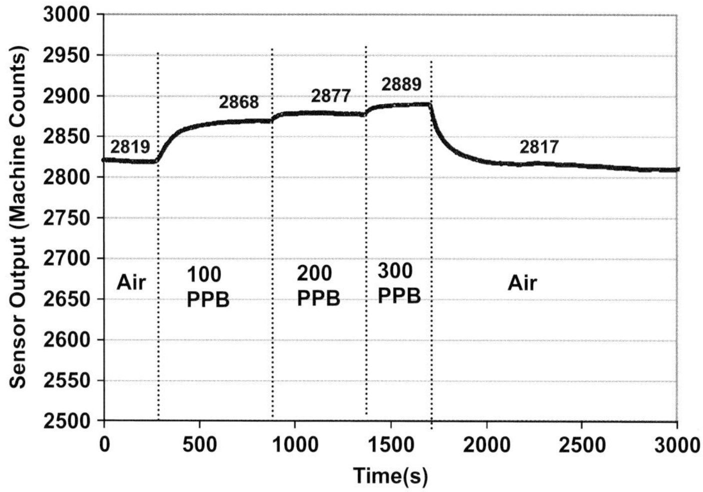

FIGURE 17.8
Response of prototype hardware system with a 10 sensor array to air, multiple NO/air concentrations, and air. The output values of the sensor system are labeled on the graph. The sensor responds to changes in NO concentration over a wide range of concentrations. The magnitude of the signal is a digital output produced by the signal conditioning hardware, which can then be converted into NO concentrations using the microprocessor and stored calibration tables.

system, the size and power consumption of the system needs to be minimized and thus miniaturization is necessary.

17.5 ASTHMA MONITORING SYSTEM MINIATURIZATION

The miniaturization of the breath monitoring system in Figure 17.7 in order to achieve a portable system requires several activities. The following will describe efforts and the present status of these activities.

17.5.1 Sensor miniaturization

In order for increased system portability as well as the capability to operate without an external power source, miniaturization of the OSU NO sensor design is necessary. The basic approach to this sensor miniaturization is to reproduce the OSU NO sensor structure in Figure 17.5 and 17.6 using MEMS-based microprocessing techniques.[7–11] For example, Figure 17.9a shows a 12 sensor array. This array has the capability of measuring low concentrations of NO, but has a surface area of approximately 10 square centimeters (cm^2). In order to operate this array at the necessary temperatures, several watts (W) of power are necessary. Figure 17.9b shows an early attempt at

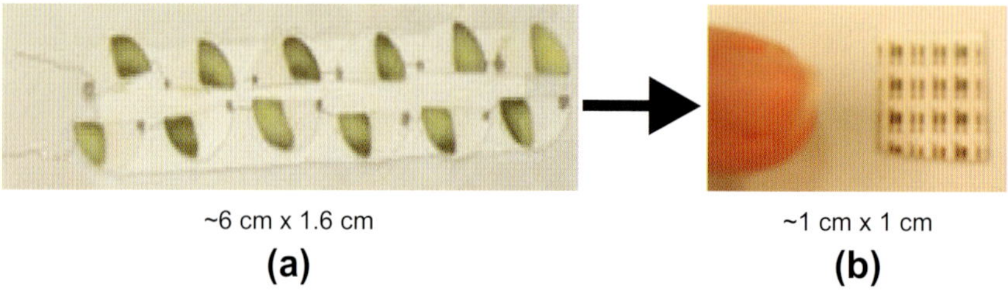

FIGURE 17.9

OSU NO sensors fabricated using two different approaches. (a) Standard fabrication approach used to produce the sensors described elsewhere.[27] (b) Basic sensor design fabricated using microprocessing techniques. The miniaturization is meant to significantly decrease size and power consumption.

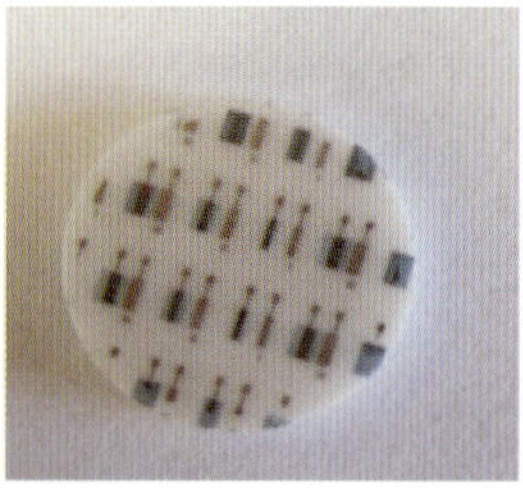

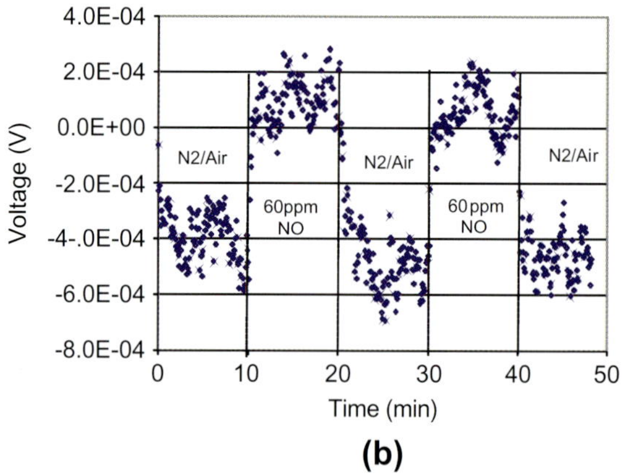

FIGURE 17.10

Miniaturization of the NO sensors. (a) The basic OSU potentiometric electrochemical cells design as shown in Figures 17.5 and 17.6 fabricated using microprocessing techniques. (b) Response of one electrode pair, that is, one working and one reference electrode to 60 ppm of NO in a 40% air/nitrogen background mixture.

miniaturizing this sensor array. Although both Figures 17.9a and 17.9b show 12 and 15 sensor arrays, respectively, the array in Figure 17.9b has a surface area of near 1 cm^2 with the power requirements for operation estimated to be less than 200 mW. Efforts using this approach are ongoing to produce miniaturized sensors based on the OSU NO sensor design.

An example of preliminary efforts toward fabrication of these microsensors is shown in Figure 17.10. Figure 17.10a shows a close-up of a number of sensors using the basic OSU potentiometric cell design deposited on a zirconia substrate. One of the advantages of these microprocessing techniques is the capability to produce a large number of sensors in a single fabrication run. Figure 17.10b shows the response of just one of those electrode pairs (reference and working electrodes) to 60 ppm of NO in a 40% air/nitrogen background mixture. While the signal has noise (on the order of 0.1 mV), and combining signals of a number of sensors would be necessary for a sensor array to have a high sensitivity needed for asthma monitoring applications, the sensor does show response to ppm levels of NO. Further maturation of the sensor processing is necessary in order to produce a sensor array of this design with the sensitivity needed for NO monitoring applications. Future work will mature these processing steps with an aim toward improved sensitivity combined with batch fabrication of interconnected sensor arrays.

17.5.2 Sensor packaging

The packaging of an OSU NO sensor has been an on-going activity of our group related to engine emissions monitoring for a number of years.[8,9,34] An example is shown in Figure 17.11 which shows a single OSU NO sensor in a package for engine emission applications. The overall approach is to integrate the sensor structure with a heater and temperature detector for temperature control. Sensors were bonded to probe assemblies using microwelds, and the probe assemblies were attached to the common manifold. Microwelding is preferred to high temperature conductive pastes because it provides a more robust physical mounting and a more reliable electrical contact for heating and signal transmission. Ceramic and glass seals for lead wires are custom designed for the test probes. This integrated sensor/temperature control system is then suspended using a platinum wire system in order to decrease heat loss.[31] Using this approach, an operational temperature of 600 C can be achieved at a power consumption near 200 mW. Power consumption at this level, and lower if possible, is an enabling property that allows a portable battery-powered system.

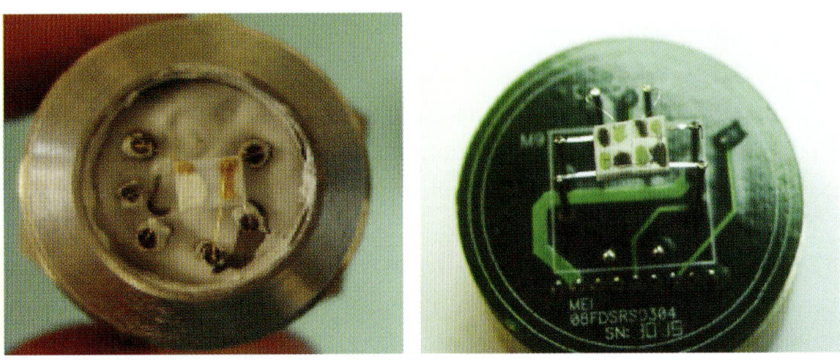

(a) (b)

FIGURE 17.11
Packaging of the OSU NO sensor system. (a) Sensor package used for engine monitoring applications including microwelds and suspended sensor structure. (b) A 4-element NO sensor for testing with a portable NO monitor.

This same approach can be used to package the OSU NO sensor. This is shown in Figure 17.11b with a 4 sensor array of the present version of the OSU NO sensor in a package structure similar to that of Figure 17.11a. This approach allows investigations of the OSU NO sensor system in a portable breath monitoring system, but perhaps without optimal sensitivity. Further work will use the miniaturized sensor structure described above (e.g. Figure 17.10) when they reach a more developed stage. In effect, a 12 microsensor array can fit in the space inhabited by a single sensor shown in Figure 17.11a, thus strongly increasing system sensitivity at nearly the same power consumption. This is the core for assembly of a complete home health asthma breath monitoring system.

17.5.3 Integrated system concept and future work

The integrated system for home breath health monitoring will be a modified and compacted version of the Exercise Breath Sensor System shown in Figure 17.2, with components including a mouthpiece for breath collection, gas handling of the exhaled breath, a sensor manifold for analysis of the breath, and visual data interface display. However, one of the major points of this chapter is that the use of Smart Sensor System technology can significantly enhance the versatility and capabilities of the core system. This type of system can be miniaturized for home, clinical, and aerospace applications, but can also enable new health care opportunities at considerably lower cost. The platform of Figure 17.1a can already accommodate 3 sensors, so not only could NO be monitored, but demonstrated capabilities exist to monitor CO_2 and O_2 or other gases as well. Thus, a single system could monitor asthma, but also parameters related to exercise physiology. It is envisioned that electronic or sensor boards could be swapped out as needed from the core exhaled breath monitoring system to allow, in a plug and play fashion, monitoring of other health parameters. For example, the NO sensor could be exchanged for an appropriate CO sensor such as that reported elsewhere[17] for monitoring of patient smoke exposure.[1] The microprocessor and associated calibration tables could be reprogrammed accordingly and the display to the user optimized to provide relevant alerts as necessary, as discussed elsewhere in a more detailed description of Smart Sensor Systems.[15] Thus, the hardware for the asthma breath monitoring system can serve as a core platform for a range of human health breath monitoring applications building toward the realization of a SBHDS.

Further, the “Lick and Stick” system already has wireless capabilities and an operational central processing node system. Information from the asthma breath monitoring system can be fed to this node, stored, and relayed as required through, for example, a web-based communication system. An asthma breath monitoring system could be combined with a “Lick and Stick” environmental monitoring system in the same home measuring, for example, CO, NO, hydrogen-bearing species, such as acetone, ethane, or pentane, and other atmospheric constituents. Such an approach would allow correlation between exhaled breath and environmental factors yielding a more whole-field approach to home health care.

The fundamental point is that much of the these core capabilities already exist; tailoring of existing technologies to breath monitoring applications is necessary to realize the potential advancements in health care and is an objective of this work. A first step toward realization of these potential advancements is integration of a portable home asthma breath monitoring system. This must be followed by extensive testing and demonstration in the laboratory, clinical studies, and home studies. Future enabling developments must also include advancements in sensor technology to allow even smaller, lower power sensor systems with both improved sensitivity and selectivity, perhaps including the use of multiparameter systems and nanotechnology.[10,15,35,36] This potential paradigm shift enabled by new technologies affects clinical applications and home health care, as well as remote applications such as space exploration.

17.6 SUMMARY AND CONCLUSION

Smart Sensor Systems combined with microsensor technology enable the capabilities to make units that are small, smart, and multifunctional. These advancements in sensor technology have the potential to revolutionize healthcare and can offer a cost advantage by adaptation into home healthcare. A system used to monitor exercise physiology has been demonstrated with this sensor approach. Efforts are underway to miniaturize NO sensor technology and integrate it into a portable Smart Sensor System based on "Lick and Stick" technology. Core capabilities already exist that have the potential to not only replace a larger system with a portable unit suitable for home healthcare, but also provide a basic platform with capabilities far beyond those of costly home healthcare analytical instrumentation. This includes measurement of species for a range of medical conditions and the development of a SBHDS.

Acknowledgments

The authors would like to acknowledge the contributions of: Dr. L. Matus, J. Wrbanek, L. Evans, and D. Litt of NASA GRC; D. Lukco of ASRC Aerospace/NASA GRC; J. Gonzalez, M. Artale, P. Lampard, and C. Hampton of Sierra Lobo, Inc.; and Dr. L. Chen of OAI.

References

1. Hunter GW, Dweik RA. Applied breath analysis: an overview of the challenges and opportunities in developing and testing sensor technology for human health monitoring in aerospace and clinical applications. *J Breath Res* 2008;**2**:037020.
2. Grob NM, Dweik RA. Exhaled nitric oxide in asthma From diagnosis, to monitoring, to screening: are we there yet? *Chest* 2008;**133**:837–9.
3. Machado RF, Laskowski D, Deffenderfer O, Burch T, Zheng S, Mazzone PJ, et al. Detection of lung cancer by sensor array analyses of exhaled breath. *Am J Respir Crit Care Med* 2005;**171**:1286–91.

4. Dweik RA. Nitric oxide reactions in the asthmatic airway. In: Marczin N, Yacoub MH, editors. *Disease markers in exhaled breath.* (Amsterdam, Netherlands): IOS Press; 2002. p. 159–66.
5. Gill M, Graff GR, Adler AJ, Dweik RA. Validation study of fractional exhaled nitric oxide measurements using a handheld monitoring device. *J Asthma* 2006;**43**:731–4.
6. Anonymous Expert panel report II: guidelines for the diagnosis and the management of asthma. *National Institutes of Health*, Bethesda; 1997.
7. Liu CC, Hesketh P, Hunter GW, Chemical Microsensors. In *ECS Interface*. Electrochemical Society Inc; 2004; p. 22–9.
8. Hunter GW, Xu JC, Liu CC, Makel DB. Microfabricated chemical sensors for aerospace applications. In: Gad-El-Hak M, editor. *MEMS: Applications (The MEMS Handbook, Second Edition)* Baton Rouge: CRC Press; 2006.
9. Hunter GW, Xu JC, Makel DB. Case studies in chemical sensor development. In: Hesketh P, editor. *Bionanofluidic Mems* New York: Springer Press; 2008. p. 197–231.
10. Stetter J, Hesketh P, Hunter GW. Sensors: Engineering Structure and Materials from Micro to Nano, in ECS Interface. 2006, Electrochemical Society Inc. p. 66–69.
11. Liu CC, O'Connor E, Strohl KP, Klann KP, Ghiurcan GA, Hunter G, Dudik L, Shao MJ. An assessment of microfabrication to sensor development and the integration of the sensor microsystem. In: Hesketh PJ, editor. *Microfabricated Systems and Mems Vi, Proceedings*. Electrochemical Society; 2002. p. 1–12.
12. Hunter GW, Xu JC, Evans LJ, Vander Wal RL, Berger GM, Kulis MJ, et al. Chemical sensors based on metal oxide nanostructures. *ECS Transactions* 2006;**3**: 199–209.
13. Dweik RA, Comhair SA, Gaston B, Thunnissen FB, Farver C, Thomassen MJ, et al. NO chemical events in the human airway during the immediate and late antigen-induced asthmatic response. *Proc Natl Acad Sci U S A* 2001;**98**:2622–7.
14. Dweik RA, Sorkness RL, Wenzel S, Hammel J, Curran-Everett D, Comhair SA, et al. Use of exhaled nitric oxide measurement to identify a reactive, at-risk phenotype among patients with asthma. *Am J Respir Crit Care Med* 2010;**181**:1033–41.
15. Hunter GW, Stetter JR, Hesketh PJ, Liu CC. Smart sensor systems. In *Interface Magazine 2011*. Electrochemical Society Inc. p. 66–69.
16. Hunter GW, Xu JC, Dungan LK, Ward BJ, Rowe S, Williams J, et al. Smart sensor systems for aerospace applications: From sensor development to application testing. *ECS Transactions* 2008;**16**:333–44.
17. Hunter GW, Xu JC, Dungan LK, Ward BJ, Dutta P, Adeyemo AD, Liu CC, Gianettino DP. Smart chemical sensor systems for fire detection and environmental monitoring in spacecraft. In *40th International conference on environmental systems*. Barcelona, Spain: Paper AIAA-2010-6307; 2010.
18. Hunter GW, Neudeck PG, Xu J, Lukco D, Trunek A, Artale M, et al. Development of SiC-based gas sensors for aerospace applications. In: Dudley M, Gouma P, Kimoto T, Neudeck PG, Saddow SE, editors. *Silicon carbide 2004—materials processing and devices*. Warrendale, PA: Mater. Res. Soc. Symp. Proc; 2004. p. J.4.4.1-12.
19. Hunter GW, Xu J, Neudeck PG, Makel DB, Ward B, and Liu CC, Intelligent chemical sensor systems For in-space safety applications. In *42nd AIAA/ASME/ SAE/ASEE joint propulsion conference & exhibit*. Sacramento, California: Paper AIAA-2006-4356; 2006.
20. Dweik RA. The promise and reality of nitric oxide in the diagnosis and treatment of lung disease. *Cleve Clin J Med* 2001; **68**: 486, 488, 490, 493.

21. Khatri SB, Ozkan M, McCarthy K, Laskowski D, Hammel J, Dweik RA, et al. Alterations in exhaled gas profile during allergen-induced asthmatic response. *Am J Respir Crit Care Med* 2001;**164**:1844–8.
22. Machado RF, Stoller JK, Laskowski D, Zheng S, Lupica JA, Dweik RA, et al. Low levels of nitric oxide and carbon monoxide in alpha 1-antitrypsin deficiency. *J Appl Physiol* 2002;**93**:2038–43.
23. Khatri SB, Hammel J, Kavuru MS, Erzurum SC, Dweik RA. Temporal association of nitric oxide levels and airflow in asthma after whole lung allergen challenge. *J Appl Physiol* 2003;**95**:436–40 discussion 435.
24. Barreto M, Villa MP, Olita C, Martella S, Ciabattoni G, Montuschi P. 8-Isoprostane in exhaled breath condensate and exercise-induced bronchoconstriction in asthmatic children and adolescents. *Chest* 2009;**135** :66–73.
25. Montuschi P, Martello S, Felli M, Mondino C, Barnes PJ, Chiarotti M. Liquid chromatography/mass spectrometry analysis of exhaled leukotriene B4 in asthmatic children. *Respir Res* 2005;**6**:119.
26. Montuschi P, Mondino C, Koch P, Ciabattoni G, Barnes PJ, Baviera G. Effects of montelukast treatment and withdrawal on fractional exhaled nitric oxide and lung function in children with asthma. *Chest* 2007; **132**:1876–1881.
27. Dweik RA, Amann A. Exhaled breath analysis: the new frontier in medical testing. *J Breath Res* 2008;**2**:030301.
28. Grob NM, Dweik RA. Exhaled nitric oxide in asthma: progress since the introduction of standardized methodology. *J Breath Res* 2008;**2**:37002.
29. Dweik RA. Nitric oxide production in the lung and its regulation by oxygen in: Marczin N, Yacoub MH. Disease markers in exhaled breath: basic mechanisms and clinical applications 2002 Amsterdam, Netherlands IOS press p.11–17.
30. Mondal SP, Dutta PK, Hunter GW, Ward BJ, Laskowski D, Dweik RA. Development of high sensitivity potentiometric NO(x) sensor and its application to breath analysis. *Sensor Actuat B-Chem* 158 2011 292–298.
31. Yang JC, Dutta PK. High temperature amperometric total NOx sensors with platinum-loaded zeolite Y electrodes. *Sensor Actuat B-Chem* 2007;**123**:929–936.
32. Szabo NF, Dutta PK. Strategies for total NOx measurement with minimal CO interference utilizing a microporous zeolitic catalytic filter. *Sensor Actuat B-Chem* 2003;**88**:168–177.
33. Yang JC, Dutta PK. Promoting selectivity and sensitivity for a high temperature YSZ-based electrochemical total NO(x) sensor by using a Pt-loaded zeolite Y filter. *Sensor Actuat B-Chem* 2007;**125**:30–39.
34. Ward BJ, Wilcher K, Hunter GW. Gas microsensor array development targeting enhanced engine emissions testing, in AIAA Infotech@Aerospace. 2010. Atlanta, Georgia: AIAA 2010–3327.
35. Hunter GW, Xu JC, Evans L, Biaggi-Labiosa A, Ward BJ, Rowe S, Makel DB, Liu CC, Dutta P, Berger GM, Vander Wal RL, The development of micro/nano chemical sensor systems for aerospace applications. In SPIE Newsroom. 2010.
36. National Nanotechnology Initiative. Nanotechnology-Enabled Sensing 2009 Workshop. 2009. Arlington, Virginia: available at <http://www.nano.gov>;2012, [accessed 28.02.12].

CHAPTER

18 VOC Analysis by SIFT-MS, GC-MS, and Electronic Nose for Diagnosing and Monitoring Disease

Claire Turner

The Open University, Department of Life, Health and Chemical Sciences, Walton Hall, Milton Keynes, MK7 6AA, UK

18.1 METHODOLOGY AND VOC ANALYSIS

Breath analysis has been around for some years now and there have been many instances of research into its potential for diagnosing disease.[1] It has many advantages over other diagnostic modalities as it is truly non-invasive, samples are very easily obtained from a conscious or unconscious patient, whether adult or child, breath is exhaled many times per minute and there is evidence that people are much happier to give a breath sample as opposed to most if not all other sample types. Breath analysis has been approved for use in a number of FDA approved tests, including enforcing drink driving legislation, asthma monitoring using nitric oxide,[2] and early detection of heart transplant rejection.[3] In addition, it has shown promise in numerous other applications, which may well lead to routinely used clinical tests. This includes the analysis of compounds produced by *Pseudomonas aeruginosa*[4] which may be used to detect infection with this organism in cystic fibrosis; diagnosis of *Helicobacter pylori* infection[5]; monitoring blood glucose in diabetes.[6] However, there are confounding factors that can affect the levels of various compounds in breath, chief among them being the mouth flora, which contribute to the VOC profile of breath.[7–9] For this reason, attention is turning to the analysis of VOCs in other sample types.

After breath, the next easiest sample to obtain is urine and for this reason, analysis of VOCs in the headspace of urine is becoming more common. In addition, urine is much easier to store than breath if analysis cannot be carried out immediately and not much urine is required to produce an acceptable sample of headspace. Urine has been shown to contain many of the VOCs in breath and blood and also seems to concentrate others, especially ketones which are easily measured in urine.[10] The concentration of urine will vary depending upon the time of day and degree of hydration of the patient[11]; hence, it is important to ensure that a method of standardization may be developed. In analyzing the soluble components of urine, the concentration of creatinine is used as a sort of internal standard against which its concentration may be normalized[12]; is there a suitable alternative VOC? Urine may be of use in monitoring the concentration

Volatile Biomarkers. http://dx.doi.org/10.1016/B978-0-44-462613-4.00018-0

of systemic VOCs; it also has value in diagnosing or monitoring localized infections or disease, such as bladder cancer.[13,14]

Blood is also relatively easy to obtain and will give information about VOCs present systemically. However, there is still a number of patients who do not like needles and for whom taking a sample of blood is a trial, so this is perhaps not so convenient as breath or urine. However, analyzing the headspace of urine (i.e. the vapor phase above the urine and not the liquid) will give a very direct snapshot of the systemic VOC profile in the body at the time and this may prove useful in diagnosing or monitoring metabolic diseases, for example.

Analyzing the headspace of fecal samples may give a good indication of the health of the gastrointestinal tract and may say something about the population of bacteria inhabiting the bowel. However, patients do not like providing samples and processing is not so easy or pleasant as the other samples discussed hitherto. Nevertheless, it has been shown[15] that much information may be obtained about the health of the bowel through analyzing the headspace of fecal samples.

The headspace of other samples may also provide some diagnostic information. This includes localized samples such as sputum, joint aspirates, bronchial lavage, gastric lavage, tissue biopsies, and other samples. The VOC profile in the headspace above such samples may provide information about localized infections or damage and may in the future be used for specialized diagnoses where speed is important.

In this chapter, we discuss some of the recent studies we have conducted analyzing the breath and VOC profile in the headspace of clinical samples from patients and healthy volunteers using selected ion flow tube mass spectrometry (SIFT-MS), thermal desorption-gas chromatography-mass spectrometry (TD/GC-MS), and electronic nose (e-nose).

18.2 HEALTHY VOLUNTEERS

If breath analysis is to be used to diagnose disease and monitor the health of individuals, it is essential to first determine the composition of a normal breath profile, i.e. which compounds are likely to be present and at what concentration. Many studies have been carried out on analyzing breath VOCs using different instruments[1,15–18] and it is clear that in determining the compounds present, particularly those present at low concentration, then GC-MS is the method of choice, particularly when combined with a concentration method such as thermal desorption (TD) or solid phase microextraction (SPME). By concentrating VOCs in breath and separating them using chromatography, hundreds of breath volatiles may be identified in breath. Although this method is powerful, it is not truly quantitative and is time consuming. It may detect hundreds of compounds, but many of those, especially those present at low concentrations, may not be endogenous and their detection and analysis may complicate the identification of true biomarkers.

An alternative method ideally suited to breath analysis is SIFT-MS, which is discussed extensively in the chapter of this book by Španěl and Smith. This is the method of choice for analyzing breath directly and is quantitative, rapid, and easy to do, requiring no additional equipment for collecting breath samples. For this reason, this method was chosen to monitor the breath profiles of 30 self-certified healthy volunteers in a 6 month longitudinal study.[7,19–21] Weekly breath samples were taken from 30 volunteers and the concentrations of the following seven compounds were determined: ethanol, methanol, isopropanol, acetaldehyde, isoprene, ammonia, and acetone. These seven compounds represent the most abundant VOCs in breath (with the exception of methane, which was not investigated in this study). The samples were taken in the morning between breakfast and lunch. Details were taken of each volunteer: their age, body mass index, diet in the previous 24 h, whether they smoked, how much alcohol they had taken, and any medication they were on. These data were used to generate distributions of the range of concentrations of these VOCs in the breath of the healthy volunteers over the time period to enable an understanding of what is "normal" to be reached.

The distributions for each of the compounds were found to be log-normal and the range in parts-per-billion by volume (ppbv), medians, and geometric standard deviations are given in Table 18.1.[7,19–21,25]

Further analysis of these data shows some interesting correlations. One of these is the increase in exhaled breath ammonia with age. We originally believed that this was due to a reduction in kidney function with age,[21] and worsening kidney function is probably a factor. However, subsequent work[8,23,24] demonstrated that much of the ammonia present on breath is generated in the mouth by mouth flora and is not systemic. A further clue for this came with an investigation into the variation of breath ethanol. It was clearly shown in this and a subsequent study[7,8] that breath ethanol increases for up to 2 h after the consumption of sweet food or a sweet drink. This is apparently down to yeast and bacteria in the mouth metabolizing the sugar and producing ethanol.

Table 18.1 The medians, geometric standard deviation, and range of seven compounds determined in the breath of 30 healthy volunteers taken weekly over 6 months.

Compound	Geometric Mean (ppb)	Geometric Standard Deviation (ppb)	Range (ppb)
Ammonia[21]	833	1.62	248–2935
Acetone[21]	477	1.58	148–2744
Methanol[20]	461	1.62	32–1684
Ethanol[7]	112	3.24	0[b]–1663
Isoprene[19]	106	1.65	0[b]–474
Propanol[21]	18	Not calculated[a]	0[b]–135
Acetaldehyde[7]	22	Not calculated[a]	0[b]–104

[a] *Data contain many values below detectable levels so GSD not calculated.*
[b] *Below detectable level (about 10 ppb).*

The behavior of ammonia and ethanol in response to the action of mouth flora demonstrates an important issue in breath analysis: it is essential to separate out the effects of mouth flora from the systemic VOCs. One way of doing this is by taking breath from the nose instead of the mouth. It was clearly demonstrated[24] that some compounds (ethanol, ammonia, HCN) are much higher when breath samples are taken through the mouth than the nose. Those compounds present on breath which appear not to be significantly influenced by mouth flora are acetone, methanol, and isoprene.

These wide-ranging studies of the trace compounds in the exhaled breath of healthy volunteers in the age range of 24–59 have been extended in later studies to analyze the breath of children and young adults.[26,27] These plus earlier studies provide a comprehensive body of data for these seven compounds, and have recently been added to by investigating the presence of HCN and sulfur-containing compounds in breath using SIFT-MS.[26,28]

18.3 ACETONE AND DIABETES

Acetone is a compound present in breath that is not significantly affected by mouth flora.[23] The longitudinal studies described above have characterized typical breath acetone concentrations for healthy (non-diabetic) people of a range of ages, body mass indices, and of both genders. These studies were single breath samples and no attempt was made to determine the blood glucose concentrations of the volunteers when they provided breath samples. However, it had been shown that the concentration of acetone is strongly influenced by diet and how hungry an individual is,[29] a result strongly confirmed by recent studies.[30] For this reason, the relationship between breath acetone and blood glucose in several healthy volunteers was investigated after fasting and for 2 h following the consumption of 75 g of glucose using SIFT-MS and a commercially available portable blood glucose meter.[31] It was found that after fasting, when blood glucose was low, acetone tended to be relatively high. Following the consumption of glucose, blood glucose increased, and breath acetone declined. The relationship was not a direct linear correlation but the general trend in healthy volunteers was that when blood glucose was low, breath acetone was relatively high and vice versa.

The earliest recorded medical histories of diabetes from ancient Greece at the time of Hippocrates note that sufferers had the smell of rotten apples on their breath. This odor is acetone. For this reason, modern breath analysis has concentrated on acetone as a biomarker of diabetes. The physiological reason for this is that acetone is one of three ketone bodies which are produced in the mammalian liver as an alternative energy source when glucose is not readily available. People who are fasting and/or performing prolonged exercise will tend to have low blood glucose (BG) levels and will, therefore, have elevated levels of ketone bodies in their blood and the fruity acetone smell on their breath. Diabetics, either lacking insulin or having insulin resistance, will not be able to use the glucose that is present, so it builds up in the blood (hyperglycemia). At the same time, the ketone bodies will increase too as the body responds to an energy shortage, and the liver will break down fat, producing the three ketone bodies acetoacetate, 3-β-hydroxybutyrate, and acetone. Acetone is more abundant in breath as it is much more volatile than the other two.

Breath acetone has been measured in diabetics, in breath,[22] and in blood.[32] As early as 1969, it was reported that breath acetone was elevated in diabetics on insulin, hypoglycemic tablets, or reducing diets, even if their blood glucose levels were near normal. Other studies also found acetone levels higher in diabetics.[33,34] Thus conventional wisdom is that acetone is markedly elevated in hyperglycemia. Is it possible, therefore, that there is a direct correlation between blood glucose and breath acetone? If this were demonstrated, it may one day be possible to replace blood glucose monitoring with a simple breath analysis. In order to achieve this, it is necessary to determine the correlation between blood glucose and breath acetone for a number of different individuals at different blood glucose concentrations and under different conditions, e.g. insulin levels, the influence of diet, alcohol consumption, etc. It is also very likely that the relationship between blood glucose and breath acetone is different in type 1 and type 2 diabetes, with levels of breath acetone being reported as being between 0.92 and 1.2 parts-per-million by volume, ppmv.[34] Much higher concentrations of acetone have been reported in type 1 diabetics (up to 20 ppmv)[6] than in type 2; however, there have been very few studies comparing the two with respect to breath acetone and other potential breath biomarkers. Some aspects of this have been carried out,[35] and it was reported that breath acetone was measured in 34 type 1 diabetics, type 2 diabetics, and 15 healthy individuals. Blood glucose (BG) concentrations were measured at the time the breath samples were taken. This study found a correlation between mean group acetone and mean group BG for the blood glucose levels when the BG levels are grouped as follows: 40–100, 101–150, 151–200, and 200–419 mg/dL. The same study also found a mean acetone concentration in type 1 diabetics at all the BG levels measured of 2.19 ppm as opposed to 0.48 ppm for the healthy volunteers.

We conducted a study[6] in which patients at lower BG concentrations (hypoglycemia) were monitored in a glucose clamp experiment where dextrose and insulin were fed intravenously. Eight type 1 diabetics were recruited; they had diabetes for a mean of 28 ± 3 years and had sub-optimal glycemic control (Hb1A level of 8.8 ± 0.4%). The patients were admitted to a clinical research facility overnight to stabilize BG concentrations using intravenous insulin. The following morning, the insulin clamp technique enabled a controlled reduction in BG concentration in steps of 40 min duration by varying intravenous dextrose while keeping insulin infusion rates constant. The nominal BG concentrations for which data were obtained were 5, 3.8, 3.3, 2.8, and 2.2 mM/L. BG levels were measured every 5 min and when stable, breath samples were taken into Nalophan bags for determination of acetone concentrations by SIFT-MS. The results of this study very clearly showed a linear relationship between breath acetone and BG concentration for all eight volunteers. However, what was also apparent was that the range of breath acetone values for baseline levels of BG (i.e. at the start of the experiment—around 5 mM/L) varied very widely from 1.0 to 21 ppmv. In all cases, breath acetone declined throughout the

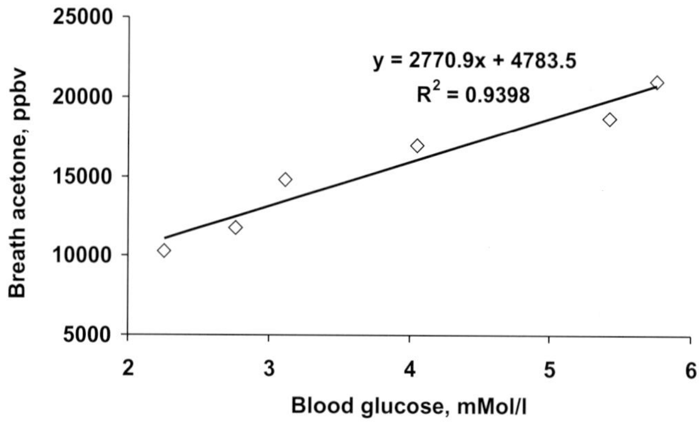

FIGURE 18.1

Glucose clamp study for one volunteer with type 1 diabetes. Acetone concentration was measured in the breath of the volunteer at clamped blood glucose concentrations.[6]

clamp (Figure 18.1, shown for one patient). This demonstrates that the absolute breath acetone concentration is not on its own an indicator of blood glucose level. What is not clear is whether this would be reproduced in the absence of a constant insulin infusion, and whether the linear correlation would be extended in the euglycemic and hyperglycemic range.[36]

Although these studies do not definitively prove a linear and reliable correlation between breath acetone and blood glucose concentrations, they do indicate a strong link, especially for type 1 diabetes. However, it is clear that as the levels of acetone vary so widely for each blood glucose concentration in each individual that a direct and absolute relationship between blood glucose and acetone does not exist. One advantage of using acetone as a potential marker of blood glucose level in diabetes is its abundance in breath, and in diabetes, as we have seen, it generally becomes much more abundant, which makes analysis much easier. Other compounds have also been proposed as potential markers of blood glucose in diabetes.[36]

While SIFT-MS is the best technique for accurate analysis of breath acetone and the presence and concentration of other breath metabolites, it is not viable as an alternative to a blood glucose monitor. The only real alternative to this is a gas sensor built into a small portable device. In our study described above, we also took breath samples by asking patients to breathe into a prototype portable device (called a Breathotron) containing a metal oxide sensor. The response of the sensor was plotted at each blood glucose concentration and results for one patient are shown in Figure 18.2. All eight patients displayed a linear correlation. A gas sensor is only semi-selective and responds to many reducing gases or vapors. However, acetone is one of the most abundant VOCs in breath even in healthy volunteers; in diabetics levels may be very high. Thus, under normal circumstances, it appears that the acetone dominates the sensor response and exactly mimics the correlation between breath acetone measured

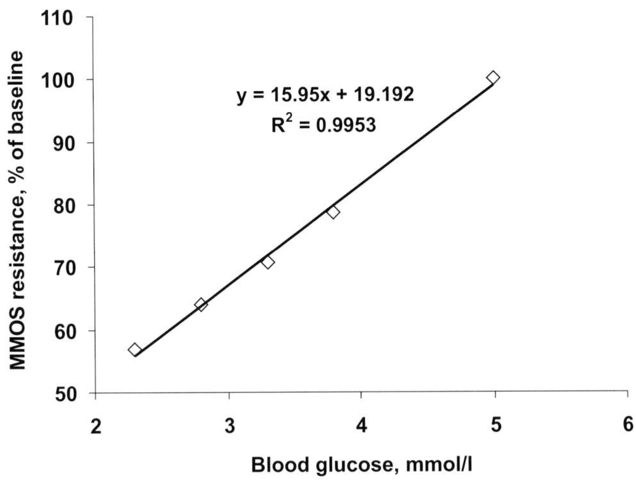

FIGURE 18.2
Response of a metal oxide sensor exposed to the breath of a volunteer with type 1 diabetes during a glucose clamp studies, plotted against blood glucose concentration.

by SIFT-MS and BG. This provides some support for the view that a gas sensor may one day be used to monitor BG.

18.4 ACETONE FROM SKIN

As acetone is a systemic VOC and unaffected by mouth flora, we wished to determine whether it could also be detected through the skin. Monitoring skin acetone may also be used to provide an estimate of BG concentration if its concentration may be accurately determined. To this end, we devised a method for monitoring VOCs from skin.[31] This involved placing the forearm inside a Nalophan bag and securing it closed. Hydrocarbon free air was then added to the bag via a tube fitting connected to the other end of the Nalophan bag. After a 10 min wait, the air from the bag was sampled directly by SIFT-MS (Figure 18.3). This method of monitoring skin VOCs was used during an oral glucose tolerance test. Volunteers were asked to fast overnight, then visit the laboratory and give a breath and skin VOC sample (as described above). Seventy five grams of glucose was given to volunteers and then breath and skin VOCs and BG concentrations were monitored at intervals for 2 h. The results demonstrated that there was a correlation between breath acetone and skin acetone (Figure 18.4), which suggests that skin may also be used to infer blood acetone concentrations. The technology exists for a portable gas sensor to monitor breath or blood glucose; however, the clinical studies have yet to be carried out to understand this relationship for many different individuals under a variety of conditions.

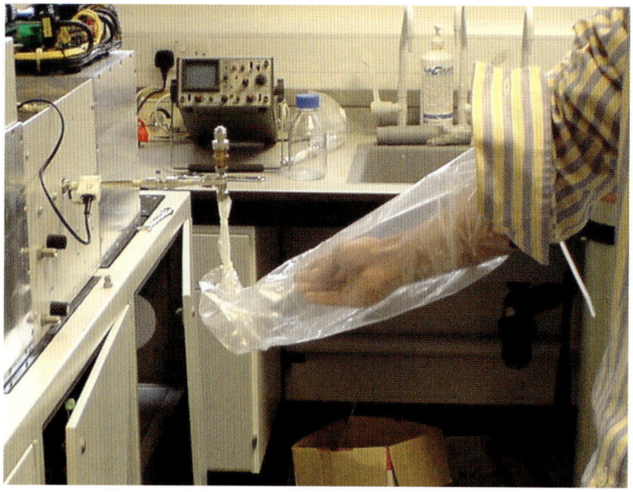

FIGURE 18.3
Taking samples of VOCs emitted from the skin by placing the forearm inside a Nalophan tube, equilibrating with hydrocarbon free air and then directly monitoring using SIFT-MS.[31]

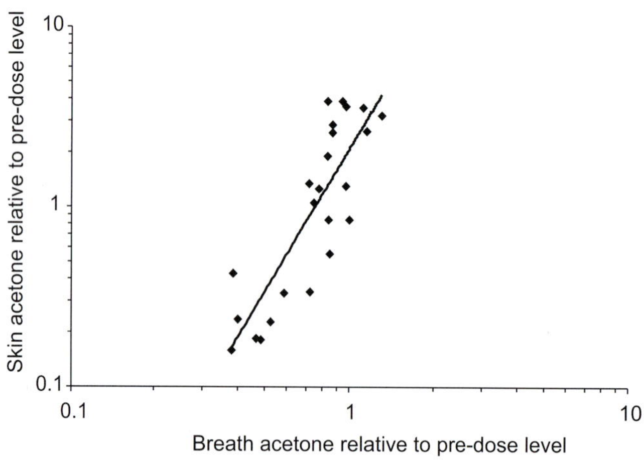

FIGURE 18.4
Correlation of skin acetone levels and breath acetone levels during an oral glucose tolerance test, each taken relative to the initial levels before glucose ingestion.[31]

18.5 TUBERCULOSIS AND OTHER INFECTIOUS DISEASES

Breath analysis is probably best suited to systemic metabolic disturbances, acute systemic infections or localized lung disease or infection. This is because the levels of VOCs present in breath are generally not very high and many are highly influenced by mouth flora. It is reasonable to suppose that breath analysis may offer a method for diagnosing pulmonary tuberculosis (TB). Indeed, ancient physicians used to throw sputum from tuberculosis sufferers onto the fire; if a particular foul odor was detected, then phthisis (the ancient name for tuberculosis) was suspected. This implies a particular profile of VOCs being present in TB sufferers.

Our own studies have tried to compare analysis of VOCs from culture of mycobacteria, headspace of serum from badgers and cattle and from breath in cattle. There have been other studies investigating the VOCs produced by mycobacteria in culture.[37,38] Such studies have found some biomarkers, including methyl phenylacetate, methyl *p*-anisate, methyl nicotinate, and *o*-phenylanisole[38] and naphthalene, among others.[37] Our own studies[39] of mycobacteria revealed different marker compounds; however, the presence of each of these compounds in the headspace above the mycobacterial cultures is highly dependent upon the medium on which the mycobacteria are grown. This implies that the same markers may not be detected *in vivo*, as the growth medium is obviously different. In addition, any VOCs produced by this slow-growing organism may be modified by the host organism, or masked by VOCs generated by the host's immune response to infection or the presence of exogenous environmental VOCs that have been absorbed by the host. So markers for TB infection should best be sought by looking at clinical samples.

18.5.1 Badgers

Samples of serum were taken from 250 badgers trapped in the wild during the UK's testing of badgers.[40] Each badger was clinically examined and serum samples were tested for TB by culture and gamma interferon tests. Serum samples of 250 μL were headspaced and the headspace analyzed using SIFT-MS. The data were carefully analyzed and it was noted that no unique individual VOC markers were found. Multivariate classification algorithms were therefore used to extract a simple TB diagnosis from the complex multivariate response provided by the SIFT-MS data, and this was the first time that such multivariate analysis has been applied to SIFT-MS data. Using partial least squares discriminant analysis (PLSDA), an accuracy of TB discrimination of approximately 88% true positive was achieved. This shows promise, but the corresponding false positive rate of 38% indicates that there is more work to do before this approach could replace the culture test. However, this demonstrates that there clearly is some difference in the VOC profile between positive and negative TB serum samples. There were no *specific* markers for TB, i.e. VOCs which were present in TB positive and absent in TB negative sample, which implies that these differences lie in the proportions of the usual VOCs present.

18.5.2 Analyzing cattle breath

The breath of cattle before and after experimental infection with the organism causing tuberculosis in cattle (*Mycobacterium bovis*) was analyzed using thermal desorption-gas chromatography-mass spectrometry (TD/GC-MS).[41] The major issue to consider with this is the use of analytical equipment to analyze highly infectious material from a category 3 pathogen, *M. bovis*. For this reason, SIFT-MS could not be used as the breath sample cannot be decontaminated and handled under category 3 containment. This is an important issue. TD/GC-MS is not rapid or direct but it offers the possibility of trapping and concentrating VOCs onto a sorbent material (Tenax TA and Carbotrap, Markes, UK), sealing the TD tubes and then heating them to 80°C to decontaminate them prior to removal from the containment area for analysis by GC-MS. Although some studies with artificial mixtures showed some reduction in concentration of VOCs of about 10%, this was deemed the best way of taking and making breath samples safe prior to analysis. Analysis of cattle samples before and after TB infection indicated some possible biomarkers for TB, i.e. VOCs present after infection but absent before infection in at least some samples. However, all these compounds were present at such low concentrations that it is impossible to be sure that they are not exogenous. Additional experiments need to be carried out on more animals to confirm this.

18.5.2.1 Electronic nose studies of cattle serum

Conducting polymer electronic nose (e-nose) was employed to investigate the headspace of serum samples from cattle experimentally infected with TB and control animals with no infection.[42] In this study the headspace of blood samples from 15 calves experimentally infected with *M. bovis* was analyzed using a Bloodhound e-nose. Five calves were vaccinated with BCG and five with another experimental TB vaccine (Ad85a) 14 weeks before and 6 weeks after infection with *M. bovis*. A further five animals were not vaccinated and acted as positive controls.

The e-nose sensor patterns were analyzed using multivariate statistics but there was little discrimination. However, looking at individual sensors, it was possible to see some differences between the groups of animals using individual sensors, but the e-nose was insufficiently robust and reliable to be able to detect the status of each individual animal. So at the population level, it was possible to show that there are some differences between positive control animals and vaccinated animals and hence some difference in VOC pattern. However, the e-noses used could not be used for the diagnosis of individual animals and, furthermore, are unable to give an indication of what the components of the VOC patterns were.

E-nose has also been used to investigate the potential for discriminating between the headspace of serum from cattle infected with an acute and rapid onset bacterial illness, *Mannheimia haemolytica*.[43] The headspace of blood samples from cattle before and after infection was analyzed using e-nose and compared with results from acute phase proteins (APPs). Significant changes of certain e-nose sensor responses allowed discrimination between samples before and after infection. The maximal changes in responses of sensitive e-nose sensors corresponded to the peak of clinical signs. Significant correlations linked decreasing responses of multiple e-nose sensors to increasing concentrations of APPs in the peripheral blood. The increased discrimination with *M. haemolytica* over TB may be down to the acute immune response, which the e-nose may be detecting.

18.6 GASTRO-INTESTINAL ILLNESS

Inflammatory bowel disease is notoriously difficult to diagnose—it can take a while to clinically differentiate between Crohn's disease (CD), ulcerative colitis (UC), and irritable bowel syndrome (IBS) among others. Therefore, we wished to investigate the potential for VOC analysis to discriminate between IBS, and the more serious CD and UC. Eighty participants (from these disease groups plus healthy volunteers) were recruited to the study, and breath samples were taken, as well as samples of blood, urine, and feces. GC-MS and SIFT-MS were used to analyze VOCs in the breath (from bag samples for SIFT-MS or TD tubes for TD-GC-MS) and the headspace of these samples. A number of multivariate statistical methods were used to analyze the data obtained and to determine whether discrimination between the disease groups was possible. Individual biomarkers for these conditions were also sought. The best results were obtained from the headspace of fecal samples, in which CD was distinguished from healthy controls with an overall classification accuracy of ~95%. Urine and blood showed some possibility as diagnostic sample matrices, but the results from breath were statistically insignificant due to the lower specificities and sensitivities. SIFT-MS showed the best discrimination results with blood, but in all cases, discrimination was much worse than when using GC-MS. Further analysis of the results has provided a list of compounds which are responsible for the discrimination in CD and UC, which, therefore, are potential markers (unpublished data).

18.7 BLADDER CANCER

In a similar way that VOCs in sputum headspace may have diagnostic potential for lung disease, VOCs in urine headspace may have diagnostic potential for bladder disease. For this reason, we analyzed urine headspace from patients with a form of bladder cancer (transitional cell carcinoma, TCC) plus three categories of controls, C1 (healthy individuals); C2 (non-cancerous condition with some abnormalities as picked up by urine dipstick); and C3 (confirmed non-cancerous urological disease).[13]

Previous studies[14] have indicated that volatile compounds specific to bladder cancer may exist in urine headspace, raising the possibility that headspace analysis could be used for diagnosis of this particular cancer. In this study, we investigated the use of a commercially available gas sensor array (e-nose) coupled with a specifically designed pattern recognition algorithm for this purpose and also analyzed the same samples using SPME/GC-MS. Using e-nose, the best diagnostic performance obtained with independent test data provided by healthy volunteers (C1) and bladder cancer patients was 70% overall accuracy (70% sensitivity and 70% specificity). When the data of patients suffering from other non-cancerous urological diseases were added to those of the healthy controls, the classification accuracy fell to 65% with 60% sensitivity and 67% specificity. While this is not sufficient for a diagnostic test, it is significantly better than random chance, leading us to conclude that there is useful information in the urine headspace. The use of SPME/GC-MS provided much better results, both in terms of sensitivity and specificity (unpublished data) and some VOC marker compounds responsible for the discrimination were identified. This method clearly has potential as a non-invasive alternative to a cystoscopy.

18.8 CONCLUDING REMARKS

VOC analysis has now been established as a highly promising technique for diagnosing disease or monitoring the condition of a patient. VOC analysis is perhaps best known with breath analysis, and there are a number of methods currently used in routine clinical practice for diagnosing disease or monitoring patients using breath analysis, such as diagnosis of *H. pylori* infection, or monitoring the condition of asthma sufferers. The concentration distributions of several common metabolites in exhaled breath of healthy individuals have been determined in order to recognize abnormal values in disease states. Certain metabolic conditions such as diabetes may also potentially be monitored by analyzing VOCs in breath or from skin and it is conceivable that such analysis may be used to reduce the reliance on blood glucose monitoring, particularly if acetone, which is usually highly abundant in diabetes, is shown to be a reliable monitor, although there are potential issues with this.[36,44] Large mass spectrometers may be replaced by nonspecific but small and portable gas sensors. These devices would not be of any use in monitoring substances present at low levels as they are insufficiently selective, but with acetone, their potential has been demonstrated.

However, there are issues with breath monitoring. The presence of interfering compounds produced by mouth flora needs to be considered, for example by taking breath from the nose. This is less convenient for the patient than mouth breathing but provides more representative data. For this reason, other samples have been headspaced for VOC analysis. Urine has been used to determine its potential for diagnosing bladder cancer, feces for diagnosing inflammatory bowel disease, and blood, serum, and sputum for tuberculosis.

Various studies have shown that mass spectrometric methods (SIFT-MS and GC-MS) are essential for identifying and quantifying individual markers. Mass spectra and chromatographic profiles may be used to look for characteristic patterns, and this may also be used to diagnose disease. This is particularly important where no unique markers at a sufficiently high concentration have been found for a condition. It is also important to know the identity of the compounds responsible for discriminating between a condition and its absence, even if they are not unique biomarkers. However, other devices such as e-nose have also been shown to play a role. Although they cannot identify individual compounds, e-nose can be used to determine whether there is a characteristic VOC profile for a particular disease condition. Small, portable gas sensors are also more convenient and cheaper for use by individuals, such as in blood glucose monitoring in diabetes. However, the use of these instruments should always be underpinned by a full understanding of the VOC profiles present, which can only be carried out through mass spectrometric methods.

Acknowledgments

The work of the following colleagues is acknowledged for their contribution to this work: Conrad Bessant, Michael Cauchi, Mark Chambers, Mark Evans, Dawn Fowler, Shu Hoashi, John Hunter, Henri Knobloch, Ruth McNerney, Mitesh Patel, Petra Reinhold, David Smith, Patrik Španěl, Andrew Spooner, Christopher Walton, Christina Weber, Carolyn Willis.

References

1. Amann A, Smith D, editors. *Breath analysis for clinical diagnosis and therapeutic monitoring*. World Scientific Singapore; 2005.
2. Massaro AF, Gaston B, Kita D, Fanta C, Stamler JS, Drazen JM. Expired nitric oxide levels during treatment of acute asthma. *Am J Respir Crit Care Med* 1995;**152**:800–3.
3. Phillips M, Boehmer JP, Cataneo RN, Cheema T, Eisen HJ, Fallon JT, et al. Prediction of heart transplant rejection with a breath test for markers of oxidative stress. *Am J Cardiol* 2004;**94**:1593–4.
4. Wang TS, Smith D, Španěl P. Selected ion flow tube, SIFT, studies of the reactions of H_3, O^+ NO^+ and $O_2^{(+)}$ with compounds released by Pseudomonas and related bacteria. *Int J Mass Spectrom* 2004;**233**:245–51.
5. Berger A. Helicobacter pylori breath tests. *BMJ* 2002;**324**:1263.
6. Turner C, Walton C, Hoashi S, Evans M. Breath acetone concentration decreases with blood glucose concentration in type I diabetes mellitus patients during hypoglycaemic clamps. *J Breath Res* 2009;**3**:046004.
7. Turner C, Španěl P, Smith D. A longitudinal study of ethanol and acetaldehyde in the exhaled breath of healthy volunteers using selected-ion flow-tube mass spectrometry. *Rapid Commun Mass Spectrom* 2006;**20**:61–8.
8. Španěl P, Turner C, Wang T, Bloor R, Smith D. Generation of volatile compounds on mouth exposure to urea and sucrose: implications for exhaled breath analysis. *Physiol Meas* 2006;**27**:N7–17.
9. Yasukawa T, Ohmori M, Sato S. The relationship between physiologic halitosis and periodontopathic bacteria of the tongue and gingival sulcus. *Odontology* 2010;**98**:44–51.
10. Wang T, Španěl P, Smith D. Selected ion flow tube mass spectrometry of 3-hydroxybutyric acid, acetone and other ketones in the headspace of aqueous solution and urine. *Int J Mass Spectrom* 2008;**272**:78–85.
11. Perucca J, Bouby N, Valeix P, Bankir L. Sex difference in urine concentration across differing ages, sodium intake, and level of kidney disease. *Am J Physiol Regul Integr Comp Physiol* 2007;**292**:R700–5.
12. Haddow JE, Knight GJ, Palomaki GE, Neveux LM, Chilmonczyk BA. Replacing creatinine measurements with specific gravity values to adjust urine cotinine concentrations. *Clin Chem* 1994;**40**:562–4.
13. Weber CM, Cauchi M, Patel M, Bessant C, Turner C, Britton LE, et al. Evaluation of a gas sensor array and pattern recognition for the identification of bladder cancer from urine headspace. *Analyst* 2011;**136**:359–64.
14. Španěl P, Smith D, Holland TA, Al Singary W, Elder JB. Analysis of formaldehyde in the headspace of urine from bladder and prostate cancer patients using selected ion flow tube mass spectrometry. *Rapid Commun Mass Spectrom* 1999;**13**:1354–9.

15. Probert CS, Ahmed I, Khalid T, Johnson E, Smith S, Ratcliffe N. Volatile organic compounds as diagnostic biomarkers in gastrointestinal and liver diseases. *J Gastrointestin Liver Dis* 2009;**18**:337–43.
16. Manolis A. The diagnostic potential of breath analysis. *Clin Chem* 1983;**29**:5–15.
17. Miekisch W, Schubert JK, Noeldge-Schomburg GF. Diagnostic potential of breath analysis—focus on volatile organic compounds. *Clin Chim Acta* 2004;**347**:25–39.
18. Smith D, Španěl P. Selected ion flow tube mass spectrometry (SIFT-MS) for on-line trace gas analysis. *Mass Spectrom Rev* 2005;**24**:661–700.
19. Turner C, Španěl P, Smith D. A longitudinal study of breath isoprene in healthy volunteers using selected ion flow tube mass spectrometry (SIFT-MS). *Physiol Meas* 2006;**27**:13–22.
20. Turner C, Španěl P, Smith D. A longitudinal study of methanol in the exhaled breath of 30 healthy volunteers using selected ion flow tube mass spectrometry, SIFT-MS. *Physiol Meas* 2006;**27**:637–48.
21. Turner C, Španěl P, Smith D. A longitudinal study of ammonia, acetone and propanol in the exhaled breath of 30 subjects using selected ion flow tube mass spectrometry, SIFT-MS. *Physiol Meas* 2006;**27**:321–37.
22. Deng C, Zhang J, Yu X, Zhang W, Zhang X. Determination of acetone in human breath by gas chromatography-mass spectrometry and solid-phase microextraction with on-fiber derivatization. *J Chromatogr B: Analyt Technol Biomed Life Sci* 2004;**810**:269–75.
23. Wang T, Pysanenko A, Dryahina K, Španěl P, Smith D. Analysis of breath, exhaled via the mouth and nose, and the air in the oral cavity. *J Breath Res* 2008;**2**:037013.
24. Smith D, Wang TS, Pysanenko A, Španěl P. A selected ion flow tube mass spectrometry study of ammonia in mouth- and nose-exhaled breath and in the oral cavity. *Rapid Commun Mass Spectrom* 2008;**22**:783–9.
25. Smith D, Turner C, Španěl P. Volatile metabolites in the exhaled breath of healthy volunteers: their levels and distributions. *J Breath Res* 2007;**1**:014004.
26. Španěl P, Dryahina K, Smith D. Acetone, ammonia and hydrogen cyanide in exhaled breath of several volunteers aged 4–83 years. *J Breath Res* 2007;**1**:011001.
27. Španěl P, Dryahina K, Smith D. The concentration distributions of some metabolites in the exhaled breath of young adults. *J Breath Res* 2007;**1**:026001.
28. Pysanenko A, Španěl P, Smith D. A study of sulfur-containing compounds in mouth- and nose-exhaled breath and in the oral cavity using selected ion flow tube mass spectrometry. *J Breath Res* 2008;**2**:046004.
29. Musa-Veloso K, Likhodii SS, Cunnane SC. Breath acetone is a reliable indicator of ketosis in adults consuming ketogenic meals. *Am J Clin Nutr* 2002;**76**:65–70.
30. Španěl P, Dryahina K, Rejskova A, Chippendale TW, Smith D. Breath acetone concentration; biological variability and the influence of diet. *Physiol Meas* 2011;**32**:N23–31.
31. Turner C, Parekh B, Walton C, Španěl P, Smith D, Evans M. An exploratory comparative study of volatile compounds in exhaled breath and emitted by skin using selected ion flow tube mass spectrometry. *Rapid Commun Mass Spectrom* 2008;**22**:526–32.
32. Tassopoulos CN, Barnett D, Fraser TR. Breath-acetone and blood-sugar measurements in diabetes. *Lancet* 1969;**1**:1282–6.
33. Owen OE, Trapp VE, Skutches CL, Mozzoli MA, Hoeldtke RD, Boden G, et al. Acetone metabolism during diabetic ketoacidosis. *Diabetes* 1982;**31**:242–8.
34. Ueta I, Saito Y, Hosoe M, Okamoto M, Ohkita H, Shirai S, et al. Breath acetone analysis with miniaturized sample preparation device: in-needle preconcentration and subsequent determination by gas chromatography-mass spectroscopy. *J Chromatogr B: Analyt Technol Biomed Life Sci* 2009;**877**:2551–6.

35. Wang CJ, Mbi A, Shepherd M. A study on breath acetone in diabetic patients using a cavity ringdown breath analyzer: exploring correlations of breath acetone with blood glucose and glycohemoglobin A1C. *IEEE Sens J* 2010;**10**:54–63.
36. Smith D, Španěl P, Fryer AA, Hanna F, Ferns GA. Can volatile compounds in exhaled breath be used to monitor control in diabetes mellitus? *J Breath Res* 2011;**5**:022001.
37. Phillips M, Cataneo RN, Condos R, Ring Erickson GA, Greenberg J, La Bombardi V, et al. Volatile biomarkers of pulmonary tuberculosis in the breath. *Tuberculosis (Edinb)* 2007;**87**:44–52.
38. Syhre M, Chambers ST. The scent of *Mycobacterium tuberculosis. Tuberculosis (Edinb)* 2008;**88**:317–23.
39. McNerney R, Mallard K, Okolo PI, Turner C. Production of volatile organic compounds by mycobacteria. *FEMS Microbiol Lett* 2012;**328**:150–6.
40. Spooner AD, Bessant C, Turner C, Knobloch H, Chambers M. Evaluation of a combination of SIFT-MS and multivariate data analysis for the diagnosis of *Mycobacterium bovis* in wild badgers. *Analyst* 2009;**134**:1922–7.
41. Turner C, Knobloch H, Richards J, Richards P, Mottram TTF, Marlin D, et al. Development of a device for sampling cattle breath. *Biosyst Eng*, 112(2), pp. 75–81.
42. Knobloch H. *Serum and urine headspace analysis using electronic nose (e-nose) technology*. Cranfield University; 2009.
43. Knobloch H, Schroedl W, Turner C, Chambers M, Reinhold P. Electronic nose responses and acute phase proteins correlate in blood using a bovine model of respiratory infection. *Sensor Actuat B: Chem* 2010;**144**:81–7.
44. Turner C. Potential of breath and skin analysis for monitoring blood glucose concentration in diabetes. *Expert Rev Mol Diagn* 2011;**11**:497–503.

PART

G

Exhaled Breath Condensate (EBC) and Particulates

CHAPTER

19

Measurement of Biomarkers of Oxidative Stress and Airway Inflammation in Exhaled Breath Condensate: Methodology and Potential Applications in Patients with COPD and Healthy Smokers

Paolo Montuschi

Department of Pharmacology, Faculty of Medicine, Catholic University of the Sacred Heart, Largo F. Vito, 1, 00168 Rome, Italy

19.1 INTRODUCTION

Exhaled breath consists of a gaseous phase, which contains volatile biomolecules (e.g. nitric oxide, carbon monoxide, and volatile organic compounds) and a liquid phase termed exhaled breath condensate (EBC). EBC is principally formed from water vapor, but also contains aerosol particles in which non-volatile biomolecules including 8-isoprostane (15-F_{2t}-isoprostane), a marker of lipid peroxidation, leukotrienes, prostaglandins, hydrogen peroxide, reactive nitrogen species, and adenosine have been detected.[1–3] Measurement of the fraction of exhaled nitric oxide (F_ENO) has been approved in the clinical setting for assessing and monitoring airway inflammation in patients with asthma,[4,5] but its utility in patients with chronic obstructive pulmonary disease (COPD) is uncertain. EBC, a non-invasive method for collecting airway secretions, is useful for studying the composition of the airway lining fluid.[6,7] Respiratory fluid represents between 0.01% and 2% of EBC by volume, as reported by one study in which EBC electrolyte concentrations were measured as a dilution marker.[8] Measurement of biomolecules in EBC is potentially useful for monitoring lung inflammation and oxidative stress, an important aspect of inflammation in patients with COPD and healthy smokers.[9] Because it is non-invasive, EBC collection might also be suitable for longitudinal studies and for monitoring the effects of pharmacological treatment in patients with COPD. Different profiles of biomarkers in EBC might reflect different aspects of lung inflammation or oxidative stress. Identification of

Volatile Biomarkers. http://dx.doi.org/10.1016/B978-0-44-462613-4.00019-2

selective biomarker profiles in EBC in patients with respiratory diseases might be of value for differential diagnosis. However, several methodological aspects have to be formally addressed before EBC analysis can be considered for application in the clinical setting. This article describes the methodology of EBC analysis, summarizes the results of the studies aimed at measuring biomarkers of inflammation and oxidative stress in EBC in patients with COPD and healthy smokers, presents advantages and limitations of EBC analysis and indicates directions for future research.

19.2 EBC ANALYSIS: METHODOLOGY

19.2.1 Experimental setup

Home-made or commercially manufactured devices for collecting EBC (Figures 19.1, 19.2, 19.3) consist of a mouthpiece with a one-way valve connected to a collecting system that is kept refrigerated.[10] Subjects are generally asked to breathe tidally through a mouthpiece connected to the condenser. Exhaled air enters and leaves the chamber through a one-way valve at the inlet and outlet while the chamber is kept closed. The EBC volume collected can be highly variable[11,12] depending on differences in type, material, and efficiency of the condenser and its temperature, turbulent airflow, and respiratory parameters and patterns. Commercially manufactured condensers include the EcoScreen® and the EcoScreen® II (Jaeger Toennies, Hoechberg, Germany) (Figures 19.2 and 19.3) and the RTube (Respiratory Research, Inc., Charlottsville, Virginia, USA).[13] The EcoScreen, which can be connected to a pneumotachograph and a computer for online recording of respiratory parameters,[14] has been used in several studies, but its production has recently been discontinued due to a

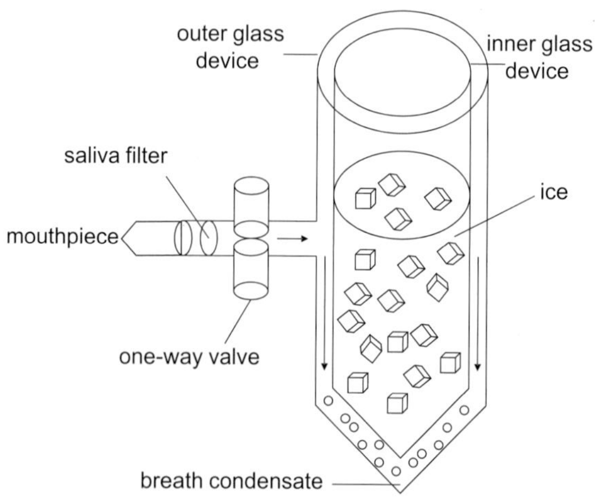

FIGURE 19.1
Schematic representation of a "home-made" device for collecting EBC.

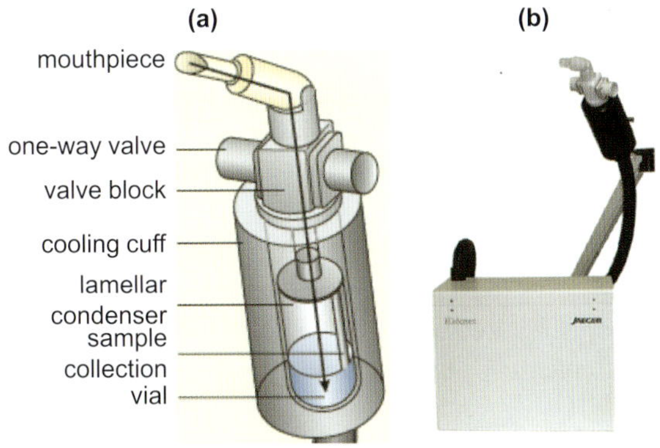

FIGURE 19.2
(a) Schematic representation of the EcoScreen® (Jaeger Toennies, Hoechberg, Germany) collecting system for EBC. The EBC is collected in the vial, as indicated by the arrow. (b) the EcoScreen® condenser. (Panel A: Modified with permission from Nature Reviews Drug Discovery, Vol. 1, Indirect monitoring of lung inflammation, Paolo Montuschi, pp. 238–242, copyright 2002 Macmillan Magazines Ltd).

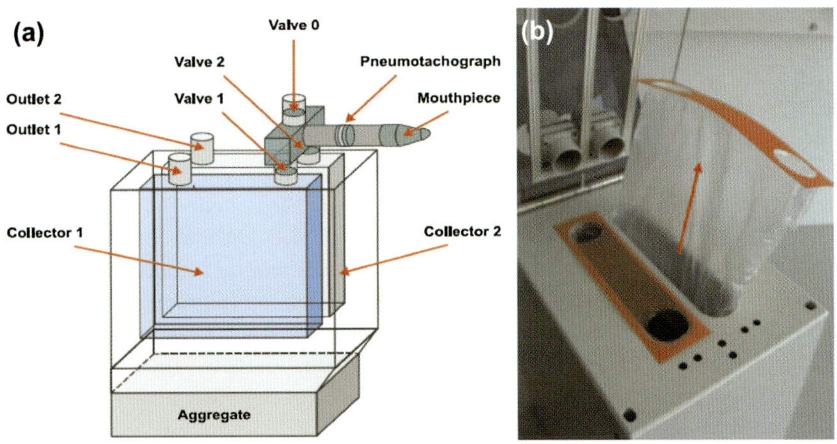

FIGURE 19.3
(a) Schematic representation of the EcoScreen® II condenser (Jaeger Toennies, Hoechberg, Germany). (b) The EcoScreen® II condenser.

manufacturer's decision. With the EcoScreen® II (Jaeger Tonnies, Hoechberg, Germany) (Figure 19.3), respiratory parameters can be measured during EBC collection. EBC derived from the airways or the alveoli can be collected at the same time into two separate collecting systems that consist of plastic bags inserted into a refrigerating system (Figure 19.3). The EcoScreen® II might be useful for studying the origin of biomarkers in the respiratory system (airways vs alveoli). Using this condenser, fractionated hydrogen peroxide measurement could differentiate hydrogen peroxide concentrations in different parts of the lung in asthmatic and healthy children.[15] The RTube consists of a disposable polypropylene tube that is connected to an exhalation valve that also serves as a syringe-style plunger to pull fluid off the condenser walls.[13] Breath is cooled down by placing an aluminum cooling sleeve over the polypropylene tube. The RTube is a portable condenser suitable for on-site collection of EBC samples. Generally, a one minute procedure can provide sufficient EBC for a pH measurement.[13] Samples collected on site are then transported to the laboratory for analysis. Using a selective sensor, hydrogen peroxide can be measured immediately after EBC sample collection.[16] The development of selective sensors for detecting specific biomolecules in the breath is currently being carried out.

19.2.2 Analytical techniques

Biomolecules in EBC in patients with COPD can be measured with different analytical techniques (Table 19.1). In most studies, immunoassays have been used.[6,17] Mass spectrometry has confirmed the presence of leukotriene (LT) B_4,[18–20] 8-isoprostane,[21,22] prostaglandin (PG) E_2,[21] aldehydes,[23] and free 3-nitrotyrosine[24,25] in EBC in healthy subjects and in patients with respiratory diseases. LTB_4[18–20] and aldehydes[23] in EBC have been measured by liquid chromatography/mass spectrometry (LC/MS). Using gas chromatography/mass spectrometry (GC/MS) LTB_4, LTD_4, and LTE_4 concentrations in EBC have been measured in healthy subjects and patients with asthma.[26] PGE_2 in EBC has been measured by GC/MS in patients with acute respiratory distress syndrome.[21] 8-Isoprostane and free 3-nitrotyrosine have been measured by both GC/MS[21,25] and LC/MS,[22,24] whereas EBC adenosine[27] and reduced glutathione[28] have been detected by HPLC.

Generally, immunoassays for biomolecules in EBC require validation with reference analytical techniques such as MS or HPLC. Two radioimmunoassays (RIAs) for 8-isoprostane and PGE_2 developed in our laboratory[29] and a commercially available enzyme immunoassay (EIA) for LTB_4[30] were qualitatively validated by HPLC. After HPLC separation, malondialdehyde (MDA) has been measured in EBC using fluorescence detection.[31]

19.2.3 Methodological aspects of EBC analysis

The priorities in EBC analysis are standardization of procedures, identification and quantitative assessment of biomolecules using “gold-standard” analytical techniques, and validation of the immunoassays for measuring biomarkers of inflammation and oxidative stress. Methodological aspects of EBC analysis have to be formally addressed[7,17]; these include the origin(s) of biomolecules; their flow- and time-dependence; organ specificity; characterization of the physical-chemical properties of different biomolecules; the use of reference indicators to estimate the biomolecule

Table 19.1 Inflammatory biomarkers in EBC in patients with COPD.

Biomarker	Method	Values	Ref.
Leukotrienes			
LTB_4	EIA	*100.6 (73.5–145.0) pg/mL	48
	EIA	§8.5 ± 0.8 pg/mL	50
	EIA	§§86.7 ± 19 pg/mL	49
LTE_4	EIA	*23.3 (9.1–31.3) pg/mL	48
Prostanoids			
PGE_2	EIA	*98.0 (57.0–128.4) pg/mL	48
PGE_2	RIA	#93.5 (84.0–105.5) pg/mL	51
$PGF_{2\alpha}$	EIA	*15.0 (10.9–19.0) pg/mL	48
PGD_2-MOX	EIA	*11.2 (8.7–15.0) pg/mL	48
Isoprostanes			
8-Isoprostane	EIA	§45 ± 3.6 pg/mL	56
	RIA	#47.9 (40.5–51.9) pg/mL	51
	EIA	¶47 (95% CI, 41–53) pg/mL	57
	EIA	§43.7 ± 2.8 pg/mL	61
	EIA	§6.0 ± 0.7 pg/mL	50
	EIA	§18.1 ± 2.0 pg/mL	58
	EIA	#14.2 [9.5–26.4]	59
Hydrogen peroxide	Spectrophotometry	§0.205 ± 0.054 μM	65
	Spectrophotometry	*2.6 (1.9–3.5) μM	74
	Spectrophotometry	§0.46 ± 0.03 μM	72
	Fluorimetry	§§0.32 ± 0.03 μM	75
	Spectrophotometry	§0.50 ± 0.11 μM	11
	Spectrophotometry	¶0.66 (0.54–0.68) μM	57
NO-derived products			
S-Nitrosothiols	Spectrophotometry	§0.24 ± 0.04 μM	82
Peroxynitrite ^^	Fluorimetry	§§7.9 ± 3.0 nM	83
Peroxynitrite ^^	Fluorimetry	§295.1 ± 52.7 nM	61
Nitrite	Spectrophotometry	§2.62 ± 0.52 μM	82
Nitrite/nitrates	Fluorimetry	§37.8 ± 3.4 μM	61
Nitrite/nitrates	Fluorimetry	^ 10.2 (2.0–93.8) μM	84
Nitrotyrosine	EIA	§24.6 ± 2.2 ng/mL	61
Hydrogen ions	pH meter	¶7.16 (7.09–7.23)	56
Aldehydes			
MDA	GC/MS/MS	§57.2 ± 2.4 nmol/l	23
Hexanal	GC/MS/MS	§63.5 ± 4.4 nmol/l	23
Heptanal	GC/MS/MS	§26.6 ± 3.9 nmol/l	23
ATP	Bioluminescence	§141 ± 44 pM	90

(Continued)

Table 19.1 (Continued)

Biomarker	Method	Values	Ref.
Cytokines			
IL-1β	Multiplex fluorescent bead immunoassay	**not available	92
IL-6	Multiplex fluorescent bead immunoassay	**not available	92
	EIA	§8.0 ± 0.1 pg/mL	94
IL-8	Multiplex fluorescent bead immunoassay	**not available	92
IL-10	Multiplex fluorescent bead immunoassay	**not available	92
IL-12p70	Multiplex fluorescent bead immunoassay	**not available	92
	Multiplex bead immunoassay	*3.75 (3.15–5.02) pg/mL	93
sICAM-1	Multiplex bead immunoassay	*22.8 (16.6–35.6) pg/mL	93
MIF	Multiplex bead immunoassay	§37.2 ± 8.9 (pg/mL)	93
RANTES	Multiplex bead immunoassay	§22.5 ± 1.7 pg/mL	93
TNF α	Multiplex fluorescent bead immunoassay	**not available	92

*Values are expressed as §mean ± SEM, §§mean ± SD, *median with the minimum to maximum range, ¶mean with 95% confidence intervals for the differences, #median with interquartile range, or ^ geometric means ± SEM. ^^ Peroxynitrite was measured using oxidation of 2′,7′-dichlorofluorescin in EBC collected at high pH.[83] ** Data are presented as figures, numerical data are not available. For some biomarkers, there is a high variability in the concentrations reported in different studies. This can be partially explained by the lack of standardization of the exhaled breath condensate method and the different analytical techniques.*

Abbreviations: ATP, adenosine triphospahte; sICAM-1, soluble intracellular adhesion molecule-1; IL, interleukin; LT, leukotriene; MDA, malondialdehyde; MIF, macrophage migration inhibitory factor; MOX, methoxime, a stable PGD_2 metabolite; NO, nitric oxide; PG, prostaglandin; RANTES, regulated on activation normal T cell expressed and secreted; TNF α, tumor necrosis factor α.

concentrations in the airway lining fluid and to normalize for variations in non-volatile compounds that can result from variations in aerosol particle dilution by water vapor[8]; the effects of temperature, humidity, and collecting setup materials on EBC analysis; possible nasal, saliva, and sputum contamination; between-day, within-day, and analytical variability; comparison of different condensers; sample stability; pre-treatment of samples; effect of different factors on biomolecule concentrations.

19.2.3.1 Origin(s) of biomolecules in EBC

The cellular source(s) of biomolecules detected in EBC, their site of production in the respiratory system compartments (airways as opposed to alveoli), and the extent to which EBC reflects the composition of airway lining fluid have to be clarified. EBC is not suitable for studying inflammatory cells in the respiratory system. However, indirect information on the cellular origin(s) of a given biomolecule can be obtained by correlating its concentrations in EBC with sputum or bronchoalveolar lavage cell counts of a given cell type. However, such correlation does not necessarily mean that the biomolecule is released by that cell type, as elevated biomolecule concentrations in EBC could also result from cell activation without increased cell counts. Establishing the cellular origin(s) of biomolecules requires invasive techniques, including bronchoscopy and bronchial biopsies.

Flow dependence might indicate whether EBC biomolecules are principally produced in the airways as opposed to the alveolar region.[32] Flow rate modifies the time available for the accumulation of a biomolecule in the respiratory tract, and, therefore, their concentrations in the EBC.[32] The fact that hydrogen peroxide concentrations in EBC depend on expiratory flow rate suggests that this biomarker of oxidative stress is principally produced in the airways.[32] Concentrations of 8-isoprostane[14,21] and PGE_2[14,21] in EBC do not depend on minute ventilation; exhaled glutathione[28] and aldehydes[23] are not flow-dependent. Taken together, these data indicate that these biomolecules are mostly produced in the alveolar region where the effect of flow rate in determining their concentrations in EBC is less important.[32] Data on flow-dependence of biomolecules are mainly derived from healthy subjects. Future studies are required to clarify the effects of different flow rates on biomolecule concentrations in EBC in patients with COPD and other lung diseases. In these patients, the greater turbulence of the airflow can increase the formation of aerosol particles that contain non-volatile biomolecules.[13] Moreover, the increased numbers and/or activation of selective cell types in different compartments of the respiratory system might have a different impact on the concentrations of a given biomarker in the EBC as compared with healthy subjects. As it is designed for differential sampling of EBC from the airways and the alveolar region, the EcoScreen® II could be a suitable tool for studying the origin of EBC biomolecules in the respiratory system.[15]

19.2.3.2 Flow-dependence of biomolecules in EBC and effect of respiratory patterns

The volume of EBC collected depends on respiratory patterns.[33,34] In healthy subjects who maintained the same tidal volume for 15 min, the mean EBC volume collected was almost twofold higher when the respiratory rate was increased from 14 breaths/min to 28 breaths/min.[33] EBC volume correlates with expiratory minute ventilation, but higher values of expiratory minute ventilation are associated with decreased condenser efficiency.[34]

In healthy non-smokers and patients with mild asthma, EBC hydrogen peroxide concentrations depend on the expiratory flow rate.[32] By contrast, exhaled MDA and glutathione do not seem to be flow-dependent, because similar concentrations were observed when EBC was collected at different flow rates.[23,28] 8-Isoprostane,[14,21]

PGE_2,[21] LTB_4,[33] LTE_4,[33] nitrite,[34] total protein,[34] and ethanol[35] values in EBC seem to be independent of respiratory patterns.

We studied the effect of ventilation patterns on 8-isoprostane and PGE_2 concentrations in EBC.[14] Fifteen healthy adults were asked to breathe at 10, 20, and 30 l/min for 10, 10, and 5 min, respectively.[14] 8-Isoprostane and PGE_2 concentrations were similar, indicating that they are not affected by expiratory minute ventilation.[14] Consistent with our findings, there was no correlation between 8-isoprostane concentrations and minute ventilation in mechanically ventilated patients.[21] Similar LTB_4 and LTE_4 values in EBC were reported in healthy subjects who were asked to breathe at 14 and 28 breaths/min maintaining the same tidal volume.[33] As airflow rate and respiratory patterns can have a different effect on the concentrations of different biomolecules in EBC, each biomolecule should be studied individually. Further studies should be undertaken to clarify the effects of respiratory patterns on the concentrations of EBC biomarkers as: (1) data on the flow-dependence of biomolecules is mostly unavailable; (2) available data on the effect of ventilation patterns on biomolecules is mainly derived from healthy subjects and they cannot be directly extrapolated to patients with COPD and other lung diseases; (3) data comparison is often difficult due to the fact that a different methodology was generally used in the published studies. Clarifying the effect of respiratory parameters on the exhaled biomolecule concentrations is relevant to the standardization of the EBC technique.

19.2.3.3 Time-dependence of biomolecules in EBC

After 20 min of tidal breathing, the mean volume of EBC collected is almost twofold higher than that collected after 10 min of tidal breathing.[23] EBC 8-isoprostane concentrations[21] and EBC pH values[36] are independent of the duration of collection, whereas exhaled hydrogen peroxide concentrations decrease with prolonged sampling time.[37] Data on MDA are conflicting. In one study, concentrations of MDA in EBC samples collected at 10 min and 20 min were similar,[23] whereas in another study MDA concentrations decreased over time.[36] The duration of EBC collection could be relevant for less stable biomolecules, including cysteinyl-LTs and hydrogen peroxide. Further studies are required to determine the effect of sampling duration on the concentrations of different biomolecules and to optimize sampling time for each of them.

19.2.3.4 Organ specificity of EBC

EBC analysis is potentially useful for providing organ-specific quantitative assessment of inflammation in the respiratory system. However, it is possible that biomolecules produced at sites remote from the lungs could circulate in plasma and enter the pulmonary epithelial lining fluids, particularly when inflammation increases capillary permeability. Studies aiming at comparing biomolecule concentrations in EBC and in biological fluids that reflect systemic production of inflammatory mediators such as plasma and urine in patients with airway inflammatory diseases could provide an answer to this question. Likewise, measuring biomolecules in EBC in patients with systemic inflammatory diseases might also be useful, although the possible concomitant systemic inflammation in patients with COPD could be a confounding factor.

19.2.3.5 Nasal and saliva contamination

Nasal and salivary contamination of EBC samples should be avoided. In healthy subjects and patients with allergic rhinitis, concentrations of adenosine, ammonia, and thromboxane (Tx) B_2 in EBC are not affected by the mode of inhalation (nasal vs oral).[27] Salivary contamination of EBC is excluded by measuring amylase concentrations that are generally undetectable,[3] even when very sensitive assays are used.[8] When amylase is detected, its concentrations in EBC are less than 0.01% of those in saliva, indicating very little, if any, salivary contamination.[8] Consistent with these findings, similar adenosine and TxB_2 concentrations were observed in EBC samples obtained through tracheostomy in mechanically ventilated patients or through the mouth in healthy subjects.[27] Nuclear magnetic resonance (NMR) spectroscopy studies[38–40] and GC/MS[41] have confirmed that EBC is not generally contaminated by saliva and that salivary contamination, if any, is not biologically significant.

19.2.3.6 Dilution reference indicators

EBC is mainly formed from water vapor, but also contains aerosol droplets in which non-volatile biomolecules are detected. Measurement of dilution reference indicators in EBC, including conductivity, urea, and electrolytes, has been proposed to normalize for inter-individual variations in droplet formation and to estimate the concentrations of non-volatile biomolecules in the airway lining fluid.[8,42] Inter-individual differences in non-volatile biomolecule concentrations could be explained, at least partially, by differences in respiratory droplet dilution. However, the selective increase of biomolecules and the lack of correlation between concentrations of biomolecules which are structurally similar are not consistent with this hypothesis.[33] The role of dilution reference indicators in EBC analysis is not defined. To normalize for possible variations in biomolecule concentrations due to differences in respiratory droplet dilution, EBC results can be expressed as a change in the concentration ratio of one biomolecule to another.

19.2.3.7 Diurnal and between-day variability of biomolecules in EBC

Within-day and between-day variability has to be established for each single biomolecule. Between-day reproducibility of RIA measurement for 8-isoprostane and PGE_2 concentrations in EBC was assessed by collecting samples on days 1, 3, and 7 from healthy subjects.[14] Specificity of these RIAs, developed in our laboratory, has been demonstrated by validation experiments with reverse phase-high performance liquid chromatography (RP-HPLC).[29] RIAs for 8-isoprostane and PGE_2 in EBC were highly reproducible, as reflected by their intraclass correlation coefficients

(0.95 and 0.90, respectively).[14] Between-day reproducibility of RIA measurement of 8-isoprostane, PGE_2, and TxB_2 in EBC, expressed as limits of agreement according to Bland and Altman analyzes,[43] was also assessed in children with asthma.[44,45] Limits of agreement for RIA measurement of these eicosanoids in EBC were as follows: 8-isoprostane, −5, and 5 pg/mL; PGE_2, −5.3 and 5.1 pg/mL; TxB_2, −3.7, and 4.7 pg/mL.[44] TxB_2 was detectable only in some children with asthma. Measurement of leukotrienes in EBC in children with asthma using commercially available enzyme immunoassays (Cayman Chemicals, Ann Arbor, MI) was less reproducible than RIAs for prostanoids as indicated by their limits of agreement.[46]

19.3 ANALYSIS OF EBC FROM PATIENTS WITH COPD AND HEALTHY SMOKERS

Several biomolecules, including leukotrienes, prostaglandins, isoprostanes, hydrogen peroxide, hydrogen ions (pH), nitric oxide-derived products and adenosine triphosphate, have been measured in EBC in healthy subjects. Some inflammatory mediators are elevated in patients with COPD.[9]

19.3.1 Eicosanoids

In most EBC studies, leukotriene and prostanoid concentrations have been measured by commercially available immunoassays that generally require validation. The limitations of this approach and the need for a robust analytical methodology for a specific and accurate quantitative assessment of eicosanoid concentrations in EBC have been discussed.[7,17,47] The presence of LTB_4,[18–20] 8-isoprostane,[21,22] and PGE_2[21] in EBC has been confirmed by mass spectrometry. RIAs for 8-isoprostane and PGE_2 have been qualitatively validated RP-HPLC.[29]

19.3.1.1 Leukotrienes

Compared with healthy smokers, LTB_4 concentrations are elevated in EBC from patients with stable COPD independent of glucocorticoid treatment.[48,49] In patients with COPD,[48] the profile of leukotrienes in EBC is different from that observed in patients with asthma.[33] Patients with asthma,[33] but not patients with COPD,[48] have elevated LTE_4 concentrations in EBC. Exhaled LTB_4 is elevated in both patients with asthma and COPD, but the latter have higher average LTB_4 concentrations in EBC.[48] In patients with COPD exacerbations, exhaled LTB_4 concentrations are decreased 2 weeks after treatment with antibiotics, an effect that persists after 2 months.[50]

19.3.1.2 Prostanoids

Compared with healthy non-smokers, PGE_2 concentrations in EBC are elevated in steroid-naïve and steroid-treated patients with stable COPD.[48] In both COPD groups of patients, exhaled PGE_2 and LTB_4 concentrations are correlated,[48] possibly indicating higher production of PGE_2, an endogenous biomolecule that may have anti-inflammatory effects in the airways, in conditions of increased lung inflammation. Non-selective cyclo-oxygenase (COX) inhibition by oral ibuprofen (400 mg q.i.d. for 2 days) reduces PGE_2 concentrations in EBC in patients with COPD,[51] an effect that might be relevant to the modulation of airway inflammation. In contrast, oral rofecoxib (25 mg once a day for 5 days), a selective COX-2 inhibitor, has no effect on PGE_2 and LTB_4 concentrations in EBC in patients with stable COPD.[51] Taken together, these results indicate that PGE_2 in EBC in patients with COPD is mainly derived from COX-1 activity.[51] The pathophysiological implications of selective and non-selective inhibition of COX for pulmonary inflammation in patients with COPD have to be established.

19.3.1.3 Isoprostanes

Isoprostanes, prostaglandin-like compounds that are produced *in vivo* independently of COX by peroxidation of arachidonic acid due to free radicals,[52] are considered among the best markers of oxidative stress.[53,54] 8-Isoprostane (15-F_{2t}-IsoP or 8-iso-$PGF_{2\alpha}$), a compound that belongs to the F_2 class of isoprostanes, is detected in EBC in healthy non-smokers.[10,21,55] 8-Isoprostane concentrations in EBC are elevated in healthy smokers and, to a greater extent, in patients with stable COPD who are current or ex-smokers.[55] Similar 8-isoprostane concentrations in current and ex-smokers with COPD might suggest that oxidant stress due to chronic smoking reaches a point at which smoking cessation has little, if any, effect.[55] Elevated exhaled 8-isoprostane concentrations in healthy smokers and patients with stable COPD has been confirmed by other authors.[56–60] 8-Isoprostane concentrations in EBC can reflect the severity of emphysema assessed by high resolution computed tomography.[58] In patients with COPD exacerbations, exhaled 8-isoprostane concentrations are higher than in healthy subjects[50,60] and are reduced after antibiotic treatment.[50] One study reported at least 10-fold lower 8-isoprostane concentrations in EBC than in sputum supernatants.[60] Inhibition of inducible nitric oxide-synthase (iNOS) by inhalation of aminoguanidine, a relatively selective inhibitor, has no effect on 8-isoprostane concentrations in EBC in patients with COPD, healthy smokers, and healthy non-smokers.[61]

19.3.2 Hydrogen peroxide

Hydrogen peroxide concentrations in EBC have been measured by spectrofluorimetry,[62–64] spectrophotometry,[11,65,66] fluorimetric assays[62–69] and chemiluminescence.[70,71] Widely variable mean hydrogen peroxide concentrations in EBC in healthy non-smoking adults have been reported (0.01 nM–0.45 μM).[62,63] Even under strictly controlled conditions, there is a high degree of between-day variability that emphasizes the fact that standardization of the experimental technique is a priority in this research area.[32]

Compared with healthy non-smokers, exhaled hydrogen peroxide concentrations are increased in healthy smokers[62] and, to a greater extent, in patients with stable

COPD.[11,64,65] In these patients, exhaled hydrogen peroxide concentrations seem independent of smoking habit.[64] Hydrogen peroxide concentrations in EBC are further elevated in patients with COPD exacerbations[65,72] and increase immediately after maximal exercise in patients with stable COPD.[73]

Data on the effect of inhaled glucocorticoids on exhaled hydrogen peroxide concentrations in patients with COPD is conflicting.[74,75] In one study, beclomethasone at a dose of 500 μg twice daily for 2 weeks had no effect on exhaled hydrogen peroxide concentrations in non-smoking patients with stable COPD.[74] In contrast, other authors reported that HFA-beclomethasone dipropionate at a dose of 400 μg twice daily for 4 weeks and fluticasone dipropionate at a dose of 375 μg twice daily for 4 weeks reduced exhaled hydrogen peroxide concentrations in patients with stable moderate COPD.[75] In patients with COPD exacerbations who required hospital admission because of lower respiratory tract infections, treatment with intravenous dexamethasone, inhaled salbutamol/ipratropium, and antibiotics, as needed, did not reduced hydrogen peroxide concentrations in EBC.[76] In contrast, other authors reported reduce exhaled hydrogen peroxide concentrations after intravenous glucocorticoid treatment in patients with COPD exacerbation.[72,77] In patients with stable COPD, oral *N*-acetyl-cysteine, which could have antioxidant effects, at a dose of 600 mg once a day for 12 months reduced hydrogen peroxide concentrations in EBC starting from 9 months.[78] In a similar study, this effect started from 15 days.[79] Despite some evidence of reduced oxidant stress, a multicenter trial has shown that *N*-acetyl-cysteine is not effective at preventing deterioration in lung function and exacerbations in patients with COPD.[80] However, it is not known whether the dose used in this study had an antioxidant effect, as the clinical pharmacology of *N*-acetyl-cysteine is not well established and measurement of biomarkers of oxidative stress was not included in this study.[80] Hydrogen peroxide concentrations in EBC in patients with COPD are transiently increased by 30 min after *N*-acetyl-cysteine was given by nebulization.[81] The clinical relevance of these findings is unknown. Large controlled studies to establish the effects of pharmacological treatment on hydrogen peroxide concentrations in EBC in patients with COPD are required. As with other biomolecules, a reliable methodology for EBC analysis is a prerequisite for planning and interpreting the results of interventional studies.

19.3.3 NO-derived products

Patients with COPD have higher *S*-nitrosothiols in EBC than healthy non-smokers,[82] but whether elevated *S*-nitrosothiols merely reflects NO production or whether they play a role in the pathophysiology of COPD has to be established. Peroxynitrite concentrations, measured using oxidation of 2′, 7′-dichlorofluorescin in EBC collected at high pH, are increased in patients with moderate to severe COPD compared with those in healthy non-smokers and smokers and are negatively correlated with forced expiratory volume (FEV_1).[83] Nitrite and nitrates are end-products of NO metabolism. EBC nitrite concentrations were found to be increased in patients with COPD.[82] Another study reported similar EBC nitrite/nitrates concentrations in patients with COPD and healthy subjects.[84] In patients with COPD, EBC nitrite concentrations are correlated with hyperinflation.[85] In healthy chronic smokers, smoking induces an acute increase in EBC nitrate, but not nitrite.[86] Peroxynitrite levels, measured using oxidation of 2′,7′-dichlorofluorescein, in patients with stable COPD are higher than those in healthy smokers and healthy non-smokers.[61,83] In a double-blind, placebo-controlled study, inhalation of aminoguanidine caused a significant, but not complete, inhibition of peroxynitrite production as reflected by its EBC concentrations in patients with COPD and healthy smokers 60 min after inhalation,[61] but no effect in healthy non-smokers.[61] Aminoguanidine reduced nitrite/nitrate concentrations in patients with COPD at 60 min and 120 min after inhalation, whereas in healthy smokers and healthy non-smokers this reduction was observed only at 60 min.[6] Inhaled aminoguanidine has no effect on EBC nitrotyrosine concentrations in patients with COPD, healthy smokers, and healthy non-smokers.[61] Taken together, this data suggests that both constitutive and inducible nitric oxide synthase (NOS) isoforms might contribute to the elevated peroxynitrite and nitrite/nitrate concentrations in EBC observed in patients with COPD.[61]

19.3.4 pH

Mean values of EBC pH, a robust and reproducible assay of airway acidity,[13,87] are lower in patients with COPD compared with patients with asthma and healthy subjects.[56] In patients with COPD, EBC pH is correlated with sputum neutrophilia and exhaled hydrogen peroxide.[56] In this study, EBC pH values in patients with COPD were similar, irrespective of treatment with inhaled glucocorticosteroids.[56] However, large controlled studies are required to establish the effects of glucocorticoids on EBC pH values in patients with COPD. EBC pH is negatively correlated with gastroesophageal reflux disease symptoms in patients with COPD and healthy controls.[88]

19.3.5 Aldehydes

Malondialdehyde (MDA), a product of lipid peroxidation, is a widely used biomarker of oxidative stress, but its reliability for assessing oxidative stress *in vivo* is questionable.[89] Using liquid chromatography-tandem mass spectrometry, EBC concentrations of MDA and other aldheydes (hexanal and heptanal) were found to be elevated in patients with COPD as compared with non-smoking subjects, whereas only MDA was elevated in patients with COPD compared with healthy smokers.[23]

19.3.6 Adenosine triphosphate

Adenosine triphosphate (ATP) has been detected in EBC in healthy non-smokers, healthy smokers, and patients with COPD exacerbation.[90] Similar ATP concentrations in EBC were reported in the three study groups.[90] Treatment improved oxygenation in patients with COPD, but did not affect EBC ATP concentrations.[90]

19.3.7 Other markers

A wide inter-individual variability (from undetectable to 1.4 mg) in the amount of total proteins in EBC, even in healthy subjects, has been reported.[91] Amylase concentrations in EBC, also measured with very sensitive analytical techniques, are generally undetectable or very low, indicating little, if any, salivary contamination.[8,41] EBC concentrations of interleukin (IL)-1β, IL-6, IL-8, IL-10, IL-12p70, and tumor necrosis factor (TNF) α, measured with a multiplex bead array test combined with flow cytometry, are elevated in patients with COPD exacerbation as compared with healthy non-smokers, healthy smokers, and patients with stable COPD.[92] Healthy smokers had increased EBC concentrations of all cytokines compared to healthy non-smokers and, with the exception of IL-1β, compared to patients with stable COPD.[92] Patients with COPD had higher IL-1β and IL-12 concentrations than healthy non-smokers.[92] One study reported elevated EBC concentrations of macrophage migration inhibitory factor, IL-12, regulated on activation normal T cell expressed and secreted (RANTES), and soluble intracellular adhesion molecule-1 in patients with moderate to severe COPD.[93] Treatment with infliximab, an anti-TNF-α antibody, had no effect on cytokine concentrations in EBC.[93] In all published studies, cytokine concentrations in EBC were measured by immunoassays. These are very sensitive and work well in buffer, but their behavior in EBC is not established.

Reported cytokines concentrations in EBC are very low and close to the lower limit of quantification of the immunoassays used. At these concentrations, the analytical variability is very high making it difficult for the interpretations of results. The presence of cytokines in EBC has not been confirmed by gold-standard analytical techniques and an accurate quantification of their concentrations in EBC has not yet been provided. These relevant issues need to be formally addressed in future studies in which a specific and robust analytical methodology should be used.

Metabolomics is the study of molecules generated by metabolic pathways and this approach has been applied to the analysis of EBC.[38–40] Metabolomic analysis of EBC by NMR spectroscopy is seen to distinguish between patients with COPD, laringectomized patients, and healthy subjects.[38] NMR spectra of saliva are very different from the corresponding EBC samples,[39,40] thus excluding significant salivary contamination.

19.4 ADVANTAGES AND LIMITATIONS OF THE EBC TECHNIQUE

In combination with measurement of volatile compounds in exhaled breath, analysis of EBC with MS techniques or NMR spectroscopy is part of a new “omics technique” called “breathomics”.[40] EBC analysis may provide insights into the pathophysiology of COPD. Measurement of surrogate markers of lung inflammation in EBC in patients with COPD might be useful for identifying subgroups of healthy chronic smokers who are more susceptible to develop COPD, molecular phenotyping, identifying those patients who are most likely to benefit from pharmacological treatment, assessing the effect of anti-inflammatory treatment, and testing new drugs for COPD.

The identification of molecular phenotypes based on profiles of inflammatory markers in EBC might be useful for a more homogeneous selection of patients to be recruited for COPD clinical trials.

The lack of standardization of EBC analysis is currently the main limitation of this technique. This explains most of the variability of the results reported in different studies and makes difficult the meaningful comparison of data obtained in different laboratories. Other current limitations of the EBC technique are: the relatively small size of the published studies which are generally single center studies; this technique is unable to provide any information on lung inflammatory cells; unable to distinguish between pulmonary rather than systemic inflammation; the origin (airways vs alveolar region) and cellular sources of most biomolecules in EBC have to be identified.

19.5 FUTURE RESEARCH

EBC analysis is currently a research tool with potential clinical applications. Standardization of EBC procedures, clarification of the methodological issues discussed above, and validation of the analytical techniques for measuring various biomolecules in EBC are the priorities in this research area. A robust analytical methodology is essential for the development of this technique. Future research should include development of reference analytical techniques (e.g. mass spectrometry) to confirm the presence of biomolecules in EBC and provide an accurate quantitative assessment of their concentrations; development of sensitive, specific, and reproducible immunoassays or validation of those already available to be used as a routine analytical tool; identification of reference values for different inflammatory biomarkers in healthy subjects and quantitative assessment of their concentrations in patients with stable COPD of different severity and in patients with COPD exacerbations; studies of diurnal and between-day variations and analytical reproducibility of measurements; large longitudinal studies in patients with COPD; studies aimed at clarifying the relationships between EBC markers and symptoms, lung function, and other indices and/or techniques for quantifying lung inflammation in patients with COPD.

Large controlled studies are required to establish the utility of EBC analysis for monitoring the response to pharmacological treatment in patients with COPD. As the behavior of different biomolecules in EBC could be different, standardized procedures have to be implemented for each individual biomolecule.

EBC technique should be part of an integrated non-invasive approach that consists of clinical assessment, lung function testing, sputum induction, and possibly breath analysis with techniques including mass spectrometry and electronic noses. Because of the relative lack of non-invasive methods for assessing and monitoring airway inflammation and therapeutic intervention in patients with COPD, further research in this area is warranted.

Acknowledgments

Supported by PRIN 2007 and Catholic University of the Sacred Heart, Fondi di Ateneo 2009–2011.

References

1. Montuschi P. Indirect monitoring of lung inflammation. *Nat Rev Drug Discov* 2002;**1**: 238–42.
2. Montuschi P, editor. *New perspectives in monitoring lung inflammation: analysis of exhaled breath condensate*. Boca Raton: CRC Press; 2005.
3. Horvath I, Hunt J, Barnes PJ, Alving K, Antczak A, Baraldi E, et al. Exhaled breath condensate: methodological recommendations and unresolved questions. *Eur Respir J*, 2005;**26**:523–48.
4. Anonymous. Recommendations for standardized procedures for the on-line and off-line measurement of exhaled lower respiratory nitric oxide and nasal nitric oxide in adults and children-1999. This official statement of the American Thoracic Society was adopted by the ATS Board of Directors, July 1999. *Am J Respir Crit Care Med* 1999;**160**:2104–17.
5. Silkoff PE, Carlson M, Bourke T, Katial R, Ogren E, Szefler SJ. The Aerocrine exhaled nitric oxide monitoring system NIOX is cleared by the US Food and Drug Administration for monitoring therapy in asthma. *J Allergy Clin Immunol* 2004;**114**:1241–56.
6. Montuschi P, Barnes PJ. Analysis of exhaled breath condensate for monitoring airway inflammation. *Trends Pharmacol Sci* 2002;**23**:232–37.
7. Montuschi P. Analysis of exhaled breath condensate in respiratory medicine: methodological aspects and potential clinical applications. *Ther Adv Respir Dis* 2007;**1**:5–23.
8. Effros RM, Hoagland KW, Bosbous M, Castillo D, Foss B, Dunning M, et al. Dilution of respiratory solutes in exhaled condensates. *Am J Respir Crit Care Med* 2002;**165**:663–69.
9. Montuschi P. Exhaled breath condensate analysis in patients with COPD. *Clin Chim Acta* 2005;**356**:22–34.
10. Montuschi P, Corradi M, Ciabattoni G, Nightingale J, Kharitonov SA, Barnes PJ. Increased 8-isoprostane, a marker of oxidative stress, in exhaled condensate of asthma patients. *Am J Respir Crit Care Med* 1999;**160**:216–20.
11. De Benedetto F, Aceto A, Dragani B, Spacone A, Formisano S, Cocco R, et al. Validation of a new technique to assess exhaled hydrogen peroxide: results from normals and COPD patients. *Monaldi Arch Chest Dis* 2000;**55**:185–88.
12. van Beurden WJ, Harff GA, Dekhuijzen PN, van den Bosch MJ, Creemers JP, Smeenk FW. An efficient and reproducible method for measuring hydrogen peroxide in exhaled breath condensate. *Respir Med* 2002;**96**:197–203.
13. Hunt J. Exhaled breath condensate: an evolving tool for non-invasive evaluation of lung disease. *J Allergy Clin Immunol* 2002;**110**:28–34.
14. Montuschi P, Varone F, Valente S, Spina D, Mondino C, Ciappi G, et al. Methodological aspects of exhaled prostanoid measurements. *Eur Respir J* 2003;**22**(Suppl. 45):18s.
15. Trischler J, Merkel N, Konitzer S, Muller CM, Unverzagt S, Lex C. Fractionated breath condensate sampling: H_2O_2 concentrations of the alveolar fraction may be related to asthma control in children. *Respir Res* 2012;**13**:14.

16. Gajdocsi R, Brindicci C, Barnes PJ, Kharitonov SA. The reproducibility and accuracy of online hydrogen peroxide bioassay and the decreasing effect of storage on H_2O_2 concentration. *Eur Respir J* 2003;**22**(Suppl. 45):279s.
17. Montuschi P. Analysis of exhaled breath condensate: methodological issues. In: Montuschi P, editor. *New perspectives in monitoring lung inflammation: analysis of exhaled breath condensate*. Boca Raton: CRC Press; 2005.
18. Montuschi P, Martello S, Felli M, Mondino C, Chiarotti C. Ion trap liquid chromatography/tandem mass spectrometry analysis of leukotriene B_4 in exhaled breath condensate. *Rapid Commun Mass Spectrom* 2004;**18**:2723–9.
19. Montuschi P, Martello S, Felli M, Mondino C, Barnes PJ, Chiarotti M. Liquid chromatography/mass spectrometry analysis of exhaled leukotriene B_4 in asthmatic children. *Respir Res* 2005;**6**:119.
20. Montuschi P. LC/MS/MS analysis of leukotriene B_4 and other eicosanoids in exhaled breath condensate for assessing lung inflammation. *J Chromatogr B: Analyt Technol Biomed Life Sci* 2009;**877**:1272–80.
21. Carpenter CT, Price PV, Christman BW. Exhaled breath condensate isoprostanes are elevated in patients with acute lung injury or ARDS. *Chest* 1998;**114**:1653–9.
22. Syslova K, Kacer P, Kuzma M, Klusackova P, Fenclova Z, Lebedova J, et al. Determination of 8-iso-prostaglandin $F_{2\alpha}$ in exhaled breath condensate using combination of immunoseparation and LC-ESI-MS/MS. *J Chromatogr B: Analyt Technol Biomed Life Sci* 2008;**867**:8–14.
23. Corradi M, Rubinstein I, Andreoli R, Manini P, Caglieri A, Poli D, et al. Aldehydes in exhaled breath condensate of patients with chronic obstructive pulmonary disease. *Am J Respir Crit Care Med* 2003;**167**:1380–6.
24. Goen T, Muller-Lux A, Dewes P, Musiol A, Kraus T. Sensitive and accurate analyses of free 3-nitrotyrosine in exhaled breath condensate by LC-MS/MS. *J Chromatogr B: Analyt Technol Biomed Life Sci* 2005;**826**:261–6.
25. Larstad M, Soderling AS, Caidahl K, Olin AC. Selective quantification of free 3-nitrotyrosine in exhaled breath condensate in asthma using gas chromatography/tandem mass spectrometry. *Nitric Oxide* 2005;**13**:134–44.
26. Cap P, Chladek J, Pehal F, Maly M, Petru V, Barnes PJ, et al. Gas chromatography/mass spectrometry analysis of exhaled leukotrienes in asthmatic patients. *Thorax* 2004;**59**:465–70.
27. Vass G, Huszar E, Barat E, Valyon M, Kiss D, Penzes I, et al. Comparison of nasal and oral inhalation during exhaled breath condensate collection. *Am J Respir Crit Care Med* 2003;**167**:850–5.
28. Corradi M, Folesani G, Andreoli R, Manini P, Bodini A, Piacentini G, et al. Aldehydes and glutathione in exhaled breath condensate of children with asthma exacerbation. *Am J Respir Crit Care Med* 2003;**167**:395–9.
29. Montuschi P, Ragazzoni E, Valente S, Corbo G, Mondino C, Ciappi G, et al. Validation of 8-isoprostane and prostaglandin E_2 measurements in exhaled breath condensate. *Inflamm Res* 2003;**52**:502–7.
30. Montuschi P, Ragazzoni E, Valente S, Corbo G, Mondino C, Ciappi G, et al. Validation of leukotriene B_4 measurements in exhaled breath condensate. *Inflamm Res* 2003;**52**:69–73.
31. Larstad M, Ljungkvist G, Olin AC, Toren K. Determination of malondialdehyde in breath condensate by high-performance liquid chromatography with fluorescence detection. *J Chromatogr B: Analyt Technol Biomed Life Sci* 2002;**766**:107–14.

32. Schleiss MB, Holz O, Behnke M, Richter K, Magnussen H, Jorres RA. The concentration of hydrogen peroxide in exhaled air depends on expiratory flow rate. *Eur Respir J* 2000;**16**:1115–8.
33. Montuschi P, Barnes PJ. Exhaled leukotrienes and prostaglandins in asthma. *J Allergy Clin Immunol* 2002;**109**:615–20.
34. McCafferty JB, Bradshaw TA, Tate S, Greening AP, Innes JA. Effects of breathing pattern and inspired air conditions on breath condensate volume, pH, nitrite, and protein concentrations. *Thorax* 2004;**59**:694–8.
35. Rickmann J, Dauletbaev N, Bratzke H, Diesel H, Hirche T, Bargon J. Breath condensate reflects different compartments of respiratory tract depending on ventilation pattern. *Am J Respir Crit Care Med* 2001;**163**:A407.
36. Larstad M, Toren K, Olin AC. Influence of sampling time on malondialdehyde levels and pH in exhaled breath condensate. *Eur Respir J* 2003;**22**(Suppl. 45):38s.
37. Svensson S, Olin AC, Toren K. Sampling time is important for the collection of hydrogen peroxide in exhaled breath condensate (EBC). *Eur Respir J* 2003;**22**(Suppl. 45):78s.
38. de Laurentiis G, Paris D, Melck D, Maniscalco M, Marsico S, Corso G, et al. Metabonomic analysis of exhaled breath condensate in adults by nuclear magnetic resonance spectroscopy. *Eur Respir J* 2008;**32**:1175–83.
39. Motta A, Paris D, Melck D, de Laurentiis G, Maniscalco M, Sofia M, et al. Nuclear magnetic resonance-based metabolomics of exhaled breath condensate: methodological aspects. *Eur Respir J* 2012;**39**:498–500.
40. Montuschi P, Paris D, Melck D, Lucidi V, Ciabattoni G, Raia V, et al. NMR spectroscopy metabolomic profiling of exhaled breath condensate in patients with stable and unstable cystic fibrosis. *Thorax* 2012;**67**:222–8.
41. Hoffmann HJ, Tabaksblat LM, Enghild JJ, Dahl R. Human skin keratins are the major proteins in exhaled breath condensate. *Eur Respir J* 2008;**31**:380–4.
42. Effros RM, Biller J, Foss B, Hoagland K, Dunning MB, Castillo D, et al. A simple method for estimating respiratory solute dilution in exhaled breath condensates. *Am J Respir Crit Care Med* 2003;**168**:1500–5.
43. Bland JM, Altman DG. Statistical methods for assessing agreement between two methods of clinical measurement. *Lancet* 1986;**1**:307–10.
44. Montuschi P, Mondino C, Pistelli R, Trove A, Ciabattoni G, Koch P. Profile of prostanoids in exhaled breath condensate in childhood asthma. *Eur Respir J* 2003;**22**(Suppl. 45):400s.
45. Baraldi E, Ghiro L, Piovan V, Carraro S, Ciabattoni G, Barnes PJ, et al. Increased exhaled 8-isoprostane in childhood asthma. *Chest* 2003;**124**:25–31.
46. Mondino C, Ciabattoni G, Koch P, Pistelli R, Trove A, Barnes PJ, et al. Effects of inhaled corticosteroids on exhaled leukotrienes and prostanoids in asthmatic children. *J Allergy Clin Immunol* 2004;**114**:761–7.
47. Montuschi P. Isoprostane, prostanoids, and leukotrienes in exhaled breath condensate. In: Montuschi P, editor. *New perspectives in monitoring lung inflammation: analysis of exhaled breath condensate*. Boca Raton: CRC Press; 2005.
48. Montuschi P, Kharitonov SA, Ciabattoni G, Barnes PJ. Exhaled leukotrienes and prostaglandins in COPD. *Thorax* 2003;**58**:585–8.
49. Kostikas K, Gaga M, Papatheodorou G, Karamanis T, Orphanidou D, Loukides S. Leukotriene B_4 in exhaled breath condensate and sputum supernatant in patients with COPD and asthma. *Chest* 2005;**127**:1553–9.

50. Biernacki WA, Kharitonov SA, Barnes PJ. Increased leukotriene B_4 and 8-isoprostane in exhaled breath condensate of patients with exacerbations of COPD. *Thorax* 2003;**58**:294–8.
51. Montuschi P, Macagno F, Parente P, Valente S, Lauriola L, Ciappi G. Effects of cyclo-oxygenase inhibition on exhaled eicosanoids in patients with COPD. *Thorax* 2005;**60**:827–33.
52. Morrow JD, Hill KE, Burk RF, Nammour TM, Badr KF, Roberts LJ. A series of prostaglandin F2 -like compounds are produced *in vivo* in humans by a non-cyclooxygenase, free radical-catalyzed mechanism. *Proc Natl Acad Sci USA* 1990;**87**:9383–7.
53. Montuschi P, Barnes PJ, Roberts LJ. Isoprostanes: markers and mediators of oxidative stress. *FESEB J* 2004:**18**:1791–800.
54. Montuschi P, Barnes P, Roberts LJ. Insights into oxidative stress: the isoprostanes. *Curr Med Chem* 2007;**14**:703–17.
55. Montuschi P, Collins JV, Ciabattoni G, Lazzeri N, Corradi M, Kharitonov SA, et al. Exhaled 8-isoprostane as an *in vivo* biomarker of lung oxidative stress in patients with COPD and healthy smokers. *Am J Respir Crit Care Med* 2000;**162**:1175–7.
56. Kostikas K, Papatheodorou G, Ganas K, Psathakis K, Panagou P, Loukides S. pH in expired breath condensate of patients with inflammatory airway diseases. *Am J Respir Crit Care Med* 2002**165**:1364–70.
57. Kostikas K, Papatheodorou G, Psathakis K, Panagou P, Loukides S. Oxidative stress in expired breath condensate of patients with COPD. *Chest* 2003:**124**:1373–80.
58. Makris D, Paraskakis E, Korakas P, Karagiannakis E, Sourvinos G, Siafakas NM, et al. Exhaled breath condensate 8-isoprostane, clinical parameters, radiological indices and airway inflammation in COPD. *Respiration* 2008;**75**: 138–44.
59. Ko FW, Lau CY, Leung TF, Wong GW, Lam CW, Hui DS, et al. Exhaled breath condensate levels of 8-isoprostane, growth related oncogene alpha and monocyte chemoattractant protein-1 in patients with chronic obstructive pulmonary disease. *Respir Med* 2006;**100**:630–8.
60. Mazur W, Stark H, Sovijarvi A, Myllarniemi M, Kinnula VL. Comparison of 8-isoprostane and interleukin-8 in induced sputum and exhaled breath condensate from asymptomatic and symptomatic smokers. *Respiration* 2009;**78**:209–16.
61. Brindicci C, Ito K, Torre O, Barnes PJ, Kharitonov SA. Effects of aminoguanidine, an inhibitor of inducible nitric oxide synthase, on nitric oxide production and its metabolites in healthy control subjects, healthy smokers, and COPD patients. *Chest* 2009;**135**:353–7.
62. Nowak D, Kalucka S, Bialasiewicz P, Krol M. Exhalation of H_2O_2 and thiobarbituric acid reactive substances (TBARs) by healthy subjects. *Free Radic Biol Med* 2001;**30**:178–86.
63. Antczak A, Nowak D, Shariati B, Krol M, Piasecka G, Kurmanowska Z. Increased hydrogen peroxide and thiobarbituric acid-reactive products in expired breath condensate of asthmatic patients. *Eur Respir J* 1997;**10**:1235–41.
64. Nowak D, Kasielski M, Antczak A, Pietras T, Bialasiewicz P. Increased content of thiobarbituric acid-reactive substances and hydrogen peroxide in the expired breath condensate of patients with stable chronic obstructive pulmonary disease: no significant effect of cigarette smoking. *Respir Med* 1999;**93**:389–96.

65. Dekhuijzen PN, Aben KK, Dekker I, Aarts LP, Wielders PL, van Herwaarden CL. Increased exhalation of hydrogen peroxide in patients with stable and unstable chronic obstructive pulmonary disease. *Am J Respir Crit Care Med* 1996;**154**:813–6.
66. Horvath I, Donnelly LE, Kiss A, Kharitonov SA, Lim S, Chung KF, et al. Combined use of exhaled hydrogen peroxide and nitric oxide in monitoring asthma. *Am J Respir Crit Care Med* 1998;**158**:1042–6.
67. Ho LP, Faccenda J, Innes JA, Greening AP. Expired hydrogen peroxide in breath condensate of cystic fibrosis patients. *Eur Respir J* 1999;**13**:103–6.
68. Jobsis Q, Raatgeep HC, Hermans PW, de Jongste JC. Hydrogen peroxide in exhaled air is increased in stable asthmatic children. *Eur Respir J* 1997;**10**:519–21.
69. van Beurden WJ, van den Bosch MJ, Janssen WC, Smeenk FW, Dekhuijzen PN, Harff GA. Fluorimetric analysis of hydrogen peroxide with automated measurement. *Clin Lab* 2003;**49**:637–43.
70. Zappacosta B, Persichilli S, Mormile F, Minucci A, Russo A, Giardina B, et al. A fast chemiluminescent method for H_2O_2 measurement in exhaled breath condensate. *Clin Chim Acta* 2001;**310**:187–91.
71. Hu Y, Zhang Z, Yang C. A sensitive chemiluminescence method for the determination of H_2O_2 in exhaled breath condensate. *Anal Sci* 2008;**24**:201–5.
72. Oudijk EJ, Gerritsen WB, Nijhuis EH, Kanters D, Maesen BL, Lammers JW, et al. Expression of priming-associated cellular markers on neutrophils during an exacerbation of COPD. *Respir Med* 2006;**100**:1791–9.
73. Mercken EM, Hageman GJ, Schols AM, Akkermans MA, Bast A, Wouters EF. Rehabilitation decreases exercise-induced oxidative stress in chronic obstructive pulmonary disease. *Am J Respir Crit Care Med* 2005;**172**:994–1001.
74. Ferreira IM, Hazari MS, Gutierrez C, Zamel N, Chapman KR. Exhaled nitric oxide and hydrogen peroxide in patients with chronic obstructive pulmonary disease: effects of inhaled beclomethasone. *Am J Respir Crit Care Med* 2001;**164**:1012–5.
75. van Beurden WJ, Harff GA, Dekhuijzen PN, van der Poel-Smet SM, Smeenk FW. Effects of inhaled corticosteroids with different lung deposition on exhaled hydrogen peroxide in stable COPD patients. *Respiration* 2003;**70**:242–8.
76. van Beurden WJ, Smeenk FW, Harff GA, Dekhuijzen PN. Markers of inflammation and oxidative stress during lower respiratory tract infections in COPD patients. *Monaldi Arch Chest Dis* 2003;**59**:273–80.
77. Gerritsen WB, Asin J, Zanen P, van den Bosch JM, Haas FJ. Markers of inflammation and oxidative stress in exacerbated chronic obstructive pulmonary disease patients. *Respir Med* 2005;**99**:84–90.
78. Kasielski M, Nowak D. Long-term administration of N-acetylcysteine decreases hydrogen peroxide exhalation in subjects with chronic obstructive pulmonary disease. *Respir Med* 2001;**95**:448–56.
79. De Benedetto F, Aceto A, Dragani B, Spacone A, Formisano S, Pela R, et al. Long-term oral N-acetylcysteine reduces exhaled hydrogen peroxide in stable COPD. *Pulm Pharmacol Ther* 2005;**18**:41–7.
80. Decramer M, Rutten-van Molken M, Dekhuijzen PN, Troosters T, van Herwaarden C, Pellegrino R, et al. Effects of N-acetylcysteine on outcomes in chronic obstructive pulmonary disease (Bronchitis Randomized on NAC Cost-Utility Study, BRONCUS): a randomised placebo-controlled trial. *Lancet* 2005;**365**:1552–60.

81. Rysz J, Stolarek RA, Luczynski R, Sarniak A, Wlodarczyk A, Kasielski M, et al. Increased hydrogen peroxide concentration in the exhaled breath condensate of stable COPD patients after nebulized N-acetylcysteine. *Pulm Pharmacol Ther* 2007;**20**:281–9.
82. Corradi M, Montuschi P, Donnelly LE, Pesci A, Kharitonov SA, Barnes PJ. Increased nitrosothiols in exhaled breath condensate in inflammatory airway diseases. *Am J Respir Crit Care Med* 2001;**163**:854–8.
83. Osoata GO, Hanazawa T, Brindicci C, Ito M, Barnes PJ, Kharitonov S. Peroxynitrite elevation in exhaled breath condensate of COPD and its inhibition by fudosteine. *Chest* 2009;**135**:1513–20.
84. Liu J, Sandrini A, Thurston MC, Yates DH, Thomas PS. Nitric oxide and exhaled breath nitrite/nitrates in chronic obstructive pulmonary disease patients. *Respiration* 2007;**74**:617–23.
85. Gessner C, Hammerschmidt S, Kuhn H, Hoheisel G, Gillissen A, Sack U. Breath condensate nitrite correlates with hyperinflation in chronic obstructive pulmonary disease. *Respir Med* 2007;**101**:2271–8.
86. Balint B, Donnelly LE, Hanazawa T, Kharitonov SA, Barnes PJ. Increased nitric oxide metabolites in exhaled breath condensate after exposure to tobacco smoke. *Thorax* 2001;**56**:456–61.
87. Vaughan J, Ngamtrakulpanit L, Pajewski TN, Turner R, Nguyen TA, Smith A, et al. Exhaled breath condensate pH is a robust and reproducible assay of airway acidity. *Eur Respir J* 2003;**22**:889–94.
88. Terada K, Muro S, Sato S, Ohara T, Haruna A, Marumo S, et al. Impact of gastro-oesophageal reflux disease symptoms on COPD exacerbation. *Thorax* 2008;**63**:951–5.
89. Halliwell B. Lipid peroxidation, antioxidants and cardiovascular disease: how should we move forward? *Cardiovasc Res* 2000;**47**:410–8.
90. Lazar Z, Huszar E, Kullmann T, Barta I, Antus B, Bikov A, et al. Adenosine triphosphate in exhaled breath condensate of healthy subjects and patients with chronic obstructive pulmonary disease. *Inflamm Res* 2008;**57**:367–73.
91. Scheideler L, Manke HG, Schwulera U, Inacker O, Hammerle H. Detection of non-volatile macromolecules in breath a possible diagnostic tool? *Am Rev Respir Dis* 1993;**148**:778–84.
92. Gessner C, Scheibe R, Wotzel M, Hammerschmidt S, Kuhn H, Engelmann L, et al. Exhaled breath condensate cytokine patterns in chronic obstructive pulmonary disease. *Respir Med* 2005;**99**:1229–40.
93. Dentener MA, Creutzberg EC, Pennings HJ, Rijkers GT, Mercken E, Wouters EF. Effect of infliximab on local and systemic inflammation in chronic obstructive pulmonary disease: a pilot study. *Respiration* 2008;**76**:275–82.
94. Bucchioni E, Kharitonov SA, Allegra L, Barnes PJ. High levels of interleukin-6 in the exhaled breath condensate of patients with COPD. *Respir Med* 2003;**97**:1299–302.

CHAPTER

20 Particles in Exhaled Air—A Novel Method of Sampling Non-Volatiles in Exhaled Air

Anna-Carin Olin

Occupational and Environmental Medicine, Sahlgrenska Academy, Box 414, SE-405 30 Gothenburg, Sweden

20.1 INTRODUCTION

It is well known that exhaled breath contains an aerosol of small particles. These particles are visualized when coughing and sneezing. Early work by Fairchild and Stampfer[1] and by Papineni and Rosenthal,[2] respectively, describe particle concentrations in exhaled breath during rest and exercise, and using different breathing modes. The main constituent of exhaled particles is water,[3] but probably all non-volatile substances that have been identified in exhaled breath condensate (EBC) are derived from these particles.

There are, however, more efficient methods of collecting particles in an aerosol than condensation, where many of the exhaled particles might pass the condenser without being collected. This was clearly indicated in early studies of EBC sampling, where two condensers were serially connected and similar levels of analytes were identified in both condensers.[4] This is one of the reasons we started to examine alternative methods of sampling non-volatiles in exhaled breath. This would not have been possible without the help and expertise of Evert Ljungström, Professor of Atmospheric Chemistry and expert in particle sampling in ambient air.

20.2 NUMBER OF PARTICLES IN EXHALED BREATH

The first step in our research was to measure the number of different-sized exhaled particles. In order to achieve this, an isolated thermostated paper box containing a reservoir for the exhaled air was constructed, and connected with this was an optical particle counter (Grimm Labortechnik GmbH, Ainring, Germany). This instrument counts all particles in the size range 0.3–4.0 μm, with a time resolution of one measurement per second. We also introduced inhalation of particle-free air for 3 min before the breath sampling in order to avoid contamination from ambient air.

From our first experiments it was clear that the number of particles exhaled during tidal breathing is very low—much lower than in the ambient air—and that there is substantial variation between subjects. With forced breathing the number of exhaled particles increased substantially (see Figure 20.1).

 http://dx.doi.org/10.1016/B978-0-44-462613-4.00020-9

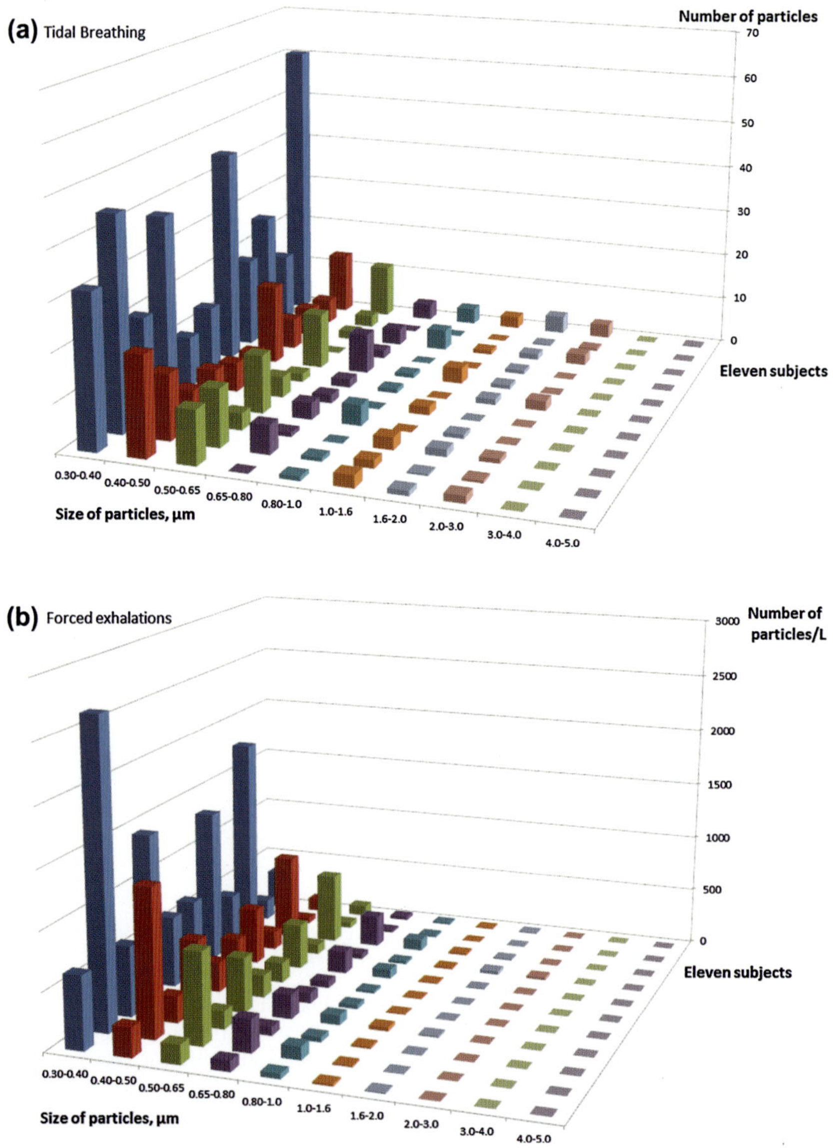

FIGURE 20.1
Number of particles in the exhaled air (PEx) of eleven healthy subjects during (a) tidal breathing, and (b) forced exhalations.

20.3 FORMATION AND ORIGIN OF EXHALED PARTICLES

The formation of endogenous particles is not yet fully understood. In a recent study we showed that by allowing for airway opening following airway closure while breathing, we increase the number of exhaled particles between 2- and 18-fold.[5] This was tested by asking 10 healthy people to use three different breathing manoeuvres where the depth of exhalation was varied from a functional residual capacity (FRC) manoeuvre, then a closing point (CP) manoeuvre to a residual volume (RV) manoeuvre. Inhalation and exhalation flows were thoroughly monitored online and all flows were kept low (200 mL/s) in order to avoid dynamic compression of the airways. The three manoeuvres were repeated ten times in all individuals. All particles of a size of 0.3–2.0 μm were counted. The main results are presented in Figure 20.2.

The number of exhaled particles increased in all subjects, varying from 2-fold to 18-fold, when breathing was performed allowing for airway closure. The results suggest that when we use this breathing manoeuvre the majority of exhaled particles originate from the very distal airways, at least in healthy people, and the amount of material sampled can be increased substantially.

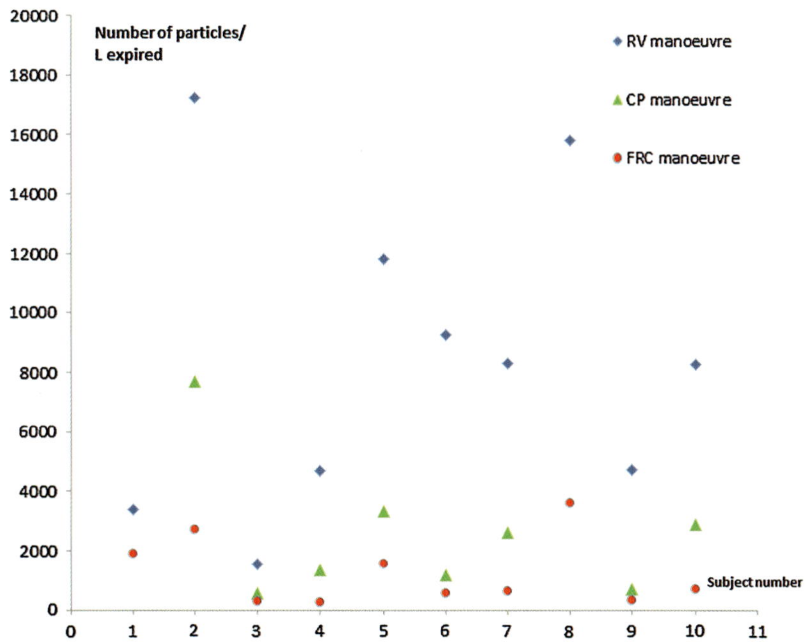

FIGURE 20.2
Number of exhaled particles (mean) after exhaling to functional residual capacity (FRC) (●), closing point (CP) (▲) and residual volume (RV) (♦).

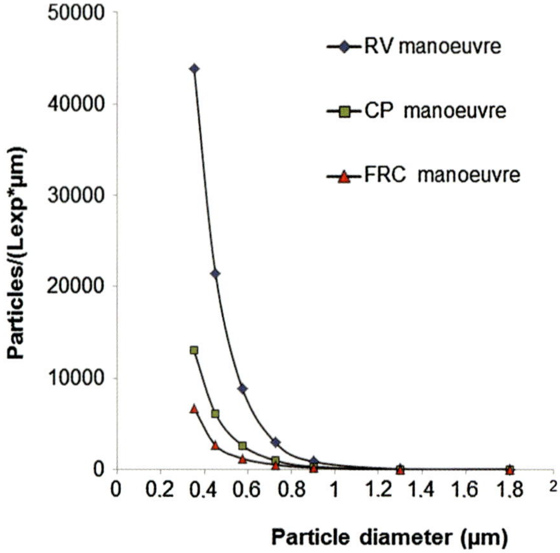

FIGURE 20.3

Particle size distributions based on average number concentrations of exhaled particles after exhaling to functional residual capacity (FRC) (▲), closing point (CP) (■), and residual volume (RV) (♦).

Examining the distribution of different-sized exhaled particles in relation to the different breathing manoeuvres showed that the profile was slightly different, with the relative increase in larger particles (>1.2 μm) being less pronounced than the increase among smaller particles (see Figure 20.3).

In a following study, which included measurement of the concentrations of even smaller particles, we showed that the number of particles with a diameter <0.05 μm was unaffected by changes in the breathing pattern, indicating that another, so far unknown, mechanism of formation is at work (see Figure 20.4).

In the same study we were able to obtain a profile of the sizes of exhaled particles down to a diameter of approximately 0.01 μm. It became clear that during tidal breathing the number of exhaled particles peaked at a diameter of around 0.07 μm, while when allowing for airway closure during breathing the peak concentration occurs for larger particles with a diameter of approximately 0.3–0.5 μm.

We have also studied the effect of breath holding, both at low and at high lung volumes, on the number of exhaled particles.[6] From these experiments it can be concluded that the number of exhaled particles increases after breath holding at low lung volumes, supporting our previous results. By contrast, the particle numbers decrease after breath holding at high lung volumes. This indicates that breath holding at low lung volumes increases airway closure and hence increases particle formation during subsequent inhalation. With breath holding at high lung volumes, when the

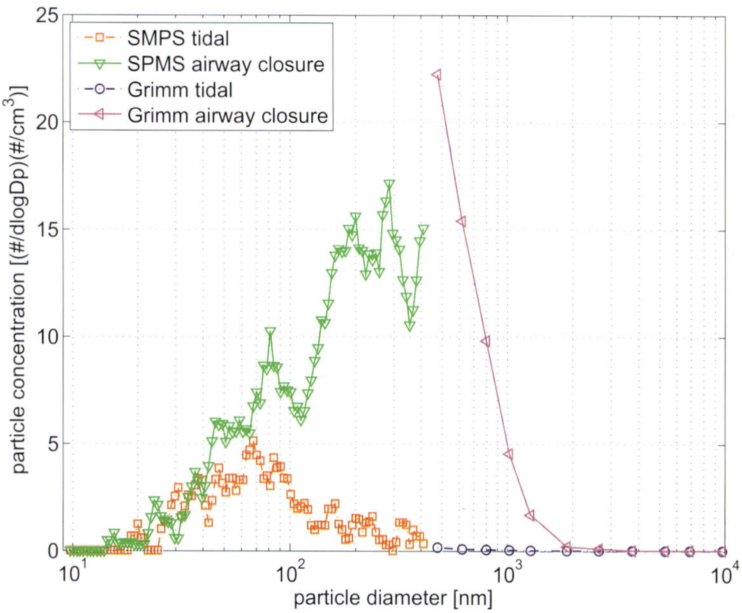

FIGURE 20.4
Particle size distributions during tidal breathing and after airway closure using two different particle counters; Scanning Mobility Particla Sizer (SMPS) system (□,▽) and Grimm dust monitor (○,◁).

particles are already formed, the longer the breath hold the longer time for the particles to sediment, and the fewer particles are found in the exhaled air.

20.4 COMPOSITION OF EXHALED PARTICLES

In an initial pilot study, the observed phospholipid composition of the collected particles was found to be similar to that of the respiratory tract lining fluid (RTLF).[7] We showed, using time-of-flight secondary ion mass spectrometry (TOF-SIMS), that particles in exhaled air (PEx) contain substantial amounts of phospholipids. TOF-SIMS is an very sensitive analytical technique that does not require any sample preparation. Only 10 exhalations were required to detect phospholipids. Phospholipids are the major component of the surfactant that covers the epithelium of the small airways.

20.4.1 Phospholipids in asthma

To examine whether the composition of phospholipids in the RTLF is altered in asthma, PEx from 15 non-smoking subjects with asthma (GINA guidelines) and 11 non-smoking healthy controls[8] were examined in an exploratory study.

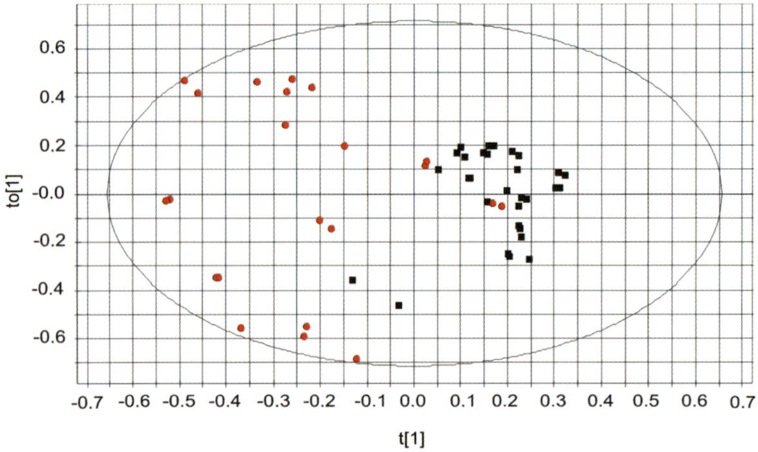

FIGURE 20.5
Orthogonal projections to latent structures discriminant analysis (OPLS-DA) score plot of positive ion spectra. ■ = subjects with asthma; ● = controls.

The participants performed ten consecutive forced exhalations with a target flow of 90% of their forced expiratory volume in 1 s (FEV_1) (±10%) into the sampling instrument. Subjects with asthma exhaled significantly lower numbers of particles than did controls ($p = 0.03$). The samples were analyzed using TOF-SIMS. Multivariate statistical analyses—principal component analysis (PCA) and orthogonal projections to latent structures discriminant analysis (OPLS-DA) of the acquired data—were used to screen TOF-SIMS spectra in order to discriminate between subjects with asthma and healthy controls. The resulting score plot showed good separation between the groups (see Figure 20.5).

Molecular ions from phospholipids (PC30:0+H, PC32:0+H, and PC32:0+Na) (positive spectra) and PG28:0, PG32:0, PG34:1, and PG36:1 (negative spectra) were stronger in all samples from subjects with asthma compared with control samples. The ratio of unsaturated to saturated phospholipids was significantly lower in samples from subjects with asthma (0.25 vs. 0.35, $p = 0.036$). These findings are from a small explorative study, but nevertheless they indicate that PEx can be used to examine alterations of the phospholipid composition of the surfactant in the small airways in respiratory disease, which finding is unique and needs further exploration.

20.4.2 Proteins

The total amount of protein in PEx has been determined to be in average 0.33 ng/L of exhaled air, which corresponds to 23% of the total mass of exhaled particles.[9]

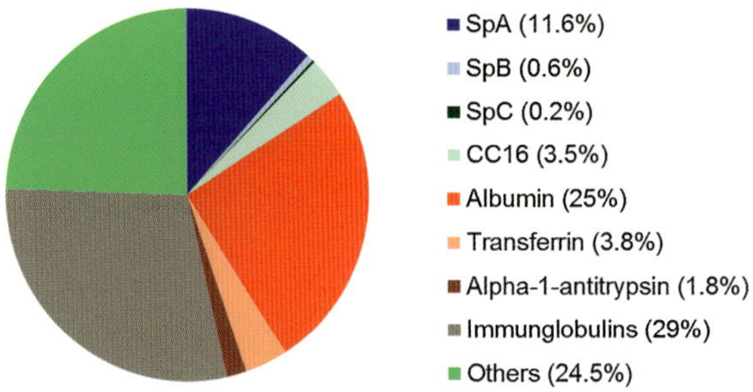

FIGURE 20.6
Relative distribution of identified proteins in particles in exhaled air (PEx). CC16 = Clara cell protein; SP-A, SP-B, SP-C= surfactant proteins A, B, C.

A shot-gun proteomic study has been performed to characterize the protein content in PEx.[9] Two pooled samples were prepared from six respectively eight healthy subjects, with a total volume of 3000 L and 4400 L of exhaled air. In total, 124 proteins were identified, the majority of which were extracellular. The single most abundant protein was albumin (25%), followed by immunoglobulins (29%) and surfactant protein A (SP-A) (11.6%). No amylase was detected, indicating the absence of salivary contamination. Figure 20.6 shows the relative distribution of the identified proteins.

20.4.2.1 Surfactant protein A in particles in exhaled air and exhaled breath condensate

Surfactant protein A (SP-A) is the most common lung-specific protein in PEx. In experimental studies and in studies of broncho-alveolar lavage (BAL), SP-A has been shown to modulate the immune response of inhaled material, having important functions in host defence, but also being of importance in keeping the surface tension high. Surfactant protein A has been suggested to be a biomarker for early chronic obstructive pulmonary disease (COPD) and acute rejection after lung transplantation. We therefore wanted to adapt and validate an enzyme-linked immunosorbent assay (ELISA) method to analyze SP-A in PEx.[10] As albumin is abundant in PEx, we simultaneously analyzed this, also with conventional ELISA. In this initial study we used an SP-A antibody[11] SP-A kindly provided by Albert Bernard. We chose to compare SP-A in PEx and EBC, the latter sampled by traditional technique using an EcoScreen.

Nine healthy subjects exhaled 100 L of breath for PEx analyses, using a breathing manoeuvre allowing airway opening, and 100 L for EBC, using tidal breathing, at two sampling sessions within 1 week. Albumin and SP-A in PEx and EBC were analyzed using ELISA. SP-A well above the detection limit was detected in all PEx samples, but only in 5/18 EBC samples. The mean concentration (± standard deviation (SD)) of SP-A was 2.6 (0.7) in PEx and 0.2 (0.4) in EBC, reported as arbitrary units. The number

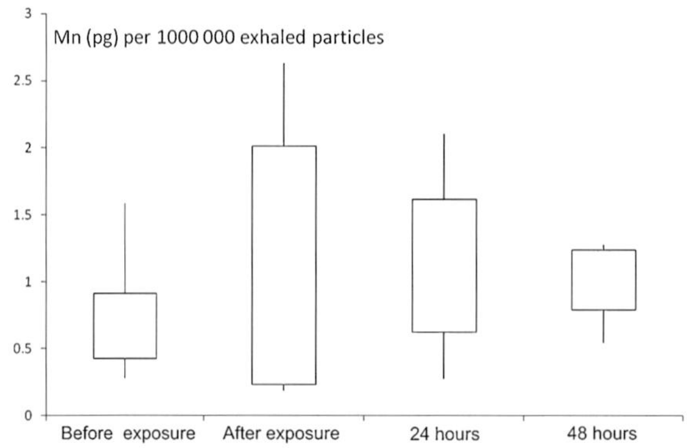

FIGURE 20.7

Manganese in particles in exhaled air (PEx) (pg/10^6 exhaled particles) before and directly after exposure and 24 and 48 h later. Box plots showing maximal, 25th and 75th percentile and minimum concentrations.

of PEx was significantly correlated to the amount of SP-A in PEx ($r_S = 0.86, p < 0.01$). The intra-individual coefficient of variation (CV) for two measurements of SP-A in PEx within 1 week was 13%.

20.5 SAMPLING OF EXHALED PARTICLES IN OCCUPATIONAL MEDICINE

In occupational medicine there is a possible utility of PEx assessment for exposure to heavy metals. In a small pilot study, five healthy subjects were exposed to welding fumes for 2 h in an exposure chamber. Particles in their exhaled air were sampled before, directly after, and 24 and 48 h after the exposure. Manganese in PEx was analyzed using inductively coupled plasma mass spectrometry (ICP-MS). Figure 20.7 shows box plots of manganese in PEx (pg/10^6 exhaled particles) giving maximal, 25th and 75th percentile and minimum levels at the different timepoints. This needs to be repeated in a larger study, once the analytical technique has been validated for the current matrix.

20.6 CONCLUSIONS

The novel PEx method opens up a new window into the distal airways and enables easy and repeated sampling of the respiratory tract lining fluid. The method provides a novel opportunity to identify and monitor pathological processes in the RTLF, most probably also reflecting pathological processes in airway epithelium. The method has high potential for use in both clinical settings and in research.

Acknowledgments

The work and findings described in this chapter are the result of joint work by the multi-disciplinary research group "Biomarkers and Airway Inflammation" at Occupational and Environmental Medicine at Gothenburg University, Sweden. Professor Evert Ljungström of the group has developed the sampling device.

References

1. Fairchild CI, Stampfer JF. Particle concentration in exhaled breath. *Am Ind Hyg Assoc J* 1987;**48**:948–9.
2. Papineni RS, Rosenthal FS. The size distribution of droplets in the exhaled breath of healthy human subjects. *J Aerosol Med* 1997;**10**:105–16.
3. Holmgren H, Bake B, Olin AC, Ljungstrom E. Relation between humidity and size of exhaled particles. *J Aerosol Med Pulm Drug Deliv* 2011;**24**:253–60.
4. Gessner C, Kuhn H, Seyfarth HJ, Pankau H, Winkler J, Schauer J, et al. Factors influencing breath condensate volume. *Pneumologie* 2001;**55**:414–9.
5. Almstrand AC, Bake B, Ljungstrom E, Larsson P, Bredberg A, Mirgorodskaya E, et al. Effect of airway opening on production of exhaled particles. *J Appl Physiol* 2010;**108**: 584–8.
6. Holmgren H, Gerth E, Ljungstrm E, Almstrand AC, Larsson P, Bake B, et al. Effects of breath holding at low and high lung volumes on amount of exhaled particles; Respir Physiol Neurbiol 2012 In Press.
7. Almstrand AC, Ljungstrom E, Lausmaa J, Bake B, Sjovall P, Olin AC. Airway monitoring by collection and mass spectrometric analysis of exhaled particles. *Anal Chem* 2009;**81**:662–8.
8. Almstrand AC, Josefson M, Bredberg A, Lausmaa J, Sjovall P, Larsson P, et al. TOF-SIMS analysis of exhaled particles from patients with asthma and healthy controls. *Eur Respir J* 2012;**39**:59–66.
9. Bredberg A, Gobom J, Almstrand AC, Larsson P, Blennow K, Olin AC, et al. Exhaled endogenous particles contain lung proteins. *Clin Chem* 2012;**58**: 431–40.
10. Larsson P, Mirgorodskaya E, Samuelsson L, Bake B, Almstrand AC, Bredberg A, et al. Surfactant protein A and albumin in particles in exhaled air. *Respir Med* 2012;**106**: 197–204.
11. Ellingsen DG, Ulvestad B, Andersson L, Barregard L. Pneumoproteins and inflammatory biomarkers in asphalt pavers. *Biomarkers* 2010;**15**:498–507.

PART

H

Volatiles of Microbial Origin: Urine, Stool and *in vitro* Cultures

CHAPTER

Challenges in the Investigation of Volatile Disease Biomarkers in Urine

21

Jae Kwak* **and George Preti*,†**

*Monell Chemical Senses Center, 3500 Market Street, Philadelphia, PA 19104, USA

†Department of Dermatology, School of Medicine, University of Pennsylvania, PA 19104, USA

21.1 INTRODUCTION

Distinctive odors or volatile profiles are often a consequence of disease. A portion of these volatile organic compounds (VOCs) are excreted into the exhaled breath and urine and through the skin, which can be recognizable by animals and analytical instruments. Because documented volatile disease biomarkers may be sampled in a non-invasive manner as they emanate from the body or body fluids, they have attracted considerable attention. As sophisticated analytical techniques coupled with comprehensive chemometric analysis become readily available, scientists are able to collect, separate, and identify numerous compounds from complex mixtures in body fluids, performing unbiased metabolic profiling of the body fluids. For example, numerous studies regarding volatile compounds in breath have been conducted and a variety of VOCs have been reported as potential disease biomarkers. While this "volatile metabolomic" approach can be useful, the chemical nature of VOCs has been neglected in many studies. Furthermore, their origin and biochemical formation are often ignored. As a result, only a few volatile disease biomarkers in breath have been reliably demonstrated (reviewed in Ref. 1).

Investigators have also looked at other body fluids such as urine, blood, and skin emanations for volatile disease biomarkers. In particular, the use of urine to identify biomarkers has several advantages. Urine samples can be easily collected and stored for long periods compared to other body fluids. In general, VOCs are present at much higher concentrations in urine than in other body fluids. Importantly, biochemical materials in urine contain rich information about individual physiological conditions. Some disease-associated VOCs released from urine have been routinely used in clinical diagnosis. For example, inborn errors of metabolism (e.g. trimethylaminuria [TMAU], phenylketonuria, maple syrup urine disease, isovaleric academia, etc.) are often accompanied by characteristic urinary odors and alteration of volatile profiles, which are due to defects in certain enzymes (see Refs. 2,3 for excellent reviews of this topic). However, as we have previously suggested for breath studies,[1]

Volatile Biomarkers. http://dx.doi.org/10.1016/B978-0-44-462613-4.00021-0

many reported disease-related volatile biomarkers in urine have not been consistently demonstrated and/or are not even mammalian metabolites. Here, we review a number of studies that have identified volatile disease biomarkers in urine, and discuss the challenges as well as potential approaches that may provide a better chance of identifying disease-associated biomarkers in urine.

21.2 DISTINCTIVE ODORS AND/OR VOLATILE PROFILES ASSOCIATED WITH DISEASES ARE PRESENT IN URINE AND DETECTED BY ANIMALS AND ANALYTICAL INSTRUMENTS

Animals can be trained to discriminate urinary odors of cancer patients from those of healthy control subjects. Dogs may detect the distinct odors in the urine samples collected from patients with bladder cancer as well as with prostate cancer,[4,5] although the dogs employed by Gordon et al.[6] did not recognize the urine odors of breast or prostate cancer patients. Mice could be trained to discriminate the urinary odors of tumor-bearing mice from those of genetically identical mice without tumors.[7]

A distinct VOC pattern of bladder cancer in urine has been detected by a gas sensor array coupled with a pattern recognition algorithm.[8] Arasaradnam et al.[9] also reported, in a preliminary communication, that an electronic nose could distinguish the volatile fingerprints in the urine samples of patients suffering from inflammatory bowel disease or diabetes from those of healthy volunteers.

Differential expressions of VOCs in urine may be indicative of the onset or progression of diseases. Robinson and Pauling[10] and Rosenberg et al.[11] observed that the volatile profiles of the urine samples collected from patients suffering from multiple sclerosis, Huntington's disease, breast cancer, Duchenne muscular dystrophy, or olivopontocerebellar degeneration differed from those of healthy control subjects. Smith et al.[12] reported that dozens of urinary VOCs were frequently detected in the urine samples collected from prostate cancer patients whereas their occurrence was rare in healthy control subject and vice versa. Matsumura et al.[7] injected two different mouse lung cancer cell lines and induced tumors in mice. Many of the urinary VOCs were down-regulated upon development of tumors. In addition, different volatile profiles were observed between groups bearing different cancer cell lines, indicating the specificity of different types of cancer cells. An elevated level of urinary formaldehyde was observed in the patients with bladder or prostate cancer.[13] A higher excretion of dimethylamine was observed in the urine samples collected from the patients with coronary artery disease or end-stage liver disease when compared to healthy controls.[14]

A number of urinary metabolites associated with diabetes have been reported, but the results are inconsistent. Earlier studies reported elevated levels of alcohols (e.g. ethanol, propanol, isobutanol, butanol, isopentanol, octanol isomers, etc.), ketones (e.g. 4-heptanone, 2-heptanone, cyclohexanone, etc.), and pyrazines in humans with diabetes.[15–20] However, Kronert et al.[21] observed that the urinary levels of isobutanol, butanol, and isopentanol were lower, but those of ethanol and propanol were higher in type 1 diabetic patients compared with healthy controls. Notably, Walker and Mills[22] and Wahl et al.[23] reported independently that some of the diabetes-associated VOCs were derived from plastic materials. For example, one of the octanol isomers (2-ethylhexanol), 4-heptanone, and 2-heptanone were metabolites of di-(2-ethylhexyl)-phthalate, a major plasticizer.[22,23] Cyclohexanone is a solvent used for plastic materials.[23,24] Wahl et al.[23] demonstrated that their increased levels in the urine samples of diabetic patients were due to the repeated exposure of plasticizers and plastic materials during hemodialysis. The authors also confirmed that non-diabetic hemodialysis patients had higher levels of 4-heptanone compared with healthy control subjects. Further to the point of mistaking exogenous VOCs for disease biomarkers, investigators employed elaborate procedures and sophisticated statistics only to report obvious exogenous compounds as biomarkers. For example, Banday et al.[25] reported that the levels of *o*-xylene and isopropyl acetate were higher, but those of 3-pentanol, dimethylstyrene, and cymol were lower in the urine samples of tuberculosis patients than in those of healthy control subjects. However, *o*-xylene, isopropyl acetate, dimethylstyrene, and cymol are likely derived from exogenous sources, not from the result of disease processes.

In most cases, differential expressions of disease-associated VOCs may not be a direct consequence, but rather a secondary effect of diseases. Although these reported biomarkers may have a diagnostic value, their origins and biochemical pathways as well as the mechanisms for their differential expressions upon manifestation of diseases are largely unknown and need to be further investigated. Also, there are likely additional disease biomarkers in urine most of which are of low volatility or totally non-volatile. Some of these can be detected through gas chromatography (GC) when derivatization techniques are employed to render them more volatile and compatible with GC analyses. Because these compounds have little to no odor or volatility, they will not be discussed here.

21.3 CHALLENGES IN MONITORING VOLATILE DISEASE BIOMARKERS IN URINE

Volatile profiles in urine are influenced by intrinsic and extrinsic factors such as genetic variation, diet, fluid intake, daily, and diurnal variations, to name a few. In addition, collection methods for VOCs and chromatographic conditions can alter urinary volatile profiles dramatically.

21.3.1 Intrinsic and extrinsic factors

Hundreds of VOCs composed of a variety of different structural classes are detected in urine.[26,27] Many of these compounds are endogenous metabolites. In particular, volatile sulfur compounds such as hydrogen sulfide and methanethiol are formed in mammalian tissues as well as by bacteria in the intestinal tract.[28] Methanethiol oxidizes spontaneously to dimethyl disulfide.[29] Many VOCs from exogenous sources are detected in urine; these may be derived from foods, flavoring agents, personal care products, and/or drugs (e.g. allyl isothiocyanate, allyl methylsulfide, safranal, carvone, limonene, menthol, phellandrene, pinene, pulegone, terpinene, terpineol, methylsalicylate, ibuprofen, etc.). The occurrence of these diet-derived VOCs varies between individuals.[18,27,30] Some obvious environmental contaminants (e.g. dichloromethane, chloroform, toluene, benzene, and naphthalene and their derivatives, styrene and its derivatives, dichlorobenzene, diisopropylether, methacrolein, methallyl cyanide, etc.) have also been reported in urine.[27] Some VOCs can be derived from both endogenous and exogenous sources. For example, while dimethylamine is produced endogenously, it is also a metabolite of ingested foods (e.g. fish and eggs) (Ref. 14, and references therein). Some VOCs in urine are produced through mammalian metabolism of exogenous compounds. For example, 4-heptanone is an endogenous metabolite of di-(2-ethylhexyl)-phthalate as mentioned above.[22,23]

The substantial contribution of genetic variation (e.g. variation in major histocompatibility complex (MHC) genes as well as variation in many other genes) to odors and volatile profiles in mouse urine has been demonstrated (reviewed in Ref. 31). However, it is not clear whether these genes also influence urinary odors and/or VOC profiles in humans. An earlier study conducted by Zlatkis et al.[15] showed that the volatile profiles in the urine samples collected from the same individuals over several months were very similar, whereas large between-individual variation was observed, suggesting that an individual's urinary VOC profile may be unique and persistent over time. The distinctive volatile profiles of different individuals may be a consequence of genetic variation, but further studies are required to examine the genetic contribution to human urinary volatile profiles.

Sex and age may influence urinary volatile profiles. Rhodes et al.[20] observed gender-dependent patterns of volatile profiles in human urine. 12 out of 90 isolated VOCs were quantitatively different between sexes. Robinson and Robinson[32] reported several age-associated VOCs (mainly ketones and furans) released from human urine.

Urinary VOCs can be subject to daily and diurnal variations. The urinary level of dimethylamine in humans was reported to be slightly different between days mainly due to dietary consumption.[14] The excretion of dimethylamine in humans and trimethylamine in mice was reported to be affected by diurnal rhythms.[14,33] Some VOCs may also be influenced by menstrual[34,35] or estrous cycles.[36] In females with certain genetic variants of TMAU, increased levels of urinary trimethylamine are found around the time of menses,[37,38] suggesting an effect of reproductive hormones on the oxidative enzyme flavin-containing monooxygenase 3 that converts odorous trimethylamine to its non-odorous oxide (trimethylamine oxide).[38]

Zlatkis and colleagues reported that dietary manipulation has a minor effect on the volatile profiles in urine.[15,30] Volunteers were asked to consume a chemically defined diet, "Vivonex," as a meal for 5 days, and then their urinary VOC profiles were examined before and after the ingestion of the diet. No major change was observed.

In contrast, we observed substantial diet effects on the urinary VOC profiles in mice and humans. We compared urinary volatile profiles of two groups of mice fed different diets. A laboratory diet was provided to one group of mice and a synthetic (chemically defined) diet to the other group. The levels of 20 out of 50 identified urinary metabolites that were significantly altered by diet variation.[39] Nevertheless, the odor profiles associated with MHC remain despite the substantial dietary variation as determined by sensor mice.[39] We also investigated the effect of diet on the volatile profiles in human urine, showing that dietary manipulation has a substantial impact on urine. We asked volunteers to consume capsules containing fish oil, garlic, turmeric, and mint for 10 days and monitored their urine for changes in VOCs. Numerous volatile compounds derived from the capsules produced abundant VOCs in urine, as measured by solid-phase microextraction followed by gas chromatography/mass spectrometry.[40] The discrepancy in the diet effect on urinary volatile profiles may be attributed to the differences in the diets used in different studies. The major ingredients of "Vivonex" are amino acids, carbohydrates, and lipids. Small amounts of VOCs might be detected in "Vivonex" since the diet is supplemented with some vitamins which may contain low levels of VOCs. In contrast, substantial amounts of numerous VOCs were detected in the capsules containing fish oil, garlic, turmeric, and mint used in our study.[40] Many of the VOCs derived from the capsules were detected in the urine samples.

VOC profiles in urine can be altered by urological diseases such as urinary tract infection,[41,42] which may be a confounding factor for the detection of other disease biomarkers. For example, when the data of patients suffering from urological diseases were included in those of the healthy subjects, the overall accuracy to detect bladder cancer samples by an E-nose dropped.[8]

While a number of urinary VOCs are shared between humans and mice, there are mouse-specific volatile and non-volatile compounds. Many of them have been identified as pheromones, mediating social communication (reviewed in Ref. 43). In particularly, mice excrete considerable amounts of "major urinary proteins (MUPs)" in urine that sequester a variety of VOCs with different chemical structures.[43,44] In addition, since MUPs are synthesized in the liver,[45] conditions which influence the health of the mouse as well as impact its liver may affect the type and amount of MUP production, leading to unexplored alterations in the uptake and release of murine urinary VOCs which are not applicable to humans. Because of the unique chemical properties of mouse urine as well as the fact that many of the disease-associated VOCs were found to be mouse-specific,[7] disease biomarkers detected in mouse urine may not be applicable to humans.

21.3.2 Analytical aspects

Urine is a complex mixture containing numerous volatile and non-volatile compounds within broad structural types and a range of polarity. Some compounds are present at high concentrations and others at much lower levels. Therefore, analytical techniques for separation and concentration of VOCs from urine should be optimized depending on the chemical nature of analytes of interest. For example, solvent extraction is recommended for extracting less volatile compounds, whereas headspace sampling techniques are frequently used for highly volatile compounds.[26,46] Among headspace sampling techniques, solid-phase microextraction and purge-and-trap sampling are the most frequently used methods. Volatile profiles in the same urine sample can be considerably different depending on which method is used. The technical details have been reviewed elsewhere.[46,47] Salts can be added to enhance the release of analytes from urine to the headspace.[26]

The pH of urine is an important factor determining the release of VOCs from urine. Both mouse and human urines are slightly acidic, ranging pH 5.8–6.7.[13,48,49] While both ionized and un-ionized forms of acidic and basic VOCs exist in intact urine, only the un-ionized forms are volatile and are found in the headspace of urine. Therefore, urinary pH may need to be adjusted to quantify analytes accurately. Ammonia and amines are best extracted under basic conditions ($pH > 10$) and organic acids under acidic conditions ($pH < 2$).[12,26,49,50] Examination of urinary esters should not be performed using basic conditions which may lead to hydrolysis.[27]

Analytical artifacts can be formed during analyses of urine. 3-Penten-2-one is produced as an artifact through aldol condensation of acetaldehyde with acetone at alkaline pH ($pH > 10$). Likewise, 3-hepten-2-one and 3-nonen-2-one can be generated through the condensation of butanal and hexanal with acetone, respectively.[51] In contrast, 3-penten-2-one was not detected in acidified urine samples in most cases.[51] Lactones may be either analytical artifacts or endogenously formed. When urine samples are analyzed using acidic conditions ($pH < 2$), hydroxy acids are converted to their corresponding lactones.[52] Lactone formation may also occur in the body as part of metabolism as well.[3] At least three cyclic vinyl ethers derived from 6-hydroxy-6-methyl-3-heptanone, a mouse-specific volatile pheromone, are formed during gas chromatographic (GC) analysis. These compounds are analytical artifacts formed via cyclization followed by dehydration in the heated GC injection port.[53,54]

Biochemical reactions in urine can occur even in frozen samples. We compared the volatile profiles between freshly collected mouse urine samples and the urine samples stored in a freezer (−20°C) for several months. When the frozen samples were defrosted and analyzed, we observed that a substantial portion of 3,4-dehydro-*exo*-brevicomin, another mouse-specific volatile pheromone,[43] appeared to be converted to its saturated compound, *exo*-brevicomin. Consequently, the unsaturated compound (3,4-dehydro-*exo*-brevicomin) was predominant in the fresh samples, whereas the saturated one (*exo*-brevicomin) was dominant in the stored samples (unpublished data). Therefore, timing and methods for storage should be carefully examined for the investigation of volatile disease biomarkers in urine.

21.4 DISCUSSION

Intrinsic and extrinsic factors, as well as analytical conditions, can influence volatile profiles in urine. In particular, exogenous compounds derived from foods and personal care products as well as environmental and analytical contaminants can become major volatile constituents in urine, thereby interfering with the detection of endogenously produced biomarkers.

How can the analytical investigation of volatile disease biomarkers productively proceed? Firstly, the impact of dietary and environmental contamination on urinary VOC profiles should be minimized and controlled. Urine donors should be encouraged to eat a bland diet and avoid foods containing aromatic spices which are likely produce more volatile metabolites (e.g. asparagus,[55] *Allium* vegetables like onions and garlic, *Brassica* vegetables including broccoli and cabbage, etc.) several days prior to and during the urine collection period as well as to record all foods, beverages and medication consumed during the study period. In addition, the medication record of patients should be reviewed to exclude volatile metabolites derived from the drugs used for treatment of diseases. Secondly, the investigation of volatile disease biomarkers should focus on the identification of biochemically relevant compounds rather than all VOCs detected from urine. Therefore, VOCs derived from exogenous sources should be excluded prior to data analysis and evaluation for biomarkers. This type of search for relevant volatile disease biomarkers is highlighted by the case of the metabolites of plastic-derived materials found in the urine of diabetic patients during early studies on biomarkers, as discussed above.[23] Finally, once candidate biomarkers are identified, their origins and biochemical pathways, as well as the mechanisms for their differential expressions, should be thoroughly investigated. As we begin to understand the biochemical mechanisms leading to alteration of endogenous VOCs in urine by disease, we should make progress in chemically defining robust, reliable and therefore practical disease-associated volatile biomarkers.

Acknowledgments

We are grateful to Dr. Gary Beauchamp for useful suggestions on an earlier version of this manuscript. This work was supported by a subcontract from UES under Air Force prime contract FA8650-10-C-6152 and by the US Army Research Office under grant number W911NF-11-1-0087. Opinions, interpretations, conclusions, and recommendations are those of the authors and not necessarily endorsed by the US Government.

References

1. Kwak J, Preti G. Volatile disease biomarkers in breath: a critique. *Curr Pharm Biotechnol* 2011;**12**:1067–74.
2. Liddell K. Smell as a diagnostic marker. *Postgrad Med J* 1976;**52**:136–8.
3. Burke DG, Halpern B, Malegan D, McCairns E, Danks D, Schlesinger P, et al. Profiles of urinary volatiles from metabolic disorders characterized by unusual odors. *Clin Chem* 1983;**29**:1834–8.
4. Willis CM, Church SM, Guest CM, Cook WA, McCarthy N, Bransbury AJ, et al. Olfactory detection of human bladder cancer by dogs: proof of principle study. *BMJ* 2004;**329**:712.

5. Cornu JN, Cancel-Tassin G, Ondet V, Girardet C, Cussenot O. Olfactory detection of prostate cancer by dogs sniffing urine: a step forward in early diagnosis. *Eur Urol* 2011;**59**:197–201.
6. Gordon RT, Schatz CB, Myers LJ, Kosty M, Gonczy C, Kroener J, et al. The use of canines in the detection of human cancers. *J Altern Complement Med* 2008;**14**:61–7.
7. Matsumura K, Opiekun M, Oka H, Vachani A, Albelda SM, Yamazaki K, et al. Urinary volatile compounds as biomarkers for lung cancer: a proof of principle study using odor signatures in mouse models of lung cancer. *PLoS One* 2010;**5**:e8819.
8. Weber CM, Cauchi M, Patel M, Bessant C, Turner C, Britton LE, et al. Evaluation of a gas sensor array and pattern recognition for the identification of bladder cancer from urine headspace. *Analyst* 2011;**136**:359–64.
9. Arasaradnam RP, Quraishi N, Kyrou I, Nwokolo CU, Joseph M, Kumar S, et al. Insights into "fermentonomics": evaluation of volatile organic compounds (VOCs) in human disease using an electronic "e-nose". *J Med Eng Technol* 2011;**35**:87–91.
10. Robinson AB, Pauling L. Techniques of orthomolecular diagnosis. *Clin Chem* 1974;**20**:961–5.
11. Rosenberg RN, Robinson AB, Partridge D. Urine vapor pattern for olivopontocerebellar degeneration. *Clin Biochem* 1975;**8**:365–8.
12. Smith S, White P, Redding J, Ratcliffe NM, Probert CSJ. Application of similarity coefficients to predict disease using volatile organic compounds. *IEEE Sens J* 2010;**10**:92–6.
13. Spanel P, Smith D, Holland TA, Al Singary W, Elder JB. Analysis of formaldehyde in the headspace of urine from bladder and prostate cancer patients using selected ion flow tube mass spectrometry. *Rapid Commun Mass Spec* 1999;**13**:1354–9.
14. Tsikas D, Thum T, Becker T, Pham VV, Chobanyan K, Mitschke A, et al. Accurate quantification of dimethylamine (DMA) in human urine by gas chromatography-mass spectrometry as pentafluorobenzamide derivative: evaluation of the relationship between DMA and its precursor asymmetric dimethylarginine (ADMA) in health and disease. *J Chromatogr B* 2007;**851**:229–39.
15. Zlatkis A, Bertsch W, Lichtenstein HA, Tishbee A, Shunbo F, Liebich HM, et al. Profile of volatile metabolites in urine by gas chromatography-mass spectrometry. *Anal Chem* 1973;**45**:763–7.
16. Liebich HM. Specific detection of volatile metabolites in urines of normal subjects and patients with diabetes mellitus using computerized mass fragmentography. *J Chromatogr* 1975;**112**:551–7.
17. Liebich HM. Gas chromatographic-mass spectrometric determination of total 4-heptanone, a new marker in diabetes mellitus. *J Chromatogr* 1983;**273**:67–75.
18. Liebich HM, Al-Babbili O. Gas chromatographic-mass spectrometric study of volatile organic metabolites in urines of patients with diabetes mellitus. *J Chromatogr* 1975;**112**:539–50.
19. Liebich HM, Al-Babbili O, Zlatkis A, Kim K. Gas-chromatographic and mass-spectrometric detection of low-molecular-weight aliphatic alcohols in urine of normal individuals and patients with diabetes mellitus. *Clin Chem* 1975;**21**:1294–6.
20. Rhodes G, Miller M, McConnell ML, Novotny M. Metabolic abnormalities associated with diabetes mellitus, as investigated by gas chromatography and pattern-recognition analysis of profiles of volatile metabolites. *Clin Chem* 1981;**27**:580–5.
21. Kronert K, Kunzel M, Reutter B, Zimmermann C, Liebich HM, Luft D, et al. Urinary excretion patterns of endogenously produced alcohols in type 1 (IDDM) and type 2

(NIDDM) diabetes mellitus compared with healthy control subjects. *Diabetes Res Clin Pract* 1990;**10**:161–5.

22. Walker V, Mills GA. Urine 4-heptanone: a beta-oxidation product of 2-ethylhexanoic acid from plasticisers. *Clin Chim Acta* 2001;**306**:51–61.
23. Wahl HG, Hong Q, Hildenbrand S, Risler T, Luft D, Liebich H. 4-Heptanone is a metabolite of the plasticizer di(2-ethylhexyl) phthalate (DEHP) in haemodialysis patients. *Nephrol Dial Transplant* 2004;**19**:2576–83.
24. Mills GA, Walker V. Urinary excretion of cyclohexanediol, a metabolite of the solvent cyclohexanone, by infants in a special care unit. *Clin Chem* 1990;**36**:870–4.
25. Banday KM, Pasikanti KK, Chan EC, Singla R, Rao KV, Chauhan VS, et al. Use of urine volatile organic compounds to discriminate tuberculosis patients from healthy subjects. *Anal Chem* 2011;**83**:5526–34.
26. Mills GA, Walker V. Headspace solid-phase microextraction profiling of volatile compounds in urine: application to metabolic investigations. *J Chromatogr B: Biomed Sci Appl* 2001;**753**:259–668.
27. Smith S, Burden H, Persad R, Whittington K, de Lacy Costello B, Ratcliffe NM, et al. A comparative study of the analysis of human urine headspace using gas chromatography-mass spectrometry. *J Breath Res* 2008;**2**:037022.
28. Tangerman A. Measurement and biological significance of the volatile sulfur compounds hydrogen sulfide, methanethiol and dimethyl sulfide in various biological matrices. *J Chromatogr B: Analyt Technol Biomed Life Sci* 2009;**877**:3366–77.
29. Hayward NJ. Effect of inoculum size on ampicillin and amoxycillin susceptibility determined by gas-liquid chromatography for members of the family Enterobacteriaceae. *J Clin Microbiol* 1986;**23**:755–9.
30. Zlatkis A, Liebich HM. Profile of volatile metabolites in human urine. *Clin Chem* 1971;**17**:592–4.
31. Kwak J, Willse A, Preti G, Yamazaki K, Beauchamp GK. In search of the chemical basis for MHC odourtypes. *Proc Biol Sci* 2010;**277**:2417–25.
32. Robinson AB, Robinson LR. Quantitative measurement of human physiological age by profiling of body fluids and pattern recognition. *Mech Ageing Dev* 1991;**59**:47–67.
33. Schaefer ML, Wongravee K, Holmboe ME, Heinrich NM, Dixon SJ, Zeskind JE, et al. Mouse urinary biomarkers provide signatures of maturation, diet, stress level, and diurnal rhythm. *Chem Senses* 2010;**35**:459–71.
34. Diskin AM, Spanel P, Smith D. Increase of acetone and ammonia in urine headspace and breath during ovulation quantified using selected ion flow tube mass spectrometry. *Physiol Meas* 2003;**24**:191–9.
35. Smith D, Ismail KM, Diskin AM, Chapman G, Magnay JL, Spanel P, et al. Increase of acetone emitted by urine in relation to ovulation. *Acta Obstet Gynecol Scand* 2006;**85**:1008–11.
36. Andreolini F, Jemiolo B, Novotny M. Dynamics of excretion of urinary chemosignals in the house mouse (Mus musculus) during the natural estrous cycle. *Experientia* 1987;**43**:998–1002.
37. Zhang AQ, Mitchell SC, Smith RL. Exacerbation of symptoms of fish-odour syndrome during menstruation. *Lancet* 1996;**348**:1740–1.
38. Shimizu M, Cashman JR, Yamazaki H. Transient trimethylaminuria related to menstruation. *BMC Med Genet* 2007;**8**:2.
39. Kwak J, Willse A, Matsumura K, Curran Opiekun M, Yi W, Preti G, et al. Genetically-based olfactory signatures persist despite dietary variation. *PLoS One* 2008;**3**:e3591.

40. Kwak J, Yi W, Willse A, Preti G, Mennella J, Steinmeyer A, et al. Effect of diet on volatile profiles of urines and sweat in humans. *Chem Senses* 2007;**32**:A122.
41. Aathithan S, Plant JC, Chaudry AN, French GL. Diagnosis of bacteriuria by detection of volatile organic compounds in urine using an automated headspace analyzer with multiple conducting polymer sensors. *J Clin Microbiol* 2001;**39**:2590–3.
42. Guernion N, Ratcliffe NM, Spencer-Phillips PT, Howe RA. Identifying bacteria in human urine: current practice and the potential for rapid, near-patient diagnosis by sensing volatile organic compounds. *Clin Chem Lab Med* 2001;**39**:893–906.
43. Novotny MV. Pheromones, binding proteins, and receptor responses in rodents. *Biochem Soc Trans* 2003;**31**:117–122.
44. Kwak J, Josue J, Faranda A, Opiekun MC, Preti G, Osada K, et al. Butylated hydroxytoluene is a ligand of urinary proteins derived from female mice. *Chemical Senses* 2011;**36**:443–52.
45. Finlayson JS, Asofsky R, Potter M, Runner CC. Major urinary protein complex of normal mice: origin. *Science* 1965;**149**:981–2.
46. Zlatkis A, Brazell RS, Poole CF. The role of organic volatile profiles in clinical diagnosis. *Clin Chem* 1981;**27**:789–97.
47. Demeestere K, Dewulf J, De Witte B, Van Langenhove H. Sample preparation for the analysis of volatile organic compounds in air and water matrices. *J Chromatogr A* 2007;**1153**:130–44.
48. Singer AG, Beauchamp GK, Yamazaki K. Volatile signals of the major histocompatibility complex in male mouse urine. *Proc Natl Acad Sci* USA 1997;**94**:2210–4.
49. Smith D, Spanel P, Holland TA, Al Singari W, Elder JB. Selected ion flow tube mass spectrometry of urine headspace. *Rapid Commun Mass Spectrom* 1999;**13**:724–9.
50. Mills GA, Walker V, Mughal H. Quantitative determination of trimethylamine in urine by solid-phase microextraction and gas chromatography-mass spectrometry. *J Chromatogr B: Biomed Sci Appl* 1999;**723**:281–5.
51. Walker V, Mills GA, Stansbridge EM. 3-Penten-2-one, a novel aldehyde adduct, is a biomarker for increased acetaldehyde in urine. *J Chromatogr B: Analyt Technol Biomed Life Sci* 2009;**877**:784–90.
52. Hoffmann KJ, Weidolf L. Identification of felodipine metabolites in rat urine. *Biomed Mass Spectrom* 1985;**12**:414–23.
53. Harvey S, Wiesler D, Novotny M. Formation of cyclic enol ethers from a labile biological precursor: an example of analytical artifacts. *J Chromatogr* 1989;**491**:27–36.
54. Novotny MV, Jemiolo B, Wiesler D, Ma W, Harvey S, Xu F, et al. A unique urinary constituent, 6-hydroxy-6-methyl-3-heptanone, is a pheromone that accelerates puberty in female mice. *Chem Biol* 1999;**6**:377–83.
55. Waring RH, Mitchell SC, Fenwick GR. The chemical nature of the urinary odour produced by man after asparagus ingestion. *Xenobiotica* 1987;**17**:1363–71.

CHAPTER

22 Volatile Organic Compounds (VOCs) Found in Urine and Stool

Ben de Lacy Costello and Norman M. Ratcliffe

Centre for Research in Biosciences, University of the West of England, Coldharbour Lane, Frenchay, Bristol BS16 1QY, UK

22.1 INTRODUCTION

VOC analysis of small molecules is a facile method for assessing stool and urine. Gut chemistry is exceedingly complex and the results of VOC analysis offer a simplified picture of the chemical soup that is in the gut. Much remains to be understood of the chemical compounds in the gut and their mode of synthesis and the difficulties are compounded by the diverse food intake of humans and the very dynamic nature of the biotransformations which are continuously occurring within the gut. This chapter gives an overview of the VOCs in urine and stool and how they change with disease. Much of the published work in this area has been undertaken using small studies and there is a risk that the correlations found between "biomarkers" and disease may not hold out in larger studies. We could question why some researchers are assessing VOCs for e.g. cancer diagnosis in urine, when as yet there are no clear biosynthetic routes established for specific cancer biomarker volatile compounds? The answer must be that this endeavor is worthwhile, because the present test(s) to identify some conditions, particularly non-benign prostate cancer and bowel cancer, are inadequate and:

i. Successful tests based on biomarker(s) would save a very large number of lives, reduce suffering, and reduce the burden on health services.
ii. Urine/stool volatile analysis is a non-invasive procedure and potentially low cost.

It is the first reason that is really important. It is worthwhile to invest in what might be a long shot if the potential returns are so significant. Furthermore analysis of VOCs in urine and stool could readily be used to assess changes in health status and also be used to assess drug metabolism and aid the development of personalized medicine.

22.2 URINE VOCs AND DISEASE

The first major study of VOCs in urine was undertaken by Mills and Walker[1] using solid-phase micro extraction (SPME), in conjunction with GC-MS to investigate the VOCs identified in the headspace above urine. They reported 103 volatile compounds

Volatile Biomarkers. http://dx.doi.org/10.1016/B978-0-44-462613-4.00022-2

associated with urine and analyzed urine from five patients with metabolic conditions. A more recent review of volatiles from urine expanded on the numbers of compounds and also included data of volatiles from elderly, apparently healthy individuals.[2] The results of a review of six studies of the volatiles found in urine from apparently healthy individuals are described by Smith *et al.* and updated herein (Table 22.1). Nine compounds were present in all studies: propanone, 2-butanone, 2-pentanone, 2-heptanone, 3-hexanone, 4-heptanone, 2,5-dimethylfuran, 2-ethyl-5-methylfuran, toluene, and so must be considered to be present with a very high degree of certainty. The source of some or all of these compounds may be the gut. A study[3] of 4-heptanone in urine strongly suggests that its presence originates at least in part from *in vivo* oxidation of the plasticizer component, 2-ethylhexanoic acid. Propanone, 2-butanone, 2-pentanone, and 2-heptanone were also found ubiquitously in the headspace of stool from a large number of samples donated from apparently healthy individuals, while a significant number of individuals were found to have 2,5-dimethylfuran, 2-ethyl-5-methylfuran, 3-hexanone and 4-heptanone.[4] The compounds so far reported in the headspace of urine and, for comparison, stool are presented (Table 22.1). The large number of ketones in urine probably, at least partially, arise from bacterial action in the gut, maybe by decarboxylation from the corresponding oxo-acids, since ketones were found at much lower concentrations in the urine of "germ free" rats.[1] Levels of the key ketone bodies, propanone (acetone), and acetoacetate have been found to vary between 1.16–14 mol/L and 1.3–15 mmol/L, respectively, in urine.[5] The ketone bodies (acetoacetate, hydroxybutyrate, propanone) are produced in the liver during periods of rapid fat oxidation, when the rate of fat breakdown exceeds the capacity of the Krebs cycle to process the resulting acetyl CoA. Propanone can be produced by the non-enzymic decarboxylation of acetoacetate and may sometimes be smelt on the urine and breath in acute diabetics. In summary (Table 22.1) the volatiles in urine cover a range of chemical classes: acids, alcohols, ketones, aldehydes, amines, *N*-heterocycles, *O*-heterocycles, sulfur compounds, and hydrocarbons. An insignificant number of esters have been reported. A large number of terpenes are described, and have been considered to be derived from foods.[1]

Little data exist on quantitative measurements of VOCs in urine. Concentrations of phenol (typically 10 mg/day excreted in urine) and *p*-cresol (typically 52 mg/day excreted in urine) have been reported to increase in urine with increasing protein intake. Their formation is considered to be due to gut microflora acting on tyrosine; anaerobic bacteria in the left colon producing phenol and aerobic bacteria in the ileum/cecum producing *p*-cresol. The relationship is complicated by fiber intake. High fibre intake with high protein resulted in a smaller increase in concentration due to decreased transit time.[6] This study was motivated by phenols being implicated in bladder and colon cancer. Normal alcohol ranges reported are 0–46 mg/24 h for ethanol, 0–300 μg/24 h for *n*-propanol and 0–18 μg/24 h for *n*-butanol, these levels approximately mirror blood serum levels.[7] Trimethylamine and 4-heptanone, were quantified as 0.5–20 μg/mL and 40–800 ng/mL, respectively, in urine.[8]

The highest concentrations in a range of urinary volatile short-chain fatty acids (and semi-volatile acids) were hippuric > glycolic > benzoic > ethanoic > 2-ketoglutaric >

Table 22.1 Compounds identified in the headspace of urine and stool samples reported in the literature.

Compound Name (Alphabetical)	CAS-Number	Urine						Stool		
		Ref. 2	Ref. 1	Ref. 95	Ref. 96	Ref. 97	Ref. 34	Ref. 4	Ref. 68	Ref. 69
Acetaldehyde*	75-07-0	p			p		p	p	p	p
Acetamide	60-35-5							p		
Acetic acid*	64-19-7		p		p		p	p	p	p
Acetic acid butyl ester	123-86-4							p		
Acetic acid ethenyl ester	108-05-4									p
Acetic acid ethyl ester*	141-78-6		p					p	p	
Acetic acid hexyl ester	142-92-7							p		
Acetic acid methyl ester	79-20-9							p		
Acetic acid pentyl ester	628-63-7							p		
Acetic acid propyl ester	109-60-4							p		
Acetic anhydride[a]	108-24-7								p	
Acetone*	67-64-1	p	p	p	p	p	p	p	p	p
Acetonitrile	75-05-8							p		
Acetophenone*	98-86-2	p						p	p	
2-Acetylfuran*	1192-62-7	p	p	p(isomer?)	p(isomer?)		p	p		
2-Acetyl-5-methylfuran	1193-79-9							p	p	
5-Acetyl-2-methylpyridine	36357-38-7							p		
p-Allyl anisole	140-67-0								p	
Allyl mercaptan	870-23-5	p								
Allyl methylsulfide	10152-76-8	p								
Ammonia[b]	7664-41-7									

(Continued)

Table 22.1 (Continued)

Compound Name (Alphabetical)	CAS-Number	Urine						Stool		
		Ref. 2	Ref. 1	Ref. 95	Ref. 96	Ref. 97	Ref. 34	Ref. 4	Ref. 68	Ref. 69
2-(Aziridin-1-yl) ethanamine	4025-37-0									p
Benzaldehyde*	100-52-7	p	p	p	p		p	p	p	p
Benzene*	71-43-2				p	p	p	p	p	
Benzene acetaldehyde	122-78-1							p	p	p
Benzene ethanethioic acid, S-methyl ester	5925-74-6							p		
Benzoic acid*	65-85-0		p					p		
Benzonitrile	100-47-0							p	p	
Benzophenone	119-61-9									p
Alpha-bergamotene	17699-05-7							p		
Benzyl alcohol*	100-51-6		p					p		
Betazole (1H- Pyrazole-3-ethanamine)	105-20-4								p	
1,3-Bis(1,1-dimethylethyl)-benzene	1014-60-4									p
2,6-Bis(1,1-dimethylethyl)-4-(1-oxopropyl) phenol	14035-34-8							p		
Butanal	123-72-8							p	p	
3-Butenal	7319-38-2				p?		p			
2,3-Butanediol	513-85-9				p		p			
2,3-Butanedione*	431-03-8	p		p	p		p	p	p	p

(Continued)

Table 22.1 (Continued)

Compound Name (Alphabetical)	CAS-Number	Urine						Stool		
		Ref. 2	Ref. 1	Ref. 95	Ref. 96	Ref. 97	Ref. 34	Ref. 4	Ref. 68	Ref. 69
Butane thioic acid, S-Methyl ester	2432-51-1							p		
Butanoic acid*	107-92-6		p					p	p	p
Butanoic acid butyl ester*	109-21-7		p					p	p	
Butanoic acid ethyl ester	105-54-4							p	p	
Butanoic acid methyl ester	623-42-7							p		
Butanoic acid propyl ester	105-66-8							p	p	
1-Butanol*	71-36-3	p	p	p	p		p	p	p	
2-Butanol	78-92-2							p		
2-butanone*	78-93-3	p	p	p	p	p	p	p	p	p
2-Butenal	123-73-9							p		
But-1-ene	106-98-9									p
2-Butoxyethan-1-ol	111-76-2		p							
2-n-Butylfuran	4466-24-4							p		
Butylisocyanate	111-36-4						p			
Butylisothiocyanate	592-82-5				p					
1-Butylpyrrole	589-33-3			p?	p		p			
α-calacorene	21391-99-1	p								
Camphene	79-92-5							p	p	
Car-2-en-4-one	28226-49-5		p							
Carbon disulfide*	75-15-0						p	p	p	p

(Continued)

Table 22.1 (Continued)

Compound Name (Alphabetical)	CAS-Number	Urine						Stool		
		Ref. 2	Ref. 1	Ref. 95	Ref. 96	Ref. 97	Ref. 34	Ref. 4	Ref. 68	Ref. 69
(+)-4-Carene	29050-33-7	p								
(−)-carveol	99-48-9		p							
Alpha-caryophyllene	6753-98-6							p		
Beta-caryophyllene	87-44-5							p		
Chlorocyclohexane	542-18-7					p	p			
Chloroform*	67-66-3		p	p	p	p		p	p	
Alpha-copaene	3856-25-5							p(isomer?)		
Cumene	98-82-8								p	
Cuminyl alcohol	536-60-7		p							
2-Cyano-2-butene	4403-61-6		p							
Cyclobutane	287-23-0									p
Cyclobutylamine	2516-34-9								p	
1,3,5-Cycloheptatriene	544-25-2									p
Cyclohexane	110-82-7								p	
Cyclohexanecarboxylic acid	98-89-5							p	p	
Cyclohexanecarboxylic acid ethyl ester	3289-28-9							p		
Cyclohexanecarboxylicacid methyl ester	4630-82-4							p		
Cyclohexanol	108-93-0					p				
Cyclohexanone*	108-94-1				p		p			

(Continued)

Table 22.1 (Continued)

Compound Name (Alphabetical)	CAS-Number	Urine						Stool		
		Ref. 2	Ref. 1	Ref. 95	Ref. 96	Ref. 97	Ref. 34	Ref. 4	Ref. 68	Ref. 69
Cyclohexene	110-83-8					p				
1,3,5,7-Cyclooctatetraene	629-20-9								p	
Cyclopentane	287-92-3							p		
Cyclopentanone	120-92-3				p		p			
Cyclopropane	75-19-4									p
Cyclopropylbenzene	873-49-4							p	p	
1-Cyclopropylethanone	765-43-5	p								
Cymenene	26444-18-8		p							
p-Cymene-8-ol	1197-01-9		p							
beta-damascenone	23726-93-4		p							
Decanal	112-31-2							p		
Decane	124-18-5							p		
1-Decanol	112-30-1							p		
2-Decanone	693-54-9							p		
Decatriene	n/a							p		
Diallyl disulfide	2179-57-9	p								
1,4-Dichlorobenzene	106-46-7					p	p			
Diethyl ether	60-29-7								p	
2,3-Dihydrofuran	1191-99-7							p		
Dihydro-2(3H)-furanone	96-48-0							p		
1,3-Dihydro-2H-indol-2-one	59-48-3									p

(Continued)

Table 22.1 (Continued)

Compound Name (Alphabetical)	CAS-Number	Urine						Stool		
		Ref. 2	Ref. 1	Ref. 95	Ref. 96	Ref. 97	Ref. 34	Ref. 4	Ref. 68	Ref. 69
Dihydro-2-(3H)-thiophenone*	1003-10-7				p		p	p		
Dihydro-5-methyl-2(3H)-furanone* (Valerolactone)	108-29-2				p		p	p		
3,6-Dihydro-4-methyl-2-(2-methyl-1-propenyl)-2H-pyran	1786-08-9	p								
2,6-Dimethyl-7-octen-2-ol* (dihydromycenol)	18479-58-8	p	p					p		
2,3-Dihydropyran	110-87-2						p			
1-(4,5-Dihydro-2-thiazolyl)-ethanone	29926-41-8							p		
1,2-Dihydro-1,1,6-trimethyl-naphthalene	30364-38-6	p								
2,2-Dihydroxy-1-phenylethanone	1075-06-5								p	
Diisopropylether	108-20-3						p			
1,3-Dimethoxybenzene	151-10-0							p		
Dimethylamine*	124-40-3	p	p						p	
2-(Dimethylamino)acetonitrile	926-64-7	p								
2,4-Dimethylbenzenamine	95-68-1							p		
4,7-Dimethylbenzofuran	28715-26-6	p								
3,3-Dimethyl-2-butanone	75-97-8	p								
2,3-Dimethyl-1-butene	563-78-0							p		

(Continued)

Table 22.1 (Continued)

Compound Name (Alphabetical)	CAS-Number	Urine						Stool		
		Ref. 2	Ref. 1	Ref. 95	Ref. 96	Ref. 97	Ref. 34	Ref. 4	Ref. 68	Ref. 69
2,3-Dimethyl-2-butene	563-79-1							p		
3,4-Dimethyldihydrofuran-2,5-dione	7475-92-5								p	
Dimethyldisulfide*	624-92-0	p	p	p	p	p	p	p	p	p
Dimethyl ether	115-10-6								p	
5,5-Dimethyl-1-ethyl-1,3-cyclopentadiene	496862-86-3	p								
2,3-Dimethylfuran	14920-89-9			p	p		p			
2,4-Dimethylfuran	3710-43-8				p	p	p			
2,5-Dimethylfuran*	625-86-5	p	p		p	p	p		p	
2,4-Dimethylheptane*	2213-23-2			p				p		
3,3-Dimethyl-2-hexanone	26118-38-7						p			
1-(1,5-Dimethyl-4-hexen-1-yl)-4-methylbenzene*	644-30-4	p						p		
6,6-Dimethyl-2-methylene-bicyclo[2.2.1]heptan-3-one	499126-76-0	p								
(Z,Z)-2,7-Dimethyl-3,5-octadiene	28980-73-6							p		
3,7-Dimethyl-1,6-octadiene (S)	2436-90-0							p		
3,7-Dimethyl-1,6-octadien-3-ol* (Linalool)	78-70-6		p					p		

(Continued)

Table 22.1 (Continued)

Compound Name (Alphabetical)	CAS-Number	Urine						Stool		
		Ref. 2	Ref. 1	Ref. 95	Ref. 96	Ref. 97	Ref. 34	Ref. 4	Ref. 68	Ref. 69
2,7-Dimethyl-oxepine	1487-99-6								p	
3,3-Dimethyloxetane	6921-35-3								p	
2,2-Dimethyl-3-pentanone	564-04-5	p								
2,4-Dimethyl-3-pentanone	565-80-0							p		
2,3-Dimethylpyrazine	5910-89-4			p	p		p			
2,5-Dimethylpyrazine	123-32-0			p(isomer?)			p			
2,6-Dimethylpyrazine	108-50-9		p		p(isomer?)		p			
2,6-Dimethylpyridine	108-48-5							p		
2,5-Dimethylpyrrole[ui]	625-84-3			p(isomer?)	p(isomer?)		p			
Dimethylsulfone	67-71-0				p		p			
Dimethylsulfoxide*	67-68-5						p	p		
Dimethyl sulfuric acid ester	77-78-1							p		
Dimethylsulfide	75-18-3							p	p	p
Dimethyltetrasulfide	5756-24-1						p			
2,4-Dimethylthiophene	638-00-6	p					p			
3,4-Dimethylthiophene	632-15-5	p								
Dimethyltrisulfide*	3658-80-8	p	p			p	p	p	p	p
4,7-Dimethylundecane	17301-32-5							p		
Di-n-butyl ether	142-96-1							p		
Dipropyldisulfide	629-19-6							p		
7,9-Di-tert-butyl-1-oxaspiro	82304-66-3									p

(Continued)

Table 22.1 (Continued)

Compound Name (Alphabetical)	CAS-Number	Urine						Stool		
		Ref. 2	Ref. 1	Ref. 95	Ref. 96	Ref. 97	Ref. 34	Ref. 4	Ref. 68	Ref. 69
(4.5)deca-6,9-diene-2,8-dione										
Dodecanal	112-54-9							p	p	
Dodecane	112-40-3							p		p
1-Dodecanol	112-53-8							p		
2-Dodecanone	6175-49-1							p		
Trans-edulan	41678-29-9		p							
Eicosane	112-95-8							p		
Ethanethioic acid, S-methyl ester	1534-08-3							p		
Ethanol*	64-17-5	p		p	p		p	p		p
4-Ethenyl-1,2-dimethylbenzene	27831-13-6							p		
1-Ethenyl-3-methoxybenzene	626-20-0								p	
(3R trans)-4-ethenyl-4-methyl-3-(1-methylethenyl)-1-(1-methylethyl)-cyclohexene	20307-84-0							p		
2-Ethenyltetrahydro-2,6,6-trimethyl-2H-pyran	7392-19-0	p								
2-Ethoxyethanol	110-80-5							p		
2-(2-ethoxyethoxy)-ethanol	111-90-0							p		p
4-Ethoxy-2-pentanone	33330-50-6				p		p			
Ethylamine	75-04-7	p								
Ethylbenzene*	100-41-4				p			p	p	

(Continued)

Table 22.1 (Continued)

Compound Name (Alphabetical)	CAS-Number	Urine						Stool		
		Ref. 2	Ref. 1	Ref. 95	Ref. 96	Ref. 97	Ref. 34	Ref. 4	Ref. 68	Ref. 69
2-Ethylbutanal	97-96-1						p			
3-Ethylcyclopentanone	10264-55-8	p								
4-Ethyl-1,2-dimethylbenzene	934-80-5							p		
Ethylene oxide	75-21-8									p
2-Ethylfuran	3208-16-0	p			p		p			
2-Ethylhexanoic acid*	149-57-5		p					p		
2-Ethyl-1-hexanol*	104-76-7		p					p	p	
1-Ethyl-1H-pyrrole	617-92-5	p								
alpha-ethylidenebenzeneacetaldehyde	4411-89-6							p		
Ethylidenecyclopropane	18631-83-9							p		
1-Ethyl-2-methylbenzene	611-14-3								p	
1-Ethyl-4-methylbenzene	622-96-8							p		
2-Ethyl-5-methylfuran*	1703-52-2	p	p	p?	p(isomer?)	p	p		p	
6-Ethyl-2-methyloctane	62016-19-7							p		
2,6,6-trimethyl-2,4-cycloheptadien-1-one (Eucarvone)	503-93-5	p	p							
Eudesma-4(14)-11-diene	17066-67-0							p		
Beta-farnesene	18794-84-8							p		
3-Furaldehyde*	498-60-2	p						p	p	
Furan	110-00-9	p	p	p?						

(Continued)

Table 22.1 (Continued)

Compound Name (Alphabetical)	CAS-Number	Urine						Stool		
		Ref. 2	Ref. 1	Ref. 95	Ref. 96	Ref. 97	Ref. 34	Ref. 4	Ref. 68	Ref. 69
2-Furanmethanol	98-00-0		p							
3-Furanmethanol	4412-91-3							p		
Furfural*	98-01-1	p		p	p		p	p		
Heptadecane	629-78-7									p
Spiro[2.4] hepta-4,6-diene	765-46-8									p
Heptanal*	111-71-7	p						p		p
2,3-Heptanedione	96-04-8							p		
Heptanoic acid	111-14-8							p	p	p
Heptanoic acid ethyl ester	106-30-9							p		
Heptanoic acid methyl ester	106-73-0							p		
1-Heptanol	111-70-6							p		
2-Heptanone*	110-43-0	p	p	p	p	p	p	p		p
3-Heptanone*	106-35-4	p			p	p	p	p		
4-Heptanone*	123-19-3	p	p	p	p	p	p	p		
2-Hexadecen-1-ol	22104-83-2									p
Hexadecanal	629-80-1									p
Hexadecane	544-76-3							p		p
1-Hexadecanol	36653-82-4							p		
Gamma-hexalactone*	695-06-7				p		p	p		
δ-hexalactone	823-22-3				p		p			
Hexanal*	66-25-1	p	p					p	p	p

(Continued)

Table 22.1 (Continued)

Compound Name (Alphabetical)	CAS-Number	Urine						Stool		
		Ref. 2	Ref. 1	Ref. 95	Ref. 96	Ref. 97	Ref. 34	Ref. 4	Ref. 68	Ref. 69
Hexane*	110-54-3						p	p	p	
2,3-Hexanedione	3848-24-6							p		
3,4-Hexanedione	4437-51-8							p		
Hexanenitrile	628-73-9							p		
Hexanoic acid	142-62-1							p	p	p
Hexanoic acid butyl ester	626-82-4							p		
Hexanoic acid ethenyl ester	3050-69-9							p		
Hexanoic acid ethyl ester	123-66-0							p		
Hexanoic acid methyl ester	106-70-7							p		
Hexanoic acid propyl ester	626-77-7							p		
1-Hexanol*	111-27-3		p					p	p	
2-Hexanone*	591-78-6	p			p		p	p	p	p
3-Hexanone*	589-38-8	p	p	p	p	p	p	p		
4-Hexen-3-one	2497-21-4							p		
1-(1H-pyrazol-4-yl)-ethanone	25016-16-4	p								
Hydrogen sulfide	7783-06-4							p		
2-Hydroxybenzaldehyde	90-02-8							p		
3-Hydroxy-2-butanone	513-86-0							p		p
1-Hydroxy-2-propanone	116-09-6							p	p	
Indane	496-11-7								p	
Indole*	120-72-9	p						p	p	p

(Continued)

Table 22.1 (Continued)

Compound Name (Alphabetical)	CAS-Number	Urine						Stool		
		Ref. 2	Ref. 1	Ref. 95	Ref. 96	Ref. 97	Ref. 34	Ref. 4	Ref. 68	Ref. 69
Ionone (alpha)	127-41-3	p								
Ionone (beta)	125-12-2		p							
Isocineole	470-67-7		p							
Isomyocorene	6876-07-9							p		
Isophorone	78-59-1		p							
4-Isothiocyanato-1-butene	3386-97-8	p								
3-Isothiocyanato-1-propene*	57-06-7	p	p	p	p	p		p	p	
Isothiocyanatocyclopropane	56601-42-4	p								
isothiocyanatomethane	556-61-6							p		
Limonene* (4-methyl-1-(1-methylethenyl)cyclohexene)	138-86-3	p			p		p	p	p	
cis-linaloloxide	n/a	p								
trans-linaloloxide	5989-33-3	p	p							
1,5,8-*p*-Menthatriene	21195-59-5								p	
p-Menth-1-en-8-ol	98-55-5							p		
Mercaptoacetone	24653-75-6							p		
Methallyl cyanide	4786-19-0	p								
Methanamide	75-12-7							p		
Methane[c]	74-82-8									
Methanethiol*	74-93-1	p	p				p	p	p	p
Methanoic acid*	64-18-6		p					p		

(Continued)

Table 22.1 (Continued)

Compound Name (Alphabetical)	CAS-Number	Urine						Stool		
		Ref. 2	Ref. 1	Ref. 95	Ref. 96	Ref. 97	Ref. 34	Ref. 4	Ref. 68	Ref. 69
Methanoic acid butyl ester	592-84-7								p	
Methanoic acid heptyl ester	112-23-2							p		
Methanoic acid pentyl ester	638-49-3							p		
Methanol[b]	67-56-1							p	p	
1-Methoxy-4-methylbenzene	104-93-8							p	p	
2-Methoxyphenol	90-05-1	p								
1-Methoxy-4-(1-propenyl)-benzene	104-46-1							p	p	
2-methoxythiophene	16839-97-7	p								
4-Methylbenzaldehyde	104-87-0							p		
2-Methylbenzofuran	4265-25-2								p	
3-Methylbutanal*	590-86-3	p			p		p	p	p	
2-Methylbutanal*	96-17-3	p			p		p	p		
Methylbutanal[ui]	n/a									p
3-Methylbutanethioc acid, S-methyl ester	23747-45-7							p		
2-Methylbutanoic acid*	116-53-0		p					p	p	
3-Methylbutanoic acid	503-74-2							p		
Methylbutanoic acid[ui]1	n/a									p
3-Methylbutanoic acid butyl ester	109-19-3							p		
2-Methylbutanoic acid ethyl ester	7452-79-1							p		

(Continued)

Table 22.1 (Continued)

Compound Name (Alphabetical)	CAS-Number	Urine						Stool		
		Ref. 2	Ref. 1	Ref. 95	Ref. 96	Ref. 97	Ref. 34	Ref. 4	Ref. 68	Ref. 69
2-Methylbutanoic acid methyl ester	868-57-5							p		
2-Methylbutanoic acid propyl ester	37064-20-3							p		
2-methyl-1-butanol*	137-32-6				p		p	p	p	
3-Methyl-1-butanol*	123-51-3				p	p	p	p		
3-Methyl-2-butanol	598-75-4						p			
3-Methyl-2-butanone	563-80-4	p		p	p		p			
2-methyl-2-butenal*	497-03-0	p						p		
3-Methyl-2-butenal	107-86-8							p		
3-Methyl-2-buten-1-ol	556-82-1							p		
3-Methyl-3-buten-1-ol*	763-32-6		p					p		
3-Methyl-3-buten-2-one	814-78-8	p								
3-Methyl-1-butyne	598-23-2	p								
1-Methyl-1,4-cyclohexadiene	4313-57-9	p								
Methylcyclohexane	108-87-2								p	
3-Methylcyclohexanone	591-24-2		p							
2-Methyl-2-cyclopentene-1-one	1120-73-6		p							
3-Methyl-2-cyclohexen-1-one	1193-18-6	p	p							
1-(2-Methyl-1-cyclopenten-1-yl)-ethanone	3168-90-9							p		

(Continued)

Table 22.1 (Continued)

Compound Name (Alphabetical)	CAS-Number	Urine						Stool		
		Ref. 2	Ref. 1	Ref. 95	Ref. 96	Ref. 97	Ref. 34	Ref. 4	Ref. 68	Ref. 69
Methylene chloride*	75-09-2	p						p	p	
1,1′-(methylene-bis(thio))bis ethane	4396-19-4						p			
3-Methylcyclopentanone	1757-42-2			p						
Methylene-cyclopropanecarbox-aldehyde	142423-24-3								p	
3-Methyleneheptane	1632-16-2							p		
4-Methylene-1-(1-methylethyl)-bicyclo[3.1.0]hexane	3387-41-5							p		
4-(1-methylethyl)-benzaldehyde	122-03-2	p								
4-(1-methylethyl)-1-cyclo-hexene-1-carboxaldehyde	21391-98-0	p								
2-Methyl-6-ethyl-pyrazine	13925-03-6			p						
2-Methylfuran*	534-22-5	p	p						p	
3-Methylfuran*	930-27-8	p						p	p	
5-Methyl-2-furancarbox-aldehyde*	620-02-0	p	p					p	p	
6-Methyl-3,5-heptadien-2-one	1604-28-0							p	p	
3-Methylheptane	589-81-1							p		
6-methyl-2,4-heptanedione	3002-23-1							p		
2-Methyl-4-heptanone	626-33-5	p								

(Continued)

Table 22.1 (Continued)

Compound Name (Alphabetical)	CAS-Number	Urine						Stool		
		Ref. 2	Ref. 1	Ref. 95	Ref. 96	Ref. 97	Ref. 34	Ref. 4	Ref. 68	Ref. 69
3-Methyl-2-heptanone	2371-19-9					p	p			
5-Methyl-2-heptanone	18217-12-4							p		
6-Methyl-2-heptanone	928-68-7									p
6-Methyl-3-heptanone	624-42-0	p			p?					
6-Methyl-5-hepten-2-one*	110-93-0	p					p	p	p	
3-Methylhexanal	19269-28-4							p		
5-Methyl-3-hexanol	623-55-2		p							
3-Methylhexan-2-one	2550-21-2					p	p			
4-Methyl-3-hexanone	17042-16-9	p	p							
5-Methyl-2-hexanone	110-12-3							p		
5-Methyl-3-hexanone*	623-56-3	p		p	p	p	p	p		
2-Methyl-2-hexenal	28467-88-1							p		
4-Methyl-5-hydroxy-hexanoic acid lactone	98354-02-0				p					
2-Methyl-3-hydroxy-2,4,4-trimethylpentylpropanoic acid ester	74367-34-3							p		
3-Methyl-1H-indole	83-34-1							p	p	
Methyl indole*[ui]	n/a				p					p
3-Methyl-1H-pyrazole	1453-58-3								p	
1-Methyl-pyrrole*	96-54-8	p		p				p		
2-Methyl-1H-pyrrole	636-41-9	p		p						

(Continued)

Table 22.1 (Continued)

Compound Name (Alphabetical)	CAS-Number	Urine						Stool		
		Ref. 2	Ref. 1	Ref. 95	Ref. 96	Ref. 97	Ref. 34	Ref. 4	Ref. 68	Ref. 69
3-Methyl-1H-pyrrole*	616-43-3		p						p	
5-Methylisothiazole	693-97-0		p							
2-Methylmercaptofuran	28588-74-1		p							
Ethyl-methylsulfide	624-89-5						p			
1-Methyl-4-(1-methylethenyl)-benzene*	1195-32-0	p						p	p	
(S)-2-Methyl-5-(1-methylethenyl)-2-cyclohexen-1-one ((+)-carvone)	2244-16-8	p			p		p			
1-Methyl-2-(1-methylethyl)-benzene* (o-cymene)	527-84-4	p				p	p	p	p	
1-Methyl-3-(1-methylethyl)-benzene*	535-77-3	p	p					p		
1-Methyl-4-(1-methylethyl)-benzene	99-87-6	p								
2-Methyl-5-(1-methylethyl)-bicyclo[3.1.0]hex-2-ene	2867-05-2							p		
1-Methyl-4-(1-methylethyl)-1,3-cyclohexadiene (alpha-terpinene)	99-86-5	p	p			p	p			
3-Methyl-6-(1-methylethyl)-2-cyclohexen-1-one	89-81-6	p	p	p	p					

(Continued)

Table 22.1 (Continued)

Compound Name (Alphabetical)	CAS-Number	Urine						Stool		
		Ref. 2	Ref. 1	Ref. 95	Ref. 96	Ref. 97	Ref. 34	Ref. 4	Ref. 68	Ref. 69
1-Methyl-4-(1-methylethyl)-1,4-cyclohexadiene* (gamma-terpinene)	99-85-4	p				p		p	p	
5-Methyl-2-(1-methylethyl)cyclohexanol* (menthol)	89-78-1	p	p					p	p	
1-Methyl-4-(1-methylethylidene)-cyclohexene*	586-62-9	p				p	p		p	
2-Methyl-5-(1-methylethyl)-phenol	499-75-2							p	p	
2-Methyl-2-[(1-methylethyl)thio]-propane	44657-76-3		p				p			
2-Methyl-5-(methylthio)furan	13678-59-6	p								
Methyl (methylthio) methyl disulfide	42474-44-2							p		
2-Methyl-1-nitropropane	625-74-1						p			
2-Methylnonane	871-83-0							p		
4-Methyloctane	2216-34-4							p		
2-Methylpentane	107-83-5							p		
3-Methylpentane	96-14-0								p	
4-Methylpentanoic acid	646-07-1							p	p	
4-Methyl-1-pentanol	626-89-1							p		

(Continued)

Table 22.1 (Continued)

Compound Name (Alphabetical)	CAS-Number	Urine						Stool		
		Ref. 2	Ref. 1	Ref. 95	Ref. 96	Ref. 97	Ref. 34	Ref. 4	Ref. 68	Ref. 69
4-Methyl-2-pentanol	108-11-2						p			
2-Methyl-3-pentanone*	565-69-5	p			p		p	p		
3-Methyl-2-pentanone*	565-61-7	p	p	p	p		p	p		
4-Methyl-2-pentanone*	108-10-1		p	p	p		p	p		p
2-Methyl-2-pentenal	623-36-9							p		
3-methyl-2-pentene	922-61-2								p	
4-Methyl-3-penten-1-ol*	763-89-3		p					p		
2-Methyl-1-penten-3-one	25044-01-3							p		
4-Methyl-3-penten-2-one*	141-79-7		p	p?	p			p		
4-Methylphenol*	106-44-5	p	p	p	p	p	p	p	p	
Methylphenol[ui]	n/a									p
1-(2-Methylphenyl)-ethanone	577-16-2								p	
2-Methyl-3-phenylpropanal	5445-77-2							p		
1-Methyl-2-piperidinone	931-20-4		p							
2-Methylpropanal*	78-84-2	p			p		p	p	p	p
2-Methyl-2-propanethiol	75-66-1						p			
2-Methylpropanoic acid	79-31-2								p	p
2-Methylpropanoic acid ethyl ester	97-62-1							p		
2-Methyl-propanoic acid, methyl ester	547-63-7							p		

(Continued)

Table 22.1 (Continued)

Compound Name (Alphabetical)	CAS-Number	Urine						Stool		
		Ref. 2	Ref. 1	Ref. 95	Ref. 96	Ref. 97	Ref. 34	Ref. 4	Ref. 68	Ref. 69
2-Methylpropanoic acid propyl ester	644-49-5							p		
2-Methyl-1-propanol*	78-83-1				p		p	p		p
2-Methyl-2-propenal* (methacrolein)	78-85-3	p						p		
2-Methyl-1-propene	115-11-7									p
2-Methyl-2-propenylbenzene*	3290-53-7				p				p	
Allyl methyl disulfide	2179-58-0		p			p	p			
p-Methylpropylbenzene (4-cymene)	1074-55-1				p		p			
Methylpropyldisulfide*	2179-60-4	p	p			p	p	p	p	
2-Methylpyrazine*	109-08-0	p(isomer?)		p(isomer?)	p(isomer?)		p	p(isomer?)		
3-Methylpyridine	108-99-6	p	p							
Methylsalicylate*	119-36-8	p						p		
2-Methyltetrahydrofuran-3-one	3188-00-9				p		p			
Methylthiirane (ethylene sulfide)	420-12-2	p	p							
4-Methylthiobutane nitrile	59121-24-3						p			
Methyl thiocyanate*	556-64-9						p		p	

(Continued)

Table 22.1 (Continued)

Compound Name (Alphabetical)	CAS-Number	Urine						Stool		
		Ref. 2	Ref. 1	Ref. 95	Ref. 96	Ref. 97	Ref. 34	Ref. 4	Ref. 68	Ref. 69
3-Methylthiophene	616-44-4							p	p	
3-(Methylthio)-propanal	3268-49-3							p		
1-(Methylthio)-propane	3877-15-4							p		
3-(Methylthio)-1-propanol	505-10-2							p		
E-1-(Methylthio)-1-propene	42848-06-6							p		
4-Methyl-undecane	2980-69-0							p		
2-Methyl-6-vinylpyrazine	13925-09-2				p?					
Beta-myrcene	123-35- 3							p	p	
Naphthalene	91-20-3							p		
2-*n*-Butylacrolein	1070-66-2	p								
N-butyl-benzenesulfonamide	3622-84-2							p		
N-methylformamide	123-39-7	p								
N,N-dimethylformamide	68-12-2	p								
N,N-dimethylurea	598-94-7	p								
Nitric oxide[b]	10102-43-9									
Nonadecane	629-92-5							p		
Nonanal*	124-19-6	p						p		
1-Nonanol	143-08-8							p		
Nonanoic acid	112-05-0		p							
2-Nonanol	628-99-9							p		
4-Nonanone	4485-09-0	p								

(Continued)

Table 22.1 (Continued)

Compound Name (Alphabetical)	CAS-Number	Urine						Stool		
		Ref. 2	Ref. 1	Ref. 95	Ref. 96	Ref. 97	Ref. 34	Ref. 4	Ref. 68	Ref. 69
2-Nonanone*	821-55-6	p		p	p		p	p		
Octadecanal	638-66-4									p
Octadecane	593-45-3									p
1-Octadecene	112-88-9									p
Octanal*	124-13-0	p			p			p		
Octane	111-65-9							p		
Octanoic acid*	124-07-2							p	p	
Octanoic acid propyl ester	624-13-5									
1-Octanol*	111-87-5		p		p		p	p	p	
3-Octanol	589-98-0							p		
2-Octanone*	111-13-7	p			p		p	p		
3-Octanone*	106-68-3			p	p		p	p		
4-Octanone	589-63-9				p		p			
(E)-2-Octenal	2548-87-0								p	
2-Octene	111-67-1							p		
3-Octene	592-98-3							p		
(E)-3-Octene	14919-01-8							p		
1-Octen-3-ol	3391-86-4							p		
3-Octen-2-one	1669-44-9							p		
Pentadecanal	2765-11-9								p	p
Pentadecane	629-62-9									p

(Continued)

Table 22.1 (Continued)

Compound Name (Alphabetical)	CAS-Number	Urine						Stool		
		Ref. 2	Ref. 1	Ref. 95	Ref. 96	Ref. 97	Ref. 34	Ref. 4	Ref. 68	Ref. 69
1-Pentadecene	13360-61-7									p
1,3-Pentadiene	504-60-9							p		
1,4-Pentadiene	591-93-5								p	
Pentanal*	110-62-3	p						p	p	
Pentane	109-66-0							p		
2,3-Pentanedione*	600-14-6				p		p	p		
3-Pentanethiol	616-31-9						p			
Pentanoic acid	109-52-4							p	p	p
Pentanoic acid butyl ester	591-68-4							p	p	
Pentanoic acid ethyl ester	539-82-2							p		
Pentanoic acid methyl ester	624-24-8							p	p	
Pentanol*	71-41-0				p		p	p	p	
2-Pentanol	6032-29-7							p	p	
3-Pentanol	584-02-1							p		
2-Pentanone*	107-87-9	p	p	p	p	p	p	p	p	p
3-Pentanone*	96-22-0	p					p	p	p	
3-Penten-2-one*	625-33-2	p		p	p			p		
2-Pentylfuran*	3777-69-3	p		p	p		p	p	p	
Alpha-phellandrene*	99-83-2	p					p	p	p	
Beta-phellandrene*	555-10-2	p				p(isomer?)		p	p	
Phenol*	108-95-2	p	p		p	p	p	p	p	

(Continued)

Table 22.1 (Continued)

Compound Name (Alphabetical)	CAS-Number	Urine						Stool		
		Ref. 2	Ref. 1	Ref. 95	Ref. 96	Ref. 97	Ref. 34	Ref. 4	Ref. 68	Ref. 69
2-Phenoxyethanol	122-99-6							p		
Phenylethyl alcohol	60-12-8							p		
Phenylglyoxal	1074-12-0								p	
2-Phenyl-4-(phenylmethylene)-1,3-propanethiol	n/a						p			
Phenylpropanoic acid methyl ester	103-25-3							p		
3-Phenylpropan-1-ol	122-97-4							p		
1-Phenyl-1-propanone	93-55-0							p		
1-Phenyl-2-propanone	103-79-7							p		
3-Phenyl-2-propenal	104-55-2								p	
Picoline	1333-41-1				p		p			
Pinane*	473-55-2		p							
Alpha-pinene	80-56-8							p	p	
Beta-pinene*	127-91-3		p	p?	p		p	p	p	
Piperidine	110-89-4		p							
2-Piperidinone	675-20-7							p		
Piperidinone[ui]	27154-43-4						p			
Propanamide	79-05-0							p		
Propanal*	123-38-6	p			p		p	p	p	

(Continued)

Table 22.1 (Continued)

Compound Name (Alphabetical)	CAS-Number	Urine						Stool		
		Ref. 2	Ref. 1	Ref. 95	Ref. 96	Ref. 97	Ref. 34	Ref. 4	Ref. 68	Ref. 69
Propanethioc acid, S-Methyl ester*	5925-75-7			p				p		
1-Propanethiol*	107-03-9	p						p		
Propanoic acid	79-09-4							p	p	p
Propanoic acid butyl ester	590-01-2							p		
Propanoic acid ethyl ester	105-37-3							p		
Propionic acid methyl ester	554-12-1							p		
Propanoic acid hexyl ester	2445-76-3							p		
Propanoic acid 2-methyl-1-(1,1-dimethylethyl)-2-methyl-1,3-propanediyl ester*	74381-40-1	p						p		
Propanoic acid 2-methyl-3-phenyl propyl ester	103-58-2								p	
Propanoic acid pentyl ester	624-54-4							p		
Propanoic acid propyl ester	106-36-5							p		
2-Propanol*	67-63-0						p	p	p	
1-Propanol*	71-23-8	p			p		p	p		
2-Propenal*	107-02-8	p						p		
Propene	115-07-1									p
2-Propen-1-ol	107-18-6									
2-Propenylidene-cyclobutene	52097-85-5									p

(Continued)

Table 22.1 (Continued)

Compound Name (Alphabetical)	CAS-Number	Urine						Stool		
		Ref. 2	Ref. 1	Ref. 95	Ref. 96	Ref. 97	Ref. 34	Ref. 4	Ref. 68	Ref. 69
Propylenesulfide	1072-43-1				p		p			
Pulegone	89-82-7	p	p							
Pyrazine	290-37-9	p			p		p			
Pyridine*	110-86-1	p	p				p	p		
3-Pyridinecarbonitrile	100-54-9							p		
Pyrrole*	109-97-7	p	p	p	p		p	p		
Pyrrole-2,5-dicarboxylic acid, 4-(2-diethylamino)ethyl, -3-methyl-, 2-ethyl ester	n/a									p
Safranal	116-26-7	p								
Santene	529-16-8		p							
Styrene*	100-42-5		p	p				p	p	
Alpha-terpineol* (4-terpinol)	10482-56-1		p		p		p		p	
Tetrachloroethane	79-34-5		p							
Tetrachloroethylene	127-18-4					p				
Tetradecanal	124-25-4									p
Tetradecane*	629-59-4	p	p					p		p
2-Tetradecen-1-ol	22104-82-1									p
4,5,6,7-Tetrahydro-3,6-dimethyl benzofuran	494-90-6	p								

(Continued)

Table 22.1 (Continued)

Compound Name (Alphabetical)	CAS-Number	Urine						Stool		
		Ref. 2	Ref. 1	Ref. 95	Ref. 96	Ref. 97	Ref. 34	Ref. 4	Ref. 68	Ref. 69
Tetrahydro-2,2-dimethyl-5-(1-methyl-1-propenyl)furan	7416-35-5	p								
Trans-tetrahydro-2,5-dimethyl-furan	2390-94-5							p		
Tetrahydrofuran	109-99-9								p	p
1,2,4,5-Tetramethylbenzene	95-93-2	p								
3,7,11,15-Tetramethyl-2-hexadecene	2437-93-6									p
2,3,6,7-Tetramethyloctane	52670-34-5							p		
2,6,10,10-Tetramethyl-1-oxaspiro [4.5]dec-9-ene	36431-72-8	p								
1-Tetrazol-2-ylethanone	51410-11-8								p	
Thioacetic acid	507-09-5						p			
Thiophene	110-02-1				p		p			
Toluene*	108-88-3	p	p	p	p	p	p	p	p	p
Tribromomethane	75-25-2								p	
Tridecanal	10486-19-8									p
Tridecane	629-50-5							p		p
1-Tridecanol*	112-70-9		p					p		p
2-Tridecen-1-ol	74962-98-4									p
Trimethylamine	75-50-3	p								

(Continued)

Table 22.1 (Continued)

Compound Name (Alphabetical)	CAS-Number	Urine						Stool		
		Ref. 2	Ref. 1	Ref. 95	Ref. 96	Ref. 97	Ref. 34	Ref. 4	Ref. 68	Ref. 69
1,2,3-Trimethylbenzene	526-73-8								p	
3,7,7-Trimethyl-bicyclo[4.1.0]hept-3-ene (3-carene)	13466-78-9							p	p	
2,2,3-Trimethylbutane	464-06-2							p		
4-(2,6,6-Trimethylcyclohexa-1,3-dienyl)but-3-en-2-one	1203-08-3	p								
2,2,6-Trimethylcyclohexanone	2408-37-9	p								
2,4,6-Trimethyldecane	62108-27-4							p		
2,6,11-Trimethyldodecane	31295-56-4							p		
2,3,5-Trimethylfuran	10504-04-8	p		p	p		p			
1,5,5-Trimethyl-6-methylene-cyclohexene	514-95-4							p		
2,3,5-Trimethylpyrazine	14667-55-1			p	p					
Undecanal	112-44-7									p
Undecane	1120-21-4							p		
2-Undecanol*	1653-30-1	p						p		
2-Undecanone	112-12-9							p		
Vinylfuran*[ui]	31093-57-9	p						p		
2-Vinylpyrazine	4177-16-6		p		p(isomer?)		p			

(Continued)

Table 22.1 (Continued)

Compound Name (Alphabetical)	CAS-Number	Urine						Stool		
		Ref. 2	Ref. 1	Ref. 95	Ref. 96	Ref. 97	Ref. 34	Ref. 4	Ref. 68	Ref. 69
Vitispirane	65416-59-3		p							
m-Xylene	108-38-3							p		
o-Xylene	95-47-6							p	p	
p-Xylene	106-42-3							p	p	
Xylene[ui]	1330-20-7		p				p			

p = present.

[a] *Very unlikely due to the reactivity of acetic anhydride with water.*

[b] *Found in urine using SIFT-MS.*[98]

[c] *Commonly presumed to be present in stool headspace.*

* *Compounds present in both urine and stool.*

[ui] *Unspecified isomer.*

n/a CAS number not available as literature did not specify isomeric form.

2-hydroxyisovaleric > lactic > 2-hydroxysobutyric > oxaloacetic > pyruvic acids, with propionic, isobutyric, butyric and 2-methylbutyric acids as minor components.[9] Interestingly there was poor correlation between urine and serum levels, with lactic > oleic > acetic > palmitic > 3-hydroxybutyric acids as the five most abundant acids in blood serum.

It has been suggested that methylamine and other short-chain aliphatic amines may play a significant role in the central nervous system disturbances observed during hepatic and renal disease.[10] To this end a quantitative method was developed for methylamine determination in the gas phase from urine. The average output was 11 mg/day with a range of 1.7–62 mg/day, with diet having a small effect, the source was considered to be mainly endogenous. Gut bacteria are likely to be implicated in the production of methylamine (probably from creatinine) as rats with no gut bacteria produced less than half the output.[10] The average daily output for dimethylamine was about 17 mg with values for the majority of the population lying within the 0.68–35.72 mg range.[11] Healthy young adults excrete about 1 mg of trimethylamine and 40 mg of trimethylamine *N*-oxide daily, although these levels are markedly influenced by diet, particularly when it contains marine fish. When marine fish is a dietary component, several 100 mg of trimethylamine *N*-oxide may be excreted.[12]

There are a very large number of compounds found in low concentrations in urine (Table 22.1). The concentrations of most of these compounds are unknown; often the source is also not known and the effects on health are not well understood.

22.2.1 Alcoholic hepatitis

A very small study involving a patient with alcohol-induced liver failure was undertaken. The patient donated a urine sample which was acidified and the headspace analyzed using SPME fibers and GCMS. Dimethydisulfide was the main component, which was considered to be due to impaired methionine catabolism and decreased clearance of sulfur compounds biosynthesized by intestinal micro organisms.[1] It should be noted that homocystinuria, an inherited disease adversely affecting the catabolism of methionine also produces large amounts of dimethyldisulfide in the headpace of urine.

22.2.2 Alcohols, short-chain fatty acids

Commercial services exist for the analysis by gas-liquid chromatography of combined blood alcohols (methanol, ethanol, propanol, butanol) and short-chain fatty acids. To the authors' knowledge, this service is not routinely undertaken for urine, although the analysis of ethanol in urine has been used extensively for law enforcement purposes. An increase in blood ethanol, following consumption of glucose and accompanied by no methanol increase and only slight increases in other alcohols, has been suggested to be due to yeast overgrowth, whereas an increase in concentration of a range of alcohols, but not ethanol, has been suggested to be due a bacterial dysbiosis in the small intestine, possibly due to malabsorption (www.biolab.co.uk/docs/gfp.pdf). Similarly, raised levels of short-chain fatty acids, with normal or nearly normal blood alcohols, suggest increased bacterial fermentation in the colon, probably due to small intestinal malabsorption (www.biolab.co.uk/docs/gfp.pdf).

The origin of urinary acids must in part be from the gut. It is well known that short-chain fatty acids are extensively synthesized by intestinal bacteria from carbohydrates and amino acids. Bovine urine acid concentrations showed a marked dependence on acid concentration changes in the gastrointestinal tract.[13] The hydrogen breath test to assess malabsorption and possible microbial overgrowth could clearly be used to supplement these tests and aid differentiation between bacterial and fungal overgrowth as fungi are not known to produce hydrogen.

Carboxylic acid profiling in urine and plasma has been successfully undertaken for the diagnosis and study of inherited metabolic disorders[9] using GCMS methods. For instance different acyl-CoA dehydrogenase deficiencies (which are inherited diseases) exist that result in volatile acids (branched or/and straight chain SCFAs, such as acetic, *n*-butanoic, isobutanoic, and *n*-octanoic acid), these being prominent in the headspace of acidified urine[1]; also see isovaleric acidemia below. Derivatization, to increase the volatility of semi- and non-volatile acids in urine has also been successfully undertaken and several methods exist.

22.2.3 Ammonia, amines, trimethylaminuria

Most ammonia in the body is produced when protein is broken down by bacteria in the intestines. The liver normally converts ammonia to urea, which is then eliminated in urine. Ammonia levels in the blood rise when the liver is not able to convert ammonia to urea. This may be caused by cirrhosis or severe hepatitis. The diagnostic symptoms of confusion, excessive sleepiness, coma, or hand tremor and the neuropsychological condition hepatic encephalopathy may be at least partially due to ammonia poisoning that can damage the liver and the brain. Gastrointestinal bleeding may cause an elevated blood ammonia level as red blood cells contain a high concentration of protein. One of the features of Reye's syndrome is also increased blood ammonia levels. Reye's syndrome leads to symptoms similar to those seen with hepatic encephalopathy, and seizures and coma may develop in severe cases. There are tests for measuring the amount of ammonia in the blood although it appears that there are no reports on urine headspace ammonia levels for conditions other than urinary tract infections.

It is well known that trimethylamine is normally formed in the human body, mainly from dietary choline of which eggs, liver, and soybeans are the main sources. A rapid and simple assay procedure employing headspace gas chromatography has been developed for the routine quantification of volatile methylamines and trimethylamine *N*-oxide present in human urine. The test was devised to aid rapid screening of patients and the diagnosis of fish odor syndrome (trimethylaminuria, an in-born error of oxidative metabolism where an odor similar to decaying fish is produced).[12] The very malodorous trimethylamine is made by bowel bacteria that break down carnitine (biosynthesized from lysine and methionine) and choline. Trimethylamine is normally absorbed from the gut and re-oxidized in the liver to the *N*-oxide, which, together with small amounts of non-oxidized trimethylamine, is excreted through the urine. In some individuals the *N*-oxidation process in the liver is impaired, resulting in reduced conversion of trimethylamine to the non odorous trimethylamine-*N*-oxide. Trimethylaminuria often has very negative psychological effect on sufferers: social exclusion that leads to depression, low self-esteem, and anxiety.

22.2.4 Cystinuria

Cystinuria is a rare genetic disease. It is characterized by the accumulation of cystine in the body. People with cystinuria cannot properly reabsorb cystine into their bloodstream during the filtering process in the kidneys. Most of the time, this excess cystine is simply excreted in the urine. The urine may have a characteristic "rotten egg" odor.

22.2.5 Diabetes

When the pancreas releases insufficient amounts of insulin or no insulin at all, the body cannot utilize glucose. The body then switches to fat to burn as fuel. Ketone bodies are a by-product of this process and when they are released they enter the bloodstream, acidify the blood, and are eventually excreted mostly in urine. The urine has a fruity smell mainly due to acetone. Rhodes et al.[14] analyzed urine from rats with diabetes and found that levels of a number of aliphatic ketones, predominantly 4-heptanone and 2-heptanone, were increased in the urine, corroborating previous studies in humans.[14,15] A ketogenic diet, which is high in fats and low in protein and carbohydrates, or fasting raises the level of ketone bodies. Home test kits are available to check ketone levels using dip stick technology and it is unlikely that VOC detector systems will have much of a role to play in this area. Abnormally increased alcohol levels, particularly ethanol, *n*-propanol, isobutanol, *n*-butanol, and isopentanol, have also been reported, using GCMS.[15]

22.2.6 Diagnosis of gastro-intestinal conditions via the analysis of VOCs from urine

Very little research has been undertaken to assess gastro-intestinal diseases by analyzing volatile compounds found in urine. Urine is a much more pleasant biological material to deal with and easier to collect than stool. However, more research needs to be undertaken to correlate stool chemistry with urine chemistry especially as the linking of volatiles in the gut to diseases is, itself, a new area of endeavor.

Recently, a series of urine and fecal samples have been obtained from patients with ulcerative colitis, Crohn's disease and diabetes, and analyzed using electronic nose technology which has shown promise in differentiating disease groups. This was a small study and further work is needed to establish these findings.[16]

22.2.7 Isovaleric acidemia (isovaleric acid CoA dehydrogenase deficiency)

Isovaleric acidemia is a genetic disease resulting in increased 3-hydroxyisovaleric acid in the urine; the acid has a cheesy, sweaty feet smell. The signs and symptoms of this disorder become apparent within a few days after birth and include poor feeding, vomiting, seizures, and lack of energy that can progress to coma. The signs and symptoms of the disorder may also appear during childhood and may come and go over time. They are often triggered by an infection or by eating an increased amount of protein-rich foods.

22.2.8 Maple syrup urine disease (MSUD)

The symptoms of maple syrup urine disease can be mental retardation. The strong smell of maple syrup in urine is associated with keto acids (keto acidosis) and can aid diagnosis. Sotolone (3-hydroxy-4,5-dimethyl-(5H)-furanone) has also been ascribed to producing the smell of the urine from patients suffering with MSUD.[17] MSUD is caused by a deficiency in enzyme activity that catalyzes de-carboxylation of 2-oxocarboxylic acids, as a step in the metabolism of the branched chain amino acids, valine, leucine, and isoleucine. As a result there is build up of the acids in tissues and urine: 2-oxoisocaproic acid from leucine, 2-oxo-3-methylvaleric acid from isoleucine, and 2-oxoisovaleric acid from valine (plus their reduced products). Most commonly, patients have less than 2% of the enzyme, and will die unless treated. The second most common form of MSUD is with enzyme activity between 8 and 15% of normal. Children with intermittent MSUD may not show any signs of the disorder until they are 12–24 months of age, usually in response to an illness or a rapid increase in protein intake. During episodes of illness or other metabolic stress, the child may develop seizures. Successful diagnosis, preferably soon after birth, followed by a restrictive diet, minimizing branched chain amino acids, minimizes symptoms. The costs of using volatile analysis and tandem mass spectrometry (MS/MS) screening for inborn errors of metabolism such as maple syrup urine disease has been reported to compare favourably with other mass screening programs.[18]

22.2.9 Methionine malabsorption syndrome (Oast-house syndrome also called Smith-Strang disease)

Methionine malabsorption is a rare genetic disease which results in reduced absorption of methionine in the small intestine. The amino acid is then available for breakdown in the large intestine by intestinal bacteria. The resulting alpha-hydroxybutyric acid causes a beer like, or yeasty odor to urine. Symptoms can be convulsions, white hair, and mental retardation among others.

22.2.10 Ovulation and volatiles

Selected ion flow tube mass spectrometry (SIFT-MS) has been used for a detailed study of the daily variations in the acetone and ammonia content of the headspace above urine from a healthy human female subject over the course of three separate menstrual cycles. Concurrent with the time of ovulation, a 3-to-12-fold increase in the level of acetone in the urine headspace was observed. The successive peaks in acetone level and the subsequent return to baseline values were mirrored by similar increases in the ammonia levels, but these were a day out of phase.[19] A larger study showed several-fold increase in urine headspace acetone levels which were detected 2–3 days after the predicted day of ovulation in 5 of 7 volunteers.[20] In the veterinary world, there would be time and cost savings in determining oestrus accurately to determine the precise insemination time; for instance, a study of 14 mares has recently reported that *p*- and *m*-cresols emitted from urine are reliable volatile chemical markers of ovulation in mares.[21]

22.2.11 Phenylketonuria

Phenylketonuria is a genetic disease resulting in reduced ability to convert phenylalanine (Phe) to tyrosine in the liver. A build up of Phe results, which is converted to phenyl acetate in the body, imparts a musty odor to urine. Other urinary metabolites include phenylacetate, phenylpyruvate, and phenethylamine. Interestingly, the major volatile component in the headspace of urine samples is considered to be benzaldehyde, a *circa* 10 fold increase compared to urine from healthy individuals. This is probably derived from benzoic acid. Optimal treatment involves maintaining blood Phe levels in a safe range while monitoring diet and cognitive development. If, however, the condition is left untreated, it can cause problems with brain development leading to progressive mental retardation, brain damage, and seizures.

22.2.12 Pre-renal azotemia and volatiles

Pre-renal azotemia is the most common form of kidney failure in hospitalized patients. Any condition that reduces blood flow to the kidney may cause it. The inability of the kidneys to filter blood results in accumulation of nitrogen-rich compounds in blood and the urine has a very strong concentrated ammonia-like smell.

22.2.13 Tuberculosis (TB)

An estimated 1.7 million people died from TB in 2009. Improving the early and effective diagnosis of TB is very important and the development of a rapid, low-cost, non-invasive method, with the potential to be administered in the field, remains a challenge. Although TB is a disease of the respiratory system, researchers have recently undertaken a study of volatile organic compounds in human urine derived from TB patients and healthy controls. VOCs were identified and quantified using GC-MS. A significant increase in the abundance of *o*-xylene and isopropyl acetate and decreased levels of 3-pentanol, dimethylstyrene, and cymol in TB patients were found compared to controls. This study suggests a possibility of using urinary VOCs for the diagnosis of human TB. Further work is required to confirm these results and to ascertain whether other conditions will lead to false positives.[22]

22.2.14 Putative volatile biomarker(s) for urinary tract cancers

The search for a relationship between VOCs and cancer is presently adventurous research. Work showing promise, even of a pilot study nature, gives confidence that this is a worthwhile area of research to further pursue. This is the case so far with urine analyzes. The impressive ability of dog olfactory senses has been tested for assessing urine for bladder cancer detection. The first study involving canines used six different dogs trained with urine from patients with bladder cancer and with those patients who were either healthy or had another disease.[23] The dogs were then tested for their ability to pick out one bladder cancer urine sample from six controls, 54 samples (27 from cancer patients) were used in training and 27 for testing. The dogs correctly selected urine from patients with bladder cancer on 22 out of 54 occasions. This gave a mean success rate of 41% compared with 14% expected by chance alone. This suggests that tumor-related volatile compounds are present in urine, imparting a characteristic odor signature.

Another proof-of-principle study was undertaken with a Belgian Malinois shepherd dog trained to recognize urine of people having prostate cancer.[24] All urine samples were frozen for preservation and heated to the same temperature for all tests. After a learning phase and a training period of 24 months, the dog's ability to discriminate prostate cancer and control urine was tested in a double-blind procedure. Urine was obtained from 66 patients referred to a urologist for elevated prostate-specific antigen or abnormal digital rectal examination. All patients underwent prostate biopsy and two groups were considered: 33 patients with cancer and 33 controls presenting negative assessments. During each test the dog was asked to signal a urine from a cancer patient among six samples containing only one urine from a cancer patient and five randomly selected controls. The dog completed all the tests and correctly designated the cancer samples in 30 of 33 cases. Of the three cases incorrectly classified as cancer by the canine, one patient was rebiopsied and prostate cancer was diagnosed. The sensitivity and specificity were both 91%. This study shows that dogs can be trained to detect prostate cancer by smelling urine with a significant success rate. It also suggests that prostate cancer gives an odor signature to urine. A study was undertaken to train one cohort of dogs on wet samples and another on dried samples. When tested, the dogs trained on liquid urine performed significantly better, suggesting that the more volatile molecules were of importance in the cancer odor signature.

There has been limited mass spectrometry studies published on volatile analyzes and cancer diagnoses from urine samples. An early study implicated formaldehyde as a putative biomarker,[25] but this work was targeted specifically at the one compound rather than the full range of VOCs present. A later study assessed the pattern of VOCs,[26] and showed differences between urine from patients known to have prostate cancer and controls; this was a small study so caution has to be expressed. It has also been reported that 1-H NMR spectroscopy and headspace SPME/two-dimensional Gas Chromatography Time-of-Flight Mass Spectrometry (SPME/GC $\times$ GC-TOF-MS) has been used to show that the metabolic composition of urine metabolites was consistently altered in lung cancer patients ($n = 71$)

relative to control subjects ($n = 54$), therefore suggesting a metabolic signature for lung cancer in urine. The compounds showing higher importance for this discrimination were mainly aldehydes (e.g. hexanal, acetaldehyde, decanal) and ketones (e.g. pentane-2,4-dione, hexan-3-one, 2,4-dimethylpentan-3-one). The authors suggest this may lead to minimally invasive detection of pulmonary carcinomas.[27]

Urinary volatile organic metabolites as potential cancer biomarkers were investigated by SPME/GC-MS.[28] A significant increase in the peak area of 2-methyl-3-phenyl-2-propenal, *p*-cymene, anisole, 4-methyl-phenol and 1,2-dihydro-1,1,6-trimethyl-naphthalene in the urine of cancer patients (oncological group: 14 leukaemia, 12 colorectal, and 7 lymphoma) versus 21 healthy individuals has been observed and the authors found, on average, a statistically significant lower abundance of dimethyldisulfide in cancer patients. Mice could be trained to discriminate between the odor of mouse urine with and without experimental tumors, demonstrating that volatile odorants are sufficient to identify tumor-bearing mice.[29] Consistent with this result, chemical analyzes of urinary volatiles demonstrated that the amounts of several compounds were dramatically different between tumor and control mice. Clearly there are other avenues of urine biomarker exploration being explored apart from volatile metabolite research. Recent examples include the search for kidney cancer biomarkers[30] and prostate cancer. [31,99] The last reference, using GCMS for urine analysis, showed sarcosine to be a putative biomarker although this now is in significant doubt.[32]

In summary there remains a need for better biomarker(s) to aid diagnosis of urinary tract cancers. The VOC research published so far is definitely promising and further work particularly with large numbers of patients is warranted.

22.3 BACTERIA PRESENT IN HUMAN URINE, CURRENT TEST METHODS, AND VOC ANALYSES

Often there is a smell from the urine passed by sufferers of urinary tract infections (UTIs), but this is insufficient for diagnosis. The volatile compounds from infected urine have previously been reported for the detection of bacteria using e.g. GC[33] and GCMS[34] and also investigated using electronic nose technology.[35,36] Much work remains to be undertaken to establish bacterial identification by detecting volatile metabolites. However, bacterial identification for bacterial disease diagnosis is an emerging area for the application of volatile sensor technologies.

UTIs include the inflammatory response and the associated signs and symptoms that result from the presence of the bacteria. In contrast, bacteriuria refers to the presence of bacteria in the urine, but this is not equivalent to UTI. Bacteriuria may be asymptomatic, particularly in elderly adults, and can be a matter of concern. UTI are a significant cause of morbidity with 3 million UTI cases each year in the USA alone.[37] Many nosocomial infections (31%) in medical intensive care units are attributable to UTI, and it is estimated that 20% of females aged 20–65 years suffer at least one episode per year. Composite data suggest that the overall expenditures for treatment of UTI among women alone in the United States were approximately \$2.47 billion in 2000, excluding spending on outpatient prescription drugs.[38] Furthermore UTI is the most frequent cause of infection among nursing home residents and the most commonly documented source of bacteremia in the elderly population.[39] It is also the most common bacterial infection in infants and young children worldwide.[40] Newborn infants are at greatest risk of urinary tract infections, and 10% of those affected will also develop bacteremia.[41] Prompt diagnosis and early treatment are highly desirable. An important aspect of testing is for asymptomatic bacteriuria, which occurs in 5–10% of pregnancies and is associated with risk of pre-term birth.[42,43] Fortunately, the prescription of the correct antibiotic(s) is normally effective in rapidly reducing bacteria levels in urine and reducing symptoms.[44] Bladder urine is normally sterile, thus the presence of pathogenic bacteria in urine is indicative of a UTI. When voided naturally, urine from men may contain small levels of bacteria from the urethra or prepuce and with women from the urethra, vulva, or vagina, but is typically less than or equal to 200 organisms/mL. Levels of 10^5 or greater are diagnostic of UTI in a symptomatic patient.[39] Levels as low as 10^3 organisms/mL may be clinically significant in symptomatic patients suspected of having UTI. Urine culture remains the reference standard for diagnosis of urinary tract infection. However, at least 18 h are needed for detection of bacterial growth on culture media by standard microbiological techniques. This can result in delayed or inappropriate treatment. Urine culture is also expensive and needs a microbiological laboratory with skilled technicians. Rapid urine tests do not identify the pathogenic organism or suitable antibiotic treatment for urinary tract infections therefore, they cannot be used as a replacement test for urine culture. However, because of the clinical importance of early diagnosis and delays with the gold-standard microbial culturing method, rapid urine tests, such as urine dipsticks and microscopy of urine for white cells and bacteria are used widely. The nitrite test strip method is very rapid. This is based on a color change when nitrite is present. Most bacteria *responsible* for UTI (e.g. *Enterobacteriaceae*) are able to reduce nitrate in the diet to nitrite, but this test will not detect the species of bacteria (including *Enteroccocus*, *Pseudomonas*, and *Staphyloccocus*) which do not reduce nitrate. Approximately 80% of uncomplicated UTI are caused by *Escherichia coli* and 20% by enteric pathogens such as *Enterococci*, *Klebsiellae*, *Proteus* spp., *coagulase* (–) *Staphylococci* and fungal opportunistic pathogens such as *Candida albicans*.[36] Recent guidelines for the management of urinary tract infections recommend microscopy for both white cells and unstained bacteria in children younger than 3 years of age and dipstick testing for children older than 3 years. However, a recent analysis of data from 95 studies in 95,703 children revealed that microscopy for white cells should not be used for the diagnosis of UTI, because microbiology laboratory facilities are needed so the results are considerably delayed compared with dipstick results and its accuracy is no better than that of dipstick.[45]

The sensitivity and specificity for nitrite-only positive dipstick is about 49% and 98%, respectively. Meta-analysis shows that dipstick testing for nitrites and leucocyte esterase when considered in an "either positive" combination, increases the sensitivity to 88%. Microscopic examination of the urine for detection of bacteria after Gram staining has been stated to be the most accurate test, compared with the reference standard urine culture, with summary estimates of sensitivity (91%), specificity (96%). The rapid dipstick tests are negative in around 10% of children with a UTI, which in most clinical settings is still unacceptably high.[45] If clinicians, patients, their families, and policy makers want to identify all children with UTIs, then it has been stated that a urine culture is always needed, irrespective of the dipstick or microscopy result.[45] Another reason for correct UTI diagnosis is that there is also the issue that UTI bacterial ureases, particularly from *Proteus mirabilis*, have been implicated as a contributing factor in the formation of urinary and kidney stones, obstruction of urinary catheters, and pyelonephritis.[46]

In summary, there remains a requirement for a low-cost, point-of-care, rapid test to preferably identify UTI and desirably the specific infecting organism(s) in urine with a very high specificity and sensitivity. The next section considers the potential for analyzes of volatiles from urine to fulfill this role.

22.3.1 Electronic nose technology for VOC detection of bacterial metabolites in urine

There have been several approaches by different research groups to tackle the problem of making a successful volatile analyzer for UTI. Bacterial urease enzymes break down urea to ammonia and this can be used as a biomarker for the presence of bacteria in urine. The potential utility of this has been established by GCMS studies which have shown that ammonia and acetic acid can occur in the urine headspace from UTI infected patients.[36,47] There have been several efforts to use VOC detection technologies to detect UTI. The first commercial system, the Osmetech Microbial Analyzer (OMA) was built predominantly for the analysis of bacteria in urine based on volatile analyzes. It was fitted with four polymer sensors that respond to different volatile organic compounds released from microorganisms in urine, particularly ethanoic acid or/and ammonia/amines. One of the justifications for use of such a system was that urine samples spiked with various microorganisms (*P. mirabilis*, *Pseudomonas aeroginosa*, *Staphylococcus aureus*, *Staphylococcus faecalis*, *C. albicans*, *E. coli*, *Klebsiella* spp.) and clinical urine samples indicated that negative samples (defined as having $<10^5$ CFU/mL^{-1}) averaged 3 ppm acetic acid, while positives (defined as being $>10^5$ CFU/mL^{-1}) ranged from 8 to 500 ppm, with the average value being approx. 100 ppm.[36] The detection of volatile organic compounds in urine by GLC was demonstrated in early studies; however, these methods were only moderately successful in detecting infected and non-infected urine and did not develop into practical diagnostic tools.[33,48,49] When bacteriuria was defined as 100,000 CFU/mL, the sensitivity and specificity of the OMA device were reported as 84% and 88%, respectively. When bacteriuria was defined as 10,000 CFU/ml, the sensitivity fell and the specificity rose, 72% and 89%, respectively.[47] They concluded that the OMA showed promise as an automated system for the rapid routine screening of urine specimens, although limitations were reported.[47] As an example of the credibility of the system; the OMA—Urinary Tract Infection Detector received pre-marketing clearance from the US Food and Drug Administration, (FDA) in 2001 (http://www.fda.gov/cdrh/510k/sumnov01.html). However, the system is no longer commercially available. The OMA simply identified UTI; later work by Pavlou et al.[50] reported the ability to identify bacterial species in urine. Agar cultures had to be used with an incubation time of 4–5 h. The volatile production patterns were then analyzed using an electronic nose system (model BH114-Bloodhound, Leeds, UK) with 14 conducting polymer sensors. Four groups could be distinguished, i.e. normal urine and infected urine with different bacteria; *E. coli*, *Proteus* spp., and *Staphylococcus* spp.[50] The experimental analysis of the data was undertaken using a neural network (NN), enabling identification of all but one of the samples correctly when compared to the culture information. This is an important advance. Further work also using the Bloodhound sensor and sophisticated pattern recognition software, reproduced Pavlou's results, giving impressive bacterial classification.[51,52] Incubation was used, (4–5 h) however this study has again shown the potential for early detection of microbial contaminants in urine samples using electronic nose technology and there remains the opportunity to improve on these results.

Identification of bacterial ureases is also of clinical utility as the enzymes, particularly from *P. mirabilis*, have been implicated as a contributing factor in the formation of urinary and kidney stones, obstruction of urinary catheters, and pyelonephritis.

Little work has been undertaken on electronic noses for cancer biomarker(s) assessment in urine. The following gives an overview of this work. The use of a commercially available gas sensor array coupled with a specifically designed pattern recognition algorithm for this purpose has been assessed for detection of volatile bladder cancer biomarkers in urine. The best diagnostic performance that was obtained with independent test data provided by healthy volunteers and bladder cancer patients was 70% overall accuracy (70% sensitivity and 70% specificity). When the data of patients suffering from other non-cancerous urological diseases were added to those of the healthy controls, the classification accuracy fell to 65% with 60% sensitivity and 67% specificity. While this is not sufficient for a diagnostic test, it is significantly better than random chance, leading the authors to conclude that there was useful information in the urine headspace but that a more informative analytical technique, such as mass spectrometry, was required if this approach were to be exploited fully. Although the specificity achieved using this method was less than that of conventional urine cytology, sensitivity was appreciably higher.[53]

22.4 STOOL VOCs AND DISEASE

There has been very little work published on the volatiles from stool and their potential for disease diagnosis. There is anecdotal information on the change of stool with various gastro-intestinal conditions, such as nurses' ability to detect diarrhea due to *Clostridium difficile*; however, little rigorous work has been undertaken.

One of the most studied areas is microbial hydrogen generation in the gut and its measurement in breath for assessing malabsorption of sugars, particularly fructose and lactose.[54] In this case the availability of selective, very sensitive hydrogen sensors has made this possible permitting hydrogen changes in the breath to be measured even though hydrogen has very low solubility in blood.

There are many gastrointestinal conditions where it would be advantageous if point-of-care, rapid analyzes could be undertaken. Presently, it can take many days for a microbiological assessment of stool to be undertaken e.g. for *C. difficile*, which is naturally a matter of concern for patients with a life threatening infection. Rapid diagnosis between the inflammatory bowel diseases (IBD), Crohn's disease, ulcerative colitis, and irritable bowel syndrome (IBS) would be beneficial for clinicians and patients if it could reduce medical examinations, which are invasive, costly, and time consuming. A rapid, accurate test for colorectal cancer would be especially desirable.

The analysis of volatile metabolites from stool offers the possibility of rapid, low-cost diagnoses of a range of disease conditions, in a non-invasive manner and, potentially, at relatively low cost. Increased knowledge of the complex chemistry and microflora of the gut from healthy individuals and those with gastro-intestinal conditions would enable more rapid progress to be made in the correlation between chemical metabolites and disease state. The following section reviews the volatile compounds that have been found in the headspace of stool which are mainly qualitative in nature.

22.4.1 Volatiles from stool

The first report of gas analysis from stool was in 1861 when Ruge reported that human rectal gas contained hydrogen, carbon dioxide, and methane, in addition to other unidentified gases.[55] Tomlin's study,[56] 130 years later drew similar conclusions. In recent years, Levitt has led the field in the study of flatus and, in 1997, published a seminal paper on its composition.[57] He confirmed Ruge's observation and went on to quantify the most abundant gases in the flatus of 16 volunteers. Typically about 70% of gas in flatus was reported to be produced intra-luminally and flatus is considered to be a mixture of hydrogen (0–50%), nitrogen (5–90%), oxygen (0–10%), carbon dioxide (10–30%), and methane (0–10%). Methane production occurs in about 50% of the healthy population, some members producing higher levels than others; methane production is correlated with methanogenic bacteria. Similarly, sulfate-reducing bacteria are responsible for the generation of pungent sulfides.[58] Thus, evidence shows that methane[59] and sulfide production[23] can be directly related to intestinal flora. It has been shown that methanogens use hydrogen to synthesise methane, by reduction of carbon dioxide and as a result high hydrogen levels and high methane levels are considered not to occur simultaneously in flatus. Thus to assess bacterial overgrowth in the small intestine both methane and hydrogen breath levels would be ideally measured. Unfortunately, while specific, extremely sensitive hydrogen sensors are commercially available, at the present time, small, low cost, sensitive, specific sensors for methane do not exist.

Patients and healthcare professionals have observed that faeces often smell abnormal during gastrointestinal disease. Identification of volatile organic compound (VOC) biomarkers from stool offers the potential for developing a method for rapid diagnosis of gastrointestinal diseases. Surprisingly little is known about VOCs found in the gut. The diagnostic and health implications of most of these compounds remain to be explored. Significant concentrations of a range of short-chain fatty acids (SCFAs),[60] branched-chain fatty acids (BCFAs), indoles[61], and phenols[62] have been observed. Fermentation of carbohydrates in the gut produces ethanoic, propionic, butanoic, pentanoic, and hexanoic acids, particularly by *Bacteroides*.[63] *In vitro* studies[64] have provided evidence that proteinacious foods also produce SCFAs via the action of bacteria such as *Clostridia* spp.; BCFAs, such as 2-methylbutanoic acid and methyl-propionic acids, are principally produced by gut microbial action on proteins via the respective branched amino acid. Volatiles such as methanethiol and ammonia are considered[65] to be derivable from methionine by the action of bacteria such as *Clostridium sporogenes*. Hydrogen sulfide and methanethiol can be damaging to the large intestinal epithelium and are also generated from sulfur-containing substances in the diet.[66] Similarly, fermentation of tyrosine and tryptophan in stool has been shown[65] to produce the VOCs phenol and indole, respectively. Phenol and *p*-cresol are considered to be produced by aerobic intestinal micro flora acting on tyrosine and the latter by anaerobic organisms.[67] Figure 22.1 describes some of the compounds shown to be produced by incubating amino acids with fresh stool.

The two most up-to-date studies state that a total of 297 and 135 different VOCs have been identified respectively by Garner et al.[4] and De Preter et al.[68] in the

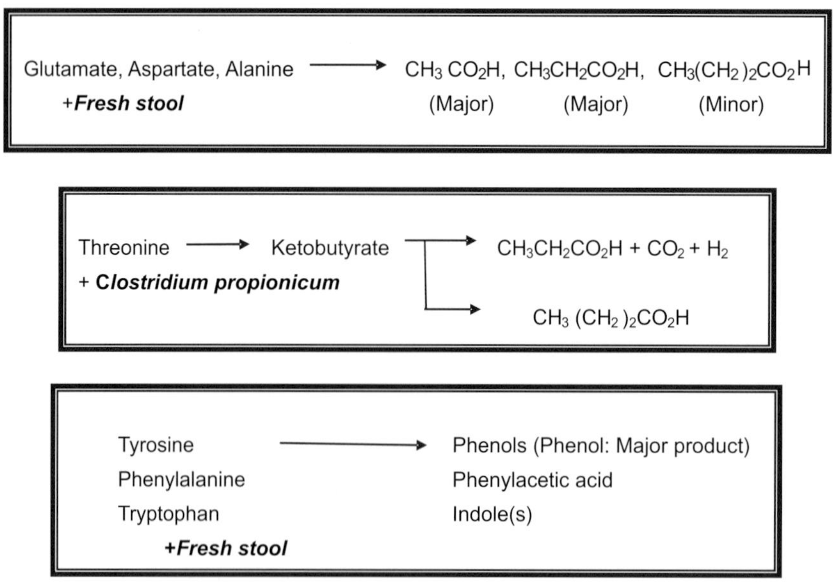

FIGURE 22.1
Origins of volatiles in the headspace of stool by incubating fresh stool/microorganisms with amino acids.

headspace of stool from apparently healthy individuals on an *ad lib* diet. These two studies show many similarities and also some differences. Typically, for each donor the number of VOCs ranged from 78 to 125 (median = 101). Interestingly, 44 compounds were stated to be common to 80% of the cohort samples.[4] Thus, many compounds were shared by asymptomatic donors. Headspace SPME has greatly facilitated the isolation and analysis of VOCs from human stool. Pioneering human fecal VOC metabolomic investigations have utilized a single SPME fiber type for analyte extraction and analysis. However Couch et al.[69] hypothesized that the varied functionality of the metabolites in the headspace of stool, dictated the use of several diverse SPME fiber coatings for more comprehensive metabolomic coverage. They evaluated eight different commercially available SPME fibers in combination with GC-FID and GCMS. This approach appears very promising; 267 peaks were found with GC-FID; the authors have yet to identify all the compounds. There is the drawback that more analysis time is required with this approach. However it is undoubtedly the case that the earlier SPME/GC-MS work is very likely to have missed compounds due to overlapping peaks or relative selectivities of the different fibers to certain volatiles. The following section describes the groups of volatiles found in stool and discusses their origins. There are not many quantitative reports of the biosynthesis of gut compounds; information when available has been included.

22.4.1.1 Acids, alcohols, and esters

Short chain fatty acids (SCFAs) in stool have been reported to be very common by a number of authors. SCFAs arise from metabolism of undigested carbohydrate, such as dietary fiber by colonic bacteria.[63] Branched chain fatty acids are also known to be abundant, and their presence can be explained from the dissimilation of amino acids in some cases.[65] SCFA are also present in appreciable concentrations *circa* 37, 12, and 13 mmol/L of stool for acetate, propionate, and butyrate, respectively.[67] Methanoic acid however is not commonly observed and neither are methyl esters. Of the other acids, benzoic acid has received some attention in quantitative rather than headspace studies of stool due its ability to inhibit the growth of fungi and some bacteria (by inhibiting glucose fermentation). Its origin is considered to be from microbial activity in the gut but mainly from diet, particularly from fruit and coffee containing quinic acid.

Alcohols were thought uncommon in adult feces.[70] However, its now known that at least 45 different alcohols can be present. Ethanol is very commonly observed. The presence of ethanol has been linked to yeast overgrowth, as well as bacterial synthesis by the Embden-Meyerhof pathway. Interestingly, specific long-chain aliphatic alcohols (C6–C20), which can inhibit the growth of various bacteria and fungi,[71] are also found. It is likely that these compounds play a role in providing an ecological niche for some bacteria by inhibiting others.

Esters were found to represent the largest group of compounds identified, (59 esters of which 4 contained sulfur, Tables 22.1 and 22.2a). An interesting readily observed feature of Table 22.2a is the similarity of the higher molecular weight compounds,

Table 22.2a A description of straight chain alcohols, short-chain fatty acids, aldehydes, alkanes, and methyl ketones found in the headspace of stool and urine of apparently healthy volunteers.

Class	Carbon Number (n)																			
	1	2	3	4	5	6	7	8	9	10	11	12	13	14	15	16	17	18	19	20
Alcohol	S	S	S	S	S	S	S	S	S	S		S	S			S				
$H{-}(CH_2)_n{-}OH$	U*	U	U	U	U	U		U					U							
Acid	S	S	S	S	S	S	S	S	U											
$H{-}(CH_2)_{n-1}{-}CO_2H$	U	U		U					U											
Aldehyde		S	S	S	S	S	S	S	S	S	S	S	S	S	S	S		S		
$H{-}(CH_2)_{n-1}{-}CHO$		U	U		U	U	U	U	U											
Alkane	S				S	S		S		S	S	S	S	S	S	S	S	S	S	S
$H{-}(CH_2)_n{-}H$						U								U						
Methyl ketone			S	S	S	S	S	S	S	S										
$H{-}(CH_2)_{n-2}{-}CO{-}CH_3$			U	U	U	U	U	U	U											

S = found in stool.
U = found in urine.
U = not detected in urine in healthy non-drinkers, easily detectable in urine of alcoholic beverage drinkers.*

Table 22.2b A description of straight chain esters found in the headspace of stool and urine of apparently healthy volunteers.

Ester		Acid Component Carbon Number (n)								
		0	1	2	3	4	5	6	7	8
Methyl	$H\text{–}(CH_2)_n - -CO_2\text{–}CH_3$		S	S	S	S	S	S		
Ethyl	$H\text{–}(CH_2)_n - CO_2\text{–}C_2H_5$		SU	S	S	S	S	S		
Propyl	$H\text{–}(CH_2)_n - CO_2\text{–}C_3H_7$		S	S	S	S	S			
Butyl	$H\text{–}(CH_2)_n - CO_2\text{–}C_4H_9$	S	S	S	SU	S	S			
Pentyl	$H\text{–}(CH_2)_n - CO_2\text{–}C_5H_{11}$	S	S	S						
Hexyl	$H\text{–}(CH_2)_n - CO_2\text{–}C_6H_{13}$		S	S						
Heptyl	$H\text{–}(CH_2)_n - CO_2\text{–}C_7H_{15}$	S								

S = found in stool.
U = found in urine.

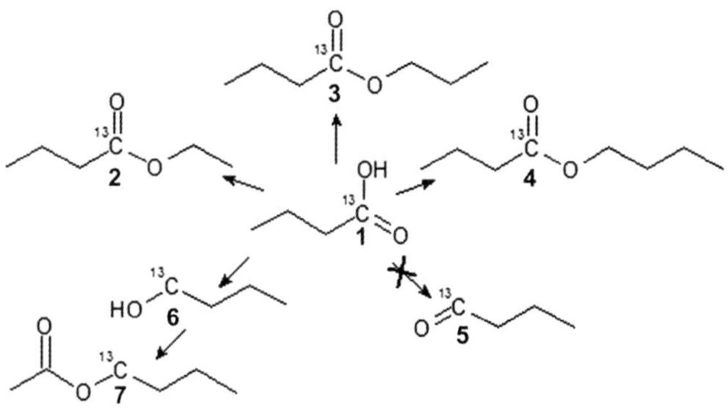

1 Butanoic acid, **2** Butanoic acid ethyl ester, **3** Butanoic acid propyl ester
4 Butanoic acid butyl ester, **5** Butanal, **6** 1-Butanol, **7** Ethanoic acid butyl ester

FIGURE 22.2
The products found from the conversion of $[1 - ^{13}C]$ butanoic acid on incubation with fresh stool from an asymptomatic donor.[4]

they either possess a long acid chain and short-chain alcohol or a short-chain acid and long-chain alcohol. This suggests that the number of esters identified is not a true picture of what is present in the stool but a limit on the method i.e. the volatility of the esters. It is very likely that a more sensitive method or better pre-concentration will significantly increase the compounds observed. An *in vitro* study using carbon labeled butanoic acid has shown the origin of some esters by stool mediated syntheses; see Figure 22.2.

22.4.1.2 Benzenoid and heterocyclic compounds

A diverse range of 56 aromatic compounds have been reported,[4] which included mono-, di-, tri-, and tetra-substituted benzenoids, mono- and di-substituted furans, and nitrogen containing derivatives of pyridine, pyrrole, and indole. Most of these have only been recently reported in faeces, although it has been established that phenolic and indole compounds arise from the metabolism of aromatic amino acids by gut bacteria.[64] It is considered that aerobic bacteria in the ileum tended to produce phenol from tyrosine, while anaerobic bacteria in the left colon produced *p*-cresol, at about 10 mg/day and 52 mg/day, respectively.[67] A wide range of furan derivatives, mainly aromatic, have been found and may be of plant origin.[72] However, this is doubtful, as furans are acid labile and would not be expected to survive the digestive process. There are many publications which have observed that alkyl furans are produced by fungi. In contrast there is a paucity of publications relating to furan biosynthesis by bacteria. Fungi are well known to be commensal organisms in the gut, which could explain the origins of furans, possibly from the metabolism of fructose. Some benzenoid compounds such as dimethylbenzenes, ethylbenzene, and toluene (constituents of petrol) may arise from air pollution.

22.4.1.3 Aldehydes and ketones

The aldehydes ethanal, propanal, butanal, and hexanal have been reported[4] in the stool of significant numbers of individuals. Ethanal is of particular interest due to its abundance and is considered to promote mutagenesis[73–75] and is associated with bowel cancer. The toxic effects of higher aldehydes have received much less attention. The origins of some aldehydes may be dietary. For instance, 2-methylpropanal, 3-methylpropanal, hexanal, nonanal, decanal, and benzaldehyde are found in potato tubers and hexanal in carrots. However, it is doubtful that these compounds would remain unchanged through the digestive system and biosynthesis by microorganisms in the gut appears more likely.

Ketones have been reported to be the second largest class of VOC found and include four of the most common compounds, of which acetone and 2-butanone probably arise from fatty acid and carbohydrate metabolism.[76] Methylketones can be produced by many species of bacteria and can also be produced by fungi from the respective alkanoic acid and undoubtedly other ketonic compounds can also be synthesised by bacteria. Some compounds of this group may have dietary origins, for example cheddar cheese contains 2-propanone, 2-butanone, and 2-heptanone, potatoes contain acetone and 2-heptanone, and 6-methyl-5-hepten-2-one is an abundant volatile in carrots.[77] The universal presence of 2,3-butanedione is interesting since it may have health implications by impacting on the growth of some bacteria and yeasts.[78] This group of compounds, and indeed other groups, are not normally the end products of metabolism by microorganisms therefore their concentrations would be expected to be continually changing in the gut.

22.4.1.4 Alkanes, alkenes, alicyclic compounds

Methane is a product of bacterial fermentation of monosaccharides, reduction of carbon dioxide, or from acetic acid. Pentane has been reported however there is little evidence of other short-chain species which could well be due to the analytical procedures used. Longer chain species have been found in small numbers in stool. Isoprene has been extracted from feces and has been investigated as a breath marker for disease.[79] Isoprene in the gut may be the result of cholesterol biosynthesis[80] and it is considered to be the most common hydrocarbon in the human body and therefore would be expected to be found in stool. Many alkenes/terpenoid compounds found are well documented as naturally occurring plant products.[81] Limonene has been reported as the most abundant of the terpenoid compounds. Most of the terpenes identified[4] are found in vegetable food stuffs and do not originate from animal products. For instance the following volatiles are present in carrots: pinene, limonene, terpinene (1-methyl-4-(1-methylethyl)-1,4-cyclohexadiene), *p*-cymene, terpinolene (1-methyl-4-(1-methylethylidene)-cyclohexene), α-caryophyllene, and humulene.[56] Copaene is found in potato extracts.[82]

22.4.1.5 Ether compounds and chloro compounds

At least seven ether compounds have been isolated from the headspace of stool. 2-ethoxyethanol commonly occurs in manufactured products like soaps and cosmetics[83] and 1,3-dimethoxybenzene is a registered food additive in Europe.[4] Similarly, it is very unlikely that chlorinated compounds found are of biological origin. Consumption of contaminated food or water is the likely source of these compounds. Chloroform may arise as a stool VOC component from several sources, it is an air contaminant and has been detected in foodstuffs.[84] Chlorination for disinfection of drinking water is another source resulting in the production of chloroform and halogenated methanes.[85] Methylene chloride is not known to occur naturally,[4] although it has been detected in the ambient air in some areas and can enter the aquatic environment from industrial waste water.

22.4.1.6 Nitrogen and sulfur compounds

Many nitrogen compounds have been reported (Table 22.1) and and are likely to arise from the diet; for instance, methylpyrazine, pyridine, and pyrrole are constituents of coffee. However, pyrrole readily polymerizes with acid and, therefore, its presence is unlikely to be dietary, as it would be unlikely to survive transit through the stomach. Ammonia results from microorganism activity. In addition increasing the amount of protein in the diet from 63 g to 136 g/day was found to increase the amount of fecal ammonia from 15 to 30 mmol/L. Interestingly, increasing the amount of fiber to the high protein diet was reported to not alter the ammonia concentration.[6] In a study of nitrogen containing compounds in the stool of 30 healthy individuals indole was the only compound found ubiquitously,[4] followed by 3-methylindole, in 73% of individuals, these compounds are well known to be produced by microbial degradation of l-tryptophan in the gut. Many compounds are present in a minority of volunteers. Allyl isothiocyanate was found to be present in 23% of cases, this compound is of particular interest due to its suspected anti-cancer properties. Its occurrence would be expected to be determined by a number of factors such as diet (cruciferous vegetables), the cooking of these vegetables, and the ability of the host's bacteria to break down sinigrin, the glucosinolate precursor.

A diverse range of sulfur compounds have been identified. For instance, methanethiol and dimethylsulfide have been commonly observed; the former is, at least in part, considered to be produced from methionine by *Clostridia* in the gut.[65] Methanethiol has a toxicity approaching cyanide and the factors controlling its concentration and biosynthesis might warrant further investigation. Methanethiol and dimethylsulfide may also be produced by methylation of hydrogen sulfide as a detoxification mechanism by mucosal thiol S-methyltransferase.[86] Dimethyldisulfide and dimethyltrisulfide have both been commonly reported in stool.[4,57,87] Hydrogen sulfide is probably most likely to occur due to the metabolism of sulfate by sulfate-reducing bacteria.[88] Sulfate, which is poorly absorbed in the small bowel, is naturally present in cruciferous vegetables (cabbage, broccoli) and nuts and as an additive in bread and beer.[88] The main sulfur-containing flatus components in healthy individuals have been quantified: hydrogen sulfide (1.06 (0.2) μmol/L), followed by methanethiol (0.21 (0.04) μmol/L) and dimethyl sulfide (0.08 (0.01) μmol/L).[88] The authors were concerned about the social aspect of pungent flatus and found in their study that hydrogen sulfide and methanethiol appeared to be principally responsible and not indole-based compounds as previously thought.

22.4.2 Metabolism of acid in stool

Long-chain alcohols may be made by reduction of the corresponding acid, for which extensive homologous series are found in the headspace of stool. This hypothesis was investigated by incubating faeces with labeled butanoic acid and it was found that labeled butanol was synthesized, which was seen to increase with incubation time.[4] No conversion to [1-^{13}C] butanal was observed. Esterification of acids was also found to readily occur in stool as evidenced by the formation of^{13}butanoic acid ethyl ester. Conversion of [1-^{13}C] butanoic acid to [1-^{13}C]butanoic acid propyl ester after 3 h incubation was also confirmed by the same method (Figure 22.2). It is likely that microbially mediated reduction of other fatty acids also occurs.

22.4.3 Specific changes in VOCs in gastrointestinal disease

This is a new area of research and there is little literature available. Headspace screening of fluid obtained from the gut during colonoscopy and breath analysis by proton transfer reaction-mass spectrometry, PTR-MS, has been undertaken as a novel approach in the diagnosis of gastrointestinal diseases. Unfortunately, no significant differences were found between patients with the inflammatory bowel diseases, IBS and IBD and healthy volunteers.[89] A later study using GCMS of the volatiles from stool of patients suffering from IBS and IBD, showed that analyzes of esters, in particular in the headspace of stool, were of particular benefit for identifying Crohn's disease.[90]

Another recent larger study of 205 individuals also using GCMS demonstrated the potential of pattern analysis of fecal VOCs in patients with IBD. This may provide a non-invasive method in the diagnosis of IBD.[91] A further recent development have been the use of a more sensitive analysis method, GC-ToFMS which has enabled breath to be analyzed for bowel conditions. It was found that 6–8 compounds could be used to classify (with 95–100% accuracy) between remission and active Crohn's disease and ulcerative colitis.[92] The sensitivity to distinguish IBD patients from controls was 100%.

The analyzes of the VOC of the stools from patients with ulcerative colitis, *C. jejuni* and *C. difficile* has been undertaken[4] and resulted in the characterization of 149, 183, and 145 compounds, respectively. Group-to-group differences were clearly observed. Some interesting differences were found among samples from the various groups. For examples, butanoic acid was common in all groups except *C. difficile* samples, in which just 41% contained butanoic acid. Butanol, however, was ubiquitous in *C. difficile* samples. *Clostridia* are known to produce alcohols such as ethanol, isopropanol, and butanol, at the expense of producing less butanoate.[93] Such observations fit with the results of the previous study.[4] Another example of the utility of stool VOC analyzes is that stool from cholera sufferers could be differentiated clearly from non-cholera controls.[94]

As a result of analysis of volatiles from the stool of neonates, it has been shown that there is promise in the early detection of necrotizing enterocolitis, which is the main reason for death in premature babies.[94]

Hydrogen production in the gut is well known and has been related to bacterial overgrowth in the small intestine. Excess hydrogen production has been used for the diagnosis of lactose and fructose malabsorption. However, the hydrogen is measured in breath and is not measured from stool, so is outside the remit of this chapter.

22.5 A COMPARISON OF VOCs FOUND IN URINE AND STOOL

The most notable feature of a comparison of VOCs found in urine and stool is that many more compounds have been reported in stool compared to urine. More compounds may exist in urine, and may simply be at too low a level to be measured, when compared to the concentrations found in stool. Tables 22.1, 22.2a, and 22.2b show the differences reported so far. It is possible that the authors have missed some data reported in the literature. The most prominent difference in chemical class were the esters, which represent the largest class in stool, with 59 compounds found compared to 2 compounds in urine. The extraction and sample treatment method has to be considered in the data analysis, hydrolysis of esters would certainly be expected if alkaline workup of the sample is undertaken. Almost certainly, the source of many urinary compounds is from the gut however in most cases there is no definitive published evidence (e.g. by using labeled compounds to prove the relationship). Some compounds are undoubtedly decreased in concentration from their journey from the gut to the bladder, e.g. benzoic acid (which occurs naturally in most berries) is derivitized (in the liver) and excreted in urine as the less-volatile hippuric acid (benzoylaminoethanoic acid). Although representatives of many chemical classes were found in the headspace of urine, in general shorter chain species were found in the headspace compared to stool. Again this is likely to be due to their presence in lower concentrations. One of the most notable differences was for straight chain hydrocarbons, (Table 22.2a), 15 compounds in stool were reported, with only 2 compounds in urine. This strongly suggests that the hydrocarbons were ingested and remained unchanged through the digestion process, their low water solubility preventing their detection in urine.

22.6 SUMMARY

Gut chemistry is very complex and this applies to omnivores in particular. The huge number of chemicals synthesized by plants enter the gut with chemical modification to a lesser or greater extent by the digestive system. The gut microorganisms are small chemical factories themselves and act on undigested food, and shed intestinal cells, bile fluids, etc. The analysis of volatiles from the headspace of both urine and stool offers a simplified assessment of the gut chemistry. There is the potential for a range of diagnoses, in a non-invasive manner. More work, with larger studies, need to be undertaken to determine the potential of volatile analyzes of urine and stool.

Acknowledgments

Thanks are due to Tanzeela Khalid and Cheryl Flynn for helping with the preparation.

References

1. Mills GA, Walker V. Headspace solid-phase microextraction profiling of volatile compounds in urine: application to metabolic investigations. *J Chromatogr B Biomed Sci Appl* 2001;**753**:259–68.
2. Smith S, Burden H, Persad R, Whittington K, de Lacy Costello B, Ratcliffe NM, et al. A comparative study of the analysis of human urine headspace using gas chromatography-mass spectrometry. *J Breath Res* 2008;**2**:037022.
3. Wahl HG, Hong Q, Hildenbrand S, Risler T, Luft D, Liebich H. 4-Heptanone is a metabolite of the plasticizer di(2-ethylhexyl) phthalate (DEHP) in haemodialysis patients. *Nephrol Dial Transplant* 2004;**19**:2576–83.
4. Garner CE, Smith S, de Lacy Costello B, White P, Spencer R, Probert CS, et al. Volatile organic compounds from feces and their potential for diagnosis of gastrointestinal disease. *Faseb J* 2007;**21**:1675–88.

5. Penders J, Fiers T, Giri M, Wuyts B, Ysewyn L, Delanghe JR. Quantitative measurement of ketone bodies in urine using reflectometry. *Clin Chem Lab Med* 2005;**43**:724–29.
6. Bone E, Tamm A, Hill M. The production of urinary phenols by gut bacteria and their possible role in the causation of large bowel cancer. *Am J Clin Nutr* 1976;**29**:1448–54.
7. Liebich HM, Buelow HJ, Kallmayer R. Quantification of endogenous aliphatic alcohols in serum and urine. *J Chromatogr* 1982;**239**:343–9.
8. Wahl HG, Hoffmann A, Luft D, Liebich HM. Analysis of volatile organic compounds in human urine by headspace gas chromatography-mass spectrometry with a multipurpose sampler. *J Chromatogr A* 1999;**847**:117–25.
9. Schatowitz B, Gercken G. Simultaneous determination of C2-C22 non-esterified fatty-acids and other metabolically relevant carboxylic-acids in biological-material by gas-chromatography of their benzyl-esters. *J Chromatogr-Biomed* 1988;**425**:257–68.
10. Mitchell SC, Zhang AQ. Methylamine in human urine. *Clin Chim Acta* 2001;**312**:107–14.
11. Zhang AQ, Mitchell SC, Smith RL. Dimethylamine in human urine. *Clin Chim Acta* 1995;**233**:81–8.
12. Zhang AQ, Mitchell SC, Ayesh R, Smith RL. Determination of trimethylamine and related aliphatic amines in human urine by head-space gas chromatography. *J Chromatogr* 1992;**584**:141–5.
13. Hradecky P. Volatile fatty-acids in urine and vaginal secretions of cows during reproductive-cycle. *J Chem Ecol* 1986;**12**:187–96.
14. Rhodes G, Holland ML, Wiesler D, Novotny M, Moore SA, Peterson RG, et al. Excretion of urinary volatile metabolites in response to alloxan induced diabetes of short duration in rats. *J Chromatogr* 1982;**228**:33–42.
15. Liebich HM, Al-Babbili O, Zlatkis A, Kim K. Gas-chromatographic and mass-spectrometric detection of low-molecular-weight aliphatic alcohols in urine of normal individuals and patients with diabetes mellitus. *Clin Chem* 1975;**21**:1294–6.
16. Quraishi MN, Harrison L, Wedlake L, Ouaret N, Maxim J, Nwokolo C, et al. Sniffing gastrointestinal diseases with an electronic e-nose – is this the future for diagnostics? *Eur J Internal Med* 2011;**22**(Suppl. 1):S78.
17. Podebrad F, Heil M, Reichert S, Mosandl A, Sewell AC, Bohles H. 4,5-Dimethyl-3-hydroxy-2[5H]-furanone (sotolone) – the odour of maple syrup urine disease. *J Inherit Metab Dis* 1999;**22**:107–14.
18. Schoen EJ, Baker JC, Colby CJ, To TT. Cost-benefit analysis of universal tandem mass spectrometry for newborn screening. *Pediatrics* 2002;**110**:781–6.
19. Diskin AM, Spanel P, Smith D. Increase of acetone and ammonia in urine headspace and breath during ovulation quantified using selected ion flow tube mass spectrometry. *Physiol Meas* 2003;**24**:191–9.
20. Smith D, Ismail KM, Diskin AM, Chapman G, Magnay JL, Spanel P, et al. Increase of acetone emitted by urine in relation to ovulation. *Acta Obstet Gynecol Scand* 2006;**85**: 1008–11.
21. Mozuraitis R, Buda V, Kutra J, Borg-Karlson AK. p- and m-Cresols emitted from estrous urine are reliable volatile chemical markers of ovulation in mares. *Anim Reprod Sci* 2012;**130**:51–6.
22. Banday KM, Pasikanti KK, Chan EC, Singla R, Rao KV, Chauhan VS, et al. Use of urine volatile organic compounds to discriminate tuberculosis patients from healthy subjects. *Anal Chem* 2011;**83**:5526–34.

23. Willis CL, Cummings JH, Neale G, Gibson GR. Nutritional aspects of dissimilatory sulfate reduction in the human large intestine. *Curr Microbiol* 1997;**35**:294–8.
24. Cornu JN, Cancel-Tassin G, Ondet V, Girardet C, Cussenot O. Olfactory detection of prostate cancer by dogs sniffing urine: a step forward in early diagnosis. *Eur Urol* 2011;**59**:197–201.
25. Spanel P, Smith D, Holland TA, Al Singary W, Elder JB. Analysis of formaldehyde in the headspace of urine from bladder and prostate cancer patients using selected ion flow tube mass spectrometry. *Rapid Commun Mass Spectrom* 1999;**13**:1354–9.
26. Smith S, White P, Redding J, Ratcliffe NM, Probert CSJ. Application of similarity coefficients to predict disease using volatile organic compounds. *Ieee Sens J* 2010;**10**:92–6.
27. Carrola J, Rocha CM, Barros AS, Gil AM, Goodfellow BJ, Carreira IM, et al. Metabolic signatures of lung cancer in biofluids: NMR-based metabonomics of urine. *J Proteome Res* 2011;**10**:221–30.
28. Silva CL, Passos M, Camara JS. Investigation of urinary volatile organic metabolites as potential cancer biomarkers by solid-phase microextraction in combination with gas chromatography-mass spectrometry. *Br J Cancer* 2011;**105**: 1894–904.
29. Matsumura K, Opiekun M, Oka H, Vachani A, Albelda SM, Yamazaki K, et al. Urinary volatile compounds as biomarkers for lung cancer: a proof-of-principle study using odor signatures in mouse models of lung cancer. *PLoS One* 2010;**5**:e8819.
30. Kim K, Aronov P, Zakharkin SO, Anderson D, Perroud B, Thompson IM, et al. Urine metabolomics analysis for kidney cancer detection and biomarker discovery. *Mol Cell Proteomics* 2009;**8**:558–70.
31. Osl M, Dreiseitl S, Pfeifer B, Weinberger K, Klocker H, Bartsch G, et al. A new rule-based algorithm for identifying metabolic markers in prostate cancer using tandem mass spectrometry. *Bioinformatics* 2008;**24**:2908–14.
32. Jentzmik F, Stephan C, Miller K, Schrader M, Erbersdobler A, Kristiansen G, et al. Sarcosine in urine after digital rectal examination fails as a marker in prostate cancer detection and identification of aggressive tumours. *Eur Urol* 2010;**58**:12–8; discussion 20–1.
33. Hayward NJ. Headspace gas-liquid-chromatography for the rapid laboratory diagnosis of urinary-tract infections caused by enterobacteria. *J Chromatogr* 1983 ;**274**:27–35.
34. Guernion N, Ratcliffe NM, Spencer-Phillips PT, Howe RA. Identifying bacteria in human urine: current practice and the potential for rapid, near-patient diagnosis by sensing volatile organic compounds. *Clin Chem Lab Med* 2001;**39**:893–906.
35. Turner APF, Magan N. Electronic noses and disease diagnostics. *Nat Rev Microbiol* 2004;**2**:161–6.
36. Persaud KC, Pisanelli AM, Evans P, Travers PJ. Monitoring urinary tract infections and bacterial vaginosis. *Sensor Actuat B – Chem* 2006;**116**:116–20.
37. Schaechter M, Medoff G, Eisenstein BI. *Mechanisms of microbial disease*. 2nd ed. Baltimore: Williams & Wilkins; 1992.
38. Griebling TL. Urologic diseases in America project: trends in resource use for urinary tract infections in women. *J Urol* 2005;**173**:1281–7.
39. Barnett BJ, Stephens DS. Urinary tract infection: an overview. *Am J Med Sci* 1997;**314**: 245–9.
40. Hellstrom A, Hanson E, Hansson S, Hjalmas K, Jodal U. Association between urinary symptoms at 7 years old and previous urinary tract infection. *Arch Dis Child* 1991;**66**: 232–4.

41. Pantell RH, Newman TB, Bernzweig J, Bergman DA, Takayama JI, Segal M, et al. Management and outcomes of care of fever in early infancy. *Jama* 2004;**291**:1203–12.
42. Millar L, DeBuque L, Leialoha C, Grandinetti A, Killeen J. Rapid enzymatic urine screening test to detect bacteriuria in pregnancy. *Obstet Gynecol* 2000;**95**:601–4.
43. Lumbiganon P, Laopaiboon M, Thinkhamrop J. Screening and treating asymptomatic bacteriuria in pregnancy. *Curr Opin Obstet Gynecol* 2010;**22**:95–9.
44. Bloomfield P, Hodson EM, Craig JC. Antibiotics for acute pyelonephritis in children. *Cochrane Database Syst Rev* 2003:CD003772.
45. Williams GJ, Macaskill P, Chan SF, Turner RM, Hodson E, Craig JC. Absolute and relative accuracy of rapid urine tests for urinary tract infection in children: a meta-analysis. *Lancet Infect Dis* 2010;**10**:240–50.
46. Jones BD, Mobley HL. Genetic and biochemical diversity of ureases of Proteus, Providencia, and Morganella species isolated from urinary tract infection. *Infect Immun* 1987;**55**:2198–203.
47. Aathithan S, Plant JC, Chaudry AN, French GL. Diagnosis of bacteriuria by detection of volatile organic compounds in urine using an automated headspace analyzer with multiple conducting polymer sensors. *J Clin Microbiol* 2001;**39**:2590–3.
48. Coloe PJ. Head-space gas liquid chromatography for rapid detection of *Escherichia coli* and *Proteus mirabilis* in urine. *J Clin Pathol* 1978;**31**:365–9.
49. Manja KS, Rao KM. Gas-chromatographic detection of urinary tract infections caused by *Escherichia coli* and *Klebsiella* sp. *J Clin Microbiol* 1983;**17**:264–6.
50. Pavlou A, Turner APF, Magan N. Recognition of anaerobic bacterial isolates *in vitro* using electronic nose technology. *Lett Appl Microbiol* 2002;**35**:366–9.
51. Kodogiannis V, Wadge E. The use of gas-sensor arrays to diagnose urinary tract infections. *Int J Neural Syst* 2005;**15**:363–76.
52. Kodogiannis VS, Lygouras JN, Tarczynski A, Chowdrey HS. Artificial odor discrimination system using electronic nose and neural networks for the identification of urinary tract infection. *IEEE Trans Inf Technol Biomed* 2008;**12**:707–13.
53. Weber CM, Cauchi M, Patel M, Bessant C, Turner C, Britton LE, et al. Evaluation of a gas sensor array and pattern recognition for the identification of bladder cancer from urine headspace. *Analyst* 2011;**136**:359–64.
54. Bate JP, Irving PM, Barrett JS, Gibson PR. Benefits of breath hydrogen testing after lactulose administration in analysing carbohydrate malabsorption. *Eur J Gastroenterol Hepatol* 2010;**22**:318–26.
55. Ruge E. Beiträge zur Kenntnis der Darmgase. *Sitzungsberichte der Kaiserlichen. Akademie der Wissenschaften* 1862;**42**:739–62.
56. Tomlin J, Lowis C, Read NW. Investigation of normal flatus production in healthy volunteers. *Gut* 1991;**32**:665–9.
57. Suarez F, Furne J, Springfield J, Levitt M. Insights into human colonic physiology obtained from the study of flatus composition. *Am J Physiol*1997;**272**:G1028–33.
58. Gibson GR, Macfarlane GT, Cummings JH. Sulphate reducing bacteria and hydrogen metabolism in the human large intestine. *Gut* 1993;**34**:437–9.
59. Miller TL, Wolin MJ. Methanogens in human and animal intestinal tracts. *Syst Appl Microbiol* 1986;**7**:223–9.
60. Cummings JH. Short chain fatty acids in the human colon. *Gut*1981;**22**:763–79.
61. Yokoyama MT, Carlson JR. Microbial metabolites of tryptophan in the intestinal tract with special reference to skatole. *Am J Clin Nutr* 1979;**32**:173–8.

62. Gostner A, Blaut M, Schaffer V, Kozianowski G, Theis S, Klingeberg M, et al. Effect of isomalt consumption on faecal microflora and colonic metabolism in healthy volunteers. *Br J Nutr* 2006;**95**:40–50.
63. Backhed F, Ley RE, Sonnenburg JL, Peterson DA, Gordon JI. Host-bacterial mutualism in the human intestine. *Science* 2005;**307**:1915–20.
64. Mortensen PB, Clausen MR. Short-chain fatty acids in the human colon: relation to gastrointestinal health and disease. *Scand J Gastroenterol Suppl* 1996;**216**:132–48.
65. Smith EA, Macfarlane GT. Dissimilatory amino acid metabolism in human colonic bacteria. *Anaerobe* 1997;**3**:327–37.
66. Pitcher MC, Beatty ER, Cummings JH. The contribution of sulphate reducing bacteria and 5-aminosalicylic acid to faecal sulfide in patients with ulcerative colitis. *Gut* 2000;**46**: 64–72.
67. Cummings JH, Hill MJ, Bone ES, Branch WJ, Jenkins DJ. The effect of meat protein and dietary fiber on colonic function and metabolism. II. Bacterial metabolites in feces and urine. *Am J Clin Nutr* 1979;**32**:2094–101.
68. De Preter V, Van Staeyen G, Esser D, Rutgeerts P, Verbeke K. Development of a screening method to determine the pattern of fermentation metabolites in faecal samples using on-line purge-and-trap gas chromatographic-mass spectrometric analysis. *J Chromatogr A* 2009;**1216**:1476–83.
69. Dixon E, Clubb C, Pittman S, Ammann L, Rasheed Z, Kazmi N. Solid-phase microextraction and the human fecal VOC metabolome. *PLoS One* 2011;**6**: e18471.
70. Lington AW, Bevan C. Alcohols. In: Clayton GD, editor. *Patty's industrial hygiene and toxicology*. New York: Wiley; 1994. p. 2585–760.
71. Elgaali H, Hamilton-Kemp TR, Newman MC, Collins RW, Yu K, Archbold DD. Comparison of long-chain alcohols and other volatile compounds emitted from food-borne and related Gram positive and Gram negative bacteria. *J Basic Microbiol* 2002;**42**:373–80.
72. Perez Locas C, Yaylayan VA. Origin and mechanistic pathways of formation of the parent furan—a food toxicant. *J Agric Food Chem* 2004;**52**:6830–6.
73. Seitz HK, Matsuzaki S, Yokoyama A, Homann N, Vakevainen S, Wang XD. Alcohol and cancer. *Alcohol Clin Exp Res* 2001;**25**:137S–43S.
74. Vakevainen S, Tillonen J, Agarwal DP, Srivastava N, Salaspuro M. High salivary acetaldehyde after a moderate dose of alcohol in ALDH2-deficient subjects: strong evidence for the local carcinogenic action of acetaldehyde. *Alcohol Clin Exp Res* 2000;**24**:873–7.
75. Homann N. Alcohol and upper gastrointestinal tract cancer: the role of local acetaldehyde production. *Addict Biol* 2001;**6**:309–23.
76. Decombaz J, Arnaud MJ, Milon H, Moesch H, Philippossian G, Thelin AL, et al. Energy metabolism of medium-chain triglycerides versus carbohydrates during exercise. *Eur J Appl Physiol Occup Physiol* 1983;**52**:9–14.
77. Kjeldsen F, Christensen LP, Edelenbos M. Quantitative analysis of aroma compounds in carrot (*Daucus carota* L) cultivars by capillary gas chromatography using large-volume injection technique. *J Agric Food Chem* 2001;**49**:4342–8.
78. Ott A, Germond JE, Chaintreau A. Vicinal diketone formation in yogurt: (13)C precursors and effect of branched-chain amino acids. *J Agric Food Chem* 2000;**48**:724–31.
79. Salerno-Kennedy R, Cashman KD. Potential applications of breath isoprene as a biomarker in modern medicine: a concise overview. *Wien Klin Wochenschr* 2005;**117**:180–6.

80. Deneris ES, Stein RA, Mead JF. Acid-catalyzed formation of isoprene from a mevalonate-derived product using a rat liver cytosolic fraction. *J Biol Chem* 1985;**260**:1382–5.
81. Kjeldsen F, Christensen LP, Edelenbos M. Changes in volatile compounds of carrots (*Daucus carota* L) during refrigerated and frozen storage. *J Agric Food Chem* 2003;**51**: 5400–7.
82. De Lacy Costello BPJ, Evans P, Ewen RJ, Gunson HE, Jones PRH, Ratcliffe NM, et al. Gas chromatography-mass spectrometry analyses of volatile organic compounds from potato tubers inoculated with *Phytophthora infestans* or *Fusarium coeruleum*. *Plant Pathol* 2001;**50**:489–96.
83. Johnson W, Expert CIR. Final report on the safety assessment of Ethoxyethanol and Ethoxyethanol Acetate. *Int J Toxicol* 2002;**21**:9–62.
84. Alloway BJ, Ayres DC. *Chemical principles of environmental pollution*. 2nd ed. London: Blackie Academic & Professional; 1997.
85. Herren-Freund SL, Pereira MA. Carcinogenicity of by-products of disinfection in mouse and rat liver. *Environ Health Perspect* 1986;**69**:59–65.
86. Weisiger RA, Pinkus LM, Jakoby WB. Thiol S-methyltransferase—suggested role in detoxication of intestinal hydrogen-sulfide. *Biochem Pharmacol* 1980;**29**: 2885–7.
87. Moore JG, Jessop LD, Osborne DN. Gas-chromatographic and mass-spectrometric analysis of the odor of human feces. *Gastroenterology* 1987;**93**:1321–9.
88. Suarez FL, Springfield J, Levitt MD. Identification of gases responsible for the odour of human flatus and evaluation of a device purported to reduce this odour. *Gut* 1998;**43**: 100–4.
89. Lechner A, Colvin HP, Ginzel C, Lirk P, Rieder J, Tilg H. Headspace screening of fluid obtained from the gut during colonoscopy and breath analysis by proton transfer reaction-mass spectrometry: a novel approach in the diagnosis of gastro-intestinal diseases. *Int J Mass Spectrom* 2005;**243**:151–4.
90. Fowler D. *Analysis of volatile marker compounds in body fluid samples from patients with gastrointestinal disease*. Cranfield: Cranfield University; 2010.
91. Ahmed I, Probert CS, Smith S, Ratcliffe N, Greenwood R. Diagnostic potential of volatile organic compounds in inflammatory bowel disease. *Gastroenterology* 2010;**138**:S763–4.
92. Bodelier A, Pierik MJ, Dallinga JW, Van Berkel J, Moonen E, Hameeteman W, et al. Analysis of volatile organic compounds in exhaled air as a non invasive biomarker for inflammatory bowel disease (IBD). *Gastroenterology* 2009; **136**:A35.
93. May T, Mackie RI, Fahey Jr GC, Cremin JC, Garleb KA. Effect of fiber source on short-chain fatty acid production and on the growth and toxin production by *Clostridium difficile*. *Scand J Gastroenterol* 1994;**29**:916–22.
94. Garner CE, Smith S, Bardhan PK, Ratcliffe NM, Probert CS. A pilot study of faecal volatile organic compounds in faeces from cholera patients in Bangladesh to determine their utility in disease diagnosis. *Trans R Soc Trop Med Hyg* 2009;**103**: 1171–3.
95. Zlatkis A, Bertsch W, Lichtenstein HA, Tishbee A, Shunbo F, Liebich HM, et al. Profile of volatile metabolites in urine by gas chromatography-mass spectrometry. *Anal Chem* 1973;**45**:763–7.
96. Zlatkis A, Brazell RS, Poole CF. The role of organic volatile profiles in clinical diagnosis. *Clin Chem* 1981;**27**:789–97.

97. Wahl HG, Hoffmann A, Luft D, Liebich HM. Analysis of volatile organic compounds in human urine by headspace gas chromatography-mass spectrometry with a multipurpose sampler. *J Chromatogr A* 1999;**847**:117–25.
98. Smith D, Spanel P, Holland TA, Al Singari W, Elder JB. Selected ion flow tube mass spectrometry of urine headspace. *Rapid Commun Mass Sp* 1999;**13**:724–9.
99. Sreekumar A, Poisson LM, Rajendiran TM, Khan AP, Cao Q, Yu J, et al. Metabolomic profiles delineate potential role for sarcosine in prostate cancer progression. *Nature* 2009;**457**:910–4.

CHAPTER

23

Volatile Organic Compounds (VOCs) Released by Pathogenic Microorganisms *in vitro*: Potential Breath Biomarkers for Early-Stage Diagnosis of Disease

Wojciech Filipiak[*,†], **Andreas Sponring**[*,†], **Anna Filipiak**[*], **Maria Baur**[*,†], **Clemens Ager**[*,†], **Helmut Wiesenhofer**[*], **Rosa Margesin**[‡], **Markus Nagl**[§], **Jakob Troppmair**[**], **and Anton Amann**[*,†]

[*] *Breath Research Institute, Austrian Academy of Sciences, Rathausplatz 4, 6850 Dornbirn, Austria*

[†] *University Clinic for Anesthesia, Innsbruck Medical University, Anichstraße 35, 6020 Innsbruck, Austria*

[‡] *Institute of Microbiology, Leopold-Franzens University of Innsbruck, Technikerstraße 25, 6020 Innsbruck, Austria*

[§] *Department of Hygiene, Microbiology and Social Medicine, Division of Hygiene and Medical Microbiology, Innsbruck Medical University, Fritz-Pregl-Straße 3, 6020 Innsbruck, Austria*

[**] *Daniel-Swarovski Research Laboratory, Department of Visceral-, Transplant- and Thoracic Surgery, Innsbruck Medical University, Innrain 66, 6020 Innsbruck, Austria*

23.1 INTRODUCTION

Apart from lung cancer, pneumonia is the most common cause of death related to infectious lung diseases. It is often characterized by the inflammation of lung parenchyma and abnormal alveolar filling with fluid (consolidation and exudation). Pneumonia can be caused by infection by bacteria, viruses, fungi, or parasites, and by chemical or physical injury to the lungs. Amongst the pathogens, *Streptococcus pneumonia* and *Haemophilus influenzae* cause community-acquired pneumonia (CAP) in children and adults by person to person transmission. In this respect, up to 60% of CAP cases are caused by *S. pneumoniae* and up to 15% by *H. influenzae*,[1] but other infections such as septicemia, meningitis or otitis, infections of the upper respiratory tract, septic arthritis or conjunctivitis might develop as a result of colonization with those microorganisms. Hospital-acquired pneumonia (HAP) is the second most common cause of nosocomial infections after urinary tract infections, which are a result of treatment in a hospital or a healthcare service unit. HAP is the primary cause of death

Volatile Biomarkers. http://dx.doi.org/10.1016/B978-0-44-462613-4.00023-4

in the intensive care unit (ICU), whereas pneumococcal pneumonia causes mortality of up to 10%,[2] even after aggressive antimicrobial treatment. Children and young adults are susceptible to lower respiratory tract infections that are typically caused by *Staphylococcus aureus*, *Haemophilus influenzae* and *Pseudomonas aeruginosa*.[1] Moreover, *P. aeruginosa* colonization of the lung is frequently found in cystic fibrosis patients, which worsens the prognosis.[3]

In HAP, the most serious complications are associated with ventilator-associated pneumonia (VAP), which occurs in patients mechanically ventilated through an endotracheal or tracheotomy tube for at least 48 h. Hence, VAP arises frequently in critically ill patients, and the risk of infection increases significantly with duration of ventilation and intensive care unit (ICU) stay. The mortality rate for VAP ranges from 24 to 76%.[4,5] VAP is often caused by pathogens such as *Staphylococcus aureus*, methicillin-resistant *Staphylococcus aureus* (MRSA), *Pseudomonas aeruginosa* and *Enterobacteriaceae* like *Klebsiella* spp. (particularly *Klebsiella pneumoniae* and *Klebsiella oxytoca*) and *Serratia* spp. Also other pathogens such as *Enterobacter* spp. (*Enterobacter aerogenes*), *Enterococcus* spp. but particularly fungi such as *Aspergillus* spp. and *Candida* spp.[4,6] are important causes of nosocomial infections and remain a serious challenge to modern medicine. *C. albicans*, which is a part of the normal human flora, but also an opportunistic pathogen, can be a serious threat to these patients. It is the predominant *Candida* species causing invasive candidiasis followed by *C. tropicalis*, *C. parapsilosis*, *C. krusei*, *C. lusitaniae* and *C. glabrata*.[7–11] Mycoses caused by *Candida* sp. range from thrush to invasive diseases like arthritis, osteomyelitis, endocarditis, endophthalmitis, meningitis or invasive candidiasis.[7] Such mycoses are often found in persons suffering from e.g. AIDS, cancer or after extensive treatment with antibiotics, corticoids and other immunosuppressive drugs.[1,12] At an advanced disease stage, VAP may develop to a multi-microbial hospital-associated infection that, according to recently published data, affects as many as 21–30% of all VAP cases.[6,13]

Early VAP diagnosis in sedated ventilated patients is very difficult since many of the typical symptoms of pneumonia will either be absent or impossible to detect. The currently applied diagnosis of VAP is time consuming and, in part, invasive due to the requirement of sputum, tracheal or bronchial secretion. Frequently, bronchoscopy, cultures of bronchoalveolar lavage (BAL), blood and serologic tests or even biopsies are needed.[14] BAL as a relatively invasive method is performed only under suspicion of pneumonia, while samples for microbiological diagnosis are collected at most twice a week. Selected samples are subjected to microbiological cultures followed by identification of the pathogens and antibiogram, which give information about antibiotic resistances and, in the case of a positive result, additionally for the next 24 h by detailed MALDI-TOF analyses of bacterial constituents (RNA, DNA, proteins). It is a major challenge that information on strain-specific antibiotic resistances are available at the earliest two days after the suspicion of pneumonia. In the meantime, a broad spectrum antibiotic therapy has to be applied, which might be disadvantageous since it may not adequately reduce a pathogen in some circumstances and by changing the normal bacterial flora it allows the pathogenic microorganisms to multiply or even lead to the development of antibiotic resistances, prolonging the ICU stay.

Therefore, it is of utmost interest to develop a method for the earlier detection of the outbreak of pneumonia in mechanically ventilated patients, preferably allowing the identification of the specific pathogens.

Monitoring the presence of pathogen-derived volatile metabolites in the alveolar air of mechanically ventilated patients seems to be a suitable method for this purpose, since it is non-invasive and the results are available immediately after measurement with no need for time-consuming cultivation of microorganisms. Moreover, breath-gas analysis already proved to be useful with critically ill patients,[15,16] children,[17] and even neonates.[18,19] In this respect, developments in analytical chemistry and concomitant improvement of sample preparation techniques such as needle trap devices[20–23] allow the detection of volatile organic compounds (VOCs) at the ppbv-pptv level. The great potential of the analysis of human breath for the diagnosis and monitoring of diseases has been demonstrated by the ^{13}C-urea breath test for the detection of *Helicobacter pylori* infection,[24,25] the nitric oxide test for asthma diagnosis[26,27] and the hydrogen breath test for the diagnosis of small intestinal bacterial overgrowth (SIBO) or fructose and lactose malabsorption.[28]

The first step toward an application of breath analysis for non-invasive diagnosis of microbial lung infections is the search for specific biomarkers of a certain microbial disease. Typically, this is done by the comparison of analyte concentration profiles obtained from healthy and diseased individuals. Nevertheless, the correct interpretation of breath-VOCs for clinical diagnosis is intricate due to multiple potential sources, such as food uptake, inspired exogenous xenobiotics (e.g. air pollutants, cigarette-smoke-derived compounds[77]) and endogenous sources. The last group comprises normal and pathological metabolic activities of organs, but also of pathogens themselves that might be highly elevated in the case of infections. Although the biochemical origin of breath-VOCs has not yet been sufficiently studied and underlying biochemical processes remain largely unknown, many compounds have been linked to diverse diseases. In addition, sensor technology has been used for the detection of microbial infections[29,30] by recording the total response of a sensor array. At this early stage of development, colorimetric sensors or electronic nose systems (EN) are often unspecific and do generate joint signal for compounds of the same chemical class.

Therefore, we postulate that *in vitro* experiments with human specimens (e.g. lung tissues), cell lines or microbial cultures applying highly specific and sensitive analytical techniques represent the most comprehensive approach to gain insight into the nature and the biochemical background of volatile metabolites, allowing appropriate interpretation of human breath data and the selection of suitable biomarkers.

Attempts to identify volatile microbial metabolites for detection and classification of virulent microorganisms have already been carried out. Encouraging results were reported, showing that some microorganisms have their own expression patterns of enzymes, producing a characteristic range of volatile metabolites.[31–34] However, the vast majority of studies on volatile organic compounds (VOCs) released from microorganisms included qualitative analyses only.[35–38] Promising from the diagnostic point of view might be the application of direct mass spectrometric techniques allowing real-time monitoring of the target analyte (potential marker). In this respect,

VOCs release from microorganisms was investigated also by means of real-time techniques, such as Selected Ion Flow Tube Mass Spectrometry (SIFT-MS)[39–45] and Proton Transfer Reaction Mass Spectrometry (PTR-MS).[46–48] However, especially in the case of quadrupole-PTR-MS the identification of VOCs is questionable since, apart from isobaric substances, several compounds may undergo fragmentation and contribute to product ions at a single mass-to-charge ratio, m/z.[49,50] The problems resulting from fragmentation under PTR-MS conditions have been discussed in details by others,[49–52] and special approaches for appropriate data interpretation have been proposed.[50] According to these works, the analytical conditions can be considerably affected by the settings of PTR-MS instruments and sample's parameters such as humidity[50] and CO_2 content.[51] However, with gas chromatography mass spectrometry (GC-MS) the unambiguous identification and quantification of more than 200 compounds in a single run is possible due to the chromatographic separation of each sample component prior to its detection. Convincing results were obtained by Scott-Thomas et al.,[53] where 2-aminoacetophenone (2-AA) was measured by GC-MS in cystic fibrosis patients (CF) as volatile biomarker produced by *P. aeruginosa*, confirming earlier studies.[54–57] Significantly higher levels of 2-aminoacetophenone were found in exhaled breath of cystic fibrosis patients colonized with *P. aeruginosa*, while the concentration of this metabolite was below the detection limit in both control groups (healthy subjects and CF patients colonized with other microbial species). However, it must be said that the large concentrations of 2-AA reported by Scott-Thomas et al. have not been seen in a combined SIFT-MS/GC-MS study of the volatiles emitted by genotypically different strains of *P. aeruginosa*.[44] Incidentally, in SIFT-MS investigations hydrogen cyanide (HCN) was investigated as a valuable biomarker of *P. aeruginosa* in the airways of CF patients.[58] Similarly, specific volatile biomarkers of *Aspergillus* spp. were detected in exhaled breath of tuberculosis patients colonized with *Mycobacterium tuberculosis*.[34,59–62]

Since most of the metabolic pathways in all microorganisms are primarily identical, the question might be whether discrimination between microbial species will ever be possible through analysis of volatile metabolites. To address this issue, *in vitro* experiments with microorganisms frequently found in VAP patients, i.e. *S. aureus, P. aeruginosa, C. albicans* and in CAP patients, i.e. *Streptococcus pneumoniae* and *Haemophilus influenzae*, were performed to investigate the release or consumption of volatile organic compounds (VOCs) from these species. A high number of diverse VOCs was generated by all investigated microorganisms in a concentration range varying from pptv to ppmv level. Clear differences in the species-specific profiles of VOCs production were observed, particularly with respect to aldehydes, which were only taken up by *P. aeruginosa* but released by *S. aureus* and to hydrocarbons, whereby entirely different VOCs of this group were released by *Streptococcus pneumoniae* and *Haemophilus influenzae*. Importantly, the *in vitro* experiments were performed under culture conditions and at microbial densities that are comparable to the *in vivo* situation in the lungs of VAP patients. Moreover, a significant release of certain metabolites was found as early as 1.5–3 h after inoculation of microorganisms, and their concentration profiles were very well correlated to the proliferation rate.

23.2 METHODOLOGY

23.2.1 Bacterial and fungal cultures

Staphylococcus aureus (ATCC 25923), *Pseudomonas aeruginosa* (ATCC 27853), *S. pneumoniae* (ATCC 49619, serotype 19F, isolated from the sputum of a 75-year-old male) and *Candida albicans* (Strain: CBS 5982) are standardized commercially available strains. *H. influenzae* were clinical isolates and a kind gift of Dr. Dorothea Orth from the Division of Hygiene and Medical Microbiology (Innsbruck Medical University). These microorganisms were deep frozen from the same stock culture and thawed and grown freshly on solid agar for each single experiment. *S. aureus*, *P. aeruginosa*, and *C. albicans* were stored for several days on Müller-Hinton agar plates.

The detailed description of the analytical protocol for the preparation of microbial cultures as well as counting of colony forming units (CFU) and optical density (OD_{600}) is given elsewhere.[63] Cultivation and headspace sampling were performed in a device of our own construction, which allows strictly controlled ventilation and VOC sampling as described in previous publications (Figure 23.1).[64–67] In each experiment, one sample containing medium with all supplements, but no bacteria or fungi, served as a control. Headspace samples for GC-MS analysis were taken 1.5, 3, 4.5, and 6 h after inoculation by *S. aureus,* 1.5, 2.25, 3, 3.75, 4.5, 5.25, 6, 24, 26, and 28 h for *P. aeruginosa,* 3, 3.7, 4.5, 6, and 7.5 h for *S. pneumoniae*, 1.5, 3, 4.5, 6, and 7.5 h for *H. influenzae* and 1.5, 3, 4.5, 6, 7.5, 26, and 28 h for *C. albicans*. These sampling time points were set according to the individual growth features of each organism.

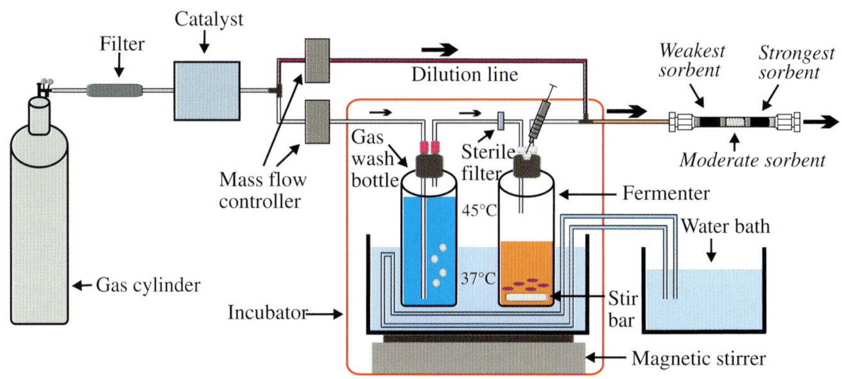

FIGURE 23.1

Overview of the system for incubation of microorganisms and headspace sampling. Carrier gas (synthetic air) was purified by a charcoal filter and a catalyst after which the main stream was divided into four independent streams adjusted by electronic mass flow controllers. The diluted headspace sample was adsorbed on a multibed sorption tube placed at the outlet of the system (at an ambient temperature of 20 °C).

23.2.2 Cultivation of microorganisms in a custom-made system for headspace sampling

Microorganisms were cultivated under strictly controlled aeration and temperature in a sophisticated custom-made system (Figure 23.1) allowing simultaneous precisely controlled headspace sample collection for subsequent GC-MS analysis. It was crucial to create culture conditions with sufficient oxygen supply and stable pH, and to avoid high shear forces. During experiments, the microorganisms were grown in suspension. Bacteria and fungi were cultivated in a 500 mL bottle (Schott AG, Mainz, Germany), which contained a magnetic stir bar with a Teflon surface (Bellco®), Vineland, NJ, Canada) that was driven by a magnetic stirrer (100 rpm). A sterile filter (Sartorius, Göttingen; Germany, hydrophobic PTFE filter, 0.2 μm pore size) was positioned at the air inlet of each culture vessel to avoid microbial contamination of the cultures. Culture vessels were placed in a water bath at a temperature of 36.5 °C. The temperature of the water bath for microbial cultures was controlled by a cooling circuit generated by a secondary water bath (Julabo, Seelbach, Germany) placed outside the incubator. To avoid condensation in transfer lines and cell culture vessels, the entire system was placed inside the incubator and held at 45 °C. The synthetic air (carrier gas), which was continuously purging the culture's headspace during microbial cultivation, had to be humidified to prevent evaporation and a concomitant change in ionic strength of the culture medium. This was done by stripping the deionised water from a gas wash bottle placed just before the culture vessel. All four independent flows of carrier gas (allowing sampling from four independent cultures) were precisely controlled by custom-made software using electronic mass flow controllers with magnetic valves (Bronkhorst, Ruurlo, Netherlands, Model: F-200CV-005-AAD-11-V Multi-Bus DMFC; measuring range 0.1–5 mL/min air).

Another important aspect was to avoid any contamination of a sample in the headspace and transfer lines. Hence, it was required to use synthetic air of high purity, i.e. 99.9995% (air 5.5, containing 5% CO_2, Linde, Stadl-Paura, Austria) for microbial cultures. The gas mixture was additionally purified by a Supelcarb™hydrocarbon trap (Supelco, Bellefonte, PA, USA) and by an inline-catalyst (Parker Zero Air Generator, Balston®, model: 75-83-220, Parker Hannifin Corporation, Haverhill, Massachusetts, USA). The contamination with laboratory indoor-air (resulting in unspecific background and irreproducible results) was prevented by sealing the atmosphere in a culture vessel with gas tight caps, ferrules and adapters at an inlet and an outlet either for aeration of cells or for VOCs sampling. For this purpose, Swagelok, Luer-lock or home-made female-male connections made of inert materials (Teflon) were used. The tightness of the system was routinely tested for all components and connections before the start of every *in vitro* experiment.

All parts of the system that were in contact with the headspace samples were made of inert materials to avoid contamination. Rubber seals, in particular, were not used because of the release of sulfur compounds (carbon disulfide, carbonyl sulphide, etc.). Similarly, all parts of the culture vessels were made of materials suitable for sterilization in a steam autoclave at 120 °C. Thus, connectors, transfer lines, caps and stir bars consisted of Teflon or an inert metal.

23.2.3 Preconcentration of VOCs on multibed tubes

Sampling on multibed sorption tubes filled with 25 mg Tenax TA (60/80 mesh), 35 mg Carboxen 569 (20/45 mesh) and 250 mg Carboxen 1000 (80/100 mesh) (arranged according to increasing sorption strength) provides high sensitivity of GC-MS analyses as shown in previous work.[64–67] The application of as strong adsorbents as carbon molecular sieves (CMS, such as Carboxen 569 and Carboxen 1000 constituting >90% of our tube's filling) yielded detection limits at the pptv level for almost all substances measured (Table 23.1). On the other hand, substantial water uptake from microbial headspace was observed for CMSs causing numerous analytical problems (e.g. decreasing adsorption efficiency, ice formation during cryo-focusing, oxidation of filament in MSD, etc.). Since commonly used techniques of water removal like drying agents and Nafion tubes result in substantial loss of low molecular polar compounds, and since dry purge of water after sample adsorption causes depletion of highly volatile VOCs, we decided to avoid excessive adsorption of water by reduction of the sample relative humidity by diluting the microbial headspace sample with dry air, as proposed by Gawrys et al.[68] For this purpose, two mass flow controllers (Bronkhorst, Ruurlo, Netherlands, measuring range: 2–100 mL/min, 0–5 mL/min) were placed inside the incubator. A dilution ratio of 1:8 (5 mL/min sample: 35 mL/min air) was optimal, providing the highest sampling efficiency (higher ratios resulted in purging of already adsorbed VOCs, while lower ratios did not reduce the adsorption of water sufficiently). Additionally, all transfer lines were made of inert material (Teflon) and kept at elevated temperature of 45 °C to prevent condensation prior to the sorption tube. Collected gas samples originating from microbial headspace had a volume of 200 mL, and the total flow through the sorption trap was 40 mL/min.

23.2.4 Thermal desorption

The injection of samples into the GC-MS was carried out by thermal desorption in a TDS3 unit equipped with a TDSA2 auto sampler (both from Gerstel, Mülheim an der Ruhr, Germany). During desorption, the carrier gas had a flow rate of 90 mL/min, whereby the initial temperature of 30 °C was elevated to 300 °C with a heating rate of 100 °C/min (held for 10 min in the splitless mode). Desorbed analytes were cryofocused at −90 °C by liquid nitrogen within the CIS-4 injector, which contained a glass liner filled with Carbotrap B (Gerstel, Mülheim an der Ruhr, Germany). Sample injection into the capillary column was induced by heating the CIS-4 injector at a rate of 12 °C/sec up to 320 °C (then hold for 2 min in the splitless mode).

23.2.5 Gas chromatography mass spectrometry (GC-MS) analysis

GC-MS analyses were carried out using a 6890N gas chromatograph equipped with a mass selective detector 5973N (both from Agilent Technologies, Waldbronn, Germany) operating at an electron impact ionization energy of 70 eV. Chromatographic separation occurred on a PoraBond Q capillary column 25 m×0.32 mm×5 μm

Table 23.1 Validation parameters for substances significantly released from bacterial and fungal cultures. Only calibrated substances are given, for which retention times were confirmed using reference materials. Ions chosen for the SIM mode are given for compounds of interest (100 ms dwell time for each m/z), otherwise data refer to measurements performed in the TIC mode. Detection limits (LOD) given in ppbv are bootsrtapped from the calibration series (described in Section 23.2.8) and do not refer to blank measurement.

	CAS	Compound	MW	t_R (min)	m/z (SIM)	LOD	R^2
Alcohols	67-56-1	Methanol	32	5.84	–	2.8	0.9575
	78-83-1	2-Methyl-propanol	74	30.39	74	4.0	0.9840
	123-51-3	3-Methyl-1-butanol	88	39.66	70, 55	5.9	0.9690
	64-17-5	Ethanol	46	12.55	–	41.2	0.9945
	71-36-3	1-Butanol	74	32.22	56, 31	3.9	0.9585
	71-23-8	1-Propanol	60	22.12	–	1.9	0.9870
Esters	141-78-6	Ethyl acetate	88	29.96	61	0.1	0.9978
	123-86-4	*n*-Butyl acetate	116	46.76	73, 56	0.2	0.9856
	123-92-2	Isopentyl acetate	130	51.68	70, 55	0.5	0.9979
	868-57-5	Methyl 2-methylbutyrate	116	44.93	56, 85	0.4	0.9426
	80-62-6	Methyl methacrylate	100	36.82	100	0.2	0.9956
	109-94-4	Ethyl formate	74	20.53	–	0.4	0.9921
	108-64-5	Ethyl isovalerate	130	50.96	70, 110	0.6	0.9866
	554-12-1	Methyl propionate	88	30.66	–	0.3	0.9762
	110-19-0	Isobutyl acetate	116	44.94	73	0.2	0.9921
	624-41-9	2-Methylbutyl acetate	130	51.45	70	0.7	0.9411
	109-60-4	Propyl acetate	102	37.13	73	0.4	0.9707
	105-54-4	Ethyl *n*-butanoate	116	46.09	–	0.2	0.9888
	105-68-0	Isoamyl propionate	144	56.67	–	0.3	0.9846

(Continued)

Table 23.1 (Continued)

	CAS	Compound	MW	t_R (min)	m/z (SIM)	LOD	R^2
Unsaturated hydrocarbons	821-95-4	1-Undecene	154	61.21	–	0.3	0.9916
	563-46-2	2-Methyl-1-butene	70	22.61	55, 70	0.4	0.9925
	513-35-9	2-Methyl-2-butene	70	23.60	55, 70	0.1	0.9981
	13688-67-0	1,10-Undecadiene	152	61.03	–	0.2	0.9732
	124-11-8	1-Nonene	126	54.28	55, 70, 126	0.2	0.9377
	872-05-9	1-Decene	140	58.63	55, 70	0.2	0.9842
	112-41-4	1-Dodecene	168	64.07	57, 70, 85	0.2	0.9867
	590-18-1	(Z)-2-Butene	56	13.55	56	1.3	0.9930
	624-64-6	(E)-2-Butene	56	13.90	56	1.3	0.9922
	115-11-7	Isobutene	56	12.56	41, 56	0.5	0.9939
	78-79-5	Isoprene	68	22.87	–	0.1	0.9926
	106-99-0	1,3-butadiene	54	12.94	54	0.5	0.9866
	563-45-1	3-Methyl-1-butene	70	20.65	–	0.3	0.9993
	646-04-8	(E)-2-Pentene	70	23.50	55, 70	0.0	0.9966
Alkanes	74-98-6	Propane	44	6.16	29, 44	0.0	0.9981
	106-97-8	Butane	58	13.70	43, 58	0.0	0.9994
	75-28-5	Isobutane	58	12.47	43	0.3	0.9960
	110-54-3	Hexane	86	33.24	–	0.0	0.9886
Ketones	67-64-1	Acetone	58	17.80	–	0.5	0.9960
	513-86-0	Acetoin	88	36.23	45, 88	3.6	0.9780
	116-09-6	Hydroxyacetone	74	30.92	–	36.2	0.8618
	821-55-6	2-Nonanone	142	60.28	–	0.2	0.9976
	107-87-9	2-Pentanone	86	36.43	86	0.2	0.9779

(Continued)

Table 23.1 (Continued)

	CAS	Compound	MW	t_R (min)	m/z (SIM)	LOD	R^2
	110-43-0	2-Heptanone	114	51.70	114	0.2	0.9626
	123-19-3	4-Heptanone	114	50.60	114	0.2	0.9701
	106-68-3	3-Octanone	128	56.78	–	0.3	0.9956
	78-93-3	2-Butanone	72	27.79	–	0.7	0.9987
	431-03-8	2,3-Butanedione	86	27.62	86	0.3	0.9787
VNCs	616-43-3	3-Methylpyrrole	81	43.06	–	0.7	0.9335
	109-97-7	Pyrrole	67	34.22	67	0.7	0.9768
VSCs	75-15-0	Carbondisulfide	76	18.02	76	0.2	0.9831
	75-18-3	Dimethylsulfide	62	18.93	47, 62	0.2	0.9892
	624-92-0	Dimethyldisulfide	94	37.66	94	0.3	0.9997
	3658-80-8	Dimethyltrisulfide	126	53.09	126	0.2	0.9591
	74-93-1	Methanethiol	48	10.30	47	1.2	0.9218
Aldehydes	100-52-7	Benzaldehyde	106	52.03	–	0.7	0.9968
	75-07-0	Acetaldehyde	44	8.92	29	2.7	0.9921
	78-85-3	Methacrolein	70	24.83	–	0.2	0.9995
	590-86-3	3-Methylbutanal	86	34.56	86	0.8	0.9697
	124-19-6	Nonanal	142	60.50	82, 98	0.2	0.9282
	66-25-1	Hexanal	100	45.72	56	0.3	0.9151
	123-38-6	Propanal	58	17.50	–	0.5	0.9983
	107-86-8	3-Methyl-2-butenal	84	41.24	84	0.3	0.9995
	107-02-8	Acrolein	56	16.52	56	0.1	0.9919
	4170-30-3	2-Butenal	70	30.87	70	0.1	0.9946

(Continued)

Table 23.1 (Continued)

	CAS	Compound	MW	t_R (min)	m/z (SIM)	LOD	R^2
	123-72-8	Butanal	72	27.35	–	0.2	0.9973
	78-84-2	2-Methylpropanal	72	25.26	72	0.2	0.9958
	124-13-0	Octanal	128	57.38	84, 100	0.2	0.9158
	497-03-0	(E)-2-Methyl-2-butenal	84	38.43	84	0.2	0.9986
	922-63-4	2-Ethylacrolein	84	33.67	84	0.4	0.9902
Acids	503-74-2	Isovaleric acid	102	45.17	60, 87	12.2	0.9799
	64-19-7	Acetic acid	60	21.19	45, 60	195.2	0.9341
Other	110-00-9	Furan	68	16.98	–	0.1	0.9957
	96-48-0	γ-Butyrolactone	86	46.47	–	14.9	0.9662
	109-92-2	Vinyl ethyl ether	72	20.89	–	0.2	0.9783

(Varian, Palo Alto, CA, USA). The following oven temperature program was used: initial 55 °C held for 6 min, then ramped 7 °C/min up to 97 °C (2 min), 2 °C/min to 110 °C (0 min), 5 °C/min to 130 °C (4 min), 5 °C/min to 160 °C (4 min), 4 °C/min to 230 °C (0 min) and 10 °C/min to 280 °C (4 min). The helium carrier gas had a constant flow rate of 1.1 mL/min. The mass spectrometer worked in a combined TIC/SIM mode. The TIC (total ion chromatogram), with an associated range set from m/z 20 to m/z 200, was used for the identification of potential target compounds. Additionally, most of compounds were quantified using SIM (selective ion monitoring) mode, with 100 ms dwell time for each ion monitored (see Table 23.1 for details). The SIM ions were chosen to ensure the best separation from neighbouring peaks; thus, they were not always the most intensive ions in the MS spectra of a particular compound. The chromatographic data processing was performed manually by the experienced analyst using the Agilent Chemstation Software (GC-MS Data Analysis from Agilent, Waldbronn, Germany).

23.2.6 Substance identification

Currently, the online computer comparison of the mass spectrum of an unknown analyte against a reference mass spectral library is used to identify the compound. The mass spectrum obtained is characteristic for a given compound; however, it is not unique—other compounds (especially isomers) could have very similar spectra. As a result, one peak in a chromatogram may be identified—with high probability—as few different compounds when based on spectra match only. Similarly, automated software (such as AMDIS) can ascribe one and the same substance to several peaks recorded on different places of the chromatogram. Therefore, we additionally used retention times based on calibration mixtures of the respective pure compounds. Peaks, for which proper identification was not possible (too low library match and no confirmation by retention time), are not investigated further.

23.2.7 Preparation of standard mixtures

The preparation of gaseous standards was performed by the evaporation of liquid compounds in 1 L glass bulbs (Supelco, Bellefonte, PA, USA). Each bulb was cleaned with methanol (Sigma-Aldrich, Steinheim, Germany), dried at 85 °C for at least 20 h with a continuous flow of purified air generated by a compressor (Breitfuss, Germany), rinsed with nitrogen 6.0 (i.e. 99.9999%, Linde, Stadl-Paura, Austria) for several minutes and subsequently evacuated using a vacuum pump (Vacuubrand, Wertheim, Germany) for at least 3 h. Liquid standards (1–10 μL, according to the desired concentration) were injected through a septum using a glass HPLC syringe (Hamilton, Bonaduz, Switzerland). After the evaporation of standards, the glass bulb was filled with nitrogen 6.0 to equalize the pressure to ambient pressure. Subsequently, the appropriate volume of the standard mixture was transferred by a gas-tight syringe (Hamilton, Bonaduz, Switzerland) into a Tedlar® bag (SKC 232 Series, Eighty Four, PA, USA), previously filled with 2 L of nitrogen (additionally purified with CMS) humidified by stripping the deionised water in a closed vessel. The trace compounds in 200 mL of this prepared gas mixture were adsorbed onto a multibed sorption tube under the same sampling conditions as described above for headspace sampling (with dilution 1:8).

23.2.8 Statistics and determination of detection limits

Statistical significance was calculated by the Kruskal-Wallis test, which is a non-parametric test to compare samples from two or more groups of independent observations.[69] This test was selected because it does not require the groups to be normally distributed and is more stable to outliers. P-values <0.05 were considered as significant. To visualize the data, results are plotted as median values with 5, 25, 75, and 95 percentiles. CFU counts are presented as mean values $\pm$ standard deviation (SD).

The limit of detection (LOD), sometimes called the minimum detectable level (MDL), is typically determined from repeated analysis of a blank and most often extrapolated to the signal-to-noise ratio S/N of 3. Such an approach may lead to too low estimates and hence to incorrect LOD values. This is particularly the case if very low concentrations of injected substances do not generate the expected (linear) detector's response, reflected in a poor correlation coefficient of the calibration curve ($r^2 < 0.9$), although there is only little (or even no) noise in the blank. Therefore, we alternatively determined the LOD individually for each substance from bootstrapping the regression line from the series of calibration measurements performed following the above-described protocol. The method of bootstrapping assumes that the calibration follows a linear relation or at least that the deviation from the linear function is smaller than the error of a measurement. In this approach, the offset parameter b of a linear function ($y = ax + b$) is important as an indicator of the noise. Taking 99% percentile of b, it can be estimated that the probability of appearance of a signal greater than noise, when the compound is not present in the sample, is less than 1%. This method is valid for various analytical techniques, and the LODs determined for such instruments as PTR-MS, SIFT-MS, SPME-GC-MS, and TD-GC-MS can be compared. However, the noise level of a measurement is not independent of signal size. Therefore, higher LODs are observed when standards of higher concentration are used for calibration (exemplified here by alcohols and acids which need to be prepared at higher concentrations due to absorption in the Tedlar® material of which sampling bags are composed).

23.3 RESULTS

Although complex gas mixtures can be analysed by several mass spectrometric techniques (like PTR-MS, SIFT-MS), the gas chromatography mass spectrometry (GC-MS) is still the gold standard amongst analytical methods in this field. It is mostly due to separation of a complex sample into individual components prior to detection allowing unequivocal identification of sampled VOCs. The identification and quantification of compounds that cannot be measured by PTR quadruple mass spectrometry (e.g. alkanes and compounds of similar proton affinity to water) is an additional advantage over direct MS techniques.

23.3.1 Limits of detection

Although sample preconcentration on as strong adsorbents as carbon molecular sieves (CMS, such as Carboxen 569 and Carboxen 1000 constituting >90% of our tube's filling) must be optimized to avoid excessive water uptake, it ensures very high sensitivity of subsequent GC-MS analysis. The above described protocol can achieve LODs at the pptv level for most substances, while the lowest LODs were observed for hydrocarbons (23 pptv for butane, 43 pptv for hexane) but also esters (86 pptv for ethyl acetate), aldehydes (118 pptv for 2-butenal), ketones (151 pptv for 4-heptanone) and sulfur-containing compounds (VSCs) (181 pptv for dimethylsulfide) (see Table 23.1 for more details). Importantly, all detection limits presented here were determined not from extrapolation of the signal-to-noise ratio to the value of $S/N = 3$ (often revealing too low values, especially .when "noise" refers to blank analysis) but individually for each analyte from bootstrapping the regression line from the series of calibration measurements performed with the above-described protocol. Consequently, the noise for LOD calculation concerns the entire analytical method (instead of pure blank measurements).

23.3.2 Linearity and precision of the sampling method

Pilot experiments with tryptic soy broth culture medium were performed to test the linearity of the sampling method (adsorption on multibed sorption tubes) and to detect a possible alteration of the applied medium headspace samples due to prolonged ventilation of cell culture vessels. First, all background peaks (impurities, artefacts, etc.) and the medium-derived compounds were determined. Thereby, 37 of a total 376 compounds were found regularly in every medium headspace, and 14 of them showed clear linearity with correlation coefficients higher than 0.9 (Table 23.2). Additionally, the so-called *matrix effect*, i.e. the existence of compounds at considerable concentration in the medium headspace not related to microbial activity (e.g. aromatics, methylated pyrazines), plays an important role in trace gas analysis. Consequently, very low levels of some VOCs released by microorganisms had to be determined in the complex matrix. Even though the headspace of the medium control showed a complex composition, linearity of the applied preconcentration method was observed over a wide volume range up to 3000 mL of the adsorbed sample. Naturally, linear relationships between collected headspace volume and measured VOCs concentration is not expected for analytes present at extremely low levels, often below detection limits, which are detectable only for large volumes collected. According to the results obtained, there was no risk of loss of analytes due to saturation of the multibed sorption tube during sample collection, even for the largest volumes of headspace, when a dry dilution technique was applied to avoid water uptake. Furthermore, the small relative standard deviations (RSD < 10% for analytes found above LOD) testify the good repeatability and good quality of the performed *in vitro* measurements with the system developed for microbial cultivation and headspace collection. Considerable deviations from low RSDs are observed for the highest sample volumes (i.e. 2000 mL and 3000 mL, see Figure 23.3), which is the consequence of a change in the composition of the medium's headspace when long adsorption time is needed

Table 23.2 Results of the linearity test for the applied sampling method (adsorption on multibed sorption tubes). Mean concentrations [ppbv] and standard deviations (%) were determined based on three independent measurements for each volume of medium headspace samples. Correlation coefficients [concentration = *f* (volume)] were determined for medium samples. Collected volumes were: 50, 100, 200, 300, 500, 1000, 2000, and 3000 mL of medium headspace. All measurements were performed in the TIC mode.

Compound	CAS	R^2 in medium	Mean 50 mL (ppbv)	RSD (%)	Mean 100 mL (ppbv)	RSD (%)	Mean 200 mL (ppbv)	RSD (%)	Mean 300 mL (ppbv)	RSD (%)	Mean 500 mL (ppbv)	RSD (%)	Mean 1000 mL (ppbv)	RSD (%)	Mean 2000 mL (ppbv)	RSD (%)	Mean 3000 mL (ppbv)	RSD (%)
2-Methyl-pyrazine	109-08-0	0.9966	2.3	8.4	6.6	2.4	15.1	4.0	24.2	1.9	41.6	0.7	85.1	1.1	161.5	8.7	237.2	6.8
2,5-Dimethyl-pyrazine	123-32-0	0.9893	6.2	7.2	12.4	2.7	25.6	0.7	55.9	3.3	96.0	1.2	200.9	1.5	325.1	12.9	464.5	8.7
1-Butanol	71-36-3	0.9878	<LOD	<LOD	7.5	3.3	14.6	3.1	19.4	0.7	31.7	2.5	56.8	1.4	113.1	17.3	157.7	12.3
2-Propanol	67-63-0	0.9868	<LOD	<LOD	2.6	8.9	7.9	2.9	10.4	2.8	19.8	0.6	41.2	3.8	84.5	19.9	117.4	11.7
1-Propanol	71-23-8	0.9794	13.3	26.7	20.0	1.8	42.3	1.8	55.9	2.2	93.5	0.7	177.0	4.8	307.6	18.7	455.0	18.8
Acetone	67-64-1	0.9705	64.2	1.4	96.4	69.9	296.7	5.6	682.2	1.6	1058.5	0.8	1843.0	1.9	2515.0	30.6	4063.4	7.9
2-Butanone	78-93-3	0.9650	4.8	2.7	9.6	0.7	21.9	6.6	41.0	3.0	64.6	1.6	105.6	3.2	211.5	30.5	311.8	22.7
Methyl methacrylate	80-62-6	0.9649	<LOD	<LOD	3.6	13.7	9.9	2.6	17.0	5.0	27.4	1.2	47.6	1.4	104.0	28.0	153.0	25.0
Benzaldehyde	100-52-7	0.9594	2.9	1.8	6.6	4.8	13.0	1.9	19.0	5.5	33.6	2.2	75.2	4.5	116.8	27.6	142.3	13.6
Ethyl acetate	141-78-6	0.9541	0.2	26.1	0.9	6.5	1.9	12.8	2.1	2.6	3.6	5.1	6.3	4.0	14.5	27.7	22.5	30.3
3-Methyl-2-butenal	107-86-8	0.9388	<LOD	<LOD	0.4	11.6	1.3	6.5	2.2	9.4	3.8	5.2	5.3	5.5	11.4	35.7	14.0	27.1
Toluene	108-88-3	0.9319	<LOD	<LOD	3.0	30.3	11.9	7.0	13.5	4.9	22.1	1.9	37.1	1.5	92.9	40.9	131.5	34.4

(Continued)

Table 23.2 (Continued)

Compound	CAS	R^2 in medium	Mean 50 mL (ppbv)	RSD (%)	Mean 100 mL (ppbv)	RSD (%)	Mean 200 mL (ppbv)	RSD (%)	Mean 300 mL (ppbv)	RSD (%)	Mean 500 mL (ppbv)	RSD (%)	Mean 1000 mL (ppbv)	RSD (%)	Mean 2000 mL (ppbv)	RSD (%)	Mean 3000 mL (ppbv)	RSD (%)
2-Methyl-propanal	78-84-2	0.9230	187.9	3.5	282.6	7.0	634.5	5.6	636.4	9.4	1043.5	2.9	1416.1	2.9	3027.0	40.3	4271.7	34.9
(Z)-2-Methyl-2-butenal	1115-11-3	0.9107	11.0	8.0	3.0	9.8	52.0	78.4	119.7	5.8	194.0	2.4	254.5	5.7	606.6	48.2	829.4	37.6
Ethanol	64-17-5	0.8977	<LOD	<LOD	102.6	14.1	306.2	17.5	311.4	26.9	523.4	37.1	1091.9	0.8	1519.7	25.0	1916.0	45.1
Methacrolein	78-85-3	0.8570	0.6	14.7	2.3	10.5	6.6	14.2	6.0	15.7	10.7	4.6	13.2	2.0	35.7	54.9	44.2	51.2
Acetic acid	64-19-7	0.8511	<LOD	<LOD	<LOD	<LOD	<LOD	<LOD	<LOD	<LOD	<LOD	<LOD	<LOD	<LOD	203.9	44.3	390.9	41.7
2-Ethylacrolein	922-63-4	0.8343	<LOD	<LOD	<LOD	<LOD	2.0	5.8	1.7	35.8	3.9	7.7	4.9	15.1	15.9	49.8	13.4	41.6
Ethyl formate	109-94-4	0.7188	<LOD	<LOD	<LOD	<LOD	<LOD	<LOD	<LOD	<LOD	<LOD	<LOD	<LOD	<LOD	<LOD	<LOD	<LOD	<LOD
(Z)-2-Butene	590-18-1	0.7144	<LOD	<LOD	<LOD	<LOD	<LOD	<LOD	<LOD	<LOD	3.5	11.8	−3.5	0.0	39.9	46.4	25.8	101.0
3-Methylbutanal	590-86-3	0.6637	92.2	3.6	69.2	23.3	323.8	5.6	333.0	2.8	166.6	9.1	132.7	10.7	431.1	104.6	783.1	55.3
Dimethyl-disulfide	624-92-0	0.6483	4.3	14.1	4.7	13.9	12.0	31.4	2.2	25.0	3.1	4.2	5.0	6.7	31.1	109.6	36.1	75.2
(E)-2-Butene	624-64-6	0.6002	<LOD	<LOD	<LOD	<LOD	<LOD	<LOD	<LOD	<LOD	3.7	17.2	6.5	22.5	33.2	74.1	14.2	105.0
Methanethiol	74-93-1	0.5168	274.8	58.3	107.6	17.5	149.0	16.6	142.1	55.5	98.0	15.9	168.8	8.1	627.1	90.9	2490.6	145.6
(E)-2-Methyl-2-butenal	497-03-0	0.4624	<LOD	<LOD	<LOD	<LOD	<LOD	<LOD	<LOD	<LOD	0.2	9.3	0.3	18.4	1.1	74.7	1.6	66.6
Ethyl isovalerate	108-64-5	0.4420	<LOD	<LOD	<LOD	<LOD	<LOD	<LOD	<LOD	<LOD	<LOD	<LOD	<LOD	<LOD	<LOD	<LOD	<LOD	<LOD

(Continued)

Table 23.2 (Continued)

Compound	CAS	R^2 in medium	Mean 50 mL (ppbv)	RSD (%)	Mean 100 mL (ppbv)	RSD (%)	Mean 200 mL (ppbv)	RSD (%)	Mean 300 mL (ppbv)	RSD (%)	Mean 500 mL (ppbv)	RSD (%)	Mean 1000 mL (ppbv)	RSD (%)	Mean 2000 mL (ppbv)	RSD (%)	Mean 3000 mL (ppbv)	RSD (%)
Butane	106-97-8	0.4395	2.7	54.8	4.2	5.3	9.5	13.6	0.0	0.0	0.9	89.7	1.3	33.3	13.6	156.3	16.8	136.2
Isovaleric acid	503-74-2	0.4391	<LOD	<LOD	<LOD	<LOD	<LOD	<LOD	<LOD	<LOD	<LOD	<LOD	<LOD	<LOD	<LOD	<LOD	24.9	109.0
Acetaldehyde	75-07-0	0.3373	117.0	51.0	138.1	20.3	424.6	19.6	563.3	3.4	751.7	4.2	477.9	125.3	1286.6	11.0	373.0	126.8
2-Methyl-1-butene	563-46-2	0.3114	<LOD	<LOD	<LOD	<LOD	<LOD	<LOD	<LOD	<LOD	<LOD	<LOD	<LOD	<LOD	<LOD	<LOD	<LOD	<LOD
Dimethyl-sulfide	75-18-3	0.1173	<LOD	<LOD	<LOD	<LOD	<LOD	<LOD	<LOD	<LOD	<LOD	<LOD	<LOD	<LOD	3.2	231.3	<LOD	<LOD
Propanal	123-38-6	−0.0154	0.7	19.7	1.1	10.5	2.7	12.9	3.2	4.7	5.5	3.9	7.0	15.8	6.1	180.9	<LOD	<LOD
Propene	115-07-1	−0.0346	<LOD	<LOD	<LOD	<LOD	<LOD	<LOD	<LOD	<LOD	<LOD	<LOD	<LOD	<LOD	<LOD	<LOD	<LOD	<LOD
Butanedione	431-03-8	−0.1609	<LOD	<LOD	2.4	182.7	15.5	205.4	<LOD	<LOD	<LOD	<LOD	<LOD	<LOD	<LOD	<LOD	<LOD	<LOD
n-Butyl acetate	123-86-4	−0.3360	<LOD	<LOD	<LOD	<LOD	<LOD	<LOD	<LOD	<LOD	<LOD	<LOD	<LOD	<LOD	<LOD	<LOD	<LOD	<LOD
Propane	74-98-6	−0.5153	2.3	78.3	4.0	38.6	9.7	29.1	3.4	36.0	2.2	31.6	<LOD	<LOD	1.2	125.4	<LOD	<LOD
Acetoin	513-86-0	−0.0424	<LOD	<LOD	<LOD	<LOD	<LOD	<LOD	<LOD	<LOD	3.8	1.0	8.2	9.9	<LOD	<LOD	<LOD	<LOD

(3000 mL of sample required 10 h of adsorption time with the applied maximum flow rate through culture vessel of 5 mL/min).

23.3.3 Reduction of the matrix effect

The aforementioned lower reproducibility for collection of large headspace volumes (2000, 3000 mL) requiring very long sampling time inevitably results from alterations in the composition of medium headspace. Therefore, investigations into the stability of headspace samples were performed. For this purpose, the headspace above three independent medium controls was purged with the continuous flow of carrier gas (synthetic air 5.6 additionally purified on carbon filter and catalyst) at the rate of 5 mL/min over one day (what refers to approx. 7000 mL of air purged). The results obtained showed that only a few compounds exhibited stable concentrations, whereas the concentrations of most VOCs decreased with time when the headspace of medium was continuously purged. Moreover, there was a tendency for the more hydrophobic (less soluble

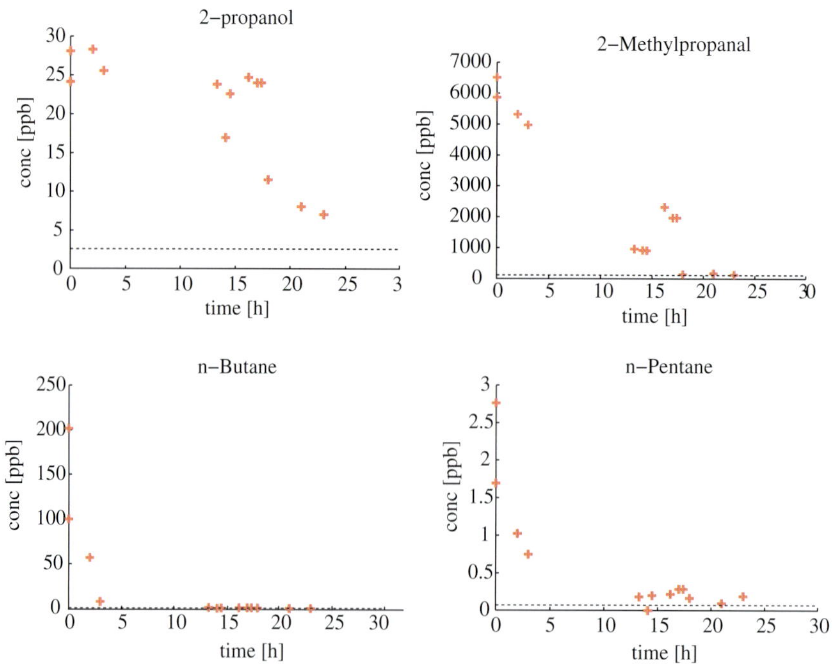

FIGURE 23.2
Stability test for the headspace of medium controls under continuous flow of carrier gas (5 mL/min). Three independent medium controls were used to verify the reproducibility of the results. The dotted horizontal line indicates the limit of detection (2.6 ppbv for 2-propanol in TIC mode; 0.7 ppbv for 2-methylpropanal in TIC mode, 0.024 ppbv for *n*-butane in SIM mode and 0.07 ppbv for *n*-pentane in SIM mode). The more hydrophobic molecules are more quickly purged from the medium headspace.

in water) compounds to be removed (purged out) faster from the medium sample than the hydrophilic compounds. For example, *n*-butane was completely purged from medium within 3 h, whereas hydrophilic 2-propanol was not completely removed from the headspace sample even after 24 h of purging (see Figure 23.2).

These results demonstrate that it is possible to decrease the background originating from the culture medium. Consequently, the matrix effect for determination of trace levels of pathogen-derived metabolites in microbial headspace can be considerably reduced. Therefore, such a "purification step" of the medium has been included in the analytical protocol for *in vitro* experiments. For this purpose, the culture medium was placed in the sampling system of our apparatus and purged at a flow rate as high as 100 mL/min over 24 h, i.e. starting one day before the actual experiment. The flow rate of carrier gas was reduced to 5 mL/min (equal to sampling flow) for 2 h before *in vitro* experiments to achieve steady state in the analyte's partition between liquid and gas phase. Otherwise VOCs accumulation in the headspace was observed in result of flow reduction. This purification procedure minimised the very high background originating from the medium (see Figure 23.3), which had been found to be the main problem in similar investigations of VOC release from lung cancer cells in our earlier studies.[64–67]

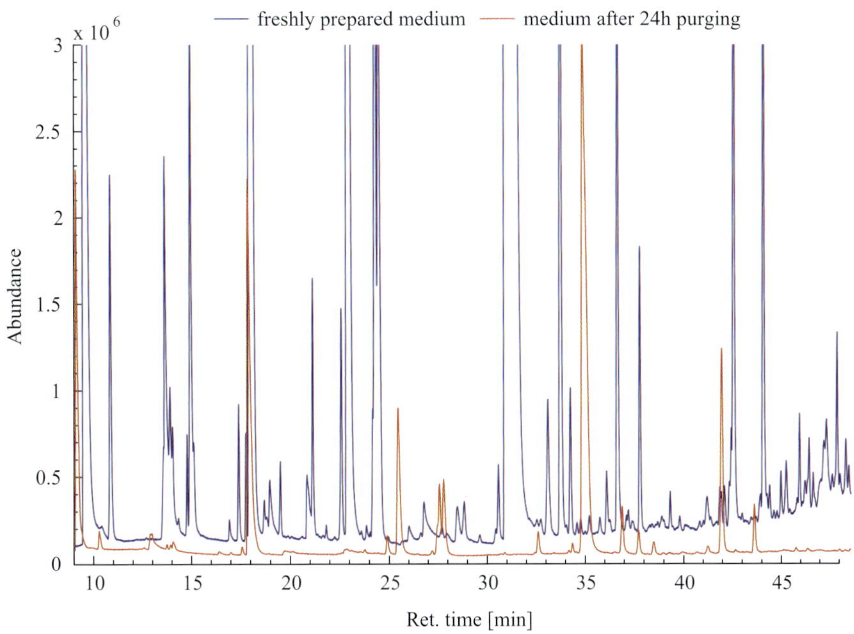

FIGURE 23.3
Decrease of background for medium control. The 24 h long purging of the medium's headspace with the carrier gas flow rate of 100 mL/min results in much lower background (red line) as compared to a headspace of the freshly prepared medium (blue line).

23.3.4 Proliferation of microorganisms

The initial microbial load in 100 mL culture medium after standardized inoculation was 10^5 CFU/ml. Importantly, bronchoscopic BAL studies typically use a diagnostic threshold of 10^4 or 10^5 CFU/ml to define both the presence of pneumonia and the etiologic pathogen.[70] Naturally, the real number of microorganisms in lungs is higher since a 10-fold dilution of BAL samples and the large surface area of the alveolar zone have to be considered. As expected from the number of microbes at time zero in our experiments, which was too low to induce turbidity, the optical density, OD_{600}, started to increase after 3 h (4.5 h for *C. albicans* and *H. influenzae*) of growth, while CFU counts began to increase after 1.5 h (3 h for *C. albicans*) (Figure 23.4).

23.3.5 Volatile metabolites released by microorganisms

The choice of appropriate analytical conditions described in previous paragraphs achieves a low background (hence reduction of matrix effect) for culture medium, good reproducibility and high sensitivity of the *in vitro* experiments. Consequently, a large number of diverse VOCs was seen to be released by each virulent microorganism in concentrations ranging from pptv to ppmv (see Tables 23.3–23.5), whereby:

- 32 VOCs from *S. aureus*, and benzaldehyde was degraded;
- 37 VOCs from *P. aeruginosa*, while additional 12 VOCs were consumed comprising 2,3-butanedione and 11 aldehydes;
- 34 VOCs from *S. pneumoniae*, with 3-methylbutanal and hexanal consumed;
- 28 VOCs from *H. influenzae*, whereby no uptake of analytes was observed;
- 26 VOCs from *C. albicans*, with degradation of 9 additional substances of diverse chemical classes (2 hydrocarbons, 1 ketone, 3 VSCs, and 3 aldehydes).

The comprehensive discussion of VOCs metabolized by bacterial species investigated in our *in vitro* study is given elsewhere.[63] In the following sections, the release and degradation of volatile substances from *C. albicans* only will be discussed in details.

Amongst 26 VOCs released by *C. albicans*, five aldehydes, four alcohols, nine esters, four ketones, two hydrocarbons, vinyl ethyl ether, and only one VSC (dimethylsulfide) had significantly elevated concentrations compared to control samples. The consumed VOCs comprised three aldehydes, three sulfur-compounds, and two hydrocarbons.

23.3.5.1 Alcohols

The current experiments revealed that under the chosen conditions *C. albicans* released four alcohols, 3-methyl-1-butanol, 2-methyl-1-propanol, 1-propanol and ethanol, all in very high concentrations with maximum levels measured after 28 h (see Table 23.3 and Figure 23.5a–c), except for ethanol that reached the plateau level after 7.5 h. Amongst them, 1-propanol was released exclusively by *C. albicans*,

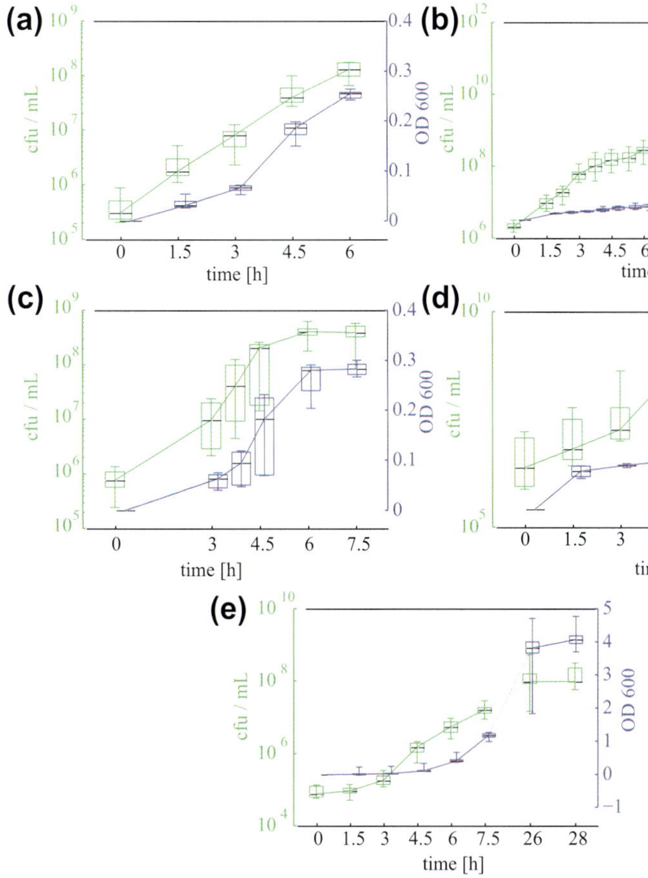

FIGURE 23.4

Growth curves of *Staphylococcus aureus* (a), *Pseudomonas aeruginosa* (b), *Streptococcus pneumoniae* (c), *Haemophilus influenzae* (d), and *Candida albicans* (e). Green boxes are the colony forming units (CFU/mL) and blue boxes are the optical densities (OD_{600}).

whereas the other three alcohols were released at least by one of the previously tested bacterial species (ethanol by *S. aureus, S. pneumoniae*, and *P. aeruginosa*, 3-methyl-1-butanol by *S. aureus* and *P. aeruginosa*, 2-methyl-1-propanol by *S. aureus*). The aforementioned alcohols were significantly released at early time points, whereby the concentration of ethanol—absent in medium control—increased immediately after fungal inoculation, yielding over 1400 ppbv after only 1.5 h of incubation and a maximum level of >26,000 ppbv after 7.5 h. Although the remaining alcohols were found at lower levels than ethanol, their concentrations were still considerably higher than those of other released compounds. In this respect, 3-methyl-1-butanol reached

Table 23.3 VOCs released or taken up by *C. albicans*. Results are presented as median concentrations [ppbv]. Values significantly higher or lower in fungi headspace samples than in medium controls ($p < 0.05$ for Kruskal-Wallis test) are indicated in italics and bold. Samples of medium headspace were collected at several different time points from inoculation at 0 h up to 28 h of microbial growth. Analytes semi-quantified to the peak area (instead of concentration) are denoted with #.

Class	Compound	CAS	Median Concentration (Respectively Peak Area)							
			Medium ($n = 15$)	1.5 ($n = 6$)	3 ($n = 6$)	4.5 ($n = 6$)	6 ($n = 6$)	7.5 ($n = 6$)	26 ($n = 4$)	28 ($n = 3$)
			Metabolites released from fungal cultures							
Alcohols	Ethanol	64-17-5	0.0	1442.0	2016.0	2900.0	12891.0	26228.0	23251.0	25168.0
	1-Propanol	71-23-8	1.3	1.3	1.3	3.7	50.4	201.4	240.6	267.2
	2-Methyl-1-propanol	78-83-1	4.5	4.5	4.5	4.5	4.5	154.7	1540.0	1833.0
	3-Methyl-1-butanol	123-51-3	2.3	2.3	2.3	25.0	227.8	490.2	2884.0	3569.0
Alde-hydes	Acetaldehyde	75-07-0	324.7	852.8	1405.0	1834.0	2477.0	4309.0	3397.0	3663.0
	Propanal	123-38-6	1.8	4.0	10.4	19.4	21.0	17.7	10.0	9.8
	2-Methylpropanal	78-84-2	113.0	164.9	202.9	328.0	1216.0	2792.0	337.0	363.0
	Methacrolein	78-85-3	1.1	1.3	1.4	2.0	5.9	6.2	1.8	2.0
	3-Methylbutanal	590-86-3	151.3	182.5	193.2	259.0	418.2	309.0	110.2	115.8
Esters	Ethyl formate	109-94-4	0.0	0.0	0.0	0.0	0.4	5.4	7.7	8.6
	Ethyl acetate	141-78-6	0.0	0.9	1.4	2.2	9.7	29.7	614.2	645.5
	Propyl acetate	109-60-4	0.2	0.2	0.2	0.2	0.2	0.2	1273.0	1402.0
	Isobutyl acetate	110-19-0	0.0	0.0	0.0	0.0	0.0	0.0	6.9	5.0
	2-Methylbutyl acetate	624-41-9	0.0	0.0	0.0	0.0	0.0	0.0	14.6	17.4

(Continued)

Table 23.3 (Continued)

Class	Compound	CAS	Median Concentration (Respectively Peak Area)							
			Medium ($n = 15$)	1.5 ($n = 6$)	3 ($n = 6$)	4.5 ($n = 6$)	6 ($n = 6$)	7.5 ($n = 6$)	26 ($n = 4$)	28 ($n = 3$)
	Isopentyl acetate	123-92-2	0.0	0.0	0.0	0.0	0.0	0.7	33.5	37.8
	Isopentyl propionate	105-68-0	0.0	0.0	0.0	0.0	0.0	0.0	0.2	0.2
	Ethyl *n*-butanoate	105-54-4	0.0	0.0	0.0	0.0	0.0	0.0	4.8	5.9
	Isopentyl *n*-butanoate #	106-27-4	0.0	0.0	0.0	0.0	0.0	0.0	6.20×10^5	7.70×10^5
Ether	Vinyl ethyl ether	109-92-2	0.0	0.0	0.4	0.8	8.6	41.5	16.4	13.9
HC	2-Methyl-2-butene	513-35-9	0.0	0.0	0.0	0.0	0.0	0.0	0.1	0.2
	(E)-2-Pentene	646-04-8	0.0	0.0	0.0	0.0	0.0	0.0	0.4	1.0
Ketones	Acetoin	513-86-0	2.0	2.0	2.0	2.0	2.0	12.1	3.8	4.9
	2-Pentanone	107-87-9	0.0	0.1	0.7	0.6	1.1	1.3	2.3	1.6
	4-Heptanone	123-19-3	0.0	0.0	0.3	1.0	1.1	1.3	0.2	0.2
	Methyl vinyl ketone #	78-94-4	0.0	2.40×10^5	2.30×10^5	2.80×10^5	2.20×10^5	0.0	0.0	0.0
VSCs	Dimethylsulfide #	75-18-3	0.0	1.4×10^3	6.90×10^2	2.0×10^3	3.0×10^3	0.0	0.0	0.0

(Continued)

Table 23.3 (Continued)

Class	Compound	CAS	Median Concentration (Respectively Peak Area)							
			Medium ($n = 15$)	1.5 ($n = 6$)	3 ($n = 6$)	4.5 ($n = 6$)	6 ($n = 6$)	7.5 ($n = 6$)	26 ($n = 4$)	28 ($n = 3$)
			Metabolites taken up (consumed) by fungal cultures							
Alde-hydes	Butanal	123-72-8	15.7	6.8	5.7	3.4	5.8	2.6	1.0	2.6
	Hexanal #	66-25-1	9.0×10^4	5.30×10^4	5.20×10^4	4.30×10^4	4.60×10^4	1.50×10^4	2.70×10^4	4.50×10^4
	Benzaldehyde	100-52-7	175.8	154.0	152.7	128.8	52.2	9.7	6.1	6.7
HC	(Z)-2-Butene #	590-18-1	6.1×10^3	8.1×10^3	7.2×10^3	8.1×10^3	8.5×10^3	0.0	0.0	0.0
	n-Butane	106-97-8	0.1	0.1	0.1	0.2	0.2	0.0	0.0	0.0
Ketone	2,3-Butanedione	431-03-8	45.9	55.5	50.5	32.2	14.8	18.1	13.6	13.1
	Methanethiol	74-93-1	12.2	12.5	10.2	2.5	1.7	0.7	0.7	0.0
VSCs	Carbon disulfide	75-15-0	0.5	0.6	1.0	0.6	0.0	0.0	0.0	0.0
	Dimethyldisulfide	624-92-0	0.4	0.4	0.1	0.1	0.1	0.1	0.1	0.1

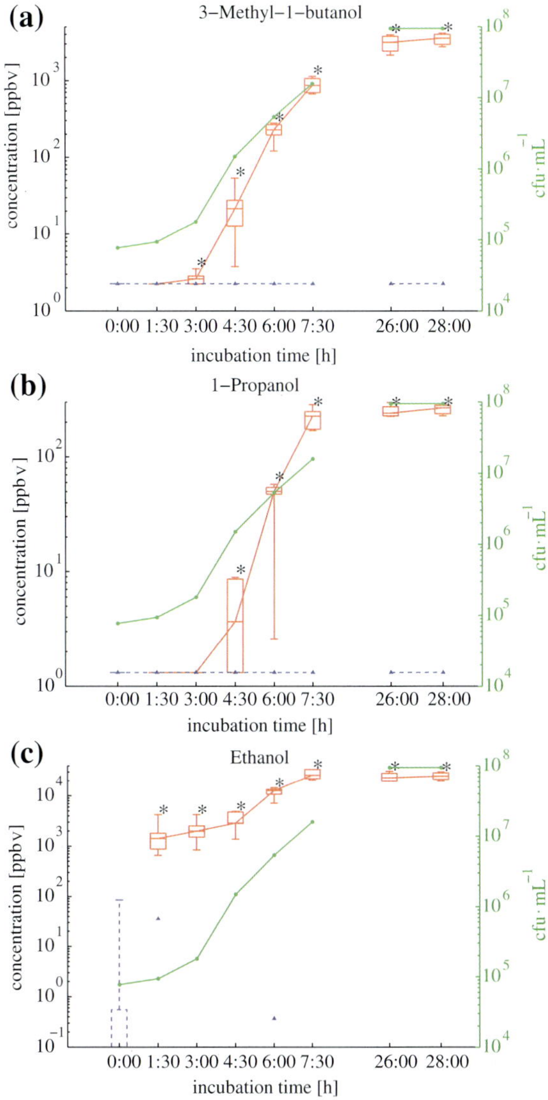

FIGURE 23.5

Exemplary alcohols released by *C. albicans*: 3-methyl-1-butanol (a), 1-propanol (b), and ethanol (c). Determined median concentrations [ppbv] for headspace over fungal cultures with 25th and 75th percentiles [ppbv] are given as red boxes with whiskers indicating 5th and 95th percentiles, while the blue box and the dotted blue line with triangles indicates the pattern for the medium control. Significant VOCs release ($p < 0.05$ for Kruskal-Wallis test) is labelled with asterisk. The green line with round points denotes the proliferation rate expressed in colony forming units [CFU/mL].

a maximum median concentration of 3569 ppbv (first significant release after 4.5 h), 2-methyl-1-propanol of 1833 ppbv (first significantly increased release after 7.5 h), and 1-propanol of 267 ppbv (first significant release after 4.5 h).

23.3.5.2 Aldehydes

Aldehydes constitute an important class of compounds produced, but also consumed by *C. albicans*, whereby five aldehydes comprising propanal, acetaldehyde, methacrolein, 2-methylpropanal, and 3-methylbutanal were released and other three (butanal, hexanal, benzaldehyde) were taken up. The highest concentrations of aldehydes were found after 7.5 h of proliferation for acetaldehyde, 2-methylpropanal, and methacrolein and at 6 h of proliferation for propanal and 3-methylbutanal. Intriguingly, the concentrations of all aldehydes had substantially decreased in the overnight experiments, i.e. after 26 h of microbial growth, which is clearly exemplified by 2-methylpropanal (Figure 23.6a). The most abundant metabolite of this group was acetaldehyde, which was significantly released immediately after inoculation, reaching over 850 ppbv at 1.5 h (in comparison to 320 ppbv found in control medium) and the maximum concentration of 4300 ppbv at 7.5 h of proliferation (Figure 23.2b). Similarly active was the metabolism of 2-methylpropanal, with its first significant release after 3 h and maximum concentration of ~2800 ppbv after 7.5 h of microbial growth. Relatively weaker release was observed for propanal and methacrolein, which reached maximum concentrations of 21 ppbv and 6.2 ppbv, respectively.

All these aldehydes were released by at least one of the previously investigated bacterial species *S. aureus*, *S. pneumoniae* and *H. influenzae*, apart from *P. aeruginosa*, for which only uptake of all aldehydes was reported. In this regard, acetaldehyde and propanal were produced by *S. aureus, S. pneumoniae*, and *H. influenzae*, methacrolein and 2-methylpropanal by *S. aureus* and *S. pneumoniae*, and 3-methylbutanal only by *S. aureus*. Importantly, three aldehydes including butanal, hexanal and benzaldehyde (Figure 23.6c) were degraded by *C. albicans*. So far, the strongest consumption of aldehydes was found for *P. aeruginosa*, which is known to metabolize diverse substances as carbon and energy sources, but three aldehydes were taken up also by *S. aureus* (benzaldehyde) and *S. pneumoniae* (hexanal, 3-methylbutanal).

23.3.5.3 Esters and ethers

Nine esters were detected at elevated concentrations in fungal cultures (Table 23.3), being the most abundant group of VOCs released by *C. albicans*. Three of them, namely ethyl formate, ethyl acetate (Figure 23.6c) and isopentyl acetate (Figure 23.7b) were gradually increasing during the course of microbial proliferation, whereas the remaining esters were not detected on the first day of the experiments and their first significant release was observed after at least 26 h of incubation. Importantly, these three esters were significantly released also by *S. aureus*.

Amongst the VOCs quantified, ethyl acetate and propyl acetate reached very high concentrations of 645 ppbv and 1402 ppbv, respectively, both 28 h after inoculation.

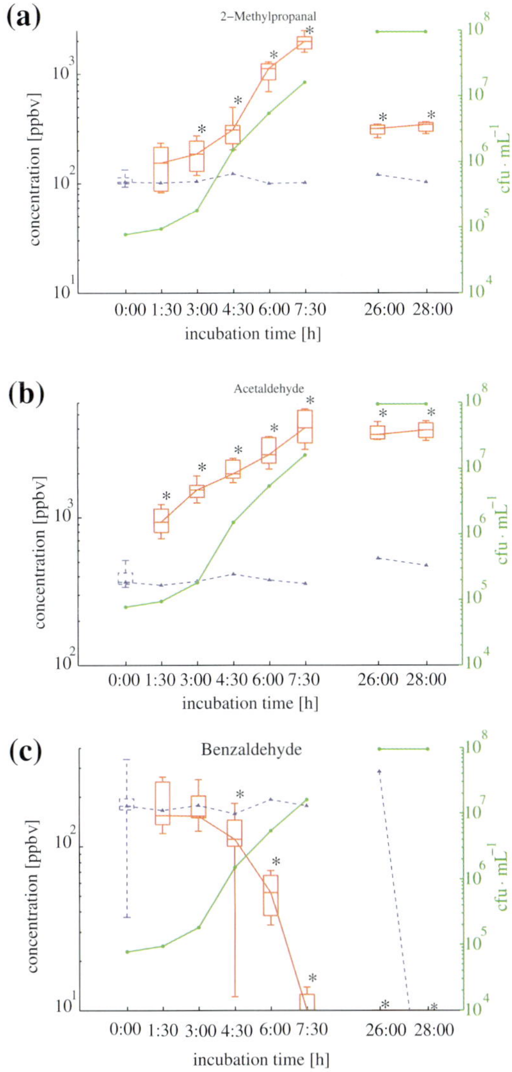

FIGURE 23.6

Exemplary aldehydes released (a–b) and taken up (c) by *C. albicans*: 2-methyl-propanal (a), acetaldehyde (b), and benzaldehyde (c). Determined median concentrations [ppbv] for headspace over fungal cultures with 25th and 75th percentiles [ppbv] are given as red boxes with whiskers indicating 5th and 95th percentiles, while the blue box and the dotted blue line with triangles indicates the pattern for medium controls. Significant VOCs release ($p < 0.05$ for Kruskal-Wallis test) is labelled with asterisks. The green line with round points denotes the proliferation rate expressed in colony forming units [CFU/mL].

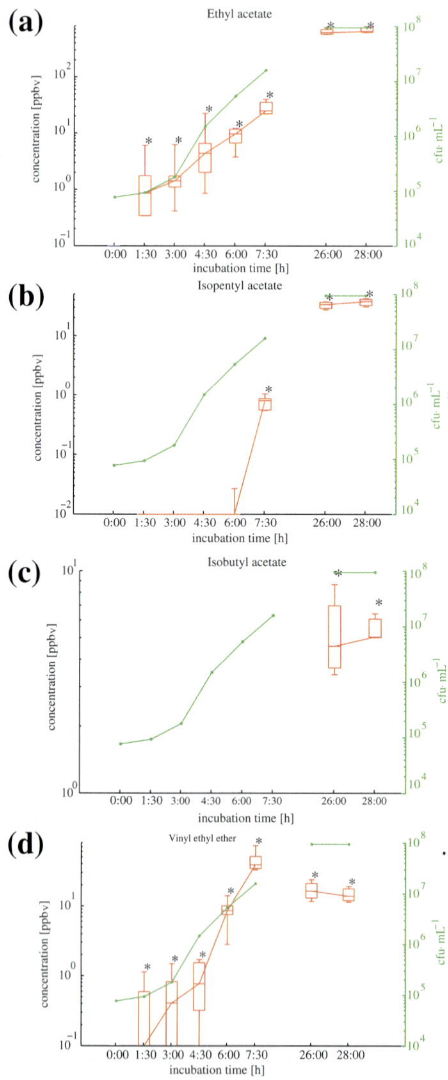

FIGURE 23.7

Release patterns of exemplary esters: ethyl acetate (a), isopentyl acetate (b) and isobutyl acetate (c), and vinyl ethyl ether (d) released by *C. albicans*. Determined median concentrations [ppbv] for headspace over fungal cultures with 25th and 75th percentiles [ppbv] are given as red boxes with whiskers indicating 5th and 95th percentiles, while the blue box and the dotted blue line with triangles indicates the pattern for medium control. Significant VOCs release ($p < 0.05$ for Kruskal-Wallis test) is indicated by an asterisk. The green line with round points denotes the proliferation rate expressed in colony forming units [CFU/mL].

Thereby, ethyl acetate was significantly increased even after 1.5 h, but propyl acetate was found at very low concentrations during the first day of the experiment (up to 7.5 h) and was significantly released at the earliest after 26 h. The other esters, isopentyl acetate (maximum 38 ppbv after 28 h), isobutyl acetate (maximum 7 ppbv after 26 h, Figure 23.7c), 2-methylbutyl acetate (maximum 17 ppbv after 28 h), ethyl formate (maximum 9 ppbv after 28 h), ethyl *n*-butanoate (maximum 6 ppbv after 28 h), and isopentyl propionate (maximum 0.2 ppbv after 26 and 28 h) were found in moderate or at very low concentrations like isopentyl propionate. However, significance was already reached at earlier time points in part (e.g. isopentyl acetate depicted on Figure 23.7b), since those esters were almost completely absent in medium controls.

Similar to *P. aeruginosa*, *C. albicans* also released several species-specific esters, namely isopentyl propionate, ethyl *n*-butanoate and diverse esters of acetic acid, such as propyl acetate, 2-methylbutyl acetate, and isobutyl acetate. In addition, isopentyl acetate released was generated at moderately high and ethyl formate at lower concentration levels (~30 ppbv and 1.9 ppbv, respectively) only by *S. aureus*. Thus, esters are potentially important volatile metabolites indicative of the presence of *C. albicans*.

Interestingly, methyl methacrylate was released in high concentration by all four bacterial species tested in this study, but not by *C. albicans*. On the other hand, *C. albicans* released ethyl acetate (commonly observed for all microbes) at much higher concentrations (645 ppbv) compared to bacteria, for which typically single ppbv (*S. aureus, P. aeruginosa*) or even sub-ppbv levels (*S. pneumoniae, H. influenzae*) were apparent.

An important metabolite, vinyl ethyl ether, was exclusively released by *C. albicans* (Figure 23.7d). Despite very low concentrations in the early phase of microbial growth (pptv level), significant differences to medium controls were detected as soon as 3 h after inoculation, because this compound was never present in medium headspace. The maximum median concentration of vinyl ethyl ether of 42 ppbv was found at 7.5 h, and it decreased overnight similar to aldehydes. Hence, vinyl ethyl ether is an important VOC related to *C. albicans*.

23.3.5.4 Ketones

As in the experiments with bacterial cultures, the release of ketones was observed for *C. albicans*, although mostly at very low concentrations (Table 23.3). 2-Pentanone and 4-heptanone were released at very low levels with highest concentrations of 2.3 ppbv after 26 h and 1.3 ppbv after 7.5 h, respectively. Apart from 2-pentanone, all ketones showed a concentration profile similar to those for aldehydes, whereby a continuous increase was observed up to 7.5 h after inoculation and a substantial decrease was found after 26 h of fungal growth. This is exemplified by methyl vinyl ketone (semi-quantified to peak area), which was not detected in the headspace samples on the second day of experiments. Importantly, it was the only ketone exclusively released by *C. albicans*. Similar to isopentyl acetate and ethyl formate, acetoin (3-hydroxy-2-butanone) was released by *C. albicans* (see Figure 23.8a) as found also for *S. aureus*, whereas concentrations determined for fungi (maximum 12 ppbv) were much

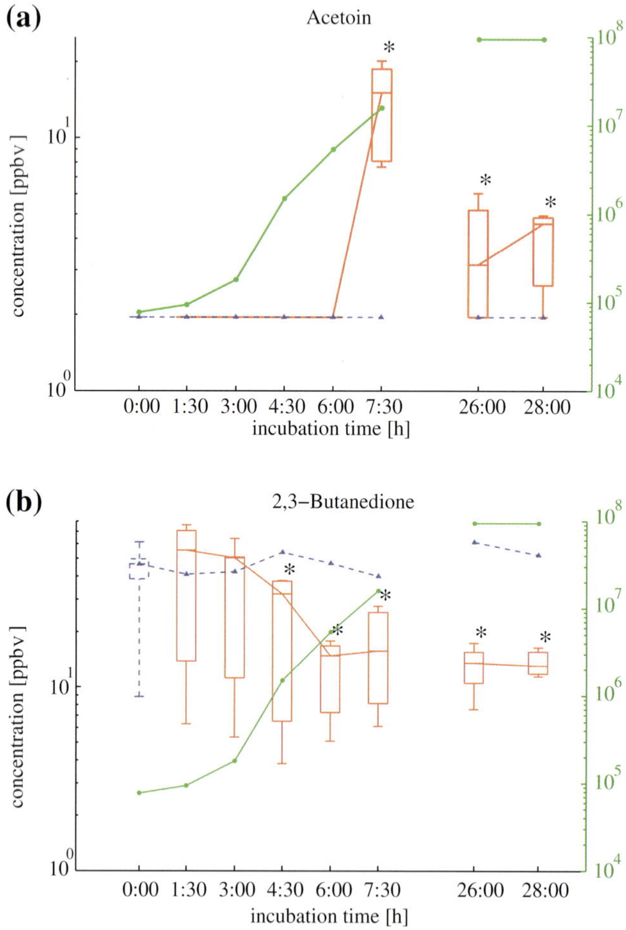

FIGURE 23.8

Exemplary ketones released (acetoin, a) or taken up (2,3-butanedione, b) by *C. albicans*. Determined median concentrations [ppbv] for headspace over fungal cultures with 25th and 75th percentiles [ppbv] are given as red boxes with whiskers indicating 5th and 95th percentiles, while the blue box and the dotted blue line with triangles indicates the pattern for medium controls. Significant VOCs release ($p < 0.05$ for Kruskal-Wallis test) is labelled with asterisks. The green line with round points denotes the proliferation rate expressed in colony forming units [CFU/mL].

lower compared to bacterial cultures (maximum 279 ppbv). The only ketone that was significantly degraded (taken up) by *C. albicans* (see Figure 23.8b) was 2,3-butanedione (see Figure 23.8b). Thereby, its levels in the medium control headspace were reduced from a median concentration of 46 ppbv to 13 ppbv after 28 h of growth.

23.3.5.5 Hydrocarbons

The only two hydrocarbons released by *C. albicans*, namely 2-methyl-2-butene and (E)-2-pentene, were detected at extremely low levels, reaching maximal concentrations of 0.2 ppbv and 1.0 ppbv, respectively, each after 28 h. These VOCs were not detected in the medium control, and (E)-2-pentene was not released by any other tested microorganism, while 2-methyl-2-butene was produced by *P. aeruginosa* and *H. influenzae*. Interestingly, two other hydrocarbons, namely *n*-butane and (Z)-2-butene showed decreasing concentrations in fungal cultures. Although the consumption of

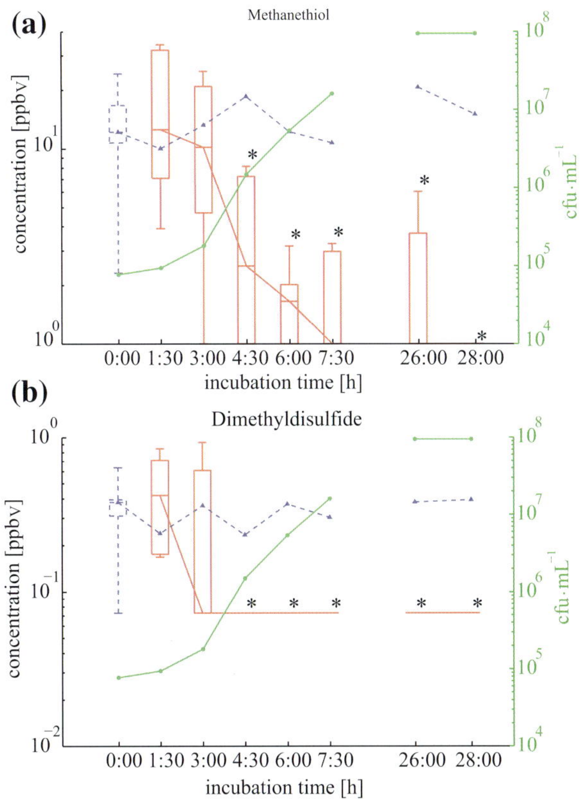

FIGURE 23.9
The uptake of volatile sulfur compounds from *C. albicans* cultures. Determined median concentrations [ppbv] of methanethiol (a) and dimethyldisulfide (b) in the headspace over fungal cultures with denoted 25th and 75th percentiles [ppbv] given as red boxes with whiskers indicating 5th and 95th percentiles, while the blue box and the dotted blue line with triangles indicates the pattern for medium controls. Significant VOCs release ($p < 0.05$ for Kruskal-Wallis test) is labelled with asterisks. The green line with round points denotes the proliferation rate expressed in colony forming units [CFU/mL].

hydrocarbons was not observed in earlier *in vitro* experiments with bacterial cultures, their uptake by *C. albicans* should be considered with care due to the trace level already found in medium controls (low pptv).

23.3.5.6 Volatile sulfur compounds (VSCs)

The *in vitro* experiments performed so far with bacterial cultures confirmed that several sulfur-containing compounds were released. In particular, methanethiol, dimethylsulfide (DMS), dimethyldisulfide (DMDS), and dimethyltrisulfide (DMTS) were released by each bacteria investigated (apart from *S. aureus*, which did not release DMS and DMTS). Therefore, unexpected profiles were observed for *C. albicans*, which released only DMS, and the remaining VSCs were consumed in the fungal cultures. Moreover, even DMS was detected at very low levels in the fungal headspace, and it was not detected after 6 h of incubation. The distinctive metabolism of sulfur-containing compounds was particularly evident for methanethiol, which was released to give headspace concentrations of hundreds of ppbv by all the mentioned bacteria (with the highest concentrations exceeding 700 ppbv for *S. aureus*), and also DMDS, whereas their concentrations decreased below the detection limits in *C. albicans* cultures (see Figure 23.9a and b).

23.4 DISCUSSION

Altogether five virulent species that often cause diseases of the respiratory tract were investigated *in vitro* to confirm the existence of pathogen-specific biomarkers. Amongst them, *Streptococcus pneumoniae* and *Haemophilus influenzae* often lead to community-acquired pneumonia, while *Staphylococcus aureus, Pseudomonas aeruginosa*, and *Candida albicans* are commonly implicated in hospital-acquired pneumonia, particularly ventilator-associated pneumonia.[1,12] For the purpose of this investigation, a special device for microbial cultivation and adsorptive headspace preconcentration, as described previously,[64] was substantially modified (see Materials and methods section), and the analytical protocol was carefully optimized, including reduction of matrix effects, selection of the best conditions for the adsorptive enrichment, and determination of the detection limits from complete calibration series (instead of blank analysis). Contrary to several earlier works,[35,36,41,71] quantification of VOCs was performed whenever possible, whereby the procedure applied allowed detection limits at the pptv level to be reached for most detected substances. Qualitative analysis by GC-MS of the majority of released metabolites was confirmed by retention times in addition to mass spectral library match, ensuring unambiguous identification. This is an important improvement, particularly in comparison to Proton Transfer Reaction Mass Spectrometry (PTR-MS) where overlapping signals (derived from different compounds) hamper both qualitative and quantitative analysis.[49,50] In addition, the dynamics of metabolites release was investigated by sampling at different phases of cultivation.

In the earlier works, relatively few VOCs were detected in the headspace of microorganisms.[40,44,71] Already known to be released from some microbes were compounds such as acetoin and methylbutanal from *S. aureus*, 1-undecene from *P. aeruginosa* or dimethyldisulfide and *iso*-pentanol from both species as reported by Zechman and co-workers.[35] However, only qualitative analyses at single incubation time point of 24 h have been performed. Similarities to our results can be found also in other studies, where the release of acetic acid from *S. pneumoniae* and *H. influenzae*, methanethiol from *S. pneumoniae* was reported by Preti and co-workers,[71] and ethanol, acetaldehyde, methanethiol and dimethylsulfide from *S. pneumoniae* was shown by Allardyce.[40]

Importantly, there are differences between our and previously published results.[35,36,40,53,57,71] Allardyce et al. reported strong release of acetic acid and acetaldehyde from *P. aeruginosa*.[40] Our results are completely opposite, whereby acetaldehyde was observed to strongly decrease from over 500 ppbv in the control medium down to 22 ppbv in the *Pseudomonas* cultures and no significant release of acids was observed for this microbe in our work. The same authors reported a decrease of acetic acid in *S. pneumoniae*, whereas an opposite concentration profile for acetic acid was observed in our experiments (released to concentrations above 1100 ppbv by microbial culture from the levels below LOD in medium control). Presumably, analytical methodologies, but especially culture conditions such as nutrient availability, could influence the VOCs metabolism in microorganisms. Therefore, standardization of these factors by application of the same culture medium and analytical protocol was an important issue in our *in vitro* studies. Although it might be insufficient to reveal the full spectrum of potential volatile metabolites, application of a single growth medium (tryptic soy broth) in all our experiments allows for the comparison of different species. This is a standard medium, ensuring fast proliferation of microorganisms (nevertheless, different supplements were applied for more fastidious strains such as *H. influenzae*). Future comparison with VOCs present in the affected patients will show whether a sufficient overlap exists. This data will also guide the development of more advanced, standardized media for *in vitro* culture.

In our experiments, we found that the patterns of VOC release from different microbes can be similar. In particular, esters such as ethyl acetate or methyl methacrylate and volatile sulfur-containing compounds like methanethiol and dimethyldisulfide are generated in biochemical pathways common for all bacteria tested. On the other hand, considerable differences were found concerning the dynamics of VOC production, and especially the uptake of volatile metabolites in the case of *P. aeruginosa*. This bacterium takes up or catabolizes (but never releases) as many as eleven aldehydes, while other microorganisms release aldehydes, for some in high concentrations. Similarly, acids were never observed to be released by *P. aeruginosa* in our study. Considering the substantial deviation in the profiles of VOCs metabolized by *P. aeruginosa* reported previously (consumption instead of production observed for other species), few more analytes might be considered as commonly released by three remaining microorganisms, namely acetaldehyde, propanal, acetic acid, 1-butanol, 2,3-butanedione, and 2-pentanone. Intriguingly, despite a higher proliferation rate of *P. aeruginosa* and *H. influenzae*, the concentrations of metabolites released by these species were lower compared to *S. aureus* and *S. pneumoniae*. A detailed list of volatile metabolites produced and degraded by tested microorganisms is given in Table 23.4.

Table 23.4 Concentrations ppbv and peak areas of volatile metabolites significantly released or degraded by all microorganisms tested *in vitro*. For each species, the median values are given for the medium control and for the highest (in case of release) or lowest (in case of uptake) median level in bacterial/fungal cultures. Analytes degraded by at least one species are denoted with[DEC], while those semi-quantified to the peak area (instead of concentration) are denoted with #. Metabolites characteristic for only one microorganism are given in bold italic, and those released by all species are given in bold. "n.d." denotes not detected. Empty fields indicate that there is no statistically significant difference between concentrations in bacteria and medium control.

	CAS	Compound	*S. aureus*		*P. aeruginosa*		*S. pneumoniae*		*H. influenzae*		*C. albicans*	
			Medium	Bacteria	Medium	Bacteria	Medium	Bacteria	Medium	Bacteria	Medium	Fungi
Alcohols	67-56-1	*Methanol*							27.96	62.0		
	78-83-1	2-Methyl-propanol	n.d.	52.6							4.5	1833.0
	123-51-3	3-Methyl-1-butanol	n.d.	210.0	<LOD	148.4					2.3	3569.0
	64-17-5	Ethanol	n.d.	11695.1	102.1	800.0	<LOD	103.1			n.d.	26228.0
	71-36-3	1-Butanol	<LOD	59.4			<LOD	56.5	<LOD	33.1		
	78-92-2	2-Butanol #			n.d.	1.5×10^4						
	71-23-8	1-Propanol									1.3	267.2
Esters	141-78-6	Ethyl acetate	n.d.	5.6	<LOD	2.9	<LOD	0.5	<LOD	0.2	n.d.	645.5
	123-86-4	*n*-Butyl acetate	n.d.	0.2								
	123-92-2	Isopentyl acetate	n.d.	1.9							n.d.	37.8
	868-57-5	Methyl 2-methylbutyrate			n.d.	1.7						
	80-62-6	Methyl methacrylate	16.0	31.9	24.8	46.7	222.0	1420.5	127.8	210.3		
	7452-79-1	Ethyl 2-methylbutyrate #			n.d.	1.8×10^5						
	2445-69-4	2-Methylbutyl isobutyrate #			n.d.	1.3×10^6						
	106-27-4	Isoamyl butyrate #			n.d.	1.4×10^6					n.d.	7.7×10^5

(Continued)

Table 23.4 (Continued)

	CAS	Compound	*S. aureus*		*P. aeruginosa*		*S. pneumoniae*		*H. influenzae*		*C. albicans*	
			Medium	Bacteria	Medium	Bacteria	Medium	Bacteria	Medium	Bacteria	Medium	Fungi
	2445-78-5	2-Methylbutyl 2-methylbutyrate #			n.d.	9.7×10^6						
	25415-62-7	Amyl isovalerate #			n.d.	6.1×10^6						
	109-94-4	Ethyl formate	n.d.	3.2							n.d.	8.6
	108-64-5	Ethyl isovalerate	n.d.	0.9								
	554-12-1	Methyl Propionate							<LOD	0.3		
	123-20-6	Vinyl butyrate #							n.d.	1.7×10^5		
	110-19-0	Isobutyl acetate									n.d.	6.9
	624-41-9	2-Methylbutyl acetate									n.d.	17.4
	109-60-4	Propyl acetate									0.2	1401.0
	105-54-4	Ethyl n-butanoate									n.d.	5.9
	105-68-0	Isoamyl propionate									n.d.	0.2
Unsaturated hydrocarbons	821-95-4	1-Undecene			0.3	317.5						
	563-46-2	2-Methyl-1-butene					n.d.	0.8				
	513-35-9	2-Methyl-2-butene			<LOD	0.7			n.d.	3.3	n.d.	0.2
	13688-67-0	1,10-Undecadiene			<LOD	6.8						
	124-11-8	1-Nonene			0.3	3.4						
	872-05-9	1-Decene			<LOD	1.2						
	112-41-4	1-Dodecene			1.9	9.5						
	590-18-1	(Z)-2-Butene^{DEC} #	n.d.	4.8			n.d.	9.8			6.1×10^3	n.d.
	624-64-6	(E)-2-Butene	1.3	13.7			n.d.	15.3				
	115-11-7	Isobutene	<LOD	21.9			0.7	31.9				

(Continued)

Table 23.4 (Continued)

	CAS	Compound	*S. aureus*		*P. aeruginosa*		*S. pneumoniae*		*H. influenzae*		*C. albicans*	
			Medium	Bacteria	Medium	Bacteria	Medium	Bacteria	Medium	Bacteria	Medium	Fungi
	78-79-5	Isoprene			<LOD	20.4			n.d.	19.7		
	106-99-0	1,3-butadiene	<LOD	4.9			<LOD	3.7				
	22370-55-4	10-Methyl-1-undecene #			n.d.	3.3×10^5						
	563-45-1	3-Methyl-1-butene							n.d.	0.748		
	646-04-8	(E)-2-Pentene									n.d.	1.0
Alkanes	74-98-6	Propane	0.91	4.9								
	106-97-8	Butane[DEC]	0.664	4.3	0.3	1.5	0.2	6.0			0.1	n.d.
	75-28-5	Isobutane					<LOD	20.4				
	110-54-3	Hexane					<LOD	1.7				
Ketones	67-64-1	Acetone					661.1	1005.6				
	513-86-0	Acetoin	<LOD	279.3							2.0	12.1
	116-09-6	Hydroxyacetone	<LOD	331.0								
	821-55-6	2-Nonanone			<LOD	22.4	n.d.	1.4				
	107-87-9	2-Pentanone			<LOD	19.3	<LOD	0.5	<LOD	0.446	n.d.	2.3
	110-43-0	2-Heptanone			n.d.	4.8						
	123-19-3	4-Heptanone			n.d.	2.1			n.d.	0.188	n.d.	1.3
	106-68-3	3-Octanone			n.d.	0.2						
	78-93-3	2-Butanone			10.1	40.6	12.0	28.6				
	108-10-1	Methylisobutylketone #			3.8×10^4	3.8×10^5						
	431-03-8	2,3-Butanedione[DEC]	22.7	68.0	77.2	5.1	88.5	255.6	56.25	115.5	45.9	13.1
	78-94-4	Methyl vinyl ketone #									n.d.	2.8×10^5

(Continued)

Table 23.4 (Continued)

	CAS	Compound	*S. aureus*		*P. aeruginosa*		*S. pneumoniae*		*H. influenzae*		*C. albicans*	
			Medium	Bacteria	Medium	Bacteria	Medium	Bacteria	Medium	Bacteria	Medium	Fungi
N-containing	5628-99-9	1-Vinyl aziridine #			n.d.	2.8×10^7						
	616-43-3	3-Methylpyrrole			<LOD	24.7						
	109-97-7	Pyrrole			1.1	49.7						
	25343-57-1	2-Acetyl-1,4,5,6-tetrahydropyridine #							n.d.	1.2×10^6		
Sulfur compounds	75-15-0	Carbondisulfide[DEC]					1.0	5.6	0195	1.8	0.5	n.d.
	75-18-3	Dimethylsulfide			<LOD	173.6	n.d.	2.6	n.d.	29.1	n.d.	# 3×10^3
	624-92-0	Dimethyldisulfide[DEC]	1.6	10.2	0.6	14.1	<LOD	6.9	0.443	5.8	0.4	0.1
	3658-80-8	Dimethyltrisulfide			<LOD	1.1	<LOD	1.3	<LOD	0.6		
	74-93-1	Methanethiol[DEC]	134.2	701.5	33.0	418.0	17.35	151.5	9.861	89.1	12.2	n.d.
	624-89-5	Ethyl methyl sulfide #							n.d.	5.0×10^5		
	31053-55-1	2-Methoxy-5-Methylthiophene #			n.d.	2.0×10^6						
	5454-45-5	3-(Ethylthio)-propanal #			n.d.	7.9×10^5			n.d.	1.9×10^5		
	24653-75-6	Mercaptoacetone #			n.d.	2.6×10^5						
	3268-49-3	3-(Methylthio)propanal #					n.d.	7.7×10^5				
	1534-08-3	S-Methyl thioacetate #							n.d.	1.4×10^6		
Aldehydes	100-52-7	Benzaldehyde[DEC]	20.6	3.8	183.9	5.7					175.8	6.1
	75-07-0	Acetaldehyde[DEC]	528.5	1417.4	515.5	23.0	465.1	1907.5	255.6	905.5	324.7	4308.0
	78-85-3	Methacrolein[DEC]	5.9	9.3	3.3	<LOD	2.2	3.0			1.1	6.2
	590-86-3	3-Methylbutanal[DEC]	317.1	4818.5	419.6	14.5	429.0	34.13			151.3	418.2
	124-19-6	Nonanal[DEC]			13.4	4.1						

(Continued)

Table 23.4 (Continued)

	CAS	Compound	*S. aureus*		*P. aeruginosa*		*S. pneumoniae*		*H. influenzae*		*C. albicans*	
			Medium	Bacteria	Medium	Bacteria	Medium	Bacteria	Medium	Bacteria	Medium	Fungi
	66-25-1	Hexanal[DEC]					1.3	<LOD			# 9×10^4	# 1.5×10^4
	123-38-6	Propanal[DEC]	4.0	14.2	2.9	<LOD	2.3	5.9	0.964	3.1	1.8	21.0
	107-86-8	3-Methyl-2-butenal[DEC]	1.5	5.7	1.3	n.d.	0.8	1.2				
	107-02-8	Acrolein[DEC]			10.0	3.6						
	4170-30-3	2-Butenal					0.5	2.0				
	123-72-8	Butanal[DEC]			24.4	<LOD	17.7	33.4	11.4	15.3	15.7	1.0
	78-84-2	2-Methylpropanal[DEC]	598.6	2044.5	181.9	4.1	173.8	304.4			113.0	2791.8
	124-13-0	Octanal[DEC]			5.4	1.3						
	1115-11-3	(Z)-2-Methyl-2-butenal	73.5	268.5								
	497-03-0	(E)-2-Methyl-2-butenal	<LOD	0.4								
	922-63-4	2-Ethylacrolein	1.7	6.5								
	90-02-8	o-Hydroxybenzaldehyde #							n.d.	2.8×10^3		
	503-74-2	Isovaleric acid	n.d.	97.4								
	64-19-7	Acetic acid	n.d.	2566.6			<LOD	1159.9	<LOD	258.3		
	110-00-9	Furan					0.3	5.1	<LOD	0.491		
	13679-41-9	3-Phenylfuran					1.1×10^6	2.1×10^6				
Aromatic	100-41-4	Ethylbenzene					4.3×10^5	1.3×10^6				
Lactone	96-48-0	gamma-Butyrolactone							<LOD	19.8		
Ether	109-92-2	Vinyl ethyl ether									n.d.	41.5

Although the origin of VOCs released by microorganisms is only partly elucidated, it is known that catabolism of amino acids yields branched-chain aldehydes.[32,72–74] In turn, aldehydes might be reduced to alcohols by alcohol dehydrogenases (e.g. 3-methylbutanal to 3-methyl-1-butanol) or oxidized to carboxylic acids by an aldehyde dehydrogenase (e.g. 3-methylbutanal to isovaleric acid) as observed for *S. aureus*. Since the aforementioned compounds were significantly released by *C. albicans* (apart from isovaleric acid) and *S. aureus*, as shown in our *in vitro* study, we presume that amino acid degradation plays an important role in these two species, especially since the culture medium consisted mainly of amino acids, peptides and glucose. Evidence supporting this hypothesis can be found also in the work of other researchers, where a link between availability of branched amino acids (e.g. valine, isoleucine) and production of branched alcohols and aldehydes was reported.[35]

Several substances (such as acetoin, 2,3-butanedione, 1-butanol, 2-propanol, acetic acid, acetaldehyde, and ethanol) significantly released by tested species are formed from pyruvate through glycolytic lactate and non-glycolytic carbohydrates fermentations.[75] The catabolism of pyruvate (depicted on Figure 23.10) seems to play an important role in the case of *S. aureus* and *C. albicans*, since the products of

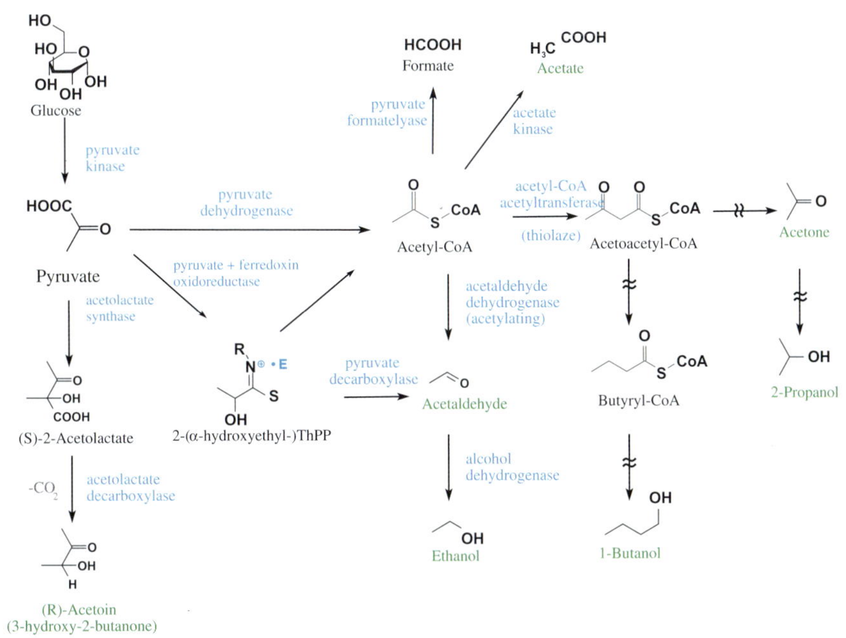

FIGURE 23.10

Simplified scheme of pyruvate metabolism via glycolytic fermentations and lactate converting fermentations, modified after Michal et al.[75] Metabolites given in green were significantly released by microorganisms in our *in vitro* study.

this metabolic pathway were found in the headspace from these species. On the other hand, the most dominant metabolic process in *P. aeruginosa* cultures is the catabolism of organic compounds such as aldehydes as carbon and energy sources. The versatile nutritional requirements of *Pseudomonas* are well known, and some of its subspecies utilize over 100 different compounds of diverse chemical classes.[2,76] Accordingly, the substantial uptake of aldehydes by *P. aeruginosa* cultures was observed, with acetaldehyde, 3-methylbutanal, 2-methylpropanal, benzaldehyde, and butanal being metabolized most efficiently.

Apart from the pathways mentioned above, the generation of volatile metabolites might be pH-dependent, whereby the final product of carbohydrate fermentation (acetic acid) lowers the pH of microbial cultures. Consequently, a greater proportion of carboxylic acids exists in the neutral form and can partition between the liquid phase of microbial culture and gaseous headspace. Lower pH might also be a result of amino acid (alkaline reaction) catabolism. For instance, leucine is converted by aminotransferase to α-ketoisocaproate and subsequently by decarboxylase to 3-methylbutanal, yielding the final product isovaleric acid after oxidation by aldehyde dehydrogenase. Since the soy broth medium used for the presented *in vitro* experiments was rich in amino acids, this metabolic pattern could play a substantial role, particularly for *S. aureus* cultures, which release high concentrations of isovaleric acid.

Formation of the same substances by the tested microorganisms prompts the expectation that at least some of these species colonize similar environments. Indeed, *Candida* albicans found in critically ill patients suffering from ventilator associated pneumonia, often coexists in infected lungs with *Staphylococcus aureus*. In accordance with this fact, multiple similarities in VOC metabolism for *S. aureus* and *C. albicans* were observed. In this respect, the same alcohols (except 1-butanol) and aldehydes (except unsaturated aldehydes) were released by *S. aureus* and *C. albicans*. Importantly, acetoin and 3-methylbutanal were generated exclusively by these two species. Furthermore, two of three esters (ethyl formate, isopentyl acetate) that gradually increased in *C. albicans* cultures during microbial proliferation were released only by *S. aureus*. On the other hand, the difference in bacterial and fungal metabolism is shown by the production of entirely different acids, hydrocarbons, ketones (apart from aforementioned acetoin, which is the product of pyruvate catabolism), and especially VSCs. This is a very important observation, since monitoring of alcohols, saturated aldehydes or esters in alveolar air of mechanically ventilated patients could be used to detect these virulent microorganisms, while isovaleric acid, propane, hydroxyacetone or unsaturated aldehydes might be used as specific biomarkers allowing differentiation of these two species. Consequently, breath analysis could be used to detect the outbreak of pneumonia in critically ill patients, who are often sedated and rarely able to communicate, and also to provide the preliminary identification of underlying pathogens.

Our *in vitro* study clearly shows that all pathogens produce species-specific metabolites or groups of metabolites allowing their identification by gas phase analysis. VOCs exclusively released by *S. aureus* and *C. albicans* comprise mostly low molecular weight analytes, while the compounds within the range of C3–C6 have

Table 23.5 Volatile species-specific biomarkers exclusively metabolized (released or degraded) by *S. aureus*, *P. aeruginosa*, *S. pneumoniae*, *H.influenzae*, and *C. albicans* cultivated *in vitro*. Degraded compounds are denoted with ^{DEC}, while numbers with arrow ($\rightarrow$) concern the initial high concentration in medium and minimum concentration in bacteria cultures. Semi-quantified compounds (to peak area) are denoted with #. Populations of healthy subjects: $n_{smokers} = 23$; $n_{non\text{-}smokers} = 32$.

Compound	CAS	Amount of Carbon Atoms in Molecule	Maximum Microbial Concentration ppbv	Time of Microbes Growth for max. Abundance (h)	Time of Microbes Growth for First Significant Increase (h)	Appearance (%) in Healthy Non-Smokers	Appearance (%) in Healthy Smokers
		VOCs characteristic for *Staphylococcus aureus*					
Isovaleric acid	503-74-2	C5	97.4	6 h	3 h	0	5.6
2-Ethylacrolein	922-63-4	C5	6.5	3 h	3 h	0	0
(Z)-2-Methyl-2-butenal	1115-11-3	C5	268.5	4.5 h	3 h	0	0
(E)-2-Methyl-2-butenal	497-03-0	C5	0.4	4.5 h	3 h	28.6	16.7
n-Butyl acetate	123-86-4	C6	0.2	6 h	4.5 h	82.1	61.1
Ethyl isovalerate	108-64-5	C7	0.9	6 h	4.5 h	0	0
Propane	74-98-6	C3	4.9	6 h	4.5 h	92.9	77.8
Hydroxyacetone	116-09-6	C3	331.0	6 h	1.5 h	0	0
		VOCs characteristic for *Pseudomonas aeruginosa*					
2-Butanol #	78-92-2	C4	1.5×10^4	26 h	26 h	0	0
Acrolein^{DEC}	107-02-8	C3	$9.9 \rightarrow 3.6$	28 h	24 h	100	100
Octanal^{DEC}	124-13-0	C8	$5.4 \rightarrow 1.3$	24 h	3.7 h	0	5.6

(Continued)

Table 23.5 (Continued)

Compound	CAS	Amount of Carbon Atoms in Molecule	Maximum Microbial Concentration ppbv	Time of Microbes Growth for max. Abundance (h)	Time of Microbes Growth for First Significant Increase (h)	Appearance (%) in Healthy Non-Smokers	Appearance (%) in Healthy Smokers
Nonanal[DEC]	124-19-6	C9	13.4 → 4.1	24 h	3 h	17.8	33.3
Methyl 2-methylbutyrate	868-57-5	C6	1.669	28 h	26 h	0	0
Ethyl 2-methylbutyrate #	7452-79-1	C7	1.8×10^5	26 h	24 h	0	0
2-Methylbutyl 2-methylbutyrate #	2445-78-5	C10	9.7×10^6	26 h	24 h	0	0
2-Methylbutyl isobutyrate #	2445-69-4	C9	1.3×10^6	26 h	24 h	0	0
Amyl isovalerate #	25415-62-7	C10	6.1×10^6	26 h	24 h	0	0
1-Nonene	124-11-8	C9	3.4	26 h	3.75 h	0	0
1-Decene	872-05-9	C10	1.2	26 h	5.25 h	0	0
1,10-Undecadiene	13688-67-0	C11	6.8	24 h	4.5 h	0	0
1-Undecene	821-95-4	C11	317.5	24 h	1.5 h	0	5.6
1-Dodecene	112-41-4	C12	9.5	26 h	6 h	0	5.6
10-Methyl-1-undecene #	22370-55-4	C12	3.3×10^5	24 h	24 h	0	0
2-Heptanone	110-43-0	C7	4.8	26 h	5.25 h	14.3	11.1
3-Octanone	106-68-3	C8	0.8	26 h	24 h	0.0	5.6
Methyl isobutyl ketone #	108-10-1	C6	3.8×10^5	26 h	3.75 h	21.4	33.3
Pyrrole	109-97-7	C4	49.7	3 h	1.5 h	92.9	88.9
3-Methylpyrrole	616-43-3	C5	24.7	24 h	5.25 h	3.6	0

(Continued)

Table 23.5 (Continued)

Compound	CAS	Amount of Carbon Atoms in Molecule	Maximum Microbial Concentration ppbv	Time of Microbes Growth for max. Abundance (h)	Time of Microbes Growth for First Significant Increase (h)	Appearance (%) in Healthy Non-Smokers	Appearance (%) in Healthy Smokers
1-Vinylaziridine #	5628-99-9	C4	2.8×10^7	2.25 h	1.5 h	0	0
Mercaptoacetone #	24653-75-6	C3	2.6×10^5	26 h	24 h	0	0
2-Methoxy-5-methylthiophene #	31053-55-1	C6	2.0×10^6	26 h	24 h	0	0
VOCs characteristic for *Streptococcus pneumoniae*							
2-Butenal	4170-30-3	C4	2.0	7.5 h	3 h	14.3	27.8
2-Methyl-1-butene	563-46-2	C5	0.8	7.5 h	6 h	64.3	66.7
n-Hexane	110-54-3	C6	1.7	7.5 h	4.5 h	100	100
iso-Butane	75-28-5	C4	20.4	7.5 h	4.5 h	100	94.4
Acetone	67-64-1	C3	1005.0	7.5 h	6 h	100	100
3-Phenylfuran #	13679-41-9	C10	2.1×10^6	7.5 h	7.5 h	0	0
Ethylbenzene #	100-41-4	C8	1.3×10^6	7.5 h	3 h	100	100
3-(Methylthio)propanal #	3268-49-3	C4	7.7×10^5	7.5 h	6 h	0	0
VOCs characteristic for *Haemophilus influenzae*							
Methanol	67-56-1	C1	62.0	7.5 h	7.5 h	100	100
o-Hydroxybenzaldehyde #	90-02-8	C7	2.8×10^3	7.5 h	3 h	0	0
Methyl Propionate	554-12-1	C4	0.3	7.5 h	7.5 h	0	0

(Continued)

Table 23.5 (Continued)

Compound	CAS	Amount of Carbon Atoms in Molecule	Maximum Microbial Concentration ppbv	Time of Microbes Growth for max. Abundance (h)	Time of Microbes Growth for First Significant Increase (h)	Appearance (%) in Healthy Non-Smokers	Appearance (%) in Healthy Smokers
Vinyl butyrate #	123-20-6	C6	1.7×10^5	7.5 h	6 h	0	0
3-methyl-1-butene	563-45-1	C5	0.7	7.5 h	6 h	3.6	0
γ-Butyrolactone	96-48-0	C4	19.8	4.5 h	1.5 h	64.3	33.3
2-Acetyl-1,4,5,6-tetrahydropyridine #	25343-57-1	C7	1.2×10^6	7.5 h	6 h	0	0
Ethyl methyl sulfide #	624-89-5	C3	5.0×10^5	6 h	4.5 h	3.6	22.2
S-Methyl thioacetate #	1534-08-3	C3	1.4×10^6	7.5 h	4.5 h	0	0
VOCs characteristic for *Candida albicans*							
1-Propanol	71-23-8	C3	267.2	28 h	4.5 h	57.1	61.1
Isobutyl acetate	110-19-0	C6	6.9	26 h	26 h	10.7	0
2-Methyl butyl acetate	624-41-9	C7	17.4	28 h	26 h	0	0
Propyl acetate	109-60-4	C5	1402	28 h	26 h	7.1	16.7
Ethyl *n*-butanoate	105-54-4	C6	5.9	28 h	26 h	0	0
Isoamyl propionate	105-68-0	C8	0.2	26 h	26 h	0	0
Vinyl ethyl ether	109-92-2	C4	41.5	7.5 h	1.5 h	0	0
(E)-2-Pentene	646-04-8	C5	0.4	26 h	26 h	3.6	22.6
Methyl vinyl ketone #	78-94-4	C4	2.8×10^5	4.5 h	1.5 h	100	100

the biggest contribution. In contrast, metabolites exclusively released by *P. aeruginosa* comprise mainly heavier esters and heavier hydrocarbons. Several compounds show significantly increased concentrations as soon as the first few hours of microbial growth. Amongst all metabolites significantly released by the tested species, nitrogen-containing compounds (VNCs) and esters are the most characteristic ones, whereby only ethyl acetate and methyl methacrylate were found for all microbes. Apart from ethyl formate, isopentyl acetate (*S. aureus* and *C. albicans*) and isopentyl butyrate (*P. aeruginosa* and *C. albicans*), the remaining esters were specific to single species (Table 23.5). Apparently, there was no ether, lactone or nitrogen-containing compound released by two different microorganisms. As such, these compounds can be used for identification of the underlying pathogen. The remaining compounds, which were released by at least two species, can still be used in exhaled breath analysis to monitor the onset of a disease (e.g. emerging pneumonia). Some of them are released at concentration levels as high as several hundreds of ppbv (e.g. methanethiol, 3-methyl-1-butanol). Nevertheless, among the tested microbes, only *S. aureus* and *C. albicans* coexist as pathogens of pneumonia.

23.5 SUMMARY

Although this study was limited to four bacterial and one fungal species, we observed characteristic VOC patterns for each microorganism, demonstrating that the approach developed (sample collection, preparation, and GC-MS analysis) is suitable to discriminate these pathogens. Hence, the results obtained encourage the continuation of this research to finally arrive at a non-invasive diagnosis of microbial lung infections. The GC-MS analysis of breath gas (after preconcentration on multibed sorption tube or needle traps) provides the results within a maximum of 2 h, while the routinely used microbiological diagnosis typically takes two days. Moreover, even real-time analysis at the bed-side could be possible in the future by mass spectrometry, e.g. PTR-MS. Even though the price of breath-diagnosis cannot be precisely estimated at this stage of the study, the overall treatment costs will surely be reduced following the early diagnosis of pneumonia (hence preventing complications) resulting from the shortening of hospitalization in the intensive care unit (where costs are as high as $\sim E3500$ per day, Innsbruck University Hospital, Austria). However, it should be clearly stated that detection of the certain VOCs in the breath of ventilated patients may result from colonization of only the conductive airways (e.g. trachea, bronchi). Thus, breath analysis will never be a stand-alone approach for the detection or verification of microbial lung infections in the clinical setting. Essential information should be gathered from currently applied—but non-specific—methods such as imaging (X-Ray, chest CT) and clinical parameters such as fever, impaired respiration or elevated level of C-reactive protein. Despite this, breath analysis might indicate the outbreak of ventilator associated pneumonia allowing the ICU personnel to intensify their efforts on a particular patient.

Acknowledgments

The research leading to these results has received funding from the Austrian Research Promotion Agency (FFG) under project no 822696, with industrial support from Roche Diagnostics Graz GmbH. We thank Dr. Horst Rüther for initiating this project and for his continuous input and support.

A.A. greatly appreciates the generous support of the government of Vorarlberg and its governor Landeshauptmann Dr. Herbert Sausgruber. M.N. appreciates support by the Austrian Science Fund (project L313-B13).

References

1. Goering R, Dockrell H, Zuckermann M, Wakelin D, Roitt I, Mims C, et al., editors. *Mims' medical microbiology*. Philadelphia: Elsevier; 2008.
2. Madigan TM, Martinko JM, Dunlap PV, Clark DP. *Brock Biology of Microorganisms*. 12th ed. In: Carlson G, Berriman L, editors. San Francisco: Pearson Education Inc; 2009.
3. Gibson RL, Burns JL, Ramsey BW. Pathophysiology and management of pulmonary infections in cystic fibrosis. *Am J Respir Crit Care Med* 2003;**168**:918–51.
4. Chastre J, Fagon JY. Ventilator-associated pneumonia. *Am J Respir Crit Care Med* 2002;**165**:867–903.
5. Thompson DS, Workman R, Strutt M. Contribution of acquired meticillin-resistant Staphylococcus aureus bacteraemia to overall mortality in a general intensive care unit. *J Hosp Infect* 2008;**70**:223–7.
6. Hidron AI, Edwards JR, Patel J, Horan TC, Sievert DM, Pollock DA, et al. NHSN annual update: antimicrobial-resistant pathogens associated with healthcare-associated infections: annual summary of data reported to the National Healthcare Safety Network at the Centers for Disease Control and Prevention, 2006–2007. *Infect Control Hosp Epidemiol* 2008;**29**:996–1011.
7. Fridkin SK, Jarvis WR. Epidemiology of nosocomial fungal infections. *Clin Microbiol Rev* 1996;**9**:499–511.
8. Edmond MB, Wallace SE, McClish DK, Pfaller MA, Jones RN, Wenzel RP. Nosocomial bloodstream infections in United States hospitals: a three-year analysis. *Clin Infect Dis* 1999;**29**:239–44.
9. Piazza O, Boccia MC, Iasiello A, Storti MP, Tufano R, Triassi M. Candidemia in intensive care patients. Risk factors and mortality. *Minerva Anestesiol* 2004;**70**:63–9.
10. Wisplinghoff H, Bischoff T, Tallent SM, Seifert H, Wenzel RP, Edmond MB. Nosocomial bloodstream infections in US hospitals: analysis of 24,179 cases from a prospective nationwide surveillance study. *Clin Infect Dis* 2004;**39**:309–17.
11. Marchetti O, Bille J, Fluckiger U, Eggimann P, Ruef C, Garbino J, et al. Epidemiology of candidemia in Swiss tertiary care hospitals: secular trends, 1991–2000. *Clin Infect Dis* 2004;**38**:311–20.
12. Brooks G, Carroll K, Butel J, Morse S. In: *Jawetz, Melnick & Adelberg's medical microbiology*. Malley J, Lebovitz H, editors. New York: The McGraw-Hill Companies; 2007.
13. Visscher S, Kruisheer EM, Schurink CA, Lucas PJ, Bonten MJ. Predicting pathogens causing ventilator-associated pneumonia using a Bayesian network model. *J Antimicrob Chemother* 2008;**62**:184–8.

14. Koulenti D, Lisboa T, Brun-Buisson C, Krueger W, Macor A, Sole-Violan J, et al. Spectrum of practice in the diagnosis of nosocomial pneumonia in patients requiring mechanical ventilation in European intensive care units. *Crit Care Med* 2009;**37**:2360–8.
15. Schubert JK, Miekisch W, Nöldge-Schomburg GFE. *VOC breath markers in critically ill patients: potential and limitations*. In: Amann A, Smith D, editors. *Breath analysis for clinical diagnosis and therapeutic monitoring*. Singapore: World-Scientific; 2005. p. 267–92.
16. Miekisch W, Schubert JK, Noeldge-Schomburg GF. Diagnostic potential of breath analysis—focus on volatile organic compounds. *Clin Chim Acta* 2004;**347**:25–39.
17. Refat M, Moore TJ, Kazui M, Risby TH, Perman JA, Schwarz KB. Utility of breath ethane as a noninvasive biomarker of vitamin E status in children. *Pediatr Res* 1991;**30**:396–403.
18. Schwarz KB, Cox JM, Sharma S, Clement L, Humphrey J, Gleason C, et al. Possible antioxidant effect of vitamin A supplementation in premature infants. *J Pediatr Gastroenterol Nutr* 1997;**25**:408–14.
19. Artlich A, Jonsson B, Bhiladvala M, Lonnqvist PA, Gustafsson LE. Single breath analysis of endogenous nitric oxide in the newborn. *Biol Neonate* 2001;**79**:21–6.
20. Mieth M, Kischkel S, Schubert JK, Hein D, Miekisch W. Multibed needle trap devices for on site sampling and preconcentration of volatile breath biomarkers. *Anal Chem* 2009;**81**:5851–7.
21. Ueta I, Saito Y, Hosoe M, Okamoto M, Ohkita H, Shirai S, et al. Breath acetone analysis with miniaturized sample preparation device: in-needle preconcentration and subsequent determination by gas chromatography-mass spectroscopy. *J Chromatogr B: Analyt Technol Biomed Life Sci* 2009;**877**:2551–6.
22. Mieth M, Schubert JK, Groger T, Sabel B, Kischkel S, Fuchs P, et al. Automated needle trap heart-cut GC/MS and needle trap comprehensive two-dimensional GC/TOF-MS for breath gas analysis in the clinical environment. *Anal Chem* 2010;**82**:2541–51.
23. Filipiak W, Filipiak A, Ager C, Wiesenhofer H, Amann A. Optimization of sampling parameters for collection and preconcentration of alveolar air by needle traps. *J Breath Res* 2012;**6**:027107.
24. Chen CP, Chen CY, Lu CL, Chang FY, Lee SD, Chu LS, et al. Infrared spectrometry based 13C-octanoic acid breath test in measuring human solid gastric emptying. *J Gastroenterol Hepatol* 2003;**18**:41–6.
25. Liao CC, Lee CL, Chiang TC, Lee SC, Huang SH, Tu TC, et al. The ^{13}C-urea breath test to detect *Helicobacter pylori* infection: a validated simple methodology with 50 mg ^{13}C-urea. *Aliment Pharmacol Ther* 2002;**16**:787–92.
26. Alving K, Weitzberg E, Lundberg JM. Increased amount of nitric oxide in exhaled air of asthmatics. *Eur Respir J* 1993;**6**:1368–70.
27. Kharitonov SA, Yates D, Robbins RA, Logan-Sinclair R, Shinebourne EA, Barnes PJ. Increased nitric oxide in exhaled air of asthmatic patients. *Lancet* 1994;**343**:133–5.
28. Eisenmann A, Amann A, Said M, Datta B, Ledochowski M. Implementation and interpretation of hydrogen breath tests. *J Breath Res* 2008;**2**:046002.
29. Hockstein NG, Thaler ER, Torigian D, Miller Jr. WT, Deffenderfer O, Hanson CW. Diagnosis of pneumonia with an electronic nose: correlation of vapor signature with chest computed tomography scan findings. *Laryngoscope* 2004;**114**:1701–5.
30. Hanson 3rd CW, Thaler ER. Electronic nose prediction of a clinical pneumonia score: biosensors and microbes. *Anesthesiology* 2005;**102**:63–8.
31. Gardner JW, Craven M, Dow C, Hines EL. The prediction of bacteria type and culture growth phase by an electronic nose with a multi-layer perceptron network. *Meas Sci Technol* 1998;**9**:120–7.

32. Marilley L, Casey MG. Flavours of cheese products: metabolic pathways, analytical tools and identification of producing strains. *Int J Food Microbiol* 2004;**90**:139–59.
33. Turner AP, Magan N. Electronic noses and disease diagnostics. *Nat Rev Microbiol* 2004;**2**:161–6.
34. Syhre M, Scotter JM, Chambers ST. Investigation into the production of 2-Pentylfuran by *Aspergillus fumigatus* and other respiratory pathogens *in vitro* and human breath samples. *Med Mycol* 2008;**46**:209–15.
35. Zechman JM, Aldinger S, Labows Jr. JN Characterization of pathogenic bacteria by automated headspace concentration-gas chromatography. *J Chromatogr* 1986;**377**:49–57.
36. Scholler C, Molin S, Wilkins K. Volatile metabolites from some gram-negative bacteria. *Chemosphere* 1997;**35**:1487–95.
37. Eriksson A, Waller KP, Svennersten-Sjaunja K, Haugen JE, Lundby F, Lind O. Detection of mastitic milk using a gas-sensor array system (electronic nose). *Int Dairy J* 2005;**15**: 1193–201.
38. Hettinga KA, van Valenberg HJ, Lam TJ, van Hooijdonk AC. Detection of mastitis pathogens by analysis of volatile bacterial metabolites. *J Dairy Sci* 2008;**91**:3834–9.
39. Carroll W, Lenney W, Wang T, Spanel P, Alcock A, Smith D. Detection of volatile compounds emitted by *Pseudomonas aeruginosa* using selected ion flow tube mass spectrometry. *Pediatr Pulmonol* 2005;**39**:452–6.
40. Allardyce RA, Langford VS, Hill AL, Murdoch DR. Detection of volatile metabolites produced by bacterial growth in blood culture media by selected ion flow tube mass spectrometry (SIFT-MS). *J Microbiol Methods* 2006;**65**:361–5.
41. Julak J, Stranska E, Rosova V, Geppert H, Spanel P, Smith D. Bronchoalveolar lavage examined by solid phase microextraction, gas chromatography–mass spectrometry and selected ion flow tube mass spectrometry. *J Microbiol Methods* 2006;**65**:76–86.
42. Chippendale TW, Spanel P, Smith D. Time-resolved selected ion flow tube mass spectrometric quantification of the volatile compounds generated by *E. coli* JM109 cultured in two different media. *Rapid Commun Mass Spectrom* 2011;**25**:2163–72.
43. Shestivska V, Nemec A, Drevinek P, Sovova K, Dryahina K, Spanel P. Quantification of methyl thiocyanate in the headspace of *Pseudomonas aeruginosa* cultures and in the breath of cystic fibrosis patients by selected ion flow tube mass spectrometry. *Rapid Commun Mass Spectrom* 2011;**25**:2459–67.
44. Shestivska V, Spanel P, Dryahina K, Sovova K, Smith D, Musilek M, et al. Variability in the concentrations of volatile metabolites emitted by genotypically different strains of Pseudomonas aeruginosa. *J Appl Microbiol* 2012;**113**:701–13.
45. Gilchrist FJ, Alcock A, Belcher J, Brady M, Jones A, Smith D, et al. Variation in hydrogen cyanide production between different strains of *Pseudomonas aeruginosa. Eur Respir J* 2011;**38**:409–14.
46. Scotter JM, Allardyce RA, Langford VS, Hill A, Murdoch DR. The rapid evaluation of bacterial growth in blood cultures by selected ion flow tube-mass spectrometry (SIFT-MS) and comparison with the BacT/ALERT automated blood culture system. *J Microbiol Methods* 2006;**65**:628–31.
47. Bunge M, Araghipour N, Mikoviny T, Dunkl J, Schnitzhofer R, Hansel A, et al. On-line monitoring of microbial volatile metabolites by proton transfer reaction-mass spectrometry. *Appl Environ Microbiol* 2008;**74**:2179–86.
48. O'Hara M, Mayhew CA. A preliminary comparison of volatile organic compounds in the headspace of cultures of *Staphylococcus aureus* grown in nutrient, dextrose and brain heart bovine broths measured using a proton transfer reaction mass spectrometer. *J Breath Res* 2009;**3**:027001.

49. Buhr K, van Ruth S, Delahunty C. Analysis of volatile flavour compounds by Proton Transfer Reaction-Mass Spectrometry: fragmentation patterns and discrimination between isobaric and isomeric compounds. *Int J Mass Spectrom* 2002;**221**:1–7.
50. Schwarz K, Filipiak W, Amann A. Determining concentration patterns of volatile compounds in exhaled breath by PTR-MS. *J Breath Res* 2009;**3**:027002.
51. Keck L, Hoeschen C, Oeh U. Effects of carbon dioxide in breath gas on proton transfer reaction-mass spectrometry (PTR-MS) measurements. *Int J Mass Spectrom* 2008;**270**:156–65.
52. Maleknia SD, Bell TL, Adams MA. PTR-MS analysis of reference and plant-emitted volatile organic compounds. *Int J Mass Spectrom* 2007;**262**:203–10.
53. Scott-Thomas AJ, Syhre M, Pattemore PK, Epton M, Laing R, Pearson J, et al. 2-Aminoacetophenone as a potential breath biomarker for Pseudomonas aeruginosa in the cystic fibrosis lung. *BMC Pulm Med* 2010;**10**:56.
54. Mann S. Über den Geruchsstoff von Pseudomonas Aeruginosa. *Arch Mikrobiol* 1966;**54**:184–90.
55. Mann S. Chinazolinderivate bei Pseudomonaden. *Arch Mikrobiol* 1967;**56**:324–9.
56. Cox CD, Parker J. Use of 2-aminoacetophenone production in identification of Pseudomonas aeruginosa. *J Clin Microbiol* 1979;**9**:479–84.
57. Labows JN, McGinley KJ, Webster GF, Leyden JJ. Headspace analysis of volatile metabolites of *Pseudomonas aeruginosa* and related species by gas chromatography-mass spectrometry. *J Clin Microbiol* 1980;**12**:521–6.
58. Enderby B, Lenney W, Brady M, Emmett C, Spanel P, Smith D. Concentrations of some metabolites in the breath of healthy children aged 7–18 years measured using selected ion flow tube mass spectrometry (SIFT-MS). *J Breath Res* 2009;**3**:036001.
59. Syhre M, Chambers ST. The scent of Mycobacterium tuberculosis. *Tuberculosis (Edinb)* 2008;**88**:317–23.
60. Syhre M, Manning L, Phuanukoonnon S, Harino P, Chambers ST. The scent of Mycobacterium tuberculosis–part II breath. *Tuberculosis (Edinb)* 2009;**89**:263–6.
61. Chambers ST, Bhandari S, Scott-Thomas A, Syhre M. Novel diagnostics: progress toward a breath test for invasive Aspergillus fumigatus. *Med Mycol* 2011;**49 Suppl. 1**:S54–61.
62. Chambers ST, Syhre M, Murdoch DR, McCartin F, Epton MJ. Detection of 2-pentylfuran in the breath of patients with Aspergillus fumigatus.*Med Mycol* 2009;**47**:468–76.
63. Filipiak W, Sponring A, Bauer M, Filipiak A, Ager C, Wiesenhofer H, et al. Molecular analysis of volatile metabolites released specifically by *Staphylococcus aureus* and *Pseudomonas aeruginosa*. *BMC Microbiol* 2012;**12**:113.
64. Filipiak W, Sponring A, Filipiak A, Ager C, Schubert J, Miekisch W, et al. TD-GC-MS analysis of volatile metabolites of human lung cancer and normal cells *in vitro*. *Cancer Epidemiol Biomarkers Prev* 2010;**19**:182–95.
65. Filipiak W, Sponring A, Mikoviny T, Ager C, Schubert J, Miekisch W, et al. Release of volatile organic compounds (VOCs) from the lung cancer cell line CALU-1 *in vitro*. *Cancer Cell Int* 2008;**8**:17.
66. Sponring A, Filipiak W, Ager C, Schubert J, Miekisch W, Amann A, et al. Analysis of volatile organic compounds (VOCs) in the headspace of NCI-H1666 lung cancer cells. *Cancer Biomark* 2010;**7**:153–61.
67. Sponring A, Filipiak W, Mikoviny T, Ager C, Schubert J, Miekisch W, et al. Release of volatile organic compounds from the lung cancer cell line NCI-H2087 *in vitro*. *Anticancer Res* 2009;**29**:419–26.

68. Gawrys M, Fastyn P, Gawlowski J, Gierczak T, Niedzielski J. Prevention of water vapour adsorption by carbon molecular sieves in sampling humid gases. *J Chromatogr A* 2001;**933**:107–16.
69. Kleinbaum D, Kupper L, Muller A, Nizam K. *Applied regression analysis and other multivariable methods*. Pacific Grove (CA): Brooks/Cole Publishing Company; 1998.
70. Guidelines for the management of adults with hospital-acquired, ventilator-associated, and healthcare-associated pneumonia. *Am J Respir Crit Care Med* 2005;**171**:388–416.
71. Preti G, Thaler E, Hanson CW, Troy M, Eades J, Gelperin A. Volatile compounds characteristic of sinus-related bacteria and infected sinus mucus: analysis by solid-phase microextraction and gas chromatography-mass spectrometry. *J Chromatogr B: Analyt Technol Biomed Life Sci* 2009;**877**:2011–8.
72. Kieronczyk A, Cachon R, Feron G, Yvon M. Addition of oxidizing or reducing agents to the reaction medium influences amino acid conversion to aroma compounds by Lactococcus lactis. *J Appl Microbiol* 2006;**101**:1114–22.
73. Morgan ME, Lindsay RC, Libbey LM, Pereira RL. Identity of additional aroma constituents in milk cultures of Streptococcus Lactis var. Maltigenes. *J Dairy Sci* 1966;**49**:15–8.
74. Yvon M, Rijnen L. Cheese flavour formation by amino acid catabolism. *Int Dairy J* 2001;**11**:185–201.
75. Michal G. *Biochemical pathways: an atlas of biochemistry and molecular biology*. 1st edn. John Wiley & Sons Inc; 1999.
76. Black JG. *Microbiology principles and explorations*. 7th edn. Hoboken (NJ, USA): Wiley; 2008.
77. Filipiak W, Ruzsanyi V, Mochalski P, Filipiak A, Ager C, Denz H, et al. Dependence of exhaled breath composition on exogenous factors, smoking habits and exposure to air pollutants. *J Breath Res* 2012;**6**:036008.

PART

I

Urban Search and Rescue Operations

CHAPTER

24 Potential Applications of Volatile Organic Compounds in Safety and Security

Agapios Agapiou[*], **Pawel Mochalski**[†,‡], **Alex Schmid**[†], **and Anton Amann**[†]

[*] *National Technical University of Athens (NTUA), School of Chemical Engineering, Sector I, 9 Iroon Polytechniou Street, 15773 Athens, Greece*

[†] *Breath Research Institute, Austrian Academy of Sciences, Rathausplatz 4, 6850 Dornbirn; Univ.-Clinic for Anesthesia, Innsbruck Medical University, Anichstr, 35, 6020 Innsbruck, Austria*

[‡] *Institute of Nuclear Physics PAN, Radzikowskiego 152, 31342 Kraków, Poland.*

24.1 INTRODUCTION

The occurrences of natural and man-made disasters are increasing worldwide. In the past decade, earthquakes have accounted for nearly 60% of all disaster-related mortalities, causing more than 780,000 deaths and directly affecting another 2 billion people.[1] In view of global urbanization and large urban center vulnerability, and bearing in mind that some of the most populous cities in the world (e.g. Tokyo, Mexico City, New York, Mumbai, Delhi, Jakarta, Shanghai, Calcutta, Los Angeles) are located in seismic zones, the probability of a near-future mega earthquake event is expected to increase.[2,3]

The survival rate of entrapped victims diminishes dramatically over time; therefore, there is a need for immediate and acute response in the disaster zone. The death toll from the last devastating earthquakes in Japan (2011), China and Haiti (2010), Indonesia and Italy (2009), reinforced the need for new, innovative technical solutions for enhancing the capabilities and intervention of urban search and rescue (USaR) teams. Until now, various instruments like geophones, telescopic or infra-red (IR) cameras, and sonars have been deployed to support the work of USaR teams. Nevertheless, the position of canines in USaR operations remains invincible and is considered as the "gold standard."[4] Search dogs exhibit excellent scenting skills and are able to search relatively large areas in a short period of time. However, they can easily get stressed, their working time is relatively short (30 min, with a subsequent break for 2 h) and their training is time-consuming and expensive.[5] These limitations have generated demands for new technical search tools, which could complement, or even replace their work.

Volatile Biomarkers. http://dx.doi.org/10.1016/B978-0-44-462613-4.00024-6

In this context, knowledge of the human scent profile and its characteristics in the disaster environment is of particular importance. Volatile species that form the human scent are released from different biological fluids (urine, blood, sweat) and tissues (skin, lungs, bowels). Their emissions depend on the conditions in the entrapment scene (confined space volume, type of collapse, temperature, humidity), the duration of entrapment, and the medical status of entrapped victims. In the entrapment environment, scent constituents are spread by air currents throughout the ruins, interact with the debris materials (brick, concrete, glass, wood, steel, etc.), and mix with environmental contaminants. Consequently, the concentration of the airborne human scent markers is low, making its detection and identification a really challenging task.

24.2 CHEMICAL ANALYSIS OF BREATH

Human breath contains numerous volatile organic compounds (VOCs) having great potential for medical, toxicological, and forensic applications. This is due to the fact that breath VOCs correlate quite well with numerous diseases, metabolic disorders, or exposure to frequently toxic environmental contaminants. Recent advances in analytical technologies resulted in novel, sensitive, and portable instruments (e.g. Ion Mobility Spectrometry, IMS[6]) allowing the on-site determination of VOCs. As a result, new challenging and exciting applications are arising in safety (exposure) and security (natural or man-made disasters) areas. Nevertheless, criteria for using breath analysis in these crucial applications need to be defined along with standardization procedures. Additionally, high throughput and sensitivity methods have to be developed.

Currently, the main applications of breath analysis in safety and security areas can be listed as follows:

- Detection of human presence.
- Identification of interactions of humans with materials and the environment leading to:
  - Illegal acts.
  - Contamination.
  - Exposure to toxics.
  - Criminal actions.
  - Law enforcement applications.

Human expired air consists of nitrogen, oxygen, carbon dioxide, water vapor, and numerous VOCs.[7–9] Although VOCs are a tiny percentage of the expired air, they can provide important medical information. VOCs in expired air are generally divided into endogenous (those that are produced in the body from metabolic processes; products of metabolism) and exogenous substances (originating outside the body). Expired air analysis is currently a field that has been widely applied in medical diagnostics, search, and rescue operations and for monitoring human exposure to specific environments. For instance, breath analysis has been applied by numerous research groups to the diagnosis of cancer by correlating the VOCs with the associated metabolic pathways leading to cancer.[8,10–14]

Chemical analysis of expired air can, in principle, provide information on human exposure to chemical, biological, and radiological agents.[15–17] Such information can be useful for identifying a person's route (sites visited) and occupation (even an illegal one), especially when unusual chemicals are present in the visited or work places (e.g. illegal drug labs, explosives). Chemical analysis of breath also has potential biometric applications. This application of expired air analysis results from the fact that vapors of chemical agents, solvents, and explosives can be inhaled by humans, distributed in human tissues, and finally detected in expired air. Furthermore, strong and fast metabolic disorders resulting from biological infection and radiological exposure produce increased quantities of certain endogenous compounds, which can be directly determined from the chemical components of the expired air.

24.3 ANALYTICAL INSTRUMENTATION; FIELD TECHNOLOGY

Until recently, the chemical analysis of breath was scientifically and technically difficult. Among the scientific problems frequently encountered were: the lack of standardization in breath sampling, difficulties in establishing reference groups, variability in the composition of expired air during day time, risk of contamination of the sample during measurements, ultra low concentrations of analytes of interest (from low ppm to low ppt), the necessity of pre-concentration, difficulties in differentiating endogenous from exogenous VOCs, difficulties in discriminating expired air components from high backgrounds, and the long time often required for analysis.

On the other hand, breath tests are non-invasive: breath can be sampled as often as it is desirable, even in real time, or during sleep,[18,19] exercise (e.g. during pedaling an ergometer[20–26]), during almost any human activity, providing unique benefits compared to conventional urine and blood sampling and testing. Note, also, the ability of sampling from handicapped people, small children, old people, critically ill patients in intensive care,[27,28] and even in the perioperative setting.[29]

Scientific and technical advances over the last 10 years have established breath testing as a very promising technique with several applications and potential for clinical applications.[30,31] Currently, several analytical techniques offer portable, selective, and very sensitive instruments ready to be used during field applications. These techniques embrace Ion Mobility Spectrometry (IMS), potentially coupled to a multicapillary column (MCC-IMS),[32–34] chemical sensors or e-nose.[35,36] In addition, various immobile or bench top, however very sensitive techniques, such as gas chromatography with mass spectrometric detection (GC-MS), proton transfer reaction mass spectrometry (PTR-MS),[37–40] selected ion flow tube mass spectrometry (SIFT-MS),[7] and laser spectrometry[41,42] are applied to breath analysis. The main advantages and disadvantages of the analytical instrumentation used in breath analysis have been summarized in Table 24.1.

Table 24.1 The main analytical techniques used in expired air analysis.

Method	Advantage	Disadvantage
PTR-MS PTR-TOF-MS	Real-time analysis, highly sensitive, no preseparation necessary, no carrier gas necessary, may use different ionization precursor ions for soft ionization (H_3O^+,NO^+,O_2^+), PTR-TOF has high mass-resolution (around 1/5000)	Large and costly instrument, mass interferences, library of compounds still to be created
SIFT-MS	Real time analysis, soft ionization, humidity measurement, uses 3 different ionization agents (H_3O^+,NO^+,O_2^+)	Uses carrier gas, less sensitive than PTR-MS
GC-MS, SPME-GC-MS TD-GC-MS	The gold standard in breath analysis. Provides full mass spectra (chemical identification) and analytical information including the type of compound, its molecular weight and retention time. SPME and TD (thermal desorption) are pre-concentration steps	Time-consuming method; for sample analysis and data interpretation, no real-time measurements are possible
IMS MCC-IMS	Low cost, portable device, with very high sensitivity, which can be improved by using different carrier gases	Difficult interpretation of results, library of compounds still to be created, insensitive to some classes of compounds
Sensor technology: Simple sensors (e.g. NO) or non-specific sensors arrays (e-noses)	Low cost, promising future, high sensitivity, use of different coatings	May need chemometric processing, suffers from cross-sensitivities
Laser spectroscopy	Measuring specific low weight volatiles, high sensitivity, promising future application with tunable lasers	Expensive due to limited production

The future analysis of VOCs is also associated with the recent developments in nanotechnology (e.g. nanotubes) and micro-machining (lab-on-a-chip concept). Promising inexpensive, fast, and widespread screening sensor arrays incorporating pattern recognition methods are expected to provide unique VOC patterns, allowing for a YES or NO identification.

24.4 FACTORS AFFECTING VOCS

Despite technological progress in the analysis of volatiles, a number of limitations still exist; these include the multiple profiles of breath analysis (as presented in Tables 24.2 and 24.3) along with the standardization of the analytical method. The developed analytical method should meet the following requirements:

- Rapid response.
- Reliability.
- Field-appliance.
- High sample throughput.
- Selectivity.
- Operation simplicity ("one-button device").

The chemical and physicochemical profile of VOC analysis strongly affects and sometimes determines a number of associated (secondary) profiles also present in expired air analysis. Safety and security barriers can be overcome by a systematic and multidisciplinary approach integrating knowledge and technology from various fields. Although VOC analysis has not been widely used in security so far, it has been used for traffic security (alcoholmeters) as a fast, reliable, field method which has received legal approval.

Chemical signatures of expired air (human indicators) can also be used in search and rescue operations for the early location of entrapped people in the ruins of collapsed buildings.[43,44] The reason for this application is that certain groups of chemical compounds found in debris air, released through breath, skin, or urine can be associated with human presence in confined spaces; this is why rescue "sniffer dogs" are usually employed. The strategy for developing an operational instrument able to

Table 24.2 The chemical and physicochemical profile of compounds in expired air, their analysis and the various factors affecting them.

Profile	Main Characteristics
Chemical	• Light gases, water, VOCs, non-condensate volatiles • Small molecules up to 250 amu • More than 200 compounds • Broad range of: H/C, O, S, N-related compounds, aromatics • May translates in markers, odor, inhaled chemicals
Physicochemical	Chemical compounds can be also looked as a range of values based on their fundamental physicochemical properties: • Boiling point • Vapor pressure • Polarity/solubility • Octanol/water partition coefficient • Henry's law constant • Acidity

Table 24.3 Secondary profiles affecting likely expired air analysis and their main characteristics.

Profile	Main Characteristics
Individual	• Gender • Age • Diseases
Environmental	• Meteorological, background emissions, interaction with background materials, field terrain (space, topography) • Takes care of sampling technique • Explains missing compounds • Explains odors • Explains concentration profiles • Differentiates among endogenous and exogenous compounds
Microbial/Radiation	• Explains how to treat samples (bio-hazards)
Transport Phenomena	• Respiration rates (flow)—Mass transfer-Laminar/ Turbulence flow • Temperature difference of human body with the environment —VOCs condensation

perform audio, video, and chemical monitoring in the ruins of collapse structures is not far away ("SGL for USaR," FP7 EC project, www.sgl-eu.org).

It appears that analysis of volatiles is a very promising method for security applications such as the identification of chemical warfare agents, monitoring and decision making in decontamination procedures, the detection of biological and radiological contaminations, the determination of recent exposure to specific chemicals (explosives, illegal drug labs, etc.), and the early location of entrapped people or dead bodies.[43,45] Applying expired air analysis in the identification of a CBRN (chemical, biological, radiological, nuclear) attack, the monitoring of decontamination processes, the tracking of illegal activities (explosives production, illegal drug labs), and the early location of entrapped people constitutes innovative promising applications. Thus, it seems that all the elements for successful application of VOC analysis in safety and security exist (i.e. basic research, applied research, critical mass, and multidisciplinary of research groups, advanced detectors—mostly detectors coping with CBRN threats; first responders are familiar with the use of IMS type detectors).

24.5 VOLATILES IN SAFETY AND SECURITY APPLICATIONS

Apart from the diagnostic potential of VOC analysis in medicine and biochemistry, a new prospective is its potential use in safety and security, based on the following questions:

Applications:

- Can VOC analysis provide information on short- or long-term exposure of humans?
- Can VOC analysis reflect on qualitative, quantitative, and temporal exposure of humans?
- Can VOC analysis discriminate humans from chemical background?
- Can exhaled breath analysis contribute to the characterization of an individual?
- Is expired air analysis a unique human fingerprint, or could it characterize a group of people, or differentiate people with respect to gender?

Technology:

- Is the present technique for volatiles' analysis in expired air sufficiently fast and accurate?
- Are fast and accurate analytical methods available suitable for field applications?

24.5.1 Safety (exposure); attack exposure

According to the US National Academy of Sciences, exposure is defined as *"an event that occurs when there is contact at a boundary between a human and the environment with a contaminant of specific concentration for an interval of time."*[46]

The various solvents extensively used in industry are considered as a potential source of human exposure. Inhalation is one of the routes of exposure along with skin absorption and ingestion. Also, breath analysis allows for exchange of compounds through the blood/breath interface.[47] This is quite crucial, especially in high-risk industries including paint, petrochemicals, and dry cleaning industries. Next to the traditional biological monitoring of blood and urine, breath analysis appears as a modern, safe, and non-invasive tool for assessing occupational exposure.

The presence of xenobiotic VOCs or the excess of endogenous VOCs in human breath can indicate recent exposure, as shown in Table 24.4. "Sick building syndrome" and bad indoor air quality are subsequent micro environments affecting modern life. Note that humans spend a lot of their time either at home or at work, indoors with exposure to a plethora of volatiles. Given below are some examples of field studies allowing the assessment of human exposure; human breath is actually serving as a surveillance tool in risk environments.

24.5.2 Safety (exposure); attack scenarios

The recent technological advancements in analysis of volatiles (new generation of small, low-cost, light-weight, mobile instruments) allows for their potential use in modern homeland security applications, including chemical, biological, and "dirty bomb" attacks. VOCs analysis seems to be of vital importance in the rapid screening of individuals in the "hot zone" in a number of security applications (attack scenarios), which are presented in Table 24.5.

Table 24.4 Breath volatiles as a result of recent exposure to xenobiotic VOCs.

VOCs	Exposure	Literature Reference
Ethanol	Ethanol ingestion	81
Alkyl halides: Vinyl chloride, Cis-1,2-dichloroethane, 1,1,1-Trichloroethylene, 1,1,1-Trichloroethane, Tetrachloroethylene, Bromodichloroethane, Bromotrichloroethane, Dichloroethane, Chloroform	Contaminated water exposure	82
Benzene, naphthalene	To jet fuel	83
Chloroform	Chlorinated municipal or swimming pool water	84, 85
Benzene	Residential benzene due to cigarette smoke	86
Tetrachloroethene	Dry cleaning	87, 88
Benzene	To coke oven industry	89
Carbon disulfide	Textile industry	90
n-Hexane, toluene butanone, acetone, toluene	Shoe manufacturing industry	87, 91
Ethyl acetate, propan-2-ol, 2-butanone, isopropyl acetate, toluene	Inks and coatings industry	87
Toluene, trimethyl benzene, tetrachloroethene, tetrahydrofuran, acetone, propan-2-ol, xylene, 2-butanone, 1-methoxy-2-propanol and *n*-hexane	Chamber studies with different solvents	87
Vinyl chloride, Cis-1,2-dichloroethene	Showering with contaminated water	92
Chloroform, bromodichloromethane	Swimming in chlorinated pool water	92
Trichloroethene	Controlled exposure chamber experiment	92
Sevoflurane, isoflurane	Post-anesthesia care units	93
Isoprene	Human presence in a private car	94

Table 24.5 Correlating attack scenarios with first responders' action.[95]

Attack Scenario	Action
Chemical	Rapid GC-MS breath screening of individuals for volatile toxins (e.g. sarin, tabun, fentanyl derivatives, anesthetics and organophosphates)
Biological[a]	Infection of individuals results in increased breath biomarkers of oxidative stress. As most pathogenic bacteria manufacture VOCs *in vitro*, their identification "fingerprint" is targeted
"Dirty Bomb"	Identification of alkanes and methylated alkanes for gamma radiation exposed victims (gamma radiation increases oxidative stress in tissues resulting in elevated in breath biomarkers, such as alkanes and methylated alkanes)

[a] *In general, there is a lack of published studies for anthrax, tularemia, or other candidate bioterrorism pathogens except for a pilot study involving Mycobacterium tuberculosis.*

24.5.3 Security (natural disasters); human life signals

24.5.3.1 Volatile human secretions

Chemical data of expired air can be used in search and rescue operations for the early location of entrapped people in the ruins of collapsed buildings.[43,48] The reason for this application is that certain groups of chemical compounds found in expired air and other biological fluids can be associated with human presence in confined spaces; canines with their superior sense are able to track these chemical human signatures. The historical employment of dogs in a variety of safety and security applications is presented in Table 24.6. Note that there are six categories of canine detection serving law enforcement: trailing, tracking, article detection (locate items deposited within a search area), substance detection, area search, and scent identification line up (use the odor scent acquired from an article of evidence to identify the suspect of a crime from a line up of scented objects).[49]

In USaR operations, first responders are mostly familiar with optical and acoustic methods. The use of field chemical detectors is limited to the identification of specific toxic compounds and chemical agents using CAM type (Chemical Agent Monitors) IMS monitoring devices. On the other hand, broader involvement appears through the use of dogs for on-site analysis. However, the detection of ultra-low concentrations in a complex background is undoubtedly a scientific and analytical challenge requiring in-depth analysis.

Human entrapment in small confined spaces, as a result of building collapse after an earthquake, leads to excretion of a variety of gases and VOCs. Concentration of VOCs are usually within the range of ppt to ppb level and potentially can increase to ppm level after hours/days of entrapment, while inorganic gases (nitric oxide, hydrogen, or ammonia) are usually excreted in ppb to ppm level.[50–52] The substances and their concentrations are strongly dependent on the entrapment time along with the victim's medical condition and the entrapment volume. The continuous human presence raises

Table 24.6 The use of dogs (*Canis familiaris* or *Canis lupus var familiaris*) as chemical detectors in different eras.[96,97]

When	Who	Where	Named
12,000 years ago	Individuals	Hunting	Hunting dogs
1800s	Individuals	Human trackers	Tracking dogs
Since World War II	Military	Location of explosives	Bomb dogs
After World War II	Civilian use	Tracking of individuals, locating drugs and bombs	Drug and bomb dogs
After World War II	Civilian use	In over 30 different applications	Chemical detector dogs
Last decade	Civilian use	Detection of flammable and ignitable liquid residues	Accelerant detector dogs
After 9/11 event in US World Trade Center	Search and rescue teams	On finding missing persons, bombing, discouraging employee drug use, termite infestation inspection, screw worm detection	Canine dogs

the concentration of VOCs originating from a variety of sources; expired air, sweat, urine, blood, faeces, vomit, gut loosing defecation, menstruation, decomposition, etc. Other chemicals present are those related to the entrapment environment, as well as, to background events in the voids. The complexity of the entrapment environment is presented in Figure 24.1. Thus, a priority filter is essential for highlighting the important sources of volatiles originating from humans.

An entrapped alive victim will continually breath, inevitably urinate, (cannot be prevented) and sweat. Perspiration strongly depends on the climate of the country; for instance in hot countries, black skins have a more odorous perspiration than white skins (except for red-haired people).[53] Blood, with fetid odor is an issue, especially when people are injured; however, blood loss is usually limited. Faeces are less frequent because constipation can sometimes last up to 10 days and vomiting depends on the individual inclinations. Menstruation is gender and age dependent and gut loosing defecation is not considered an important source. A decomposition odor is also present if the victim suffers from crush syndrome, but this becomes more intense in the case of dead victims.[45,54,55] In this context, the most promising source of human presence markers is expired air,[48] sweat, and urine.

Environmental conditions (temperature, relative humidity) and processes of diffusion, absorption, and dilution can strongly affect the identification of the human chemical signature of a victim, as the substances might interact with the entrapment environment.[56] The quick penetration of debris is important for their selection, since the fundamental physicochemical properties actually guide and determine the volatiles' fate.[56,57] Physicochemical properties predict and enhance the knowledge

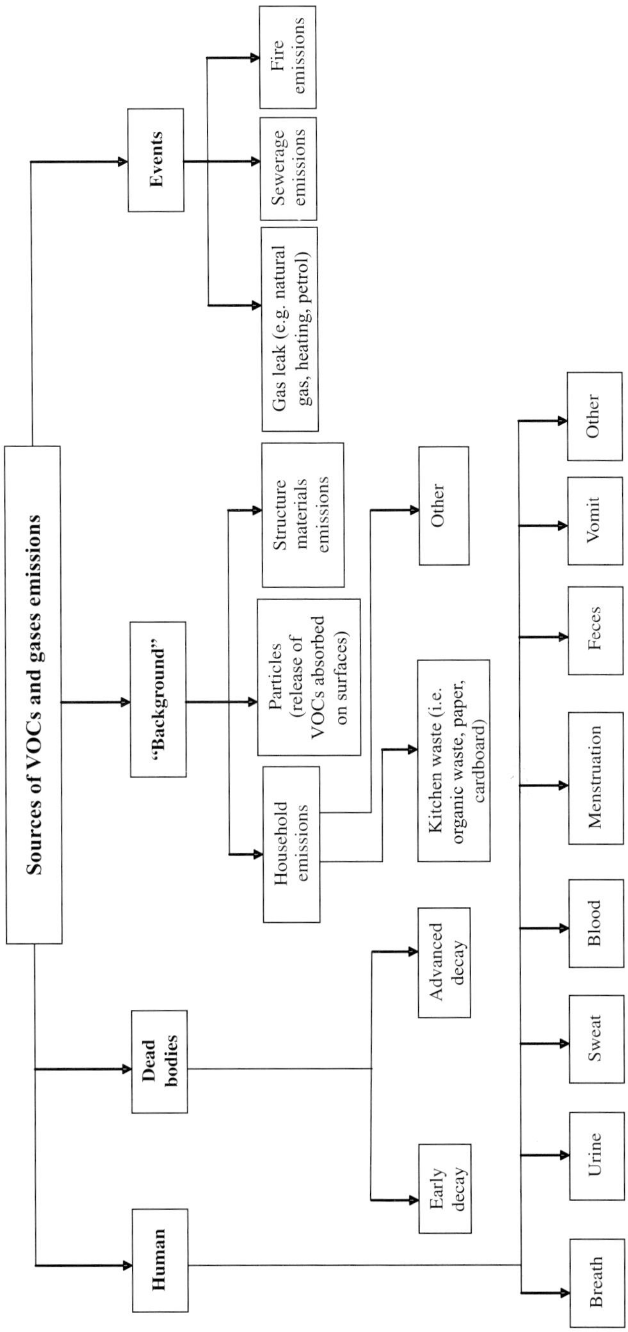

FIGURE 24.1
The complexity of the entrapment environment.

regarding the distribution and fate of volatile substances in the environment; boiling point, vapor pressure, polarity/solubility, octanol/water partition coefficient (giving an indication for the fat/blood partition coefficient), Henry's constant, acidity.

As the production of these chemicals is transient and not constant, their location presupposes that a suitable sensor is located in the appropriate space and time for the specific target chemical. Moreover, the focus on a few marker compounds is an issue, as important information might be hidden in the whole matrix. Therefore, selection criteria need to be established and developed based on the following principles: limit of detection, unique in relation to the source, concentration range in a specific time frame (as concentration is changing due to entrapment time frame), overall detection, selection of physicochemical properties that interact with the background, strength of odor.

24.6 URINE AS A POTENTIAL SOURCE OF MARKERS OF HUMAN PRESENCE

24.6.1 Background

Urine, next to breath and skin, is an important reservoir of human scent constituents. This is due to the kidneys' pre-concentration capabilities and the fact that the entrapped individuals are usually dehydrated. Therefore, urine can be an important source of volatile markers of entrapped victims. Until now, more than 230 VOCs belonging to different chemical classes (e.g. aldehydes, ketones, furans, pyrroles, terpenes, sulfur-containing compounds) have been detected in human urine.[43,58–62] Consequently, this plenitude and high abundance of urine-borne species is of particular importance for USaR operations. The ideal urine-borne markers should be relatively volatile, non-reactive, and omnipresent at relatively high concentrations to ensure their detection in the proximity of entrapped persons. Consequently, only spontaneously released molecular species from human urine can be suitable as markers of human presence. In the context of USaR operations, VOCs emitted by human urine hold almost an equally important status as compounds emitted by exhaled breath.

The selection of potential urine-borne markers of human presence requires firstly their reliable identification, secondly the investigation of their interactions with common debris materials, and finally, the evaluation of the suitability of currently available portable, field techniques (e.g. IMS, aspiration-type IMS (AIMS)) for their detection and monitoring.

24.6.2 Volatiles spontaneously emitted by human urine

Due to its high versatility, sensitivity, selectivity, and advanced identification, headspace solid-phase micro-extraction gas-chromatography-mass spectrometry (HS-SPME-GC-MS) is considered as the gold standard for the analysis of VOCs emitted from biological samples. Therefore, the analysis of VOCs spontaneously emitted by unmodified human urine (no addition of salt, or adjustment of pH, or heating) was performed using this technique. The headspace analysis of urine samples obtained from a group of 20 healthy volunteers (9 females, 11 males, median age 29.5 years) resulted in the detection of a total of 148 volatile compounds.[63] 33 VOCs were found to be omnipresent (with incidence higher than 80%). Their incidence and properties, which can govern their behavior in the entrapment scene, are presented in Table 24.7. A typical chromatogram of HS-SPME-GC-MS urine analysisis presented in Figure 24.2. The most represented chemical class was ketones with 10 representatives (acetone, 2-butanone, 3-methyl-2-butanone, 3-methyl-2-pentanone, 2-pentanone, 4-methyl-2-pentanone, 2-heptanone, 4-heptanone, 3-hexanone, 3-penten-2-one). The other well-represented class was aldehydes including seven representatives (propanal, 2-methylpropanal, 2-methyl-butanal, 2-methyl-2-butenal, pentanal, hexanal, and octanal). Within the remaining species there were seven sulfur-containing compounds (methanethiol, dimethyl sulfide (DMS), dimethyl disulfide (DMDS), dimethyl trisulfide (DMTS), allylisothiocyanate, isothiocyanocyclohexane, and dimethyl sulfone ($DMSO_2$)), one aromatic hydrocarbon (toluene), two esters (methyl acetate and ethyl acetate), one terpene (carvone), and four heterocyclics (furan, 3-methyl-furan, pyrrole, and 1-methyl-pyrrole). Aliphatic hydrocarbons were represented only by isoprene, which is considered the most prominent hydrocarbon, contained in blood and emitted in breath[20–22,26] and urine, but not from human skin.[64] Incidentally, isoprene in comparatively high concentrations is typical for humans; elephants release isoprene in relation to musth[65] and in rat, rabbit, dog, ewe, and cow the isoprene concentrations in breath or urine are very small.[66,67] In general, 17 compounds were found in all urine samples and further five substances in all samples except one. The 33 omnipresent urine species can be considered as an initial library of urine-borne potential markers of human presence.

24.6.3 Evolution of urine scent profile during sample storage at room temperature

The evolution of urine scent profile was investigated in the course of 4 days of storage. The urine samples of 20 volunteers were stored in closed glass vials at room temperature and analyzed after certain periods of time (12, 24, 48, 72, and 96 h) to observe possible changes in the headspace composition.

Headspace levels of 26 compounds omnipresent in urine were found to increase in the course of the experiment. After 4 days the increase ranged from 36% for isothiocyanocyclohexane to 525% for DMTS (median 252%). The highest rise after 96 h was noted for three sulfur species: methylmercaptan (322%), DMDS (499%), and DMTS (525%). Concentrations of 10 ketones were found to increase, with the highest gain observed for 3-hexanone (383%) and 4-methyl-2-pentanone (365%). Among the remaining compounds there were three aldehydes (2-methylpropanal, 2-methylbutanal, and pentanal), three heterocyclic compounds (1-methyl-pyrrole, furan, and 3-methylfuran), carvone, and two esters (methyl acetate and ethyl acetate). The emission of only three compounds, namely $DMSO_2$, octanal, and isoprene, decreased within 4 days of storage. However, the drop was prominent only for $DMSO_2$ (49%). Levels of all remaining compounds (acetone, toluene, DMS, propanal, pyrrole, and allylisothiocyanate), did not exhibit a pronounced trend or remained stable. Exemplary trends in the headspace levels of

Table 24.7 Incidence and selected chemical properties of VOCs released by unmodified human urine. Physicochemical data are from Refs [98–100]. Reproduced with kind permission by Informa Healthcare from Ref. [63].

Compound	CAS	Henry's Law Constant (25°C) (mol × dm^{-3} × atm^{-1})	Octanol-Water Partition Coef. Log Kow (25°C)	Incidence *n*/20 (%)	LOD (ng/mL)	RSD (%)	Possible Origin
Methanethiol	74-93-1	0.3	0.78*	20/20 (100)	300	10	Systemic, bacterial
Furan	110-00-9	0.2	1.34	20/20 (100)	15	2	Smoking, dietary
Propanal	123-38-6	13	0.59	19/20 (95)	5	6	Systemic, dietary
Acetone	67-64-1	28	−0.24	20/20 (100)	6	2	Systemic, dietary
Dimethyl sulfide	75-18-3	0.48	0.92*	19/20 (95)	10	11	Systemic, dietary
Methyl acetate	79-20-9	10	0.18	18/20 (90)	14	7	Dietary, environmental
Isoprene	78-79-5	0.028	2.42	18/20 (90)	20	4	Systemic
2-Methylpropanal	78-84-2	10	0.74*	20/20 (100)	16	2	Systemic, dietary
2-Butanone	78-93-3	20	0.29	20/20 (100)	4	1	Dietary
3-Methylfuran	930-27-8	0.1	1.91*	20/20 (100)	12	7	Smoking, dietary
Ethyl acetate	141-78-6	7	0.73	16/20 (80)	12	7	Dietary, environmental
Pyrrole	109-97-7	10	0.75	19/20 (95)	50	2	Systemic, dietary
3-Methyl-2-butanone	563-80-4	7	0.84	20/20 (100)	4	4	Dietary
2-Methylbutanal	96-17-3	5	1.23*	20/20 (100)	11	3	Systemic, dietary
2-Pentanone	107-87-9	16	0.91	20/20 (100)	5	2	Dietary
Pentanal	110-62-3	6	1.31*	19/20 (95)	19	4	Systemic, dietary
1-Methyl-pyrrole	96-54-8	5.15*	1.21	16/20 (80)	—	8	Systemic, dietary
Dimethyl disulfide	624-92-0	1	1.77	20/20 (100)	4	6	Dietary, bacterial
2-Methyl-2-butenal	1115-11-3	45	1.15*	17/20 (85)	15	8	Systemic, dietary
3-Penten-2-one	625-33-2	30	0.52	19/20 (95)	2	5	Dietary

(Continued)

Table 24.7 (Continued)

Compound	CAS	Henry's Law Constant (25°C) (mol × dm^{-3} × atm^{-1})	Octanol-Water Partition Coef. Log Kow (25°C)	Incidence *n*/20 (%)	LOD (ng/mL)	RSD (%)	Possible Origin
Cyclopentanone	120-92-3	100	0.38	15/20 (75)	7	7	Dietary
Toluene	108-88-3	0.15	2.73	20/20 (100)	2	5	Environmental
4-Methyl-2-pentanone	108-10-1	2.2	1.31	20/20 (100)	2	5	Dietary
Allylisothiocyanate	57-06-7	4.15	2.15*	16/20 (80)	—	8	Dietary
3-Methyl-2-pentanone	565-61-7	10	1.16*	20/20 (100)	5	3	Dietary
3-Hexanone	589-38-8	10	1.24*	20/20 (100)	2	2	Dietary
Hexanal	66-25-1	5	1.78	17/20 (85)	14	4	Systemic, dietary
Dimethyl sulfone	67-71-0	50000	−1.41	17/20 (85)	—	12	Environmental
3-Methylcyclopentanone	1757-42-2	20*	1.05*	13/20 (65)	—	7	Dietary
4-Heptanone	123-19-3	7	1.73*	20/20 (100)	7	1	Environmental, dietary
2-Heptanone	110-43-0	7	1.98	20/20 (100)	7	3	Environmental
Dimethyl trisulfide	3658-80-8	0.54*	1.87*	18/20 (90)	20	12	Dietary
Octanal	124-13-0	2	2.78*	17/20 (85)	6	5	Systemic, dietary
Isothiocyanocyclohexane	1122-82-3	0.177*	3.57*	20/20 (100)	—	8	Environmental
Carvone	2244-16-8	13*	2.71	17/20 (85)	15	3	Dietary

Values with asterisk are not experimental data, but calculated using EPI suite software (by US environmental protection agency).

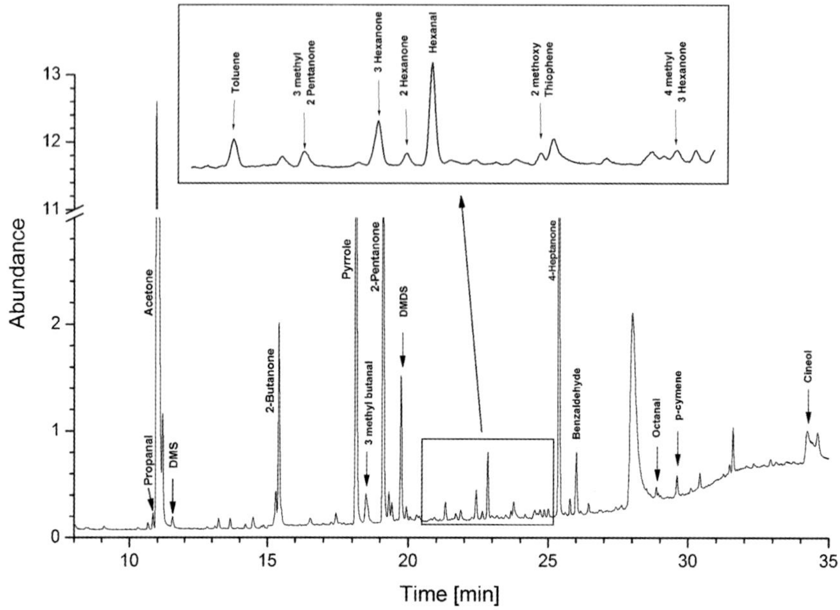

FIGURE 24.2

A typical chromatogram of HS-SPME-GC-MS urine analysis.

DMDS, 3-hexanone, $DMSO_2$, and acetone within the course of the experiment are presented in Figure 24.3. These changes in VOCs emission are most probably the result of different factors, such as intense bacterial activity, pH changes, modifications in urine characteristics (affecting for example the VOCs solubility, presence of photo-chemically degradable VOCs), and decomposition of urine constituents (e.g. proteins). The increase of the emission rate can be considered as a favorable effect improving the VOCs detection. In this context, species like sulfur compounds (DMDS and DMTS) and some ketones (e.g. 3-hexanone, 2-heptanone, 4-heptanone, 4-methyl-2-pentanone), showing the highest increase in emission during storage, seem to be attractive markers of human urine.

24.6.4 Preliminary study on interactions of urine-borne species with common debris materials

Knowledge about the emission/evolution of urine volatiles and their behavior in the presence of typical debris materials is a critical factor for the selection of urine markers of human presence. Brick and concrete, along with wood, are considered as the basic building materials in household dwellings of modern societies. The type of construction material is considered a critical factor in mortality and injury patterns during an earthquake.[68] The various gas detectors used for the determination of the chemical

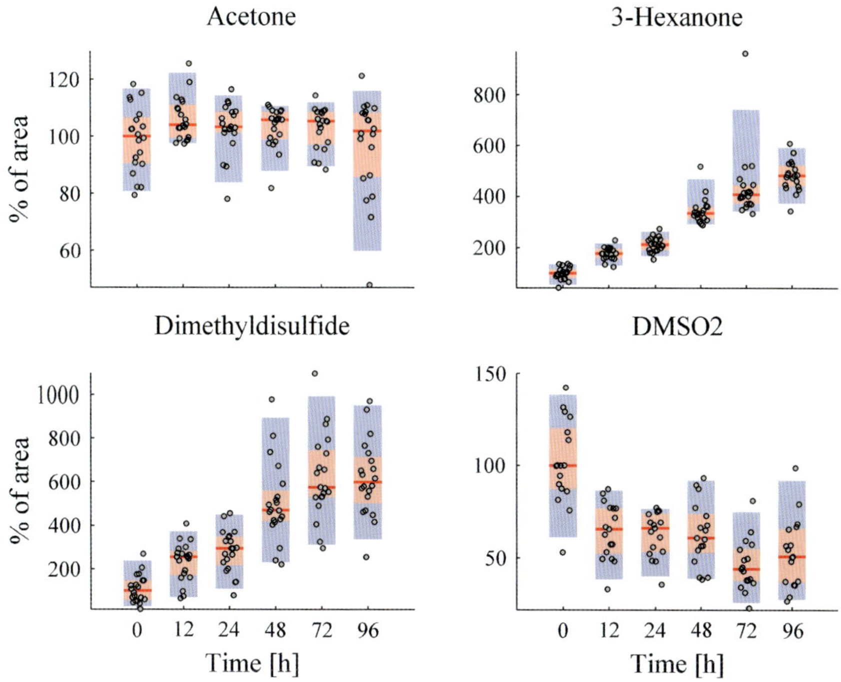

FIGURE 24.3
Evolution profile of selected urine VOCs after 72 h of sample storage at room temperature. Reproduced with kind permission by Informa Healthcare from Ref. [63].

species in ambient air are actually detecting the gas phase concentration of these molecular species, neglecting the importance of the liquid and/or adsorbed phase, where these species are accumulated in higher concentrations. Critical factors for the absorption are considered to be the indoor environmental conditions, the household materials, and the states of matter (e.g. solid, powder, liquid).

Two basic building materials, brick and concrete, were investigated using HS-SPME-GC-MS and an in-house made filling chamber mimicking the entrapment scene.[69] The filling chamber used during the experiments is presented in Figure 24.4. The chamber consisted of a closed stainless steel cylinder having internal diameter 102 mm and height 123 mm (capacity of 950 mL). Steel gauze (mesh size 1 mm) divided the chamber into two parts (41 and 82 mm high). The top part was loosely packed with tested debris materials having grain diameters of 4–8 mm and layer thickness 4 cm (approximately 350 g). The lid covering the chamber was equipped with a septum-less head (Gerstel, Germany) enabling the SPME extraction from the headspace of the chamber. The whole chamber was continuously purged with high purity synthetic air, at a flow rate of 10 mL/min, to mimic air currents spreading scent constituent in the entrapment scene. 5 mL of urine obtained from four volunteers (age range: 32–41 years, non-smokers)

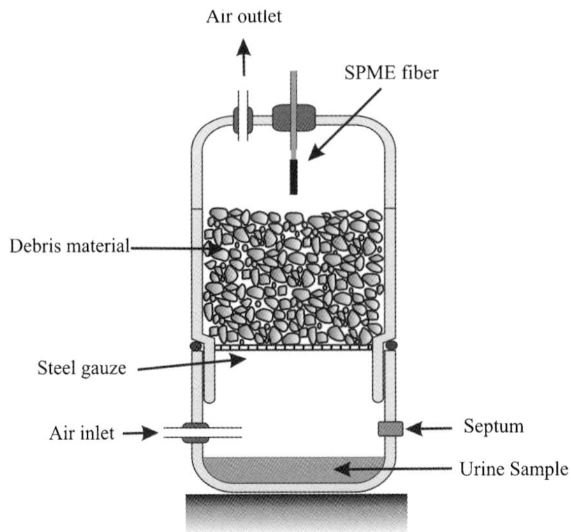

FIGURE 24.4
Sketch of the filling chamber. Reproduced with permission from Ref. [69].

were placed below the material layer on the onset of the experiment. The urine vapors in the headspace were monitored every 36.4 min for 24 h. In the context of USaR operations, this sampling rate was recognized as satisfactory, as the canine operating time is around 30 min. Three experiments were performed per volunteer, comprising both an empty chamber and chamber filled with brick or concrete. The empty chamber measurements were treated as reference points for the assessment of the interaction of urine vapors with the selected tested materials.

A total of 22 urine-borne compounds was detected in the headspace of the chamber. The identified compounds per volunteer are displayed in Table 24.8. The most represented chemical class was ketones with 11 representatives and aldehydes with five species. Among the remaining compounds, there were two furans, two volatile sulfur compounds, one nitrile, and one heterocyclic. Six compounds (acetone, 2-butanone, 2-pentanone, 4-heptanone, pyrrole, and DMS) were observed in all experiments. For the majority of compounds, the maximal observed concentration did not exceed 10 ppb. Only three species (acetone, 2-butanone, 2-pentanone) exhibited higher abundances. Particularly high concentrations ranging from 300 to 630 ppb were noted for acetone. The high abundance of this compound is consistent with other studies, implying high levels of acetone in different biological fluids and tissues.[8,47,70] The observed maximal concentrations varied strongly between volunteers. This reflects the individual urine composition, which mainly depends on the individual's diet and nutritional status, medical condition, environmental exposure, and dehydration.

Exemplary permeation profiles obtained for selected urine-borne species are presented in Figure 24.5. The typical concentration profile consists of two parts: an increase

Table 24.8 Incidence and maximal concentrations of urine-borne species obtained for different tested materials investigated within this study. For DMS and DMDS, the values in brackets refer to the second observed peak. Empty cells indicate that the respective VOC was not detected. Reproduced with permission from Ref. [69].

VOC	Volunteer 1 (ppb)			Volunteer 2 (ppb)		
	No Filling	Brick	Concrete	No Filling	Brick	Concrete
Furan	1.1 ± 0.1	1.3 ± 0.1	1 ± 0.1			
Propanal	3.2 ± 0.3	2.8 ± 0.3	3.3 ± 0.3	5.45 ± 0.44	3.15 ± 0.25	3 ± 0.25
Acetone	544 ± 25	551 ± 26	262 ± 12	296 ± 14	249 ± 12	182 ± 9
DMS	1.6 ± 0.1 (0.73)	1.9 ± 0.2	1.9 ± 0.2	1.34 ± 0.1	1.65 ± 0.1	1.36 ± 0.1
2-Methyl propanal	1.9 ± 0.1	2 ± 0.1	1.1 ± 0.1			
2,3-Butanedione	2.3 ± 0.1	4.8 ± 0.3	4 ± 0.2	1.99 ± 0.1	1.15 ± 0.06	1.45 ± 0.08
2-Butanone	46 ± 2	36 ± 2	15 ± 1	5.25 ± 0.24	4.15 ± 0.2	4.36 ± 0.2
2-Methyl furan	0.84 ± 0.06	0.85 ± 0.06	0.61 ± 0.05			
Pyrrole	11.6 ± 0.8	9.3 ± 0.7	5.4 ± 0.4	4.5 ± 0.3	4 ± 0.3	2.5 ± 0.2
3-Methyl-2-butanone	1.7 ± 0.2	1.3 ± 0.1	0.73 ± 0.06			
2-Pentanone	73 ± 4	34.4 ± 2	7.7 ± 0.5	18.5 ± 1	10.7 ± 0.6	5.4 ± 0.3
n-Pentanal						
DMDS	12.7 ± 0.7 (18.1)	6.3 ± 0.4	6.9 ± 0.4			
3-Penten-2-one	0.8 ± 0.03	0.6 ± 0.02	0.2 ± 0.01			
4-Methyl-2-pentanone	1 ± 0.02	0.4 ± 0.02	0.35 ± 0.02	0.29 ± 0.02	<LOQ	<LOQ
3-Methyl-2-pentanone	0.6 ± 0.02	0.24 ± 0.01				
3-Hexanone	2.5 ± 0.15	0.94 ± 0.06	<LOQ			
n-Hexanal	5.5 ± 0.2	4.5 ± 0.2	6.57 ± 0.2			
4-Heptanone	13.1 ± 1	2.7 ± 0.2	0.46 ± 0.03	1.74 ± 0.1	0.42 ± 0.03	0.21 ± 0.03
2-Heptanone	2 ± 0.14	0.92 ± 0.07	0.61 ± 0.05	0.54 ± 0.05	0.29 ± 0.04	0.56 ± 0.05
n-Octanal						

(Continued)

Table 24.8 (Continued)

VOC	Volunteer 3 (ppb)			Volunteer 4 (ppb)		
	No Filling	Brick	Concrete	No Filling	Brick	Concrete
Furan						
Propanal	3.34 ± 0.3	5.1 ± 0.4	3.9 ± 0.31			
Acetone	515 ± 24	551 ± 26	220 ± 10	627 ± 30	590 ± 28	348 ± 16
DMS	0.69 ± 0.06 (0.35)	1 ± 0.08	0.68 ± 0.06	0.69 ± 0.06	0.8 ± 0.06	0.6 ± 0.05
2-Methyl propanal						
2,3-Butanedione				7.8 ± 0.4	8.5 ± 0.4	10.1 ± 0.5
2-Butanone	26 ± 1	21 ± 1	14 ± 1	6.35 ± 0.3	5.7 ± 0.3	2.5 ± 0.2
2-Methyl furan						
Pyrrole	6.58 ± 0.46	6.1 ± 0.4	1.66	4.94 ± 0.12	4.5 ± 0.3	2.7 ± 0.2
3-Methyl-2-butanone	0.78 ± 0.06	0.7 ± 0.06	0.4 ± 0.03	0.4 ± 0.03	0.35 ± 0.03	0.25 ± 0.03
2-Pentanone	70 ± 4	28 ± 2	5 ± 0.3	13 ± 1	7 ± 0.4	1.2 ± 0.07
n-Pentanal	5.7 ± 0.5	6.6 ± 0.6	6 ± 0.6			
DMDS	1.97 ± 0.1 (0.94)	1.1 ± 0.06	1 ± 0.06			
3-Penten-2-one						
4-Methyl-2-pentanone						
3-Methyl-2-pentanone	0.35 ± 0.01	0.3 ± 0.01	<LOQ			
3-Hexanone	0.72 ± 0.04	0.4 ± 0.03	0.3 ± 0.03	0.4 ± 0.03	<LOQ	<LOQ
n-Hexanal	5.65 ± 0.2	7.24 ± 0.3				
4-Heptanone	21 ± 1.5	11.2 ± 1	1.6 ± 0.1	1.6 ± 0.1	0.65 ± 0.05	<LOQ
2-Heptanone	1.71 ± 0.14	0.4 ± 0.05	0.3 ± 0.04			
n-Octanal	4.62 ± 0.3	2.1 ± 0.2	<LOQ			

**<LOQ = the analyte was detected (measured above the LOD) but not quantified (being below the LOQ).*

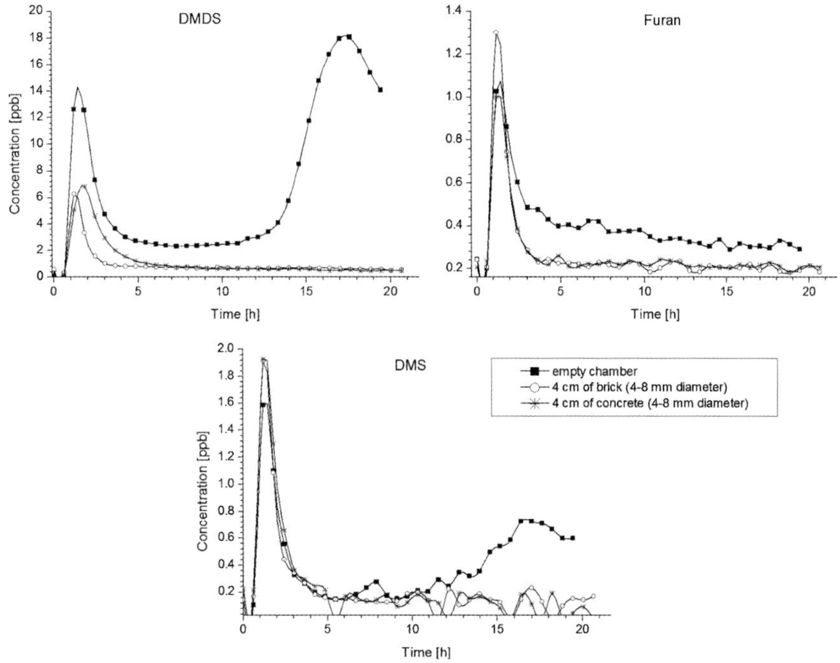

FIGURE 24.5
Permeation profiles of selected urine-borne species in the filling chamber headspace. Reproduced with permission from Ref. [69].

following the urine injection and an exponential decay. The profiles were compared using the time constant (τ), representing the removal rate of urine vapor constituents from the chamber, peak times (time of the experiment after which the maximal concentration appears) and the concentration value at the peak maximum. The average values of the aforementioned characteristics calculated for observed species are presented in Tables 24.8 and 24.9.

The influence of tested materials on the permeation profiles of analytes under study depended on their chemical class and molecular mass. The profiles of furans were found not to be affected significantly by both building materials. The values of time constants and maximal observed concentration were similar during respective experiments proving weak interactions with the surface of brick or concrete. These compounds were also relatively quickly removed from the chamber and their concentrations within 4–5 h returned to background levels. This feature seems to result from their relatively poor solubility, thus imposing their quick transfer from urine into the gas phase. Consequently, the usefulness of the compounds from this group can be limited during USaR operations.

The sulfur compounds behaved in a similar way to furans when the chamber was filled with brick or concrete. The concentration peaks related to the urine injections did

Table 24.9 The average time constants (τ) represent the removal rate of the urine-borne compounds from the chamber and the average peak times (times of the maximal concentrations measured from the time of the urine injection). Reproduced with permission from Ref. [69].

VOC	Time Constant (h)			Peak Time (h)		
	No Filling	Brick	Concrete	No Filling	Brick	Concrete
Acetonitrile	4.40	2.72	12.00	0.19	0.80	0.19
Furan	1.93	0.75	0.94	0.19	0.19	0.19
Propanal	5.42	7.59	6.67	0.39	0.80	2.35
Acetone	2.50	2.45	5.17	0.19	0.60	1.61
DMS	1.12	0.93	0.91	0.19	0.19	0.53
Propanal, 2-methyl-	2.48	1.50	21.60	0.80	0.80	1.40
2,3-Butanedione	n.c.	n.c.	n.c.	0.73	1.40	14.75
2-Butanone	2.02	2.24	10.37	0.19	0.60	2.62
2-Methyl furan	1.90	1.08	1.10	0.19	0.19	0.19
Pyrrole	4.53	5.20	7.38	0.65	1.25	2.62
3-Methyl-2-butanone	1.67	2.50	17.04	0.39	1.40	3.53
2-Pentanone	1.62	2.45	n.c.	0.19	1.20	4.84
n-Pentanal	n.c.	n.c.	n.c.	0.80	1.40	11.72
DMDS	0.89	0.76	1.36	0.19	0.19	1.10
3-Penten-2-one	2.91	n.c.	n.c.	0.80	1.40	7.47
2-Pentanone, 4-methyl-	2.43	3.59	n.c.	0.49	1.71	9.59
2-Pentanone, 3-methyl-	1.55	3.04	n.c.	0.19	1.71	11.72
3-Hexanone	1.23	3.13	n.c.	0.19	1.71	12.64
n-Hexanal	2.09	n.c.	n.c.	0.80	2.92	4.44
4-Heptanone	1.38	4.81	n.c.	0.19	2.82	19.21
2-Heptanone	1.41	17.93	n.c.	0.19	3.43	19.21
n-Octanal	n.c.	n.c.	n.c.	1.40	11.12	20.83

n.c. = not calculated.

not last longer than 5 h. Interestingly, for DMS and DMDS, two peaks were observed. The additional peaks of these substances can evidence the additional production of these compounds during the urine ageing (e.g. from decomposition of proteins).

Ketones exhibited longer residence times and stronger interactions with the debris materials. Whereas acetone did not interact significantly with the brick, the observed maximal concentrations of 2-butanone, 2-pentanone, and 4-heptanone were respectively 20%, 50%, and 75% reduced after passing through the 4 cm brick layer than in empty chamber experiments (Table 24.8). The increase of τ values ranged from 50% for C5 ketones to 250% for C7 ketones. The presence of concrete as the adsorption material modified the profiles of ketones even stronger. Generally, the concentration peaks were much wider without a clearly marked maximum point and exhibited strong tailing. This was also manifested by the huge shift of the peak times, especially visible for heavier ketones. For example, peak maxima of C7 ketones occurred 6–8 times later in concrete than in brick. The time constants for the compound removal behaved in a similar manner; however, due to the 24 h limitation of the experimental time, the τ values of heavier ketones could not be calculated. The observed maximal concentrations were substantially decreased and this drop again showed molecular mass dependence.

Aldehydes were found to be much less abundant and omnipresent in the headspace of the chamber than ketones. Their profiles did not show such pronounced peaks and their concentrations after urine injection remained elevated, usually until the end of the experiment. Interestingly, the maximal concentrations of aldehydes were relatively unaffected by the presence of the tested materials. Propanal, the only omnipresent representative of this family, exhibited weak interactions with debris materials showing similar behavior to furans or sulfur species.

Although the applied experimental set-up is a very simplified model of an entrapment scene and cannot incorporate all possible factors, several valuable pieces of information can be drawn from the urine-borne species behavior during permeating through the tested debris materials. The emission of volatiles from urine is not constant and shows a concentration profile with an initial peak related to the urination, which next is washed out by the prevailing air currents. Since urine is a limited reservoir of species, their long-term presence in the surroundings of the entrapped person will strongly depend on their physicochemical properties. Compounds well soluble in urine are released longer than compounds exhibiting low solubility. Consequently, the latter disappear relatively quickly from the victim's surroundings, showing relatively low usefulness for USaR operations. All this means that urine is considered as rather an impulse type source of potential markers of human presence. The time of the appearance of this "impulse" and its strength depends on the individual's characteristics like physiological status, diet, or dehydration. The exact time of this occurrence is difficult to predict; however, it should appear within 12 h after entrapment. Brick was found to be a much less adsorptive for urine-borne species than concrete. Although concrete reduces the observed levels of compounds under study, it prolongs their presence in the debris. More polar analytes (e.g. ketones) showed relatively long residence times. This feature is particularly desired in the context of USaR operations. Members of this chemical family also demonstrated the highest levels in the chamber headspace. Consequently, their detection in the surroundings of the entrapped person is easier. Acetone in this context holds a distinguish status since its levels were especially high. This compound accounted for 80–90% of the total mass of volatiles emitted by the urine. Bearing in mind that acetone is also a major constituent of human breath and is emitted in significant amounts from human skin,[40,70] it can be considered as a powerful biomarker of human presence.

The proposed experimental set-up is considered as a useful model for studying the interaction of human urine VOCs with the debris materials; nevertheless, it cannot cover all possible parameters. In real situations, different factors (e.g. humidity, temperature, indoor or outdoor contaminants) and hazardous events in the ruins (e.g. smoldering fires) will influence the released scent profile.

24.7 EVALUATION OF IMS-BASED PORTABLE TECHNOLOGIES FOR THE DETECTION OF URINE-BORNE HUMAN SCENT CONSTITUENTS

24.7.1 Background

Despite its high versatility and sensitivity, GC-MS is a time- and effort-consuming technique and can usually be applied only in laboratory conditions by well-trained personnel. Consequently, the application of this technique in the field operations is very limited, or even impossible. Nevertheless, the development of an on-site powerful GC-MS instrument is a nearby target. Thus, for the detection, localization and monitoring of biomarkers of human presence in the disaster scene, portable, rapid, highly sensitive and selective instruments need to be used. In addition, the field techniques should be simple and not energy demanding. IMS seems to meet these requirements as is considered a leading technology in safety and security applications. Ion mobility spectrometers can be quite small and handheld and provide excellent detection limits for numerous classes of compounds.[32,71–73] In terms of principle of operation, there are three different types of ion mobility spectrometers: conventional, aspiration, and DMS (differential mobility spectrometers) or FAIMS (field asymmetric waveform IMS).[6] The first two types of IMS have been tested for their suitability in the detection/identification of urine-borne volatiles; an ion mobility spectrometer with β-radiation source (^{63}Ni) combined with a multi-capillary column (MCC-IMS) and an AIMS.

24.7.2 Application of the MCC-IMS for the detection and quantification of urine-borne volatiles

24.7.2.1 Suitability of MCC-IMS for detecting urine volatiles

An ion mobility spectrometer provided by ISAS (Institute for Analytical Sciences, Dortmund, Germany) with a radioactive ionization source (^{63}Ni) and multi-capillary polar column (OV-5, Multichrom, Novosibirsk, Russia) was used as a prototype instrument. The study was focused exclusively on the omnipresent urine volatiles identified by the HS-SPME-GC-MS technique, as described in Section 24.6.2. Only 23 compounds from the original set of 33 potential urine markers could be detected.[56] Their identification parameters (retention time and ion mobility) are presented in Table 24.10. To further verify the suitability of the MCC-IMS for the detection of urine volatiles, a series of measurements of urine headspace was performed. A study involving a group of 30 healthy volunteers (20 males, 10 females, mean age: 22–53 years) resulted in the detection of 14 compounds from the group of omnipresent urine species; however, only seven of them: acetone, propanal, 3-methyl-2-butanone, 2-methylpropanal, 4-heptanone, 2-heptanone, and octanal were found in all samples. Further two, hexanal and 2-pentanone were detected in at least 80% of samples. A typical 3D IMS chromatogram from human's urine headspace is shown in Figure 24.6. Despite having some limitations in detecting some important families of urine-borne volatiles like hydrocarbons, sulfur compounds, or aromatics, MCC-IMS exhibited excellent

Table 24.10 MCC-IMS identification parameters of promising urine-borne markers of human presence. Reproduced with kind permission of Springer Science+Business Media from Ref. [56].

Name	CAS	Proton Affinity ($kJ\ mol^{-1}$)	Peak	Retention Time (s)	Reduced Ion Mobility ($cm^2/V \times s$)
1,2,3-Trimethylbenzene	526-73-8	No data	I	41.4	1.77
2-Butanone	78-93-3	827.3	I	5.5	1.87
			II	5.6	1.61
2-Heptanone	110-43-0	No data	I	19.1	1.57
			II	19.1	1.47
			III	19.1	1.23
2-Methyl-1-propanal	78-84-2	797.3	I	6	1.74
			II	6	1.66
2-Methyl-2-butenal	1115-11-3	843.7	I	8	1.81
			II	8	1.48
2-Methylbutanal	96-17-3	No data	I	7	1.66
2-Pentanone	107-87-9	832.7	I	8	1.77
			II	8	1.58
			III	11.3	1.46
3-Hexanone	589-38-8	843.2	I	9	1.71
			II	8	1.44
			III	9	1.38
3-Methyl-2-butanone	563-80-4	836.3	I	5	1.8
			II	5	1.65
			III	5.5	1.48
3-Methyl-2-pentanone	565-61-7	No data	I	8.3	1.69
			II	8.3	1.35
3-Methylfuran	930-27-8	854	I	9	1.87
			II	8.3	1.79
			III	8.3	1.57
			IV	8.3	1.45
3-Penten-2-one	625-33-2	864.7	I	8	1.71
			II	8	1.44
			III	8.9	1.38

(Continued)

Table 24.10 (Continued)

Name	CAS	Proton Affinity (kJ mol^{-1})	Peak	Retention Time (s)	Reduced Ion Mobility (cm^2/V × s)
4-Heptanone	123-19-3	845	I	15.6	1.62
			II	18	1.48
			III	14.5	1.27
4-Methyl-2-pentanone	108-10-1	No data	I	8.9	1.7
			II	6.7	1.55
			III	7.8	1.36
Acetone	67-64-1	812	I	5.6	1.78
Dimethyl disulfide	624-92-0	815.3		In RIP area	
Dimethyl sulfide	75-18-3	830.9		In RIP area	
Dimethyl sulfone	67-71-0	No data		In RIP area	
Dimethyl trisulfide	3658-80-8	No data		In RIP area	
Ethyl acetate	141-78-6	835.7	I	6.6	1.49
Furan	110-00-9	812		Not detected	
Hexanal	66-25-1	No data	I	8.5	1.56
			II	10	1.29
Isoprene	78-79-5	826.4	I	44.2	1.65
			II	44.2	1.53
Methanethiol	74-93-1	773.4		Not detected	
Methyl acetate	79-20-9	821.6	I	6.1	1.68
Octanal	124-13-0	No data	I	43	1.4
			II	43	1.35
			III	43	1.11
Pentanal	110-62-3	796.6	I	6.7	1.64
			II	6.7	1.41
Propanal	123-38-6	786	I	1.1	1.75
			II	1.1	1.71
Pyrrole	109-97-7	875.4	I	45.5	1.21
Toluene	108-88-3	784		Not detected	

RIP = Reactant Ion Peak.

time response and sensitivity for the dominant volatiles classes such as ketones and aldehydes. Within this context, IMS can be considered as a promising technique for the detection, localization, and monitoring of VOCs evolved from human urine for the purpose of locating entrapped victims in collapsed buildings.

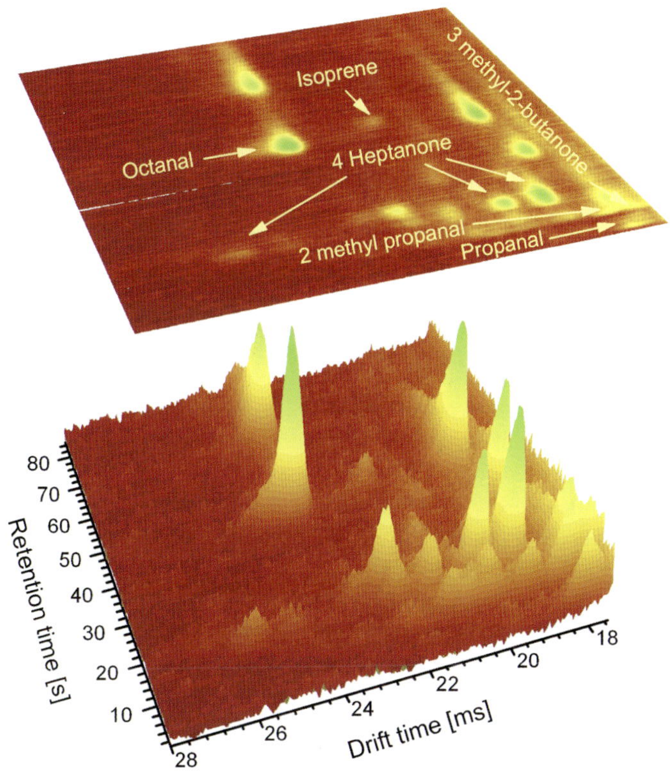

FIGURE 24.6
Three-dimensional (contour) IMS chromatogram of a headspace human urine sample.

24.7.2.2 Detection of urine-borne markers permeating through the debris material by MCC-IMS

The detection of urine-borne compounds permeating through the debris material continued with additional tests using the MCC-IMS technique.[56] The tests were performed using a filling chamber similar to the one described in Section 24.6.4; but without the closing lid. The chamber was filled with mimicking debris quartz sand. Different grain sizes were tested: 0.5–1 mm, 1–2 mm, and 4–8 mm. Small urine samples (1 mL) were placed below the quartz layer and the air composition above the quartz sand was studied by MCC-IMS. Samples were drawn every 10 min for 180 min. For each grain size, five experiments with the urine of each volunteer (20 males, 10 females, age range: 22–53 years) were performed.

Eleven compounds were identified in the air above the quartz layer during the 3-hours-long experiment. Ketones were represented by acetone, 3-methyl-2-butanone, 2-heptanone, and 4-heptanone. Another well-represented class was aldehydes, also with four representatives: propanal, 2-methylpropanal, hexanal, and octanal. Within the remaining species there was one aliphatic hydrocarbon (isoprene), one aromatic hydrocarbon (1,2,3-trimethylbenzene), and one heterocyclic compound (pyrrole). Two compounds, namely 3-methyl-2-butanone and octanal, were detected during all experiments and further two, acetone and 2-heptanone, in all samples but one. The shortest permeation times were noted for acetone, propanal, 3-methyl-2-butanone, and octanal. In general, the permeation times increased with the decrease of the quartz grain size.

24.7.2.3 Quantitative monitoring of selected urine volatiles (2-heptanone and n-octanal) permeating through the debris material by MCC-IMS

The permeation behavior of two selected promising urine markers, 2-heptanone and *n*-octanal, was quantitatively examined under different scenarios of quartz grain size (2–4 mm and 4–8 mm) and quartz layer thickness (4, 8, and 12 cm) using a MCC-IMS and the filling chamber described in section 24.6.4.[57] The vapors released by the urine sample (1 mL) of a healthy volunteer (female, 23 years old, non-smoker) were measured in the headspace above the quartz layer every 15 min for 12 h. The exemplary time profile of 2-heptanone in the headspace of different quartz layers exposed to the urine vapors is presented in Figure 24.7. The limits of detections (LODs) provided by the applied IMS instrument were 0.25 and 3 ppb for 2-heptanone and *n*-octanal, respectively. The maximum headspace concentrations noted during the experiments varied from 7 to 16 ppb for 2-heptanone and from 15 to 40 ppb for *n*-octanal. However, it must be noted that these values strongly depend on the sample composition, which in turn depends

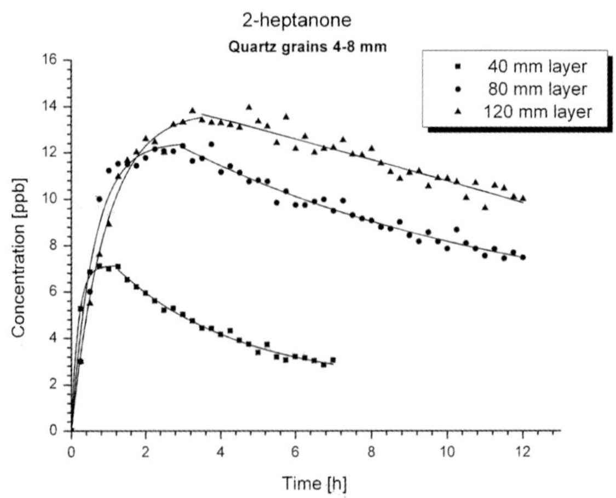

FIGURE 24.7
Time profiles of 2-heptanone in the headspace of 4–8 mm quartz sand. Reproduced with kind permission of Springer Science+Business Media from Ref. [57].

Table 24.11 Permeation time constants for 2-heptanone and *n*-octanal. Reproduced with kind permission of Springer Science+Business Media from Ref. [57].

Grain Size (mm)	Quartz Layer (mm)	Time Constant τ (min)	
		2-Heptanone	*n*-Octanal
	40	27	125
2–4	80	95	361
	120	202	854
	40	11	38
4–8	80	36	112
	120	55	271

on individual's diet, medical status, environmental exposure, and dehydration. Similar to the profiles obtained from the HS-SPME-GC-MS studies, the IMS profiles of 2-heptanone consisted of two parts: an exponential growth that asymptotically approaches a maximum limit and an exponential decay. The exponential growth parts were used to calculate the time constants (τ) representing the permeation rate of urine vapor constituents through the quartz layer. 2-heptanone was found to penetrate through quartz layer by approximately a factor 4 faster than *n*-octanal (see Table 24.11). The increase of quartz thickness had a significant influence on permeation time. The two-fold and three-fold increase of thickness of 4–8 mm quartz increased the 2-heptanone time constants, 3 and 5 times respectively. The analogous factors for *n*-octanal amounted to 3 and 7.

24.7.2.4 MCC-IMS technique; conclusions

The MCC-IMS technique can be considered as a very useful tool for the investigation of the behavior of urine-borne markers in the voids of collapsed buildings. Despite of some limitations in detecting some classes of compounds like hydrocarbons, sulfur, or aromatic compounds, it provides excellent detection and quantification of the most abundant ketones and aldehydes evolving from urine. The short analysis time (approx. 5 min) and the low LODs (low ppb–ppt range) make this technique an interesting candidate for the field analytical technique to be used during USaR operations. In this context, its potential for miniaturization can be regarded as a crucial factor. In addition to this, IMS can be particularly useful in cases when searching dogs and rescuers are exposed to toxic agents,[74] helping to detect such potential risks.

24.7.3 Application of the VAMPPI multi-sensor detector for the detection of selected volatile constituents of human scent

24.7.3.1 The VAMPPI chemical detector

VAMPPI chemical detector, a slightly modified version (display- and power board-free) of the commercial ChemPro 100i (Environics Oy, Mikkeli, Finland), a commercial multi-sensor chemical detector, developed for the detection and identification of chemical warfare agents (CWA) and toxic industrial chemicals (TICs). It contains an aspiration-type IMS and six gas sensors (two ceramic semi-conductor sensors named SC1, SC2, three micro hot-plate metal oxide sensors named MOS1, MOS2, MOS3, and a field effect transistor, FET) along with temperature, pressure, and flow control units.[75–77] The ion mobility cell (IMCell) with eight pairs of electrodes is the basic component of the AIMS. During analysis, the gas sample is continuously pumped through the IMS body where it is ionized by the radiation source ^{241}Am. The resulting clusters of ions flow through the IMCell where a perpendicular electric field induces their collisions with the electrodes. Ion separation depends on their ion mobility, which in turn is a function of their size, shape, and charge.[73] The collisions of ion clusters with electrodes generate the measured current (pA). By changing the polarity of the electric field in short time intervals, both negative and positive ions can be measured. Consequently, the measurement result is a pattern - a chemical profile - containing pA values in 16 channels (8 electrodes in positive, 8 in negative modes). The shape of the pattern does not depend on the concentration and is a basis for qualitative analysis, whereas, the intensity of the signal (sum of the electrode currents) provides the quantitative information.[77]

The VAMPPI is a handheld, fast, sensitive detector that has already been successfully applied for the early detection of toxic chemicals, pesticides, and microbial VOCs.[75–79]

24.7.3.2 AIMS profiles of selected constituents of human scent

The applicability of IMS for the identification of VOCs emitted from the human body has been validated using several selected compounds commonly released from human tissues and biological fluids: five ketones (acetone, 2-butanone, 2-pentanone, 3-methyl 2-butanone, 4-heptanone), four aldehydes (propanal, pentanal, hexanal, octanal), isoprene, ethanol, and DMDS. The aforementioned species are commonly emitted from human tissues and fluids (e.g. blood, urine, breath, skin).[43,56,58,59,70,80]

An exemplary relative AIMS pattern obtained for 4-heptanone at concentration of 500 ppb is presented in Figure 24.8. The patterns of volatiles under study obtained for single-compound calibration mixtures were normalized to create response vectors $C = (C_1, C_2, \ldots, C_{14})$ and build a reference library of compounds. Normalized response vectors were compared to assess if it is possible to distinguish between different compounds. The comparison algorithm counted the number of channels n in the analyzed vector with signal in a range of $\pm 5\%$ of the corresponding values of the library reference vector to create a similarity index (SI). Two patterns/vectors were considered as similar when the SI was higher than 80%.

Apart from isoprene, all compounds produced measurable IMCell responses; however, the intensities of these responses were substance specific. For aldehydes and ketones the responses were comparable and ranged from 40 to 80 pA. For DMDS and ethanol the signal was substantially weaker. Aldehydes and ketones produce distinct patterns and, consequently, it is possible to differentiate species belonging to these classes. The SI values of homologous compounds do not differ significantly. For instance, SI between 2-butanone and 2-pentanone amounts to 93% and between acetone

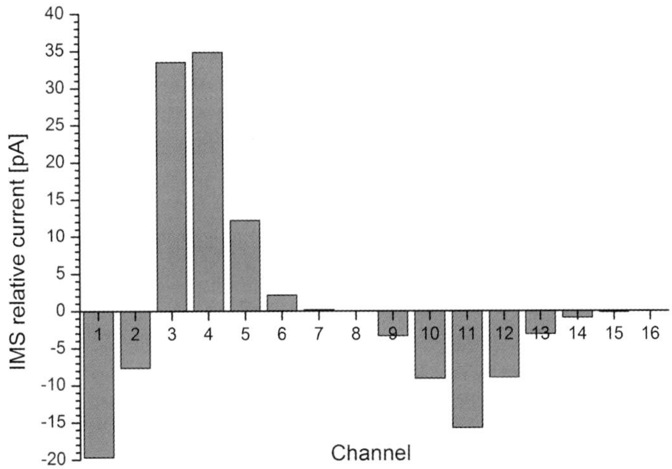

FIGURE 24.8
AIMS pattern for 4-heptanone at a concentration of 500 ppb.

and 2-butanone to 79%. On the other hand, ethanol and DMDS generated very similar pattern to ketones; thus differentiation between these species is limited. Nevertheless, the comparison of SI values obtained for the test substance (acetone) at different concentrations (25–3000 ppb) showed good agreement between profiles (SI > 79%).

In high concentration, the AIMS show saturation response, which is typical of IMS devices.[73] As a consequence, the sensitivity of the IMCell drops substantially with the concentration increase. However, the concentrations used during experiments were still within the dynamic range of the device and thus the calibration graphs allowed quantitative measurements. The relative standard deviations (RSDs) of measurements ranged from 1.5% to 5%. In general, ketones exhibited better LODs (average 9 ppb) than aldehydes (average 26 ppb), whereas, LODs of ethanol and DMDS were higher and amounted to 55 and 87 ppb, respectively.

24.7.3.3 Application of the AIMS to the detection of acetone in human breath

Most breath species exhibit very low concentration levels (ppt—several ppb) that is much below the observed detection limits of the tested AIMS unit. However, two volatiles, isoprene and acetone hold a distinguished status, since their breath abundances can be especially high. In the present study, the AIMS was also used for screening real human breath samples. Since, isoprene failed to generate signal in VAMPPI, the major breath constituent which could contribute to the breath AIMS pattern was acetone. The breath samples of ten volunteers (7 males, 3 females, age range: 25–35 years) were analysed by VAMPPI and for validation reasons by SPME-GC-MS. The samples profiles obtained exhibited the highest similarity to acetone (79–85%), confirming the decisive signal contribution of this compound. Quantitative analysis based on the AIMS signal provided an acetone concentration range of 290–540 ppb, which can be considered as typical for breath samples. For acetone concentrations below 500 ppb, good agreement was observed between AIMS and SPME-GC-MS concentrations, with differences ranging from 3 to 14% (mean: 9%). However, for higher concentrations, pronounced deviations were observed (33–70%). The latter clearly results from the decrease in sensitivity of the AIMS response.

24.7.3.4 AIMS; conclusions

To sum up, the AIMS can be considered as very promising technology for the detection, identification, and monitoring of volatiles in biological samples. The main advantages of this technique are its good sensitivity (ppb), rapid response (<5 s), portability, and low analysis cost. It is particularly attractive for studies in field conditions, where easy-in-use, portable, and sensitive instruments need to be applied. Despite its relative limited analytical power, AIMS was able to quantitatively determine breath acetone which is considered to be a strong indicator of human presence. However, further success of these type of analyzers for targeted application will depend on progress towards improving the sensitivity, selectivity and resolution of aspiration-type IMS, making possible quantitative analysis on low ppb or even ppt level (i.e. use of multivariate data analysis techniques).

24.8 SPATIO-TEMPORAL MEASUREMENTS OF VOCS

Dogs are the paradigmatic searching support in USaR missions. The olfactory capabilities of canines are outstanding and incomparable in their *real-time* capabilities, sensitivity, and the potential to detect concentration gradients. Nevertheless, dogs get tired after about 30 min of intensive searching and require a 2-hour-resting period, to be repeated after each deployment. Hence, the goal of mimicking the capabilities of dogs with miniaturized sensing systems could, if achieved, indeed support the efforts of the rescue personnel.

At present, the potentiality of source detection and plume tracking using analytical technology is just poorly understood. Therefore, as a first step, the spatiotemporal expansion of selected VOCs was monitored in indoor experiments (indoor expansion of evaporated compounds).

Plume monitoring and identification of the releasing source are an important step in the development of hand-portable devices for USaR operations.

Figure 24.9 shows an experimental setup used for indoor diffusion experiments of volatile compounds. Various compounds (e.g. acetone) can be injected into a source pan covered by different debris materials (e.g. brick split or quartz gravel). Above the source pan, green and red flags on strings hanging from the ceiling, mark the probing sites. The composition of the continuously sampled air was determined by proton-transfer-reaction time-of-flight mass spectrometry (PTR-TOF-MS). Figure 24.10 shows the observed temporal concentration progression and regression curve of acetone vapor

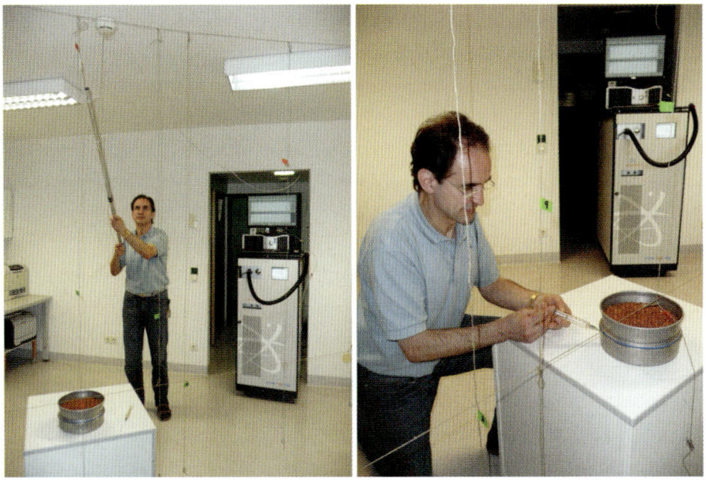

FIGURE 24.9
Experimental setup for spatiotemporal monitoring of selected volatile compounds. On the center foreground of the left panel, the acetone source pan covered by brick split is visible. Above the source pan, green and red flags on strings hanging from the ceiling, mark the probing sites; in the background, a proton-transfer-reaction time-of-flight mass spectrometer (PTR-TOF-MS) for air analysis is visible. On the right panel, acetone is injected into the source pan through a septum on the side wall.

permeated through quartz gravel (grain diameter 3–8 mm), under various experimental conditions.

PTR-TOF-MS seems to be the best available technique for this kind of experiments. Its capability of real-time measurements and its enhanced mass resolution of 1/5000 are a big advantage in comparison with quadrupole PTR-MS or with Ion Mobility Spectrometry. Therefore, plume detection experiments have been performed and were partly successful. Nevertheless, the experimental design needs to be improved, in order to establish a valid *search protocol* which *always* works, independent of the prevailing conditions (depending on non-equilibrium situations and indoor air currents). In the near future, a variety of VOCs of particular interest will be examined under different conditions for identifying their concentration gradients in time and space.

24.9 REAL-TIME MEASUREMENT OF EXHALED BREATH AND SKIN EMANATIONS

The range of measurement techniques employed for gas analytical investigations is extremely diverse and each method has its specific strengths and weaknesses.[80] Often, one actually combines the strengths of different analytical methods, such as PTR-MS,

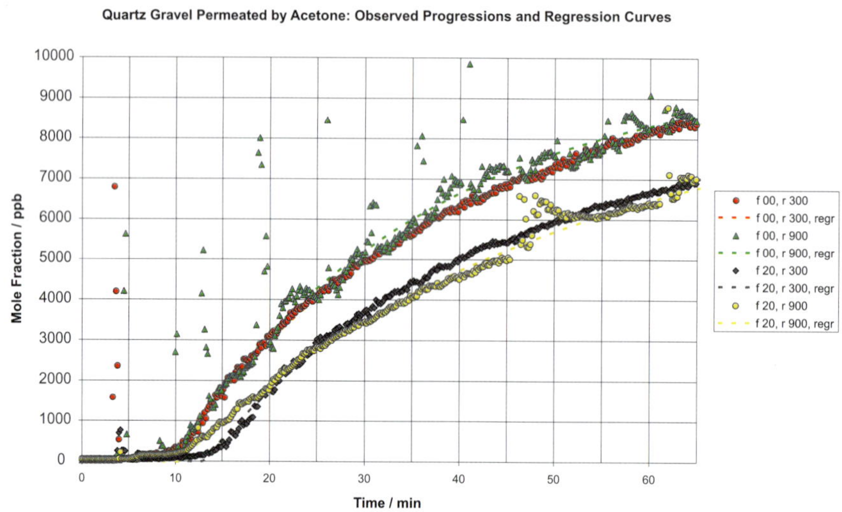

FIGURE 24.10

Observed temporal concentration progression and regression curve of acetone vapor permeating through quartz gravel (grain diameter 3–8 mm) under various experimental conditions specified on the right of diagram; the filling thickness "f" is given in mm, as well as, the distance "r" from the source, "regr" is for regression curve.

SIFT-MS, GC-MS, and IMS. PTR-MS and SIFT-MS can measure in *real-time* (similar to dogs), but cannot provide as much detailed information as GC-MS. On the other hand, IMS has the potential to be miniaturized.

A major advantage of PTR-MS is its *real-time* capability,[18,20–26]. Figure 24.11 presents a *real-time* measurement of acetone and isoprene in exhaled breath for a full night of one single volunteer. In the upper panel, the hypnogram of the volunteer is shown in blue color, with sleep stages N1, N2 (light sleep stages), N3 (deep sleep stage), and REM (rapid eye movement sleep, the sleep phase in which dreams occur). Panel 2 shows the concentration of acetone and panel 3 shows the concentration of isoprene. Isoprene is a specific example of a compound, which concentrations may increase by a factor of 4–5 while exerting effort (e.g. exercising on a stationary bicycle[21,26]). The present figure shows peaks in isoprene concentration related to movements of the limbs. Since isoprene is related to cholesterol metabolism, interesting medical applications are expected in the future.[26]

A chamber for volunteers, which is based on a body plethysmography (subject chamber from Ganshorn, Niederlauer, Germany) is presented in Figure 24.12. The chamber is adapted for separate (possibly parallel) measurements of exhaled breath

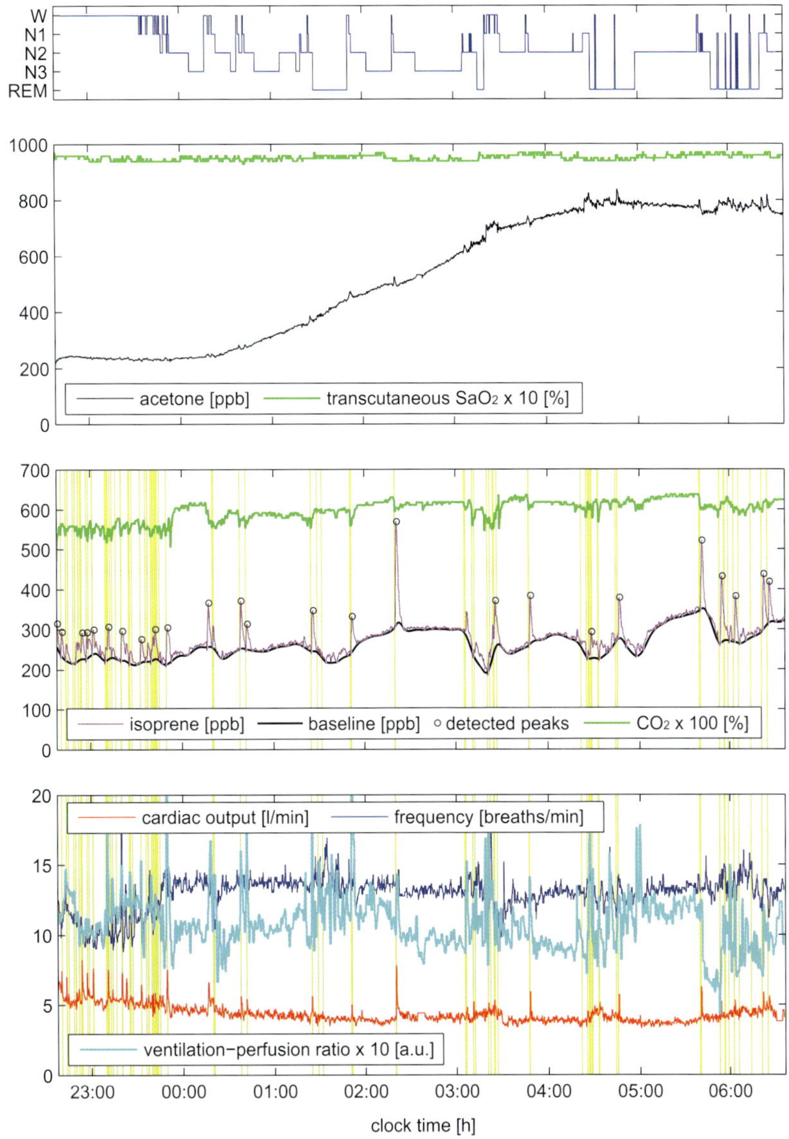

FIGURE 24.11
Measurements of acetone and isoprene concentration in exhaled breath of a single test subject during a full night. The associated hypnogram is given in the first panel (W="wake", sleep stages N1, N2, N3, and "rapid eye movement sleep" REM). The acetone concentration is shown in panel 2 and the concentration of isoprene in panel 3. The acetone concentration is relatively smooth whereas the isoprene concentration shows characteristic peaks, related to movements of the limbs. Breathing frequency and cardiac output are shown in panel 4. Vertical lines in panel 3 and 4 depict leg movements as measured by anterior tibialis electromyography. Reproduced with permission of IOP from Ref. [19].

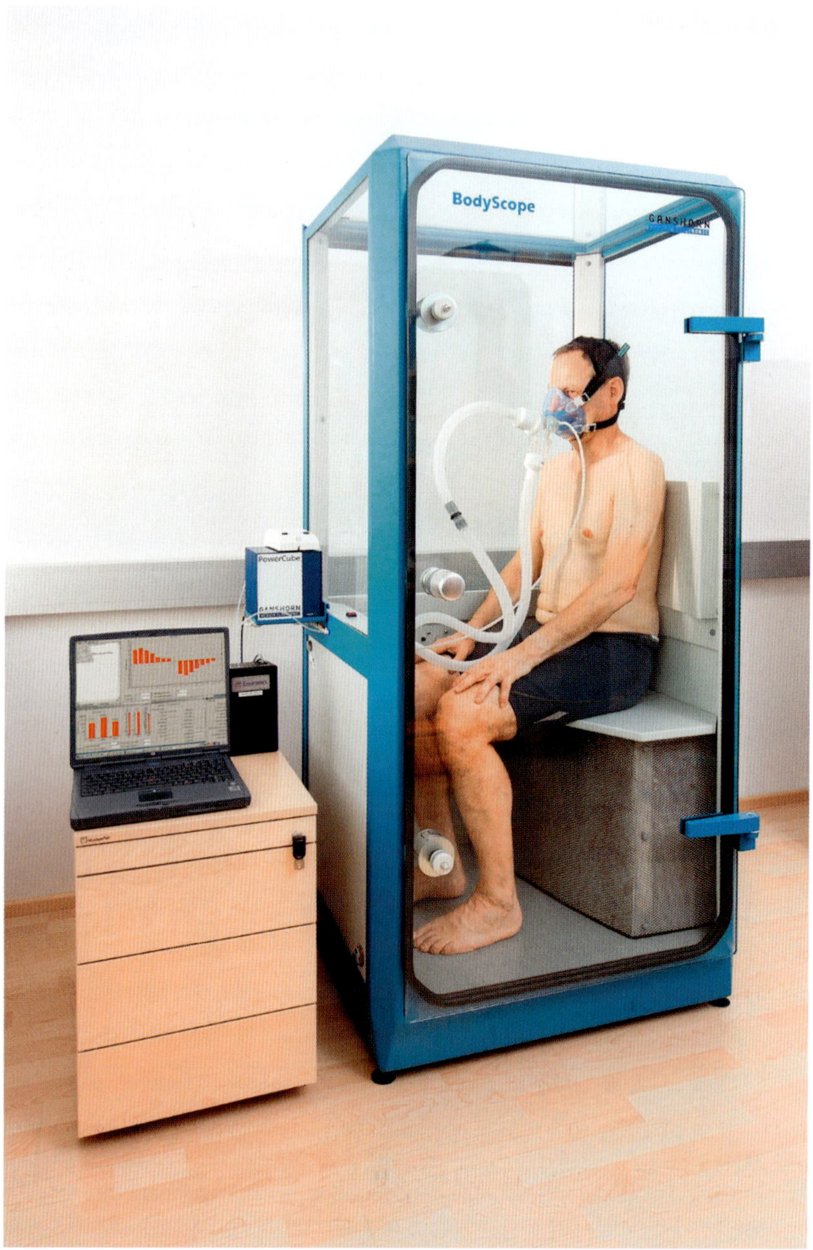

FIGURE 24.12

Chamber for volunteers based on a body plethysmography chamber from Ganshorn (Niederlauer, Germany). Exhaled breath and skin metabolites can separately be investigated and analyzed by direct mass-spectrometric techniques (PTR-MS or SIFT-MS).

and skin emanations. Breathing occurs via a tube leading to the outside of the chamber, where exhaled breath can be collected. Skin emanations are accumulated in the chamber, collected through a separate tube and measured by PTR-TOF-MS or by GC-MS.

Compounds emanating through skin include methanol, ethanol, propanal, acetone, ethyl ether, methacroleine, 2-methyl-propanal, butanal, 2-butanone, 3-methyl-butanal, 2-methyl-butanal, 3-methylbutenal, 2-pentanone, pentanal, 2-methylhept-5-en-3-one, sec-butyl acetate, toluene, 4-methyl-2-pentanone, hexanal, butyl acetate, 1-octene, 1-methoxy-2-propyl acetate, benzaldehyde, 1-nonene, 6-methyl-5-hepten-2-one, 2-pentyl-furan, octanal, nonanal, decanal, allylmethylsulfide, and methylpropylsulfide. Isoprene is not emitted through the skin, or at much lower concentrations than in exhaled breath.

24.10 A HIT LIST OF COMPOUNDS FOR URBAN SEARCH AND RESCUE OPERATIONS

Many different compounds could potentially be used in USaR operations. However, some compounds of particular importance are:
2-heptanone, 4-heptanone, 2-butanone, acetone, 2-pentanone, pyrrole, propanal, ammonia, isoprene, and carbon dioxide.

The choice of compounds is always a trade-off between their advantages and disadvantages. In the present hit list, all compounds apart from isoprene and carbon dioxide can be detected by IMS. 2-Heptanone, for example, has been investigated with respect to its permeation properties and compared to *n*-octanal; it penetrates debris much faster than octanal and is chemically comparatively stable. Higher aldehydes like nonanal and decanal can also be detected by IMS, but are not released in such high concentrations as propanal. As previously mentioned, the use of more than one compound (chemical pattern) might facilitate search operations; therefore combining the concentration profiles of these compounds is essential for the detection of entrapped persons.

24.11 SUMMARY

Analysis of VOCs released from human tissues and biological fluids such as breath, skin, or urine opens exciting research possibilities in safety and security areas. Canines have been successfully employed for this task due to their unique scenting skills; hence it should be possible to mimic and adopt their field capabilities. In this context, the potential of breath analysis is evaluated taking into consideration the recent technological progress in analytical chemistry. Next to expired air, human urine holds an almost equal role in this field. Therefore, modern state-of-the-art analytical technologies (e.g. GC-MS, IMS, and PRT-MS) have been employed for studying human urine as a potential source of human scent constituents and its interaction with debris material. Initially, the analysis of urine samples was performed using HS-SPME-GC-MS and resulted in the selection of 33 omnipresent VOCs. The evolution of urine scent profile during 4 day long sample storage at room temperature was next investigated. Headspace levels of 26 ubiquitous urine compounds were found to increase in the course of the experiment; among them sulfur compounds (DMDS and DMTS) and some ketones (e.g. 3-hexanone, 2-heptanone, 4-heptanone, 4-methyl-2-pentanone), showed the highest increase in emission during storage. The interactions of urine-borne species with common debris materials (brick and concrete) were studied using an in-house made filling chamber and HS-SPME-GC-MS. Permeation profiles of 22 volatile species, released from human urine samples, were quantitatively monitored over the selected debris materials for a time period of 24 h. The influence of debris materials on the permeation profiles of analytes depended on their fundamental physicochemical properties. In general, concrete demonstrated a stronger interaction with urine species than brick, affecting the observed concentrations and residence times of released volatiles in the chamber.

Two different types of ion mobility spectrometers have been examined for their suitability in the detection/identification of omnipresent urine-borne species; a MCC-IMS with β-radiation source (^{63}Ni) and an AIMS. The MCC-IMS study resulted in the selection of seven urine species, namely acetone, propanal, 3-methyl-2-butanone, 2-methylpropanal, 4-heptanone, 2-heptanone, and octanal, which were detected in all samples. Among them, four compounds, 3-methyl-2-butanone, octanal, acetone, and 2-heptanone, were found to permeate relatively quickly through the sand layers and during all experiments. Two very promising urine-borne species, 2-heptanone and *n*-octanal, were then examined under the influence of two crucial operational factors; the size of quartz grain and the quartz layer thickness. 2-heptanone was found to penetrate through quartz layer by approximately a factor 4 faster than *n*-octanal. The two-fold and three-fold increase of quartz sand thickness lengthened the permeation times on average 3 and 7 times for *n*-octanal and 3 and 5 times for 2-heptanone. In the same content, the AIMS was used for the screening of selected human-origin VOCs. The results indicated the potentiality of AIMS in discriminating the investigated analytes. Additionally, qualitative and quantitative measurements of real human breath samples were performed. The acetone levels measured by AIMS correlated quite well with the SPME-GC-MS results, showing huge potential of AIMS for the detection of breath acetone in low concentrations exhaled air samples; however, high deviations were observed for concentrations above 500 ppb. Finally, the enhanced analytical power of a PTR-TOF-MS was presented in different real-time measurements; in the spatiotemporal monitoring of acetone over a source pan covered by debris materials, in monitoring exhaled breath acetone and isoprene during a full night sleep and in the simultaneously monitoring of breath VOCs and skin emanations employing a body plethysmography chamber.

The current status of volatiles analysis is quite promising; the science is mature, the technology is expanding, industry is interested, and medical community is welcoming breath testing. As technology rapidly changes and transforms, no one can really predict the scientific advancements that can occur in the next decades; nevertheless, it is certain that breath chemical analysis will be involved, offering unique and challenging revolutionary applications in crucial areas, including safety, and security. Thus, as was successfully stated by Dr. R. Mukhopadhyay in a relevant editorial in Analytical Chemistry (2004), "don't waste your breath."

Acknowledgments

The research leading to these results has received funding from the European Community's Seventh Framework Program (FP7/2007-13) under grant agreement No. 217967 ("SGL for USaR" project, Second Generation Locator for Urban Search and Rescue Operations, www.sgl-eu.org). We appreciate funding from the Austrian Federal Ministry for Transport, Innovation and Technology (BMVIT/BMWA, project 836308, KIRAS). We greatly appreciate the generous support of the government of Vorarlberg and its governor Landeshauptmann Dr. Herbert Sausgruber.

References

1. United States Geological Survey. *Earthquakes hazard program: earthquakes with 1000 or more deaths from 1900–2010*. <http://earthquake.usgs.gov/earthquakes/world/world_deaths.php> [accessed 30.6.2012].
2. United Nations International Strategy for Disaster Reduction. *Earthquakes caused the deadliest disasters in the past decade*. Press release on 28 January 2010. <http://www.unisdr.org/archive/12470> [accessed 30.6.2012].
3. Bartels SA, VanRooyen MJ. Medical complications associated with earthquakes. *Lancet* 2012;**379**:748–57.
4. Ferworn A. Canine augmentation technology for urban search and rescue. In: Helton WS, editor. *Canine ergonomics: the science of working dogs*. Boca Raton: CRC Press; 2009. p. 205–44.
5. Wong J, Robinson C. Urban search and rescue technology needs: identification of needs. Federal Emergency Management Agency (FEMA) and the National Institute of Justice (NIJ); 2004 [Document number 207771].
6. Borsdorf H, Mayer T, Zarejousheghani M, Eiceman GA. Recent developments in ion mobility spectrometry. *Appl Spectrosc Rev* 2011;**46**:472–521.
7. Spanel P, Smith D. Progress in SIFT-MS: breath analysis and other applications. *Mass Spectrom Rev* 2011;**30**:236–67.
8. Bajtarevic A, Ager C, Pienz M, Klieber M, Schwarz K, Ligor M, et al. Noninvasive detection of lung cancer by analysis of exhaled breath. *BMC Cancer* 2009;**9**:348.
9. Ligor M, Ligor T, Bajtarevic A, Ager C, Pienz M, Klieber M, et al. Determination of volatile organic compounds in exhaled breath of patients with lung cancer using solid phase microextraction and gas chromatography mass spectrometry. *Clin Chem Lab Med* 2009;**47**:550–60.
10. Phillips M, Altorki N, Austin JH, Cameron RB, Cataneo RN, Greenberg J, et al. Schmitt Prediction of lung cancer using volatile biomarkers in breath. *Cancer Biomark* 2007;**3**:95–109.
11. Amann A, Corradi M, Mazzone P, Mutti A. Lung cancer biomarkers in exhaled breath. *Expert Rev Mol Diagn* 2011;**11**:207–17.
12. Wehinger A, Schmid A, Mechtcheriakov S, Ledochowski M, Grabmer C, Gastl GA, et al. Lung cancer detection by proton transfer reaction mass-spectrometric analysis of human breath gas. *Int J Mass Spectrom* 2007;**265**:49–59.
13. Peng G, Hakim M, Broza YY, Billan S, Abdah-Bortnyak R, Kuten A, et al. Detection of lung, breast, colorectal, and prostate cancers from exhaled breath using a single array of nanosensors. *Br J Cancer* 2010;**103**:542–51.
14. Hakim M, Billan S, Tisch U, Peng G, Dvrokind I, Marom O, et al. Diagnosis of head-and-neck cancer from exhaled breath. *Br J Cancer* 2011;**104**:1649–55.

15. Pleil JD, Stiegel MA, Sobus JR. Breath biomarkers in environmental health science: exploring patterns in the human exposome. *J Breath Res* 2011;**5**:046005.
16. Pleil JD, Stiegel MA, Sobus JR, Liu Q, Madden MC. Observing the human exposome as reflected in breath biomarkers: heat map data interpretation for environmental and intelligence research.*J Breath Res* 2011;**5**:037104.
17. Pleil JD, Stiegel MA, Madden MC, Sobus JR. Heat map visualization of complex environmental and biomarker measurements. *Chemosphere* 2011;**84**:716–23.
18. Amann A, Telser S, Hofer L, Schmid A, and Hinterhuber H. Exhaled breath gas as a biochemical probe during sleep. In: Amann A, Smith D. editors. *Breath analysis for clinical diagnosis and therapeutic monitoring.* Singapore: World Scientific; 2005. p. 305–16.
19. King J, Kupferthaler A, Frauscher B, Hackner H, Unterkofler K, Teschl G, et al. Measurement of endogenous acetone and isoprene in exhaled breath during sleep. *Physiol Meas* 2012;**33**:413–28.
20. King J, Koc H, Unterkofler K, Mochalski P, Kupferthaler A, Teschl G, et al. Physiological modeling of isoprene dynamics in exhaled breath. *J Theor Biol* 2010;**267**:626–37.
21. King J, Kupferthaler A, Unterkofler K, Koc H, Teschl S, Teschl G, et al. Isoprene and acetone concentration profiles during exercise on an ergometer. *J Breath Res* 2009;**3**:027006.
22. King J, Mochalski P, Kupferthaler A, Unterkofler K, Koc H, Filipiak W, et al. Dynamic profiles of volatile organic compounds in exhaled breath as determined by a coupled PTR-MS/GC-MS study. *Physiol Meas* 2010;**31**:1169–84.
23. King J, Unterkofler K, Teschl G, Teschl S, Koc H, Hinterhuber H, Amann A. A mathematical model for breath gas analysis of volatile organic compounds with special emphasis on acetone. *J Math Biol* 2011;**63**:959–99.
24. King J, Unterkofler K, Teschl G, Teschl S, Mochalski P, Koc H, et al. A modeling-based evaluation of isothermal rebreathing for breath gas analyses of highly soluble volatile organic compounds. *J Breath Res* 2012;**6**:016005.
25. King J, Unterkofler K, Teschl S, Amann A, Teschl G. Breath gas analysis for estimating physiological processes using anesthetic monitoring as a prototypic example. *Conf Proc IEEE Eng Med Biol Soc* 2011;**2011**:1001–4.
26. Koc H, King J, Teschl G, Unterkofler K, Teschl S, Mochalski P, et al. The role of mathematical modeling in VOC analysis using isoprene as a prototypic example. *J Breath Res* 2011;**5**:037102.
27. Schubert JK, Miekisch W, Geiger K, Noldge-Schomburg GF. Breath analysis in critically ill patients: potential and limitations. *Expert Rev Mol Diagn* 2004;**4**:619–29.
28. Scholpp J, Schubert JK, Miekisch W, Geiger K. Breath markers and soluble lipid peroxidation markers in critically ill patients. *Clin Chem Lab Med* 2002;**40**:587–94.
29. Boshier PR, Cushnir JR, Mistry V, Knaggs A, Spanel P, Smith D, et al. On-line, real time monitoring of exhaled trace gases by SIFT-MS in the perioperative setting: a feasibility study. *The Analyst* 2011;**136**:3233–7.
30. Modak AS. Stable isotope breath tests in clinical medicine: a review. *J Breath Res* 2007;**1**:014003.
31. Modak AS. Single time point diagnostic breath tests: a review. *J Breath Res* 2009;**4**:017002.
32. Westhoff M, Litterst P, Freitag L, Baumbach JI. Ion mobility spectrometry in the diagnosis of sarcoidosis: results of a feasibility study. *J Physiol Pharmacol* 2007;**58**(Suppl. 5):739–51.
33. Ruzsanyi V, Baumbach JI, Sielemann S, Litterst P, Westhoff M, Freitag L. Detection of human metabolites using multi-capillary columns coupled to ion mobility spectrometers. *J Chromatogr A* 2005;**1084**:145–51.

34. Westhoff M, Litterst P, Freitag L,Urfer W, Bader S, Baumbach JI. Ion mobility spectrometry for the detection of volatile organic compounds in exhaled breath of patients with lung cancer: results of a pilot study. *Thorax* 2009;**64**:744–48.
35. Machado RF, Laskowski D, Deffenderfer O, Burch T, Zheng S, Mazzone PJ, et al. Detection of lung cancer by sensor array analyses of exhaled breath. *Am J Respir Crit Care Med* 2005;**171**:1286–91.
36. Mazzone PJ. Analysis of volatile organic compounds in the exhaled breath for the diagnosis of lung cancer. *J Thorac Oncol* 2008;**3**:774–80.
37. Kushch I, Arendacka B, Stolc S, Mochalski P, Filipiak W, Schwarz K, et al. Breath isoprene-aspects of normal physiology related to age, gender and cholesterol profile as determined in a proton transfer reaction mass spectrometry study. *Clin Chem Lab Med* 2008;**46**:1011–8.
38. Kushch I, Schwarz K, Schwentner L, Baumann B, Dzien A, Schmid A, et al. Compounds enhanced in a mass spectrometric profile of smokers' exhaled breath versus non-smokers as determined in a pilot study using PTR-MS. *J Breath Res* 2008;**2**:026002.
39. Schwarz K, Filipiak W, Amann A. Determining concentration patterns of volatile compounds in exhaled breath by PTR-MS. *J Breath Res* 2009;**3**:027002.
40. Schwarz K, Pizzini A, Arendacka B, Zerlauth K, Filipiak W, Schmid A, et al. Breath acetone-aspects of normal physiology related to age and gender as determined in a PTR-MS study. *J Breath Res* 2009;**3**:027003.
41. Short LC, Frey R, Benter T. Real-time analysis of exhaled breath via resonance-enhanced multiphoton ionization-mass spectrometry with a medium pressure laser ionization source: observed nitric oxide profile. *Appl Spectrosc* 2006;**60**:217–222.
42. Lee PS, Schreck RM, Hare BA, McGrath JJ. Biomedical applications of tunable diode laser spectrometry: correlation between breath carbon monoxide and low level blood carboxyhemoglobin saturation. *Ann Biomed Eng* 1994;**22**:120–5.
43. Statheropoulos M, Sianos E, Agapiou A, Georgiadou A, Pappa A, Tzamtzis N, et al. Preliminary investigation of using volatile organic compounds from human expired air, blood and urine for locating entrapped people in earthquakes. *J Chromatogr B Analyt Technol Biomed Life Sci* 2005;**822**:112–7.
44. Huo R, Agapiou A, Bocos-Bintintan V, Brown LJ, Burns C, Creaser CS, et al. The trapped human experiment. *J Breath Res* 2011;**5**:046006.
45. Statheropoulos M, Agapiou A, Spiliopoulou C, Pallis GC, Sianos E. Environmental aspects of VOCs evolved in the early stages of human decomposition. *Sci Total Environ* 2007;**385**:221–7.
46. Committee on advances in assessing human exposure to airborne pollutants NRC, National Academy of Sciences. *Human Exposure Assessment for Airborne Pollutants: Advances and Opportunities.* Washington, DC: National Academies Press; 1991.
47. O'Hara ME, Clutton-Brock TH, Green S, Mayhew CA. Endogenous volatile organic compounds in breath and blood of healthy volunteers: examining breath analysis as a surrogate for blood measurements. *J Breath Res* 2009;**3**:027005.
48. Statheropoulos M, Mikedi K, Agapiou A, Georgiadou A, Karma S. Discriminant Analysis of volatile organic compounds data related to a new location method of entrapped people in collapsed buildings of an earthquake. *Anal Chim Acta* 2006;**566**:207–16.
49. Eckenrode BA, Ramsey SA, Stockham RA, Van Berkel GJ, Asano KG, Wolf DA. Performance evaluation of the Scent Transfer Unit (STU-100) for organic compound collection and release. *J Forensic Sci* 2006;**51**:780–9.

50. Gustafsson L. Exhaled nitric oxide: how and why we know it is important. In: Amann A, Smith D, editors. *Breath analysis for clinical diagnosis and therapeutic monitoring*. Singapore: World Scientific; 2005.
51. Eisenmann A, Amann A, Said M, Datta B, Ledochowski M. Implementation and interpretation of hydrogen breath tests. *J Breath Res* 2008;**2**:046002.
52. Dryahina K, Smith D, Spanel P. Quantification of methane in humid air and exhaled breath using selected ion flow tube mass spectrometry. *Rapid Commun Mass Sp* 2010;**24**:1296–1304.
53. Hart R. Human body odor. *Nexus* 1980;**1**.
54. Statheropoulos M, Spiliopoulou C, Agapiou A. A study of volatile organic compounds evolved from the decaying human body. *Forensic Sci Int* 2005;**153**:147–55.
55. Statheropoulos M, Agapiou A, Zorba E, Mikedi K, Karma S, Pallis GC, et al. Combined chemical and optical methods formonitoring the early decay stages of surrogate human models. *Forensic Sci Int* 2011;**210**:154–63.
56. Rudnicka J, Mochalski P, Agapiou A, Statheropoulos M, Amann A, Buszewski B. Application of ion mobility spectrometry for the detection of human urine. *Anal Bioanal Chem* 2010;**398**:2031–8.
57. Mochalski P, Buszewska M, Agapiou A, Statheropoulos M, Buszewski B, Amann A. Preliminary investigation of permeation profiles of selected head-space urine volatiles (2-heptanone, *n*-octanal) using IMS. *Chromatographia* 2012;**75**:41–6.
58. Mills GA, Walker V. Headspace solid-phase microextraction profiling of volatile compounds in urine: application to metabolic investigations. *J Chromatogr B: Biomed Sci Appl* 2001;**753**:259–68.
59. Smith S, Burden H, Persad R, Whittington K, de Lacy Costello B, Ratcliffe NM, et al. A comparative study of the analysis of human urine headspace using gas chromatography-mass spectrometry. *J Breath Res* 2008;**2**:037022.
60. Spanel P, Smith D, Holland TA, Al Singary W, Elder JB. Analysis of formaldehyde in the headspace of urine from bladder and prostate cancer patients using selected ion flow tube mass spectrometry. *Rapid Commun Mass Spectrom* 1999;**13**:1354–9.
61. Wahl HG, Hoffmann A, Luft D, Liebich HM. Analysis of volatile organic compounds in human urine by headspace gas chromatography-mass spectrometry with a multipurpose sampler. *J Chromatogr A* 1999;**847**:117–25.
62. Wang T, Spanel P, Smith D. Selected ion flow tube mass spectrometry of 3-hydroxybutyric acid, acetone and other ketones in the headspace of aqueous solution and urine.*Int J Mass Spectrom* 2008;**272**:78–85.
63. Mochalski P, Krapf K, Ager C, Wiesenhofer H, Agapiou A, Statheropoulos M, et al. Temporal profiling of human urine VOCs and its potential role under the ruins of collapsed buildings. *Toxicol Mech Methods* 2012;**22**:502–11.
64. Turner C, Parekh B, Walton C, Spanel P, Smith D, Evans M. An exploratory comparative study of volatile compounds in exhaled breath and emitted by skin using selected ion flow tube mass spectrometry. *Rapid Commun Mass Sp* 2008;**22**:526–32.
65. Rasmussen LE, Perrin TE. Physiological correlates of musth: lipid metabolites and chemical composition of exudates. *Physiol Behav* 1999;**67**:539–49.
66. Cailleux A, Cogny M, Allain P. Blood isoprene concentrations in humans and in some animal species. *Biochem Med Metab Biol* 1992;**47**:157–60.
67. Burger BV. Mammalian semiochemicals. *Top Curr Chem* 2005;**240**:231–78.

68. Coburn AW, Spence RJS, Pomonis A. Factors determining human casualty levels in earthquakes: mortality prediction in building collapse. In: Bernal A. editor. *Proceedings of the 10th World conference on earthquake engineering*. Rotterdam: Balkema; 1992. p. 5989–94.
69. Mochalski P, Agapiou A, Statheropoulos M, Amann A. Permeation profiles of potential urine-borne biomarkers of human presence over brick and concrete. *Analyst* 2012;**137**:3278–85.
70. Gallagher M, Wysocki CJ, Leyden JJ, Spielman AI, Sun X, Preti G. Analyses of volatile organic compounds from human skin. *Br J Dermatol* 2008;**159**:780–91.
71. Matz LM, Hill Jr HH. Evaluation of opiate separation by high-resolution electrospray ionization-ion mobility spectrometry/mass spectrometry. *Anal Chem* 2001;**73**:1664–9.
72. Lawrence AH. Characterization of benzodiazepine drugs by ion mobility spectrometry. *Anal Chem* 1989;**61**:343–9.
73. Eiceman GA, ZK. *Ion mobility spectrometry*. Boca Raton: CRC Press Taylor and Francis Group; 2005.
74. Murphy LA, Gwaltney-Brant SM, Albretsen JC, Wismer TA. Toxicologic agents of concern for search-and-rescue dogs responding to urban disasters. *J Am Vet Med Assoc* 2003;**222**:296–304.
75. Nousiainen M, Perakorpi K, Sillanpaa M. Determination of gas-phase produced ethyl parathion and toluene 2,4-diisocyanate by ion mobility spectrometry, gas chromatography and liquid chromatography. *Talanta* 2007;**72**:984–90.
76. Rasanen RM, Hakansson M, Viljanen M. Differentiation of air samples with and without microbial volatile organic compounds by aspiration ion mobility spectrometry and semiconductor sensors. *Build Environ* 2010;**45**:2184–91.
77. Utriainen M, Karpanoja E, Paakkanen H. Combining miniaturized ion mobility spectrometer and metal oxide gas sensor for the fast detection of toxic chemical vapors. *Sensor Actuat B – Chem* 2003;**93**:17–24.
78. Rasanen RM, Nousiainen M, Perakorpi K, Sillanpaa M, Polari L, Anttalainen O, et al. Determination of gas phase triacetone triperoxide with aspiration ion mobility spectrometry and gas chromatography-mass spectrometry. *Anal Chim Acta* 2008;**623**:59–65.
79. Tuovinen K, Kolehmainen M, Paakkanen H. Determination and identification of pesticides from liquid matrices using ion mobility spectrometry. *Anal Chim Acta* 2001;**429**:257–68.
80. Amann A, Spanel P, Smith D. Breath analysis: the approach towards clinical applications. *Mini Rev Med Chem* 2007;**7**:115–29.
81. Mason MF, Dubowski KM. Breath-alcohol analysis: uses, methods, and some forensic problems–review and opinion. *J Forensic Sci* 1976;**21**:9–41.
82. Mashir A, Dweik RA. Exhaled breath analysis: the new interface between medicine and engineering. *Adv Powder Technol* 2009;**20**:420–5.
83. Egeghy PP, Hauf-Cabalo L, Gibson R, Rappaport SM. Benzene and naphthalene in air and breath as indicators of exposure to jet fuel. *Occup Environ Med* 2003;**60**:969–76.
84. Aggazzotti G, Fantuzzi G, Righi E, Tartoni P, Cassinadri T, Predieri G. Chloroform in alveolar air of individuals attending indoor swimming pools. *Arch Environ Health* 1993;**48**:250–4.
85. Jo WK, Weisel CP, Lioy PJ. Routes of chloroform exposure and body burden from showering with chlorinated tap water. *Risk Anal* 1990;**10**:575–80.
86. Wallace LA, Pellizzari ED, Hartwell TD, Sparacino C, Whitmore R, Sheldon L, et al. The team (Total Exposure Assessment Methodology) study: personal exposures to toxic

substances in air, drinking water, and breath of 400 residents of New Jersey, North Carolina, and North Dakota. *Environ Res* 1987;**43**:290–307.

87. Dyne D, Cocker J, Wilson HK. A novel device for capturing breath samples for solvent analysis. *Sci Total Environ* 1997;**199**:83–9.
88. Droz PO, Krebs Y, Nicole C, Guillemin M. A direct reading method for chlorinated hydrocarbons in breath. *Am Ind Hyg Assoc J* 1988;**49**:319–24.
89. Drummond L, Luck R, Afacan AS, Wilson HK. Biological monitoring of workers exposed to benzene in the coke oven industry. *Br J Ind Med* 1988;**45**:256–61.
90. Campbell L, Jones AH, Wilson HK. Evaluation of occupational exposure to carbon disulphide by blood, exhaled air, and urine analysis. *Am J Ind Med* 1985;**8**:143–53.
91. Periago JF, Morente A, Villanueva M, Luna A. Correlation between concentrations of *n*-hexane and toluene in exhaled and environmental air in an occupationally exposed population. *J Appl Toxicol* 1994;**14**:63–7.
92. Pleil JD, Lindstrom AB. Exhaled human breath measurement method for assessing exposure to halogenated volatile organic compounds. *Clin Chem* 1997;**43**:723–30.
93. Gruber G, Lirk P, Amann A, Keller C, Schobersberger W, Hoffmann G, et al. Neopterin as a marker of immunostimulation: an investigation in anaesthetic workplaces. *Anaesthesia* 2002;**57**:747–50.
94. Bjorkqvist S, Spetz A, Ramnas O, Petersson G. Isoprene from expired air inside a private car. *Sci Total Environ* 1997;**207**:63–7.
95. Phillips M, Breath testing – a new tool in homeland security. In: *Fourteenth international conference on-site analysis and homeland security*. Arlington, Virginia, Washington, DC, USA: Abstracts Book; February 20–23, 2006. p. 98–9.
96. Furton KG, Myers LJ. The scientific foundation and efficacy of the use of canines as chemical detectors for explosives. *Talanta* 2001;**54**:487–500.
97. Lorenzo N, Wan T, Harper RJ, Hsu YL, Chow M, Rose S, et al. Laboratory and field experiments used to identify Canis lupus var. familiaris active odor signature chemicals from drugs, explosives, and humans. *Anal Bioanal Chem* 2003;**376**:1212–24.
98. Sangster J. In: Fogg PGT, editor. *Octanol-water partition coefficients: fundamentals and physical chemistry*. Wiley series in solution chemistry. Chichester, New York, Weinheim, Brisbane, Singapore, Toronto: John Wiley & Sons Ltd.; 1997.
99. Sander R. *Compilation of Henry's law constants for inorganic and organic species of potential importance in environmental chemistry (Version 3)*; 1999. <http://www.henrys-law.org>.
100. Lide DR, editor. *CRC handbook of chemistry and physics*. 90th ed. Taylor and Francis; 2010.

Index

C

F

G

H

N

O

P

T

U

V

W

X